W9-AQQ-185

SECOND EDITION

TECHNICAL WRITING

A READER-CENTERED APPROACH

SECOND EDITION

TECHNICAL WRITING

A READER-CENTERED APPROACH

PAUL V. ANDERSON
MIAMI UNIVERSITY, OHIO

HARCOURT BRACE JOVANOVICH COLLEGE PUBLISHERS

FORT WORTH PHILADELPHIA SAN DIEGO NEW YORK ORLANDO AUSTIN SAN ANTONIO

TORONTO MONTREAL LONDON SYDNEY TOKYO

For my family—
Margie, Christopher, Rachel

Cover: ''Manifold,'' a computer-generated artwork by Melvin L. Prueitt, Computer Graphics Group, Los Alamos National Laboratory.

Copyright © 1991, 1987 by Harcourt Brace Jovanovich, Inc.

All rights reserved. No part of this publication may be reproduced or transmitted in any form or by any means, electronic or mechanical, including photocopy, recording, or any information storage and retrieval system, without permission in writing from the publisher.

Some material derived from *Business Communication: An Audience-Centered Approach* by Paul V. Anderson, copyright © 1989 by Harcourt Brace Jovanovich, Inc.

Requests for permission to make copies of any part of the work should be mailed to: Permissions Department, Harcourt Brace Jovanovich, Inc., 8th Floor, Orlando, Florida 32887.

ISBN: 0-15-589682-2

Library of Congress Catalog Card Number: 90-82910

Printed in the United States of America

PREFACE

When preparing this new edition of *Technical Writing: A Reader-Centered Approach,* I began with the same belief that guided my work on the first edition—namely, that as teachers we all desire to prepare our students to become confident, flexible, and resourceful writers. At work they will encounter such a wide variety of writing situations that no single formula or short series of recipes we might teach will serve them in all situations. Instead, we must equip them to think for themselves as they approach each new communication. We must help them gain the knowledge, skills, and strategies necessary to size up each situation, to create a thoughtful, realistic plan for writing, and then to carry out that plan with skill and flair.

However, this edition differs from the first in many substantial ways. I have gathered suggestions and ideas for improving the book from many sources, including students and teachers who have used it, recently published research about technical writing and its teaching, and my own classroom experimentation and ongoing studies of writing in the workplace. Before previewing some of these improvements and refinements, I want to emphasize that I have preserved the features that students and instructors found most useful in the first edition. The book continues to provide thorough instruction in the writing process, to take a reader-centered approach, to organize most chapters around easy-to-understand, easy-to-remember guidelines, and to provide numerous annotated examples of writing done at work. In addition, I continue to address students directly in a manner that I hope is helpful, respectful, and encouraging.

NEW CHAPTERS AND DISCUSSIONS

Among the most significant improvements in this edition is the inclusion of three new chapters. They treat the following important topics:

- **Persuasion.** Though often neglected in textbooks, persuasion is an integral element of almost every technical communication. Chapter 5 tells students how to construct arguments, reason soundly, and build a persuasive persona. Additional advice about how to influence readers' attitudes and actions appears throughout the book.
- **Using the library.** Chapter 6 introduces students to the resources and strategies that support thorough, efficient library research. The chapter covers both printed and computerized bibliographic aids.

- **Writing collaboratively.** Collaborative writing is very common in the workplace. Chapter 26 provides students with practical suggestions for contributing effectively to group writing projects both in school and on the job.

I also have added short discussions of many other important topics. Three of the most noteworthy are the following:

- **Writing at work versus writing in school.** Presented in Chapter 1, this discussion helps students appreciate the rationale for much of the advice given throughout the book.
- **Sexist and discriminatory language.** A special section helps students recognize this language's sometimes subtle forms, explains some objections to these usages, and teaches practical strategies for avoiding them.
- **Word processors and desktop publishing.** Discussions scattered through the book introduce students to the use of these important technologies when drafting their prose, creating visual aids, and designing page layouts.

SHARPENED FOCUS, BRISKER PACE

For this edition, I have reworked discussions throughout the book to make them more pointed and direct. Most notably, I have condensed the first eighteen chapters of the first edition into ten. Thus, even with the three new chapters just described, this edition has fewer chapters than the first. The brisker pace that results should be especially helpful to instructors who teach on a quarter calendar rather than a semester one.

NEW LEARNING AND WRITING AIDS

I've added case-study problems and planning worksheets as aids. Suitable for class discussions, homework assignments, or graded projects, the case-study problems provide students with rich opportunities to apply the book's advice to the complex writing situations that arise at work.

The six planning worksheets cover instructions, proposals, and several types of reports. While developing these writing aids, I have tried to avoid giving students the impression that the worksheets provide surefire formulas for imitation. I believe that I have succeeded in designing worksheets that students will instead view as guides that will help them creatively adjust conventional writing patterns to suit their particular readers, purposes, and circumstances.

TEACHING SUPPORT

For the second edition, I have revised the *Instructor's Manual*. The new manual covers such topics as course design, classroom activities, and grading criteria. It includes a sample syllabus with more than forty pages of day-by-day teaching suggestions. It also provides master copies of worksheets and overhead transparencies.

Appendix C of the textbook includes writing assignments and case-study problems designed to support and reinforce the principles taught in the *Instructor's Manual*.

CONCLUSION

In preparing this edition, I have striven to provide a textbook that you will find teachable and that your students will find informative, realistic, and helpful. I have greatly benefited from the advice of other teachers, and I continue to welcome your suggestions and comments.

Paul V. Anderson

ACKNOWLEDGMENTS

I take great pleasure in this opportunity to thank the many people who generously gave advice and assistance while I was working on this second edition.

First, I would like to express my gratitude to Jim Clark, Bill Hardesty, Nancy Kersell, Jean Lutz, and Gil Storms—all faculty colleagues at Miami University who have provided invaluable help in many ways. In addition, I am grateful to former Miamian Marian Winner, now Dean of Libraries at Northern Kentucky University.

I also have greatly benefited from excellent reviews by the following faculty at other colleges: Frank Devlin, Salem State College; Steve Driggers, Auburn University; Glenda Hudson, California State University, Bakersfield; Debra Journet, University of Louisville; Gloria Kitto Lewis, Wayne State University; Jan Spyridakis, University of Washington; and Charie Thralls, Iowa State University. I thank them for their insightful comments and excellent suggestions.

I also wish to thank the many present and former students who have helped me with this edition. In particular, I want to express my gratitude to the following people for their special contributions: Ed Bedinghaus (now with Cincinnati Bell Information Systems), Katie Feller, Teresa Newman, J. B. Thompson (now with Oxford Associates), Kris Tyeryar (now with the Ohio River Valley Water Sanitation Commission), Diane Rawlings, Paul Schafer, Patrick Willis, and Mike Zerbe. I am also grateful to the many teaching assistants in Miami University's master's degree program in technical and scientific communication who have shared their ideas and teaching experiences with me.

For help in preparing the manuscript for this edition, I am especially indebted to Trudi Nixon for the immense amount of work she did on various drafts. I also wish to thank Kathy Fox, Jacalyn Kearns, Betty Marak, and Virginia Tobeson for their assistance with the manuscript.

Also, I wish to extend thanks to the very helpful people at Harcourt Brace Jovanovich who worked with me. I am particularly grateful to Bill McLane, whose enthusiasm and assistance have been invaluable. I also have been fortunate to have the expert support of Karen Allanson, Lynne Bush, Niamh Foley-Homan, Kay Kaylor, Steve Lux, Linda Miller, Cindy Robinson, and Lisa Werries.

Finally, I want to thank my family for their encouragement, kindness, and good humor while I worked on this new edition.

CONTENTS

CHAPTER 8   Using Six Patterns of Development   246

CHAPTER 9   Beginning a Communication   278

CHAPTER 10   Ending a Communication   300

I

INTRODUCTION

1

WRITING, YOUR CAREER, AND THIS BOOK

F rom the perspective of your professional career, one of the most valuable subjects you will study in college is writing.

Why? Imagine what your working days will be like once you begin your career. You will spend much of your time using the special knowledge and skills you are now learning in college, especially in your major department. You will search out answers to questions asked by your co-workers, develop recommendations requested by your boss, and solve problems faced by your customers. Furthermore, you will probably generate many good ideas on your own. Looking around, you will discover ways to make things better or to do them less expensively, to solve problems that have stumped others, or to bring about improvements that others haven't even begun to dream about.

YOUR TWO ROLES AT WORK

Yet all the knowledge you possess and all the ideas you generate will be useless to your employer unless you communicate them to someone else. Consider, for example, the situation of Sarah, a recent college graduate who majored in metallurgy. Sarah has just analyzed a group of pistons that broke when used in an experimental automobile engine. No matter how skillfully she conducted this analysis, Sarah's work can help her employer only if Sarah communicates the results to someone else, such as the engineer who must redesign the pistons.

Similarly, Larry, a newly hired dietitian, must communicate to make his work valuable to his employer, a large hospital. Larry has devised a way to reorganize the operation of the hospital's kitchen that will save money and provide better service to the patients. However, his insight will benefit his employer only if he communicates his recommendations to someone else, such as the kitchen director, who has the power to implement them.

Like Sarah and Larry, you will be able to make your work valuable to your employer only if you communicate effectively. Consequently, at work you will play two distinct roles. First, you will be a specialist, applying the knowledge and skills you learned in your major. As a specialist you will generate information and ideas that will be *potentially* useful. Second, you will be a communicator. In this role you will share the results of your specialized activities with co-workers, customers, and other people who rely on you.

Very often your two roles will blend together. As you plan and prepare a communication, you will not only create the message you will deliver, but also develop and refine your ideas and insights about your topic. When that happens, your communicating work and your specialized work will become different aspects of the same activity.

WRITING WILL BE CRITICAL TO YOUR SUCCESS

At work, much of your communicating will take place in writing. Numerous surveys indicate that if you are at all like the typical college graduate, you can expect to spend about 20 percent of your time at work writing.[1] That comes to

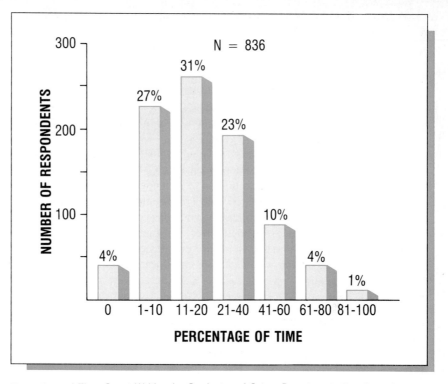

Figure 1–1 Percentage of Time Spent Writing by Graduates of Seven Departments that Send Students to Technical Writing Courses

From Paul V. Anderson, "What Survey Research Tells Us about Writing at Work," in *Writing in Nonacademic Settings*, ed. Lee Odell and Dixie Goswami (New York: Guilford Press, 1985), 18.

one day out of every five-day work week! For example, Figure 1–1 shows how much time is spent writing by graduates of seven departments that send students to technical writing courses—departments ranging from systems analysis and chemistry to home economics and office administration.[2] In fact, notice that 38 percent of the respondents spend *more* than 20 percent of their work time writing. Fifteen percent spend more than 40 percent. Similar responses were found in another survey of 245 people listed in *Engineers of Distinction*; these successful engineers report devoting nearly one-fourth (24 percent) of their time to writing.[3]

Besides enabling you to perform your job, writing well can bring you many personal benefits, including recognition in the form of praise, raises, and promotions. In many organizations, employees rarely work directly with the upper-level managers who have an important influence on decisions about pay and promotions. In such a company, your memos, reports, and other writing may be the only evidence they have of your good work as *either* a specialist or a communicator. Also, writing is an important responsibility of managers, who must convey a wide variety of messages to those above and below them in the organizational hierarchy. Consequently, employers look for writing ability when considering people for advancement.

It's not surprising, then, that 94 percent of the graduates from seven departments that send students to technical writing classes reported that the ability to "write well" (not just write, but *write well*) is of at least "some" importance to them. Furthermore, more than half—58 percent—said that it is of "great" or "critical" importance. (See Figure 1–2.) In the survey of people listed in *Engineers of Distinction,* 89 percent said that writing ability is considered when a person is being evaluated for advancement, and 96 percent said that the ability to communicate on paper has helped their own advancement.

In addition to bringing you recognition, writing well at work can bring you personal satisfaction. Most importantly, writing well enables you to make a personal impact. Perhaps you will design a new product or service you believe your employer should offer. Maybe you will seek an increased budget to support an important project you are heading or the department you manage. Or maybe you will want your employer to adopt a slightly more expensive manufacturing process because it would make working conditions substantially safer for employees. To succeed in any of these endeavors, you will need to influence other people's decisions and actions, most likely through writing.

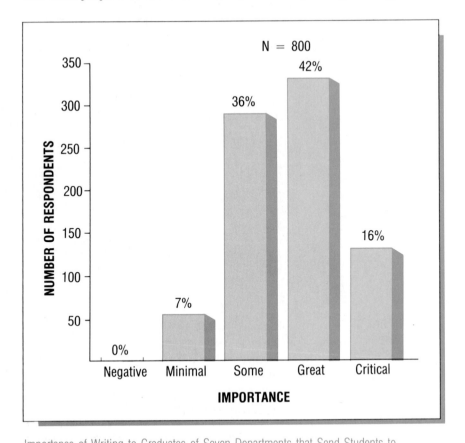

FIGURE 1–2.   Importance of Writing to Graduates of Seven Departments that Send Students to Technical Writing Courses (Percentages do not add to exactly 100 because of rounding)

From Paul V. Anderson, "What Survey Research Tells Us about Writing at Work," in *Writing in Nonacademic Settings,* ed. Lee Odell and Dixie Goswami (New York: Guilford Press, 1985), 19.

Of course, altering product lines, budget allocations, or manufacturing processes aren't the only ways you find satisfaction at work. In many cases, your good writing will simply bring you the pleasure of having helped your co-workers, customers, or other people by sharing things you know, find out, or create.

WRITING AT WORK DIFFERS FROM WRITING AT SCHOOL

This book will teach you how to succeed when you write at work. Of course, you already possess much knowledge about writing that will be indispensable to you on the job. The book will help you build on that solid foundation. However, to write successfully at work, you will also need to develop new writing skills and even new ways of thinking about writing. That's because on-the-job writing differs in some very fundamental ways from the writing done in school. Some of the key differences are summarized in Figure 1–3 and explained in the following paragraphs.

Purpose

One important difference involves your purpose for writing. As a student, you communicate for *educational* purposes. Instructors ask you to write term papers and take written exams in order to help you learn the course material and to give you a chance to demonstrate your mastery of a subject. In contrast, as an employee you will communicate for *instrumental* purposes. That is, most of your communications will be designed to help your employer achieve practical, business objectives, such as improving a product, increasing efficiency, and the like.

This difference in purpose has profound impact on the kind of writing needed. Consider just one example. At school, where your aim is to show how much you know, one of your major writing strategies is probably to say as much as you can about your subject. At work, however, your communications should usually include *only* the information your readers need—no matter how much more you know. Extra information would only clog your readers' path to what they need, thereby decreasing their efficiency and increasing their frustration.

Audience

Another important difference between writing at school and at work involves your audience. As Figure 1–4 illustrates, at school you almost always address a single person, your instructor. In contrast, at work you will often create communications that address a wide variety of people who may differ in terms of such significant variables as familiarity with your subject, the use they will

Area	School	Work
Purpose	Educational: to learn and show what you know	Instrumental: to make things happen
Audience	Simple: one person (an instructor)	Complex: often large groups of people with different needs, desires, and uses for your information
Types of Communications	Term papers, exams	Letters, memos, formal reports, proposals, instructions
Ownership	You own your communications	Your employer owns them
Conditions for Creation	Few distractions if you desire to avoid them; deadlines often flexible	Many unavoidable distractions, inflexible deadlines
Social and Political Considerations	Usually not pertinent	Complex social and political considerations often affect what you can say and how you can say it

FIGURE 1–3   Key Differences between Writing at School and Writing at Work

make of your information, and the kinds of professional and personal concerns they will bring to your communication. Consider, for example, the report in which Larry will present his recommendations for improving the hospital kitchen. His report may be read by his supervisor, Helen Swayze, who will want to know what measures she will have to take in order to follow his recommendations; by the vice president for finance, Marty DuBois, who will want to verify the cost estimates that Larry includes; by the director of purchasing,

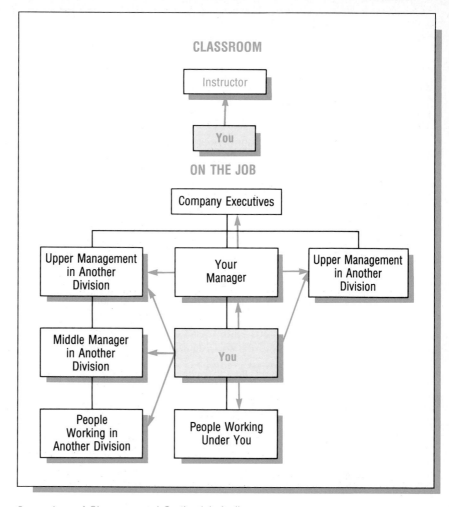

Figure 1–4   Comparison of Classroom and On-the-Job Audiences

Russell Chan, who will want to know about the new equipment he needs to order; by the director of personnel, Ginny Martinelli, who will want to learn whether she will need to rewrite any job descriptions; and by members of the labor union, who will want assurances that the new work assignments will treat them fairly. Writing for such large and diverse audiences requires skills that are not needed when writing only to your instructor.

Types of Communications

To address their audiences, people at work write different types of communications than are prepared in school. Instead of term papers and exams, they write such things as memos, business letters, instructions, project proposals, and progress reports. Each type of on-the-job communication has its own conventions. To write successfully at work, you will need to learn how to construct these kinds of communications.

Ownership

A fourth difference between writing at school and at work concerns the "ownership" of a writer's work. When you write at school, you and your instructor think of the resulting communication as *yours*. At work, however, your communications will belong only *partly* to you. They will also belong to your employer. That's because writing is part of your job, part of what you are paid to do. Furthermore, what you write represents not only you but also your department or your employer. If you write a letter or report to a customer, the customer views it as an official communication from your employer. If you write a proposal, your employer will get the contract—or lose it. To assure that their employees' communications represent the company effectively, many employers require employees to submit drafts to managers who must grant approval—and may dictate changes—before the communications can be sent.

Two other situations, both fairly common at work, also affect a writer's ownership of what he or she writes. First, employees often work on committees that write reports, proposals, and other documents collaboratively. The final version is the group's product, not that of any individual in the group. Second, people often write communications that are sent under somebody else's name. For instance, it is common for departmental reports to be signed by the head of the department, even though they are written by a staff member.

To succeed on the job, you will need to learn to write under circumstances in which your employer claims ownership of your communications and other people exert great power over what you say and how you say it.

Conditions for Creation

In addition, at work people write under much different conditions than at school. As a student, you can probably set aside blocks of uninterrupted time to write your papers. And you can probably find a quiet place to work, though that may mean walking to the library or a study area. In contrast, as an employee, you will be subject to constant distractions. People may drop by, the phone may ring, and your schedule may be peppered with meetings and appointments that leave only brief periods for writing. Despite these distractions, you will be expected to write quickly and efficiently.

Also, your employer will expect you to complete your writing on time, even though assignments sometimes will be made on very short notice. Of course, deadlines occur at school, too, but projects are usually assigned well in advance and you may be able to obtain an extension if you need one. Furthermore, even if you miss a deadline at school, the penalty is relatively mild— at least compared to the penalties for missed deadlines on the job. For example, if a company is preparing a proposal or a sales document, the communication must reach the client on time or it may not be considered at all—no matter how good it is. Employers sometimes advise employees that "It's better to be 80 percent complete than 100 percent late." Missing the deadline could cost your employer thousands, even millions, of dollars.

Social and Political Considerations

To add to the difficulty of writing in a busy, distracting environment under the pressure of deadlines, at work you will often need to consider social and political factors that don't arise when you are writing in school. To begin, you must always keep in mind your professional and social relationships with your readers: manager and subordinate, customer and supplier, co-worker and co-worker. These will influence what you say and how you say it. In addition, you will need to consider the culture of your organization—the way it perceives and presents itself. For example, your organization might be formal and conservative or informal and innovative. Whatever the culture of the organization that employs you, it is likely to influence the way you are expected to write. Finally, you will need to keep in mind the practical politics in your organization. While all individuals, departments, and divisions share the same overall organizational goals and all pursue the same overall mission, business organizations are full of competition and, at times, of intrigue as individuals and groups compete for recognition, power, and money.

AT WORK, WRITING IS AN ACTION

From the preceding discussion, it should be clear that at work you will be writing for purposes and under circumstances much different from those in college. The best way to begin preparing for writing in your career is to consider what, in general, you will want to accomplish with it. As explained earlier, writing at work almost always has an instrumental purpose: you write to make something happen. This fact is contrary to the view some people hold about writing at work. They think of writing as a passive function. They are active, they believe, only when they are playing their *specialist* role. They think that when they are writing they are merely recording or transporting information.

Nothing could be farther from the truth.

To write is to act. When you write, you exert your powers to bring about some specific result. For example, consider Larry's situation. As a dietetics *specialist,* Larry's action is to devise a plan for improving the efficiency of the kitchen. His action as a dietetics *communicator* is to write a proposal that *persuades* the decision-makers in the hospital to adopt his plan. Similarly, Sarah's action as a metallurgical *specialist* is to test the pistons to learn why they failed. As a metallurgical *communicator,* she constructs tables of data, composes sentences, and performs a broad range of other writing activities in order to present her results in a way that helps the engineer redesign the pistons so they work satisfactorily. Figure 1–5 provides examples of changes you might desire to bring about through your writing at work.

The most important thing for you to remember about the ''communication acts'' you will perform at work is that they are *social* actions. Every communication you write will be an interchange between particular, individual people:

The Way Things Are Now	You Act (You Write)	The Way You Want Things to Be
You work for a manufacturing plant. The plant manager doesn't realize that the company is using an outdated process for a key operation.	➡	The plant manager decides to investigate a newly developed process because he has read a technical report you wrote.
Your employer, a bank, is installing a new computer system for use by the tellers.	➡	The tellers perform their jobs efficiently and flawlessly using the new computer system because they have read a manual you wrote.
The manager of your department wants to decide whether or not to purchase a certain piece of equipment. You have the information the manager needs in order to make the decision.	➡	Your manager chooses wisely based on information in a feasibility report you wrote.
You work for a company that franchises fast-food restaurants featuring salads and other healthful meals. It is tempted to begin a rapid expansion.	➡	The company decides it would overextend itself if it expanded so rapidly. It makes this decision after reading a report in which you showed the reasons for caution.

FIGURE 1–5   Examples of Changes You Can Bring About through Writing

you and your readers. You will be a supervisor telling a co-worker what you want done, an adviser trying to persuade your boss to make a certain decision, the expert helping another person operate a certain piece of equipment. Your reader will be an experienced employee who has some hesitancies about your request, a manager who has been educated to ask certain questions when making a decision, a machine operator who has a particular sense of personal dignity and a specific amount of knowledge of the equipment to be operated. Even when you are writing to groups of people, your communications will establish an individual relationship between you and each person in the group. Each person will read with his or her own eyes, react with his or her own thoughts and feelings.

Of course, as a communicator your relationship, and thus your interactions, with your individual readers will be shaped by all the elements of the business context described above: the swirl of personal aims and organization goals, the cooperation and competitiveness, corporate styles and individual preferences, and the incessant demand for rapid, effective action. At bottom, however, will always be you and each individual member of your audience.

THE MAIN ADVICE OF THIS BOOK: THINK CONSTANTLY ABOUT YOUR READERS

The observation that writing is a social action leads to the main advice of this book: when writing, think constantly about your readers. Think about what they want from you—and why. Think about the ways you want to affect them and the ways they will react to what you have to say. Think about them in all the ways you would if they were standing right there in front of you while you talked together.

You may be surprised that this book emphasizes the personal dimension of writing more than such important characteristics as clarity and correctness. Although clarity and correctness are important, they cannot, by themselves, ensure that something you write at work will succeed.

To succeed, the communications you write at work must affect in specific ways the individual people you are addressing. For instance, if Larry's proposal for modifying the hospital kitchen explains the problems created by the present organization in a way that his *readers* find compelling, if it addresses the kinds of objections his *readers* will raise to his recommendation, if it reduces his *readers'* sense of being threatened by having a new employee suggest improvements to a system that they set up, then it may succeed.

On the other hand, if Larry's proposal fails to persuade his readers that the kitchen's present organization creates significant problems from their point of view, if it leaves some key questions unanswered, if it makes Larry seem like a pushy person who has overstepped his appropriate role, it will not succeed. It won't matter how "clear" the writing is, or how "correct." A communication can be perfectly clear, perfectly correct, and yet be utterly unpersuasive, utterly ineffective.

Similarly, consider Sarah's report on the faulty pistons. If she writes it so the engineer who must redesign the pistons can find the information he needs quickly and in a readily usable form, her report will achieve its purpose. It will do what Sarah created it to do. However, if the engineer has to hunt through the report for the needed information and must convert it into another form before he can use it, then Sarah's report will be ineffective—no matter how clearly and correctly it is written.

Throughout the rest of this book, you will find advice about how to create on-the-job communications that affect your readers in the ways you intend. To lay the groundwork for that advice, the remaining pages in this chapter do the following:

● Explain three important facts about what occurs while people read.
● Describe two basic strategies for keeping your readers in mind while writing.
● Preview several effective writing techniques widely used on the job.

THREE IMPORTANT FACTS ABOUT HOW PEOPLE READ

As you attempt to write in ways that will elicit the desired response from your readers, you will find it very helpful to know three important facts about how people interact with the messages they read: readers create meaning, readers' responses are shaped by the situation, and readers react on a moment-by-moment basis. Each point is discussed below.

Readers Create Meaning

Perhaps the most important fact about reading is that instead of *receiving* meaning (as is often thought), people interact with the message to *create* meaning. First, we create meaning from the individual words we see on the page. Consider, for instance, the amount of knowledge we must possess and apply to read even such a simple sentence as "It's a dog." Keep in mind that before children learn to read, they can't make any sense out of those three printed words (*it's, a, dog*). To them, the words are indecipherable black marks against the page's white background. Remember, too, the word *dog* has several meanings, so readers cannot know the meaning of the sentence without knowing its context. Is it the answer to the question, "What kind of animal is Jim's new pet?" or to the question, "What do you think of the new computer made by ABC Corporation?"

Furthermore, when reading, we not only create meaning from the individual words and sentences, but also we build those smaller meanings into larger structures of knowledge. These structures are not merely the memory of the words we've read but our own creations. Here's a quick way to demonstrate that to yourself. Imagine that someone has asked you to explain the meaning of the following statement from earlier in this chapter: "To write is to act." First write down a sentence you could use to answer that request. Then try to find a sentence in this book that exactly matches yours. Most likely, you won't be able to find one. The sentence you wrote is not one that you memorized; it expresses the meaning you *created* when reading the text.

Readers' Responses Are Shaped by the Situation

A second important fact about readers is that their responses to a communication are shaped by the total situation surrounding the message—including such factors as their purpose for reading, their perceptions of the writer's aims, their personal interests and stake in the subject discussed. and their past relations with the writer.

Imagine, for instance, you have been asked to select the brand of desktop computer to be purchased for fifty offices at your employer's corporate headquarters and as part of your research you read a report stating that the ABC computer is "a dog." Given your purpose, you might begin immediately—even before reading the report's next sentence—to consider other information and opinions you have obtained about that product. Do others agree that the computer is a dog? Do the data you have gathered support that assessment? Your response to the statement also would be shaped by your knowledge that the writer is a computer specialist in your own company—or a salesperson for one of ABC's competitors. And it would be shaped by any public pronouncements you have made on the subject. For example, if you had declared a similar view, you might feel satisfaction and pleasure, but if you had stated that you believed the ABC computer to be a superior product, you might react with embarrassment or anger.

The full range of situational factors that can affect a reader's response are too numerous and varied to list here. The key point for you to remember is that in order to predict how a reader might respond to something you write, you must understand the situation in which the person will read your communication.

Readers React on a Moment-by-Moment Basis

The third important fact about readers is that they react to communications on a moment-by-moment basis. When we are reading a humorous novel, we chuckle as we read a funny sentence or paragraph. We don't wait until we have finished the entire book. Likewise, on the job, people react to each part of a memo, report, or other business communication as soon as they come to it. The following demonstration illustrates this point and explains its significance to you as a writer.

Imagine you are the manager of the personnel department in a factory. A few days ago you met with Donald Pryzblo, who manages the data processing department, to discuss a problem: recently, the company's computer has been issuing some payroll checks for the wrong amount.

The two of you discussed the problem because your department and Mr. Pryzblo's work together to create each week's payroll. First, your clerks collect a time sheet for each employee. Then they transfer the information on those sheets to time tickets, which they forward to the clerks in Pryzblo's department. His clerks enter the information into a computer, which calculates each employee's pay and prints the payroll checks. The whole procedure is summarized in the following diagram:

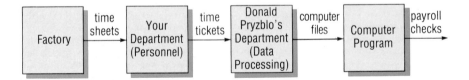

When meeting with Pryzblo, you proposed a solution that he did not like. Because you are both at the same level in your company's organization, neither one has power to impose suggestions on the other. In today's mail you find a memo from Pryzblo. Your task in this demonstration is to read the memo *very slowly*—so slowly that you can focus on something that normally happens too rapidly for you to notice, namely, the way you react, moment-by-moment, to each statement you read.

Here is the procedure you should use to slow down your reading. First, cover the memo entirely with a sheet of paper. Then slide the paper down the page, stopping as you expose each new sentence. Just after you have read each sentence, write down your thoughts and feelings (in your role as manager of the personnel department). In this demonstration, there are no "right" or "wrong" answers. You are not being tested.

Ready? Take out a sheet of paper, slip it quickly over Figure 1–6 on page 18 and begin to slide it down Pryzblo's memo. Read each sentence and then record your reaction. When you are finished, turn back here and read the next section.

Discussion of the Demonstration

Finished? Now look back over the notes you made while reading the memo in your role as manager of the personnel department. Most likely you made notes throughout the memo. For example, most people who participate in this demonstration react to the quotation marks that surround the word *errors* in the first sentence. And they react immediately; they don't wait until they have finished reading the entire memo. The word *insinuated* in the second sentence likewise draws an immediate reaction from most readers. (They laugh if they forget to play the role of personnel manager; most cringe if they remember to play the role.)

The fact that readers respond to communications on a moment-by-moment basis is important to you as a writer because their reaction when reading one sentence affects their reactions to the sentences they read afterwards. Consider, for example, what happens as most people read through Donald Pryzblo's memo. Playing the role of the personnel manager, they start to become defensive the moment they see the quotation marks around the word *errors,* or at least by the time they come to the word *insinuated*. After they have read

Pryzblo's third paragraph, their defensiveness has hardened into a grim determination to resist any suggestion that Pryzblo might make. By the time they reach the last sentence of the memo, they have little interest in accepting the recommendation it presents.

A smaller group of readers is more even-tempered. Instead of becoming defensive, they become skeptical. Their skepticism is triggered by the first two sentences, where they interpret the quotation marks around the word *errors* and the use of the word *insinuated* as signs that Pryzblo is behaving emotionally rather than intellectually. Consequently, they decide to evaluate his statements very carefully. When they read the accusation that personnel department clerks are miscopying, they immediately want to know what evidence Pryzblo has to support his claim. When they read the next sentence, which fails to provide that evidence, they are disappointed and also disinclined to go along with any suggestions Pryzblo might make.

Thus, even though different readers react to the first few sentences of this memo in different ways, their early reactions shape their reactions to later sentences, including the recommendation. As a result, even though Pryzblo's recommendation seems sensible enough in itself, no reader I've met feels like accepting it, at least not until Pryzblo provides more information.

It is possible, of course, that Pryzblo had some other purpose in writing. For example, maybe he wanted to inflame the personnel manager into making a rash and foolish response that would get him or her in trouble with their mutual boss. Then his memo might have worked very well. Even then, the basic points of this demonstration would remain unchanged: people react to the things they read on a moment-by-moment basis and their reactions at one moment shape their responses to what they read next.

The preceding discussion of how people read should explain why it is so important for you to follow the main advice of this book—"think constantly about your readers." Each reader will create his or her own response to your communication. A person will respond differently under one set of circumstances than under another. What you say in one or two early sentences can affect the outcome of an entire communication. To write effectively, you must predict your readers' likely responses to your message and design your communication accordingly. And to do that, you must keep your readers—their needs and goals, feelings and situation, preferences and responsibilities—foremost in mind throughout your work on a communication.

TWO STRATEGIES FOR KEEPING YOUR READERS IN MIND

Unfortunately, when writing you have so many things to think about that you can easily lose sight of your readers. To prevent this from happening, you can develop a reader-centered writing process, and you can "talk" with your readers. These two strategies are explained in the following paragraphs and referred to throughout the rest of the book.

Make notes on
your thoughts
and feelings
here.

NOTE: Remember to cover this memo. Read it one sentence at a time, noting your thoughts and feelings after reading each sentence and before going on to the next one.

October 10, 19—

TO: Your Name, Manager, Personnel Dept.

FROM: Donald Pryzblo, Manager, Data Processing
                  Dept.

SUBJECT: INCORRECT PAYROLL CHECKS

I have been reviewing the "errors" in the computer files.

Contrary to what you insinuated in our meeting, the majority of these errors were made by <u>your</u> clerks. I do not feel that my people should be blamed for this. They are correctly copying the faulty time tickets that your clerks are preparing.

You and I discussed requiring my computer operators to perform the very time-consuming task of comparing their computer entries against the time sheets from which your clerks are miscopying.

My people do not have time to correct the errors made by your people, and I will not hire additional help for such work.

Would you
accept this
recommendation?
Why or why
not?

I recommend that you tell your clerks to review their work carefully before giving it to the computer operators.

FIGURE 1–6   Memo for Demonstration

Develop a Reader-Centered Writing Process

The first strategy for keeping your readers in mind throughout your work at writing is to develop a strategic, reader-centered writing process. Overall, your writing process is the set of activities you perform when you prepare a message. Although these activities are quite varied, they may be classified into five groups:

- Defining your communication objectives
- Planning your communication
- Drafting your message
- Evaluating your draft
- Revising your draft in light of what you learned from the evaluation

When presented in a list like this, the activities seem to have a linear relationship with one another, as if a writer should proceed through them in order. That is not the case. When preparing a communication, successful writers typically jump around the activities many times. For instance, while planning or drafting, you may gain an additional insight into your situation that causes you to refine your objectives. Or, while drafting one part of a communication, you may think of a way to improve the organization or phrasing of another part. You may also develop more ideas about your topic, and those developments may require you to modify your plans, your draft, and even your communication objectives. By dancing among the various activities, you can take advantage of the additional insights and new ideas you gain about what you want to say and how you want to say it.

No matter how much dancing you may do, however, you need to take your readers into consideration when performing each of these activities. In Chapters 3 through 19 you will find detailed advice about how to do that. Even now, however, you can begin executing your writing process in a reader-centered way. First, when defining your objectives (whether for the whole communication or a small part of it), focus on what you want to happen *while your reader is reading*. For example, instead of saying, "My purpose in this section is to describe the two companies from which we might buy supplies," say "My purpose is to present information about the two suppliers in a way that lets my reader quickly compare them on a point-by-point basis using the specific criteria that are important to my reader." Unlike the first version, the second suggests what information you should include about the suppliers (the information related to your readers' selection criteria) and also how you should organize that information (into a point-by-point comparison).

After you have defined your objectives in terms of what you want to happen while your readers read, refer to your objectives continually so you can benefit from the practical insights they provide about how to write your message. Writers sometimes forget to do this. For instance, when planning, they concentrate on creating a "logical" organization without considering whether it will be useful or persuasive to their readers. Similarly, when evaluating a draft, they concentrate on spelling and punctuation without asking whether their readers will respond to their message in the desired way. Avoid such mistakes. Throughout all your work on a communication, refer to your reader-centered objectives. If you make them the foundation on which the rest of your work rests, you will be well on your way to using a reader-centered writing process.

Talk with Your Readers

The second strategy for keeping your readers in mind is to "talk" with them when writing. Imagine that your communication is a conversation in which you make a statement and your audience responds. Write each sentence, each paragraph, each chapter, to create the interaction—the conversation—with your readers that will bring about the final result you desire.

When following this strategy, it is crucial that you talk *with* your readers, not *to* them. When you talk *to* other people, you are like an actor reciting a speech: you stick to your script without regard to how your audience is reacting to your words. When you talk *with* others, you adjust your statements to fit their reactions. Does someone squeeze his brows in puzzlement? You explain the point more fully. Does someone twist her hands impatiently? You abbreviate your message. Are your listeners unpersuaded by one argument? Then you abandon it and try another. Do they ask to hear more? You provide additional information.

Consider how much Donald Pryzblo could have benefited from "talking" with the personnel manager as he drafted his memo. After writing the first sentence, Pryzblo would have seen the manager wrinkle his or her forehead as the manager saw the quotation marks around *errors*. After writing the second sentence, Pryzblo would have seen the manager's jaw tighten and heard the manager exclaim, "What proof do you have that my clerks are making the mistakes?" Seeing these things, Pryzblo would have known that he only could persuade the manager to accept his recommendation if he changed his draft.

Even better, Pryzblo might have thought about the manager's reactions *before* drafting. Then he might have thought, "I want to begin this memo in a way that will make the personnel manager feel open-minded about my recommendation." With that insight he could have devised an opening that would promote open-mindedness, such as beginning with a friendly statement (not accusations) and then saying that he wanted to work with the personnel manager to solve the problem of the payroll checks in a way that will best serve the interests of their mutual employer.

This strategy of imagining your readers in the act of reading works for any kind of communication, not just ones (like Pryzblo's) intended primarily to persuade. Figure 1–7 shows how another writer benefited from "talking" with her reader while writing a set of instructions. The strategy works so well because it enables you to anticipate your reader's moment-by-moment responses to your message and to write accordingly.

Marty's Talk with Her Reader

Marty, a young engineer, was writing instructions for calibrating an instrument for testing the strength of metal rods. One of her original instructions read like this:

```
15. Check the reading on Gauge E.
```

Marty then imagined how a typical user of her instructions would react after reading that instruction. She saw the reader look up and ask, "What should the reading be?" So Marty told the reader to look for the correct reading in the Table of Values.

Marty then imagined that when the reader looked at the Table of Values, the reader discovered that the value on Gauge E was incorrect. The reader asked, "What do I do now?" So Marty revised again.

In the end, her instruction read as follows:

```
15.  Check the reading on Gauge E to see if it corresponds with
     the appropriate value listed in the Table of Values (see
     Appendix IV, page 38).
     • If the value is incorrect, follow the procedures for
       correcting imbalances (page 28).
     • If the value is correct, proceed to step 16.
```

FIGURE 1–7   How One Writer "Talked" with Her Reader

SOME READER-CENTERED WRITING TECHNIQUES YOU CAN START USING NOW

In the preceding section, you read about two general strategies you can use to write successfully on the job: develop a reader-centered writing process and imagine you are talking with your readers as they read your communications. What will your communications look like if you use those strategies? Of course, every one will be different because each will be created to achieve a unique set of objectives in a unique situation. However, many communications written at work share important features, which are described later in this book. Some of the most common and useful are introduced briefly in the following paragraphs.

This introduction has two purposes. The first is to help you begin using these features right now, even before you read the more detailed explanation of them in later chapters. The second is to help you see how the common features of on-the-job writing are based on the purposes of writers and the needs of readers. Consequently, the features are organized around three goals that people usually have when writing on the job.

- **Help readers focus on key information.** Because they are very busy, most readers at work want to find key information quickly. One way writers satisfy that desire is to state their recommendations, conclusions, or other main points at the beginning of their communications, often in the first paragraph. Also, to help their readers scan through for particular facts, writers often use *headings, topic sentences,* and *lists.* In lists, writers generally make the most important points stand out by placing them first. All of these techniques are illustrated in Figure 1–8, which shows a memo in which the writer, Frank Thurmond, answers his reader's question about whether their company can use a new, inexpensive plastic to make the containers for a household oven cleaner it manufactures.

  At work, writers also help their readers focus on key information by eliminating information the readers would find irrelevant to their needs. For example, Thurmond carefully selected only the test results he knew would be important to his reader, a person who must decide whether or not to use the plastic for packaging the oven cleaner. If Thurmond were writing to someone else—say, another researcher who wanted to understand the chemical reaction between the cleaner and the plastic—Thurmond would have included a different set of test results.

- **Tell readers how things are relevant to them.** When readers at work pick up a communication, their first question is often, ''Why should I read this? How does it relate to my needs or responsibilities?'' Consequently, at work writers often begin with a sentence that explicitly tells the readers how the information in the communication is relevant to them.

  Similarly, as people at work read through a communication, they want more than mere facts; they want to know how the facts are significant to them. As a result, writers provide many statements that explain relevance. For instance, at the bottom of the first page of his memo, Thurmond not only states that the oven cleaner attacks the plastic but also explains why the attack is a problem: it deforms the bottles so that they might fall off the shelves in retail stores. (See Figure 1–9.)

- **Use a straightforward style.** Because readers desire to understand the things they read as quickly as possible, people who want to create reader-centered communications generally say things as plainly and directly as they can. Three techniques they use for achieving that style follow.

  1. **Trim away unnecessary words.** For example, they write ''because'' instead of ''because of the fact that,'' and they write ''now'' instead of ''at this point in time.''

2. **Put action in verbs rather than in other parts of speech.** For example, they say "We *conclude*" rather than "We reached the *conclusion*." (See the first paragraph of Figure 1–9.)

3. **Use verbs in the active voice rather than the passive.** In sentences in the active voice, the actor—the person or thing performing the action—is in the subject position of the sentence. For example, in the first sentence of Thurmond's memo, *we* is the actor and also the subject: "*We* have completed . . ." In contrast, when a sentence is in the passive voice, the subject of the sentence is separate from the actor, and the subject is acted on by the actor. For instance, in the passive version of the sentence just quoted ("The tests . . . were completed by *us*."), *us* is still the actor but no longer the sentence subject. Figure 1–10 highlights the lively, active verbs Thurmond used.

Do not misunderstand the preceding list. You cannot write successfully simply by using the techniques presented there. Successful writing is an effective interaction between you and your readers. What matters is how you construct this person-to-person interchange. Some or all of the techniques listed above may *help* you in that effort, but they also can be used in a thoughtless and mechanical way that is no help at all. Headings can be written in a way that assists the reader or that misleads and confuses. An introductory explanation of relevance can be written in a way that is compelling or that misses the mark. In an attempt to write in a straightforward manner, a writer might achieve a style that is crisp and vigorous or one that is telegraphic and puzzling. To work, each of these techniques—and the many others you will learn about in this book—must be skillfully attuned to the particular readers and situation you are addressing.

WHAT LIES AHEAD IN THIS BOOK

To help you learn how to create an effective interchange with your readers in every communication you write on the job, this book offers several kinds of information and advice.

First, to provide you with a detailed sample of a reader-centered writing process, the next chapter leads you through writing a resume and job application letter.

Then, Chapters 3 through 19 present a detailed discussion of each of the activities of the writing process, beginning with the definition of your objectives, and proceeding through planning, drafting, evaluating, and revising.

Chapters 20 through 25 introduce you to the general frameworks—called superstructures—used for constructing six types of communications that will be important to you in your career: reports (four types), proposals, and instructions. Although these patterns aren't recipes you can simply imitate to create

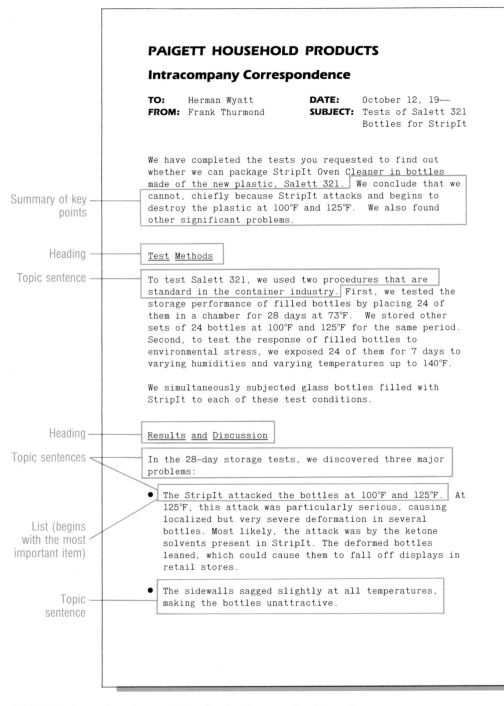

Summary of key points

Heading

Topic sentence

Heading

Topic sentences

List (begins with the most important item)

Topic sentence

PAIGETT HOUSEHOLD PRODUCTS

Intracompany Correspondence

TO: Herman Wyatt   **DATE:**   October 12, 19—
FROM: Frank Thurmond   **SUBJECT:** Tests of Salett 321
                                                                  Bottles for StripIt

We have completed the tests you requested to find out whether we can package StripIt Oven Cleaner in bottles made of the new plastic, Salett 321. We conclude that we cannot, chiefly because StripIt attacks and begins to destroy the plastic at 100°F and 125°F. We also found other significant problems.

Test Methods

To test Salett 321, we used two procedures that are standard in the container industry. First, we tested the storage performance of filled bottles by placing 24 of them in a chamber for 28 days at 73°F. We stored other sets of 24 bottles at 100°F and 125°F for the same period. Second, to test the response of filled bottles to environmental stress, we exposed 24 of them for 7 days to varying humidities and varying temperatures up to 140°F.

We simultaneously subjected glass bottles filled with StripIt to each of these test conditions.

Results and Discussion

In the 28-day storage tests, we discovered three major problems:

- The StripIt attacked the bottles at 100°F and 125°F. At 125°F, this attack was particularly serious, causing localized but very severe deformation in several bottles. Most likely, the attack was by the ketone solvents present in StripIt. The deformed bottles leaned, which could cause them to fall off displays in retail stores.

- The sidewalls sagged slightly at all temperatures, making the bottles unattractive.

FIGURE 1–8   Strategies for Helping Readers Focus on Key Information

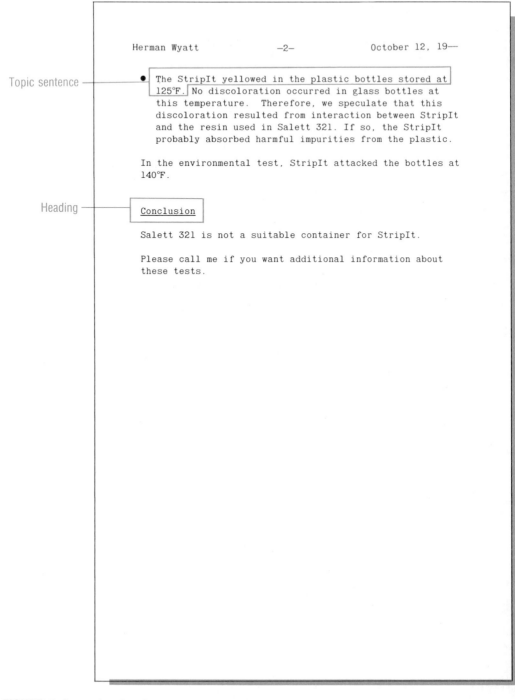

Topic sentence

Herman Wyatt   —2—   October 12, 19—

The StripIt yellowed in the plastic bottles stored at 125°F. No discoloration occurred in glass bottles at this temperature. Therefore, we speculate that this discoloration resulted from interaction between StripIt and the resin used in Salett 321. If so, the StripIt probably absorbed harmful impurities from the plastic.

In the environmental test, StripIt attacked the bottles at 140°F.

Heading

Conclusion

Salett 321 is not a suitable container for StripIt.

Please call me if you want additional information about these tests.

FIGURE 1–8   *(continued)*

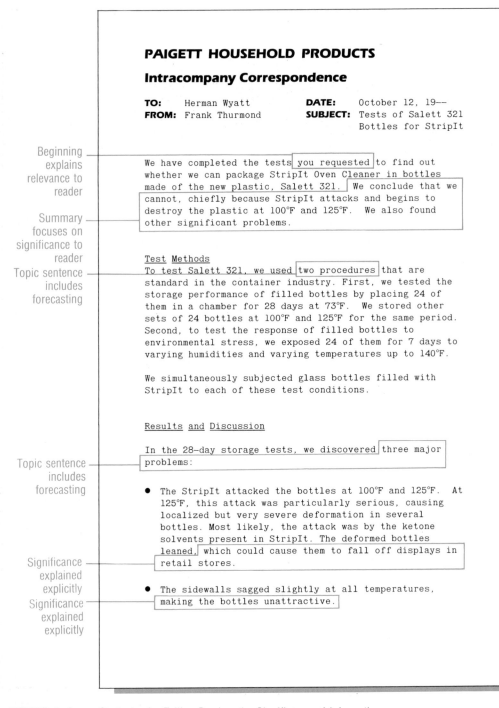

Beginning explains relevance to reader

Summary focuses on significance to reader

Topic sentence includes forecasting

Topic sentence includes forecasting

Significance explained explicitly

Significance explained explicitly

PAIGETT HOUSEHOLD PRODUCTS

Intracompany Correspondence

TO: Herman Wyatt **DATE:** October 12, 19—
FROM: Frank Thurmond **SUBJECT:** Tests of Salett 321 Bottles for StripIt

We have completed the tests you requested to find out whether we can package StripIt Oven Cleaner in bottles made of the new plastic, Salett 321. We conclude that we cannot, chiefly because StripIt attacks and begins to destroy the plastic at 100°F and 125°F. We also found other significant problems.

Test Methods
To test Salett 321, we used two procedures that are standard in the container industry. First, we tested the storage performance of filled bottles by placing 24 of them in a chamber for 28 days at 73°F. We stored other sets of 24 bottles at 100°F and 125°F for the same period. Second, to test the response of filled bottles to environmental stress, we exposed 24 of them for 7 days to varying humidities and varying temperatures up to 140°F.

We simultaneously subjected glass bottles filled with StripIt to each of these test conditions.

Results and Discussion

In the 28-day storage tests, we discovered three major problems:

● The StripIt attacked the bottles at 100°F and 125°F. At 125°F, this attack was particularly serious, causing localized but very severe deformation in several bottles. Most likely, the attack was by the ketone solvents present in StripIt. The deformed bottles leaned, which could cause them to fall off displays in retail stores.

● The sidewalls sagged slightly at all temperatures, making the bottles unattractive.

FIGURE 1–9 Strategies for Telling Readers the Significance of Information

Herman Wyatt   —2—   October 12, 19—

- The StripIt yellowed in the plastic bottles stored at 125°F. No discoloration occurred in glass bottles at this temperature. Therefore, we speculate that this discoloration resulted from interaction between StripIt and the resin used in Salett 321. If so, the StripIt probably absorbed harmful impurities from the plastic.

In the environmental test, StripIt attacked the bottles at 140°F.

Conclusion

Salett 321 is not a suitable container for StripIt.

Please call me if you want additional information about these tests.

FIGURE 1–9   *(continued)*

PAIGETT HOUSEHOLD PRODUCTS

Intracompany Correspondence

TO: Herman Wyatt   **DATE:**   October 12, 19—
FROM: Frank Thurmond   **SUBJECT:** Tests of Salett 321
                                             Bottles for StripIt

We have completed the tests you requested to find out
whether we can package StripIt Oven Cleaner in bottles
made of the new plastic, Salett 321. We conclude that we
cannot, chiefly because StripIt attacks and begins to
destroy the plastic at 100°F and 125°F. We also found
other significant problems.

Test Methods

To test Salett 321, we used two procedures that are
standard in the container industry. First, we tested the
storage performance of filled bottles by placing 24 of
them in a chamber for 28 days at 73°F. We stored other
sets of 24 bottles at 100°F and 125°F for the same period.
Second, to test the response of filled bottles to
environmental stress, we exposed 24 of them for 7 days to
varying humidities and varying temperatures up to 140°F.

We simultaneously subjected glass bottles filled with
StripIt to each of these test conditions.

Results and Discussion

In the 28-day storage tests, we discovered three major
problems:

- The StripIt attacked the bottles at 100°F and 125°F. At
  125°F, this attack was particularly serious, causing
  localized but very severe deformation in several
  bottles. Most likely, the attack was by the ketone
  solvents present in StripIt. The deformed bottles
  leaned, which could cause them to fall off displays in
  retail stores.

- The sidewalls sagged slightly at all temperatures,
  making the bottles unattractive.

FIGURE 1–10   Lively, Active Verbs

Herman Wyatt   —2—   October 12, 19—

- The StripIt (yellowed) in the plastic bottles stored at 125°F. No discoloration occurred in glass bottles at this temperature.   Therefore, we (speculate) that this discoloration resulted from interaction between StripIt and the resin used in Salett 321. If so, the StripIt probably (absorbed) harmful impurities from the plastic.

In the environmental test, StripIt (attacked) the bottles at 140°F.

<u>Conclusion</u>

Salett 321 is not a suitable container for StripIt.

Please (call) me if you want additional information about these tests.

FIGURE 1—10   *(continued)*

an effective communication, they are very helpful. Each represents a generally successful way of structuring messages in some situation frequently encountered at work.[4]

Chapters 26 through 28 provide advice about communicating in special circumstances. Chapter 26 focuses on meeting the special challenges involved with writing on teams in which several people work together to produce a single document. Chapters 27 and 28 explain how to use the ''reader-centered'' approach when you present your message orally.

Finally, Appendix A describes the conventional formats for business communications so that you can write letters that *look* like business letters, memos that look like business memos, and so forth. Appendix B tells how to write footnotes, references, and bibliographies.

A WORD ABOUT GUIDELINES

Many chapters in this book offer guidelines, which are brief summary statements designed to make the book's suggestions as easy as possible for you to remember and use. As you read the guidelines, keep two things in mind.

First, they look deceptively simple. Indeed, you could probably memorize most of them very quickly. What is difficult about the guidelines is learning how to apply them in particular situations. To do that, you need to learn the rationale for them and understand the detailed techniques for employing them. Both take careful reading of the chapters, together with concentrated thought and practice.

Second, as you read the guidelines, keep in mind that they are merely that: guidelines. They are not rules. Although you can usually succeed by following them, every one has exceptions. In some circumstances, two guidelines may even seem to conflict with one another. Thus, to apply the guidelines successfully, you need to use good sense and creativity, guided always by your thoughts of your readers and the specific ways you want to affect them.

To help you learn how to apply these guidelines to actual communications, the book describes many sample situations and shows many sample letters, memos, reports, and other communications. These reflect *typical* business concerns and practices. However, what's typical is not what's universal. As explained earlier, every communication situation is unique. The readers and circumstances you encounter in your job will certainly differ to some extent—and may differ substantially—from those described here. In fact, you may end up working for a boss or client whose regulations, values, or preferences are very *untypical*. To write successfully then, you may need to *ignore* one or more of the guidelines explained here.

However, no matter what situation you find yourself in, the overall advice taught in this book will guide you to success: size up your situation by focusing on your readers and the way you want to affect them, then use your objectives to shape your message in a way that will bring about the result you desire. If you learn how to apply that advice to the typical situations described here, you will have mastered the strategies needed to write successfully in any untypical situations you may encounter in your career.

CONCLUSION

Writing will play a very crucial role in your career. To help you succeed, this book offers numerous suggestions about how to write effectively on the job. As mentioned earlier, however, you cannot become a successful writer simply by memorizing these suggestions. You must also learn to apply them thoughtfully and creatively. This requires practice—practice that is accompanied by knowledgeable and sensitive reactions to your efforts. Such reactions, unfortunately, are something a textbook cannot provide. To learn whether or not your way of applying this book's advice is working, you must turn to other people. Hence, the value of this book will depend largely on your openness to the insights and advice provided by your instructor and your classmates as they review your efforts. With their help, you should be able to learn how to communicate your ideas and your special knowledge in ways that will make good things happen for you and your readers.

In your efforts to prepare yourself to write successfully in your career, I wish you good luck.

EXERCISES

1. Imagine two situations in which you will write on the job. For each, explain what purposes you will have for writing and what purposes your readers will have for reading. If you have written as an intern, a co-op student, or a regular employee, you may describe two of those situations.

2. Find a communication written by someone who has the kind of job you want. Explain its purpose from the points of view of both the writer and the readers. Describe some of the writing strategies the writer has used to achieve those purposes.

3. a. Find a piece of writing that you believe to be ineffective. (You might look for an unclear set of instructions or an unpersuasive advertisement for some business or technical product.) Write a brief analysis of three or four "reading moments" in which you interact with the text in a way that inhibits the author's desired results.

   b. Now analyze an effective piece of writing. This time, write about three or four "reading moments" in which you interact with the text in a way that helps the author bring about the desired result.

4. **a.** To demonstrate the extent to which readers create the meaning of the things they read, find five words in this chapter that would take on a different meaning if they were used in another sentence. In each case, write down the sentence from this book and another in which the word's meaning is different.

   b. Find one paragraph on page 168 whose main point is explicitly stated and another paragraph whose main point is obvious even though it isn't stated explicitly. Explain how you derived the meaning of the latter paragraph.

CASE   **HELPING YVONNE'S LITTLE BROTHER**

As you walk across campus this morning, you hear someone call your name. Turning around, you see a student running toward you. It's Eddie Ware, younger brother of Yvonne, your good friend from high school. When you last saw Yvonne, she told you that Eddie would attend your college, but this is the first time you have seen him on campus.

"Hi, Eddie," you say. "How are you?"

"Hi," he replies, a little self-conscious. You smile to reassure him. You don't know him well and suspect that he's worried that he's intruding.

After you and Eddie exchange a few snippets of news, Eddie asks, "You know Professor Bersani, don't you?"

"Yeah. She's great. She's my adviser."

"Well, she's mine too," Eddie says, and then he explains that he's chosen the same major as you. "Professor Bersani said she knows you. In fact, she said you might be able to help me."

"I'll try," you reply.

"When we were talking about what courses I should plan to take before I graduate, she said that I should work especially hard on improving my writing. I'd shown her my high-school grades. In English, they aren't good."

You hold your breath. Does he want you to tutor him?

Eddie continues, "I told her that I planned to work hard in *all* my classes, including English. But she said I should use one of my electives to pick up an extra writing class. But that's not what I came to college for: I want to learn my major. When I told her that, she said that maybe before I finish choosing my classes I should learn more about the writing I will do on the job once I graduate. When I asked how I could learn that, she suggested I talk to an older student. I told her earlier that I knew you, and she said that you would be the perfect person to ask."

"That's flattering," you think to yourself.

After talking more with Eddie, you learn that through his conversation with Professor Bersani he has developed the following questions:

- Will writing be an important part of my job?
- In what ways will my writing affect my success?
- How much will I write?
- What kinds of communications will I write?
- Who will my readers be?
- What sources will I have for the information and ideas I include in my communications?

By the time you finish learning what Eddie wants to know, you have to rush off. "I'll call you later with the answers," you promise.

"I don't have a phone," Eddie says with some embarrassment. "I'm trying to cut costs. But here's my address. Can you drop me a note in case I don't run into you again soon?"

"Sure," you agree. For Yvonne's brother, you'll write a note. Besides, he's a nice kid.

Your Assignment

Help Eddie by contacting two people who work in your field. Then write Eddie a note in which you report the answers you obtained to his questions. Some alternative ways to answer Eddie's questions are:

- Ask an instructor in your major department. If possible, find an instructor who has worked off-campus in the kind of job you want.
- If you have already worked in the kind of position you want as an intern, co-op student, or regular employee, base your answers on what you observed.

In your memo, give the names and positions of the persons with whom you spoke.

CASE   **SELECTING THE RIGHT FORKLIFT TRUCK**

It has been two weeks since you received this assignment from your boss, Mickey Chelini, who is the Production Engineer at the manufacturing plant that employs you. "We've been having more trouble with one of our forklift trucks," he explained. You are not surprised. Some of those jalopies have been breaking down regularly for years.

"And Ballinger's finally decided to replace one of them," Mickey continues. Ballinger is Mickey's boss and the top executive at the plant. His title is Plant Manager.

Notes on Forklifts

Present Forklift. The present forklift, which is red, moves raw material from the loading dock to the beginning of the production line and takes finished products from the Packaging Department back to the loading dock. When it moves raw materials, the forklift hoists pallets weighing 600 pounds onto a platform 8 feet high, so that the raw materials can be emptied into a hopper. When transporting finished products, the forklift picks up and delivers pallets weighing 200 pounds at ground level. The forklift moves between stations at 10 mph, although some improvements in the production line will increase that rate to 15 mph in the next two months. The present forklift is easy to operate. No injuries and very little damage have been associated with its use.

Electric Forklift. The electric forklift carries loads of up to 1,000 pounds at speeds up to 30 mph. Although the electric forklift can hoist materials only 6 feet high, a 2-foot ramp could be built beneath the hopper platform in three days (perhaps over a long weekend, when the plant is closed). During construction of the ramp production would have to stop. The ramp would cost $600. The electric forklift is $17,250 and a special battery charger costs $1,500 more. The lift would use about $2,000 worth of electricity each year. Preventive maintenance costs would be about $300 per year, and repair costs would be about $800 per year. While operating, the electric forklift emits no harmful fumes. Parts are available from a warehouse 500 miles away. They are ordered by phone and delivered the next day. The electric forklift has a good operating record, with very little damage to goods and with no injuries at all. It comes in blue and red.

Gasoline Forklift. The gasoline forklift is green and carries loads of up to 1 ton as rapidly as 40 mph. It can hoist materials 12 feet high. Because this forklift is larger than the one presently being used, the company would have to widen a doorway in the cement wall separating the Packaging Department from the loading dock. This alteration would cost $800 and would stop production for two days. The gasoline forklift costs $19,000 but needs no auxiliary equipment. However, under regulations established by the Occupational Safety and Health Administration, the company would have to install a ventilation fan to carry the exhaust fumes away from the hopper area. The fan costs $870. The gasoline forklift would require about $1,800 of fuel per year. Preventive maintenance would run an additional $400 per year, and repairs would cost about $600 per year. Repair kits are available from the factory, which is 17 miles from our plant. Other owners of the forklift have incurred no damage or injuries during the operation of this gasoline forklift.

FIGURE 1–11 Notes for Use with Case

"What finally happened to make him decide that?" you ask. "I thought he was going to keep trying to repair those old things forever."

"Actually, he wants to replace one of the newer ones that we bought just two years ago," Mickey replies. "That particular forklift was manufactured by a company that has since gone bankrupt. Ballinger's afraid we won't be able to get replacement parts. I think he's right."

"Hmmmm," you comment.

"Anyway," Mickey says, "Ballinger wants to be sure he spends the company's money more wisely this time. He's done a little investigation himself and has narrowed the choice to two machines. He's asked me to figure out which one is the best choice."

You can see what's coming. You've had a hundred assignments like this before from Mickey.

"I'd like you to pull together all the relevant information for me. Don't make any recommendation yourself; just give me all the information I need to make my recommendation. Have it to me in two weeks."

"Will do," you say, as you start to think about how you can squeeze this additional assignment into your already bulging schedule.

Your Assignment

It's now ten days later. You've gathered the information shown in Figure 1–11. First, plan your final report by performing the following activities:

- List the specific questions your boss will want your report to answer. *Note:* your boss does *not* want you to include a recommendation in your report.
- Underline the facts in your notes that you would include in your report; put an asterisk by those you would emphasize.
- Decide how you would organize the report.
- Explain which techniques from the list on pages 21 through 23 you would use when writing this report. What other things would you do to assist your readers?

Second, imagine that you have been promoted to Mickey's job (Production Engineer). Tell how you would write the report that you would send to Ballinger regarding the purchase of a new forklift truck. It must contain your recommendation.

2

EXAMPLES OF READER-CENTERED WRITING: RESUMES AND LETTERS OF APPLICATION

I magine you are sitting at your desk at work preparing to write something. Maybe it's a report for your employer or a letter to a client. You know the result you want to bring about: to persuade someone to accept your recommendation, to help someone learn the answer to a question, to enable someone to perform a new procedure. Furthermore, you remember reading in the first chapter of this book that you should take a reader-centered approach to writing, one in which you think constantly about your readers.

But you are a little puzzled. You ask, ''How am I supposed to think about my readers, and how will my thoughts about my readers help me put words down on paper?''

The rest of this book answers those questions. It presents many suggestions for thinking about your readers and for using the knowledge you gain about them to help you write effectively on the job. As explained in Chapter 1, much of this advice is organized around the five major activities of the writing process: defining objectives, planning, drafting, evaluating your draft, and revising. The purpose of this chapter is to provide you with a brief overview of that process so you can see how these five activities fit together and how to put your thoughts about your readers to good, practical use as you perform each of them.

THE PERSUASIVE AIM OF THIS CHAPTER

To introduce you to the writing process, this chapter shows how you can perform each of the five major activities of writing while creating your resume and letter of application for employment.

Why were these particular communications chosen for the demonstration? Partly to be of immediate, practical help to you. Your resume and job application letter will play important roles in your career. You may even need them right now to apply for a summer job, part-time job, internship, co-op position, or full-time employment.

Beyond that, the resume and letter of application readily illustrate the value of thinking constantly about your readers: they make it especially evident that your success depends entirely on the impression your communications make on your audience.

When you finish studying this chapter you should know two things: how to write a resume and letter of application and how to follow a reader-centered writing process. In addition, you should see *why* it is worthwhile to use a reader-centered writing process, one in which you keep your readers—not yourself, not your subject matter—primarily in mind at every step.

WRITING YOUR RESUME

In this chapter, you will walk through the writing process twice, first with the resume and then, more rapidly, with the job-application letter.

Defining Your Objectives

The first activity of writing, defining objectives, is especially important whether you are writing a resume or any other job-related communication. When defining objectives, you tell what you want your communication to do. Thus your objectives form the basis for all your other work at writing. Together they are the foundation you build on, the goal you strive for, the standard against which you measure your success as you perform the other four writing activities.

To take a reader-centered approach to defining objectives, you need to look at three things: the final result you desire, the people who will read your communication, and the specific way you want your communication to affect those people as they read your message. With a resume the final result you desire is obvious: to get you a job. However, the other two topics require some careful thought. The following paragraphs explain what you should consider as you think about the readers of your resume and the way you want your resume to affect them.

Understanding Your Readers

The first step in understanding your readers is to learn who they will be. To find out who will read your resume, you need to learn how your resume will be used by the employers you apply to. Typically, employers recruit new employees in two stages, each with a different group of readers.

In the first stage, employers try to attract applications from as many qualified people as possible. They do this by maintaining a personnel office that accepts and screens applications; by interviewing students on college campuses; and by placing advertisements in job catalogs (like the *College Placement Annual*), professional journals, and newspapers. At this stage of recruiting, resumes are usually read by people who work in the personnel office. The main objective of these readers is usually to screen out applicants who are obviously unqualified.

To understand these first-stage readers, you may find it helpful to draw an imaginary portrait of one of them. Imagine a man who has just sat down at his desk to read a stack of twenty-five to fifty new applications that arrived in today's mail. He doesn't have enough time to read all of the applications thoroughly, so, with jacket off and sleeves rolled up, he sorts quickly through the pile for the two or three persons who might merit additional consideration. He quickly finds reasons to disqualify most applicants; only occasionally does he read a full resume. As you write and revise your resume, keep in mind that it must quickly attract and hold this man's attention. It must make an immediate and positive impact if it is to stand out from the many resumes heaped on his desk.

In the second stage of recruiting, employers carefully scrutinize the qualifications of the most promising applicants. Often this involves visits by the candidates to the employer's site. The second-stage readers of your resume include especially the manager of the department you would work for, but also people you would work with in that and other departments.

To represent your reader in this stage, you might imagine the head of the department you wish to work for. This person is shorthanded and wants rapidly to fill one or more openings. She has waited a week for the personnel office to forward resumes from the most promising applicants. When they finally arrive, she clears a space on her crowded desk and hopes no one will interrupt before she can read them. Her needs are very specific and she knows precisely the qualifications she seeks. As you write and revise your resume, keep in mind that it must persuade this reader that you have the specific abilities, education, and experience she wants.

Of course, some job searches vary from the two-stage recruiting procedure just described. If you interview at a campus placement center, you will probably hand your resume to the company recruiters at the same time you meet them. If you already have a full-time job, you may be able to bypass the personnel office when you apply for your next position. In almost all job searches, however, there are readers who scan an applicant's resume quickly and others who use the resume to carefully scrutinize the applicant's qualifications. For example, the recruiter you meet in a campus placement center will have a great desire to find your key qualifications very quickly. If your interview is successful, the person will probably circulate your resume to others who will study it in great detail.

Deciding How You Want Your Resume to Affect Your Readers

After you have identified your readers, you should determine how your resume will need to affect them if it is to result in the job offer you are seeking. More precisely, you should define how you want your resume to affect your readers *while they read it*. Why? As Chapter 1 explained, that's the only time your resume has any direct influence on your readers. Of course, your readers may later recollect what your resume says, and be influenced by that information—but even what they recollect will depend on what happens as they read.

To determine how you want your resume—or any other communication—to affect your readers, you can think about two things:

- The way you want your communication to alter your readers' attitudes
- The task you want to help your readers perform while they read

The insights you gain by defining your purpose in these ways will help you immensely throughout your work at writing your resume, as will become clear later in this chapter.

Altering Your Audience's Attitudes   To define the way you want a communication to alter your readers' attitudes, first determine how your audience feels before reading what you are now writing, and then decide how you want them to feel after reading it. Your readers' attitude toward you now, before reading your resume, is neutral because they don't know anything about you.

You want them to come to believe as they read that you are highly qualified for the job you seek.

Once you have described your readers' present and desired attitudes, try to find out things about your readers that will help you plan a strategy for persuading them to change their attitudes in the way you specified. To begin, find out what will appeal to your audience—what they want. In the case of your resume, that means determining what qualities employers look for in the people they hire. By showing your readers that you have these qualities, your resume will favorably alter their attitude about you.

As common sense will tell you, employers want to hire people who are:

- Capable. Through their training and their ability to learn, applicants must be able to perform the tasks the employer assigns them.
- Responsible. Applicants must be trustworthy enough to use their training and abilities to benefit the organization.
- Pleasant. Applicants must be able to interact compatibly with other employees on the job.

Of course, these qualities are stated only generally. The readers of your resume will look for them in much more specific terms. Instead of asking, ''Is this applicant capable?'' they will ask, ''Can this person analyze geological samples?'' ''Program computers in COBOL?'' ''Manage a clothing store?'' and so on.

How can you learn exactly what qualities your readers will look for? You might ask professors in your major department or visit people who hold the kind of job you desire. The information you obtain is information you can use. It tells you what things you can say about yourself to persuade your readers that *you* are sufficiently capable, responsible, and pleasant for the job.

Helping Your Readers Perform Their Tasks   While part of your resume's purpose is to alter your readers' attitudes, another part is to enable your readers to perform the work of reading. Different kinds of communication involve different tasks. When you know what those tasks are, you can write your communication in a way that will help your readers perform them easily.

When reading your resume, your readers' primary task will be to seek the answers to the questions they ask when determining whether or not an applicant is capable, responsible, and pleasant:

- What exactly does this person want to do?
- What kind of education does this person have for the job?
- What experience does the person have in this or a similar job?
- What other activities has the person engaged in that have helped prepare him or her for the job?
- How can I get more information about this person's qualifications?

Knowing that your readers will be looking for the answers to these questions tells you a great deal about what to include in your resume. Knowing that your readers initially will devote only a brief time to reading your resume lets you know that you should help your readers find that information quickly.

In sum, when defining your objectives, you should strive to learn about your readers and say as precisely as possible how you want your communication to affect them while they are reading it. The following four sections show how you can use the insights you gain by defining your objectives to help you plan, draft, evaluate, and revise your resume.

Planning

When you plan, you decide what to say and how to organize your material. In addition, you should find out any relevant expectations that your readers have about your communication, because those expectations may limit the choices you make concerning content and organization.

Deciding What to Say

Your definition of your resume's purpose provides you with direct help in determining what to say. In essence, your resume is a persuasive argument whose purpose is to convince your readers to hire you. Like any persuasive argument, it has these two elements: a claim and the evidence that supports the claim. Your definition of objectives tells you what the (implicit) claim in your resume should be: that you are the kind of capable, responsible, and pleasant person that employers want to hire.

Furthermore, your objectives can help you identify some of the specific facts you can mention as evidence to support that claim about yourself. Your objectives do that by alerting you to the kinds of questions your readers will be asking while they read your resume. When you answer them, you have an opportunity to present evidence that you possess the key characteristics your readers seek in employees. Figure 2–1 shows how you can match your *implicit* persuasive claims to the readers' questions. Later in this chapter you will find detailed advice about the evidence you might provide to support your claims.

Organizing Your Material

When planning a communication, you need to decide not only what you will say but also how you will organize your material. Your reader-centered objectives help you in that task, too. For example, your definition of your resume's objectives tells you that you need to organize your material in a way that will emphasize the major evidence that you are qualified for the job you seek—and it helps you determine which evidence is the major evidence. Similarly, your portrait of your readers makes clear that you must organize so that your impatient, rapidly reading audience finds your most persuasive information quickly.

Conventional Resume Topics	Usual Questions By Readers	Persuasive Claims To Make
Objective	What, exactly, do you want to do?	I have a sense of direction and I want a job that you have.
Education Honors	Do you have the required knowledge?	I have the required knowledge, perhaps more than you expected.
Work Experience Activities	Do you have experience in this or a similar job? What responsibilities have you had?	I have related experience in which I learned things directly relevant to the job I want. I was trusted by my employers.
Interests	Do your interests show that you would be a pleasant and compatible employee?	I am an interesting, well-rounded person who can work well with others.
References	How can I get more information about your qualifications?	Important people with knowledge of my business qualifications will attest that I would make a good employee.

FIGURE 2–1   Matching Persuasive Claims to Readers' Questions

More than one organizational pattern can be used to achieve those
tives. Most resumes are organized around the applicant's experiences.
the topics listed in Figure 2–1 serve also as organizational categories:
tional experiences, work experiences, and so on. Figure 2–2 shows a
organized in this way.

However, some individuals choose to organize a substantial part
resume around their accomplishments and abilities. Such a resume i
times called a *functional resume* because it emphasizes the functions a
the applicant can perform. Generally, people who use functional resun

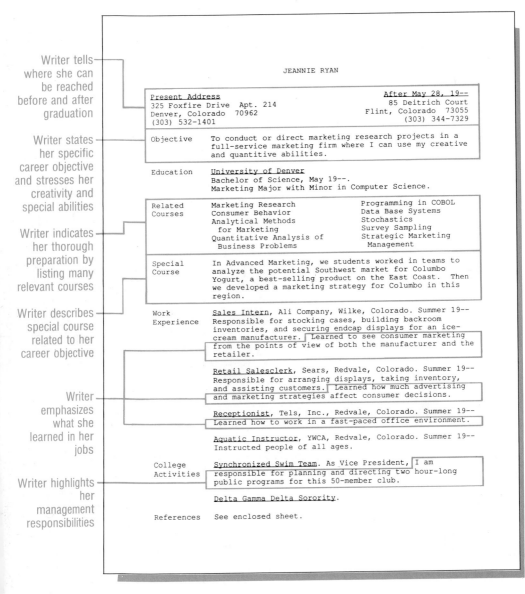

Writer tells where she can be reached before and after graduation

Writer states her specific career objective and stresses her creativity and special abilities

Writer indicates her thorough preparation by listing many relevant courses

Writer describes special course related to her career objective

Writer emphasizes what she learned in her jobs

Writer highlights her management responsibilities

JEANNIE RYAN

<u>Present Address</u>	<u>After May 28, 19--</u>
325 Foxfire Drive Apt. 214	85 Deitrich Court
Denver, Colorado 70962	Flint, Colorado 73055
(303) 532-1401	(303) 344-7329

Objective   To conduct or direct marketing research projects in a full-service marketing firm where I can use my creative and quantitive abilities.

Education   <u>University of Denver</u>
Bachelor of Science, May 19--.
Marketing Major with Minor in Computer Science.

Related   Marketing Research   Programming in COBOL
Courses   Consumer Behavior   Data Base Systems
                   Analytical Methods   Stochastics
                     for Marketing   Survey Sampling
                   Quantitative Analysis of   Strategic Marketing
                     Business Problems   Management

Special   In Advanced Marketing, we students worked in teams to
Course   analyze the potential Southwest market for Columbo
                 Yogurt, a best-selling product on the East Coast. Then
                 we developed a marketing strategy for Columbo in this
                 region.

Work   <u>Sales Intern</u>, Ali Company, Wilke, Colorado. Summer 19--
Experience Responsible for stocking cases, building backroom
                 inventories, and securing endcap displays for an ice-
                 cream manufacturer. Learned to see consumer marketing
                 from the points of view of both the manufacturer and the
                 retailer.

                 <u>Retail Salesclerk</u>, Sears, Redvale, Colorado. Summer 19--
                 Responsible for arranging displays, taking inventory,
                 and assisting customers. Learned how much advertising
                 and marketing strategies affect consumer decisions.

                 <u>Receptionist</u>, Tels, Inc., Redvale, Colorado. Summer 19--
                 Learned how to work in a fast-paced office environment.

                 <u>Aquatic Instructor</u>, YWCA, Redvale, Colorado. Summer 19--
                 Instructed people of all ages.

College   <u>Synchronized Swim Team</u>. As Vice President, I am
Activities   responsible for planning and directing two hour-long
                 public programs for this 50-member club.

                 <u>Delta Gamma Delta Sorority</u>.

References   See enclosed sheet.

FIGURE 2–2   Conventional Resume

extensive work experience with on-the-job accomplishments to talk about. Figure 2–3 shows an example of a functional resume.

Whichever organizational pattern you use, you still must decide the order in which you will present your material. If you think about your readers in the act of reading your resume, you will see that you should make your name and your professional objective prominent. By making your name prominent, you tell your readers at the outset who you are and you help them relocate your

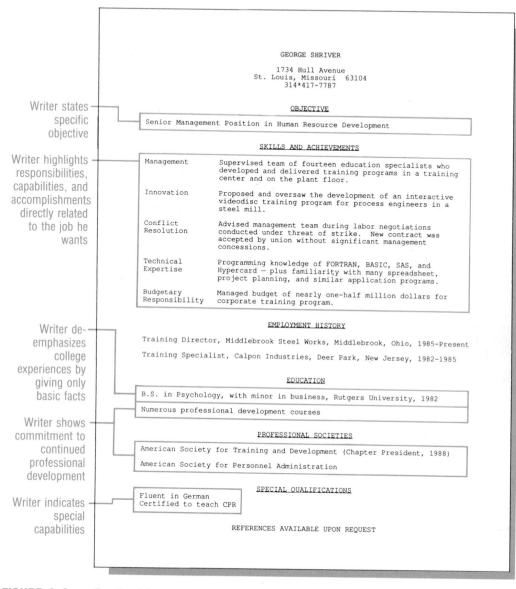

Writer states specific objective

Writer highlights responsibilities, capabilities, and accomplishments directly related to the job he wants

Writer de-emphasizes college experiences by giving only basic facts

Writer shows commitment to continued professional development

Writer indicates special capabilities

GEORGE SHRIVER

1734 Hull Avenue
St. Louis, Missouri 63104
314*417-7787

OBJECTIVE

Senior Management Position in Human Resource Development

SKILLS AND ACHIEVEMENTS

Management	Supervised team of fourteen education specialists who developed and delivered training programs in a training center and on the plant floor.
Innovation	Proposed and oversaw the development of an interactive videodisc training program for process engineers in a steel mill.
Conflict Resolution	Advised management team during labor negotiations conducted under threat of strike. New contract was accepted by union without significant management concessions.
Technical Expertise	Programming knowledge of FORTRAN, BASIC, SAS, and Hypercard — plus familiarity with many spreadsheet, project planning, and similar application programs.
Budgetary Responsibility	Managed budget of nearly one-half million dollars for corporate training program.

EMPLOYMENT HISTORY

Training Director, Middlebrook Steel Works, Middlebrook, Ohio, 1985-Present

Training Specialist, Calpon Industries, Deer Park, New Jersey, 1982-1985

EDUCATION

B.S. in Psychology, with minor in business, Rutgers University, 1982

Numerous professional development courses

PROFESSIONAL SOCIETIES

American Society for Training and Development (Chapter President, 1988)

American Society for Personnel Administration

SPECIAL QUALIFICATIONS

Fluent in German
Certified to teach CPR

REFERENCES AVAILABLE UPON REQUEST

FIGURE 2–3   Functional Resume

resume quickly if they want to refer to it again. By announcing your professional objective prominently, you help your readers judge your qualifications for that particular job as they read the rest of your resume. If you are writing a conventional resume, you can provide the desired prominence by placing your name and professional objective on the top. If you are designing an unconventional resume, you may place your name along the bottom or side.

After stating your professional objective, you should organize your remaining material by following one of the most basic strategies for writing at work: put the most important information first. This will ensure that your hurried readers come to the most important information quickly. In a resume, the most important material is that which describes your most impressive qualification. If you are like most college students, that qualification will be your education. However, if you have had substantial work experience, your accomplishments and proven skills may eclipse your education and so should go first. From there, you should arrange your information in decreasing order of importance. By convention, references come last.

Finding Out What's Expected

A final task involved with planning a communication is to consider the expectations your audience or other people may have about it so you can take these expectations into account when designing your message. Sometimes these expectations are requirements, such as page limits and regulations about what to include and how your communication should look. Often expectations arise from custom, reflecting conventional rather than required ways of doing things.

Applying for jobs is a highly ritualized activity in our society and many conventions about resumes exist. However, these conventions vary from one field to another, so the only way to deal with them effectively is again to follow the main advice of this book: learn about your particular readers and create your communication accordingly. For example, people in some fields, such as chemistry and engineering, take a very conservative approach to resumes. To them, a resume should be typed neatly on white, buff, or light gray paper with the applicant's name, address, and phone number at the top. In contrast, people in other fields, such as advertising, are accustomed to seeing highly unconventional resumes, perhaps printed on pink paper—or on balloons or pop-up boxes. For them, even typed resumes may incorporate creative designs by, for example, placing the applicant's name at the bottom or running it up the side in large, ornate letters.

This chapter's advice about resumes reflects widespread conventions. Keep in mind, however, that the conventions in your field may be different—and that you may have to do some investigating to learn whether or not that is the case. Also, remember that creative variations from convention can sometimes be effective, if they are carried out thoughtfully and purposefully.

Drafting

When you draft, you transform your plans—your notes, outlines, and ideas—into a communication. For your resume, as for many communications you will create at work, you must not only draft the prose but also design the visual appearance of your message. As the following paragraphs explain, your description of your readers and the way you want to affect them can help you do both effectively.

Drafting the Prose

While you draft the prose for your resume, keep in mind your imaginary portraits of your readers. Remember that your purpose is to enable those people to locate the answers to their questions about you (listed in Figure 2–1) and to present those answers in a way that will persuade the readers you are capable, responsible, and pleasant. The following paragraphs suggest some ways you can do so.

Professional Objective   When you state your professional objective, you answer your readers' question, "What, exactly, do you want to do?" Your answer can be extremely important to the success of your resume. In response to a survey, personnel officers of the 500 largest corporations in the United States reported that the most serious problem they find with resumes and letters of application is the applicants' failure to specify their job and career objectives.[1]

That feeling by personnel officers may surprise you. "After all," you may reason, "once I present my qualifications, shouldn't an employer be able to match me to an appropriate opening?" The answer to that question lies in your imaginary portrait of your reader—the harried manager trying to work his way rapidly through a large number of applications. That person does not have time to act as your personal employment agency, trying to match you with a job. Furthermore, if you fail to describe your employment objectives, you may seem aimless—pleasant enough, talented enough, but not motivated enough to compete successfully with the many applicants who have precisely defined their goals.

What should your professional objective look like? By convention, such statements are one or two sentences long and are usually general enough that the writer could send them, without alteration, to many prospective employers. If you follow this convention, you would not say, for example, "I want to work in the process control department of Channing Electronic's Fort Lauderdale plant." Instead you would make a more general statement, such as, "I want to work in process control in a mid-sized electronics company in the southeastern United States."

This does not mean, however, that you need to develop a single professional objective that you can send to *all* the employers you might contact. Imagine, for instance, that you are still deciding whether to go into retail sales or business-to-business sales. You could then write two resumes, each with a professional objective suited to one of those two fields. (Of course, you might make other changes, too, to tailor each resume to its professional objective.) It's even possible to create a professional objective designed to appeal to one particular employer, for instance an employer for whom you have already worked as a summer intern. However, according to convention you would still state the objective so you could send the resume to similar companies as well.

Consequently, the challenge you face when writing your professional objective is to be neither too general nor too specific. You have struck the proper balance if you could send the same resume to several companies and if your readers can see that you want to work in their particular kind of *organization* (a manufacturing plant, a pharmaceutical company), and in some particular *department* (quality control, new product development).

Education   When describing your education, you provide evidence that you are capable of performing the job you have applied for. The basic evidence is your college degree, so you should name your college and your degree, and the date of your graduation. But you may also be able to say several other persuasive things:

- If your grades are good, mention them.
- If you have earned any academic honors, tell what they are.
- If you had any special educational experience, such as a co-op assignment or internship, describe it.

Also, keep in mind that the title of your degree does not necessarily alert your readers to your specific qualifications for the job. For that reason, you may want to describe some of the courses you have taken that relate directly to that job.

By looking at the resumes of Jeannie Ryan (Figure 2–2), Ramon Perez (Figure 2–4), and Sharon Pollock (Figure 2–5), you can see how three very different people have elaborated on the way their educations qualify them for the jobs they want. Jeannie is a senior who has gone to college full-time (except summers) since graduating from high school. Like the other two, she lists her school, her major, and her date of graduation. Because her grades are somewhat below average, she does not mention her grade point average. Nevertheless, she makes a very good impression by listing courses she has taken in two areas—marketing management and analytical methods—that are critical to the kind of job she wants. Furthermore, she describes in detail a course in which she worked on a major project that closely resembles the kind of task she will have on the job.

Unlike Jeannie, Ramon has completed most of his college study in the evening while also holding a full-time job. Nevertheless, he earned excellent grades and therefore lists his grade point average. In addition, he describes his honors in a separate section, thereby making them more prominent than they would have been if he had merely included them with the other material under the heading "Education."

Sharon's much different approach to the "Education" section is one typically used by experienced employees: she lists nothing other than the basic facts of her school, major, and year of graduation. Because she graduated fourteen years ago, she answers the readers' questions about her capabilities by focusing on her work experience rather than on the details of her education.

Work Experience   By describing your work experience, you can present additional evidence about your capabilities and also show you are responsible.

Ramon Perez

16 Henry Street
Brooklyn, New York 11231
Days: (212) 374-7631
Home: (718) 563-2291

Professional Objective	A position as a systems analyst where I can use my knowledge of computer science and business to develop customized systems for financial institutions
Education	**New York University.** *B.S. in Computer Science.* *December 19--* GPA 3.4 overall; 3.7 in major
	Computer classes include artificial intelligence and expert systems, computer security, data communication, deterministic systems, stochastics, systems design
	Business classes include accounting, banking, finance, economics, business law
	Worked full-time while completing last half of course work
Honors	Dean's List three times Golden Key National Honor Society
Related Work Experience	**Miller Health Spas,** New York City, 1988-Present. *Data Entry Clerk*
	• Helped convert to a new computerized accounting system • Served on a four-person team that wrote user documentation for the new system • Trained new employees • Earned Employee of the Month Award twice
	Meninger Bank, New York City, 1986-1987. *Teller*
	• Performed all types of daily, night-deposit, and bank-by-mail transactions • Proved the vault, ordered currency, and handled daily cash flow • Learned what financial computer systems look like from the tellers' viewpoint
Activities	**Juvenile Diabetes Foundation,** 1986-Present. *Volunteer*
	• Helped design a major fund-raising event two years in a row • Successfully solicited contributions from sponsors • During the second year, the event achieved its highest level of contributions ever

References		
Professor Mildred Dobrick Finance Department New York University New York, NY 12234 (212) 254-9642	Professor R. Theodore Berger Communication Department New York University New York, NY 12234 (212) 254-7539	Wilson Meyerhoff Senior Accountant Miller Health Spas 3467 Broadway New York, NY 12232 (212) 671-9007

FIGURE 2–4   Resume of a Person Who Completed College While Working Full-Time (Prepared with Desktop Publishing)

Sharon Pollock

8965 Portage Drive
Baton Rouge, LA 70811
504/569-4230

Objective   A position as Corporate Manager of Drilling Operations for an international oil company.

Summary   Fourteen years of experience in a variety of drilling operations in both the United States and overseas. Steady progression in management and budgetary responsibilities. Extensive record of success in introducing technical innovations that improved drilling performance.

Experience and Achievements

Pilot Petroleum Company

Regional Drilling Engineer   1986-Present

Responsible for Pilot's drilling operations throughout the entire Gulf of Mexico Region. Coordinate comprehensive drilling programs for both exploration and new oil field development. Prepare one-year and three-year drilling plans for corporate head-quarters. Prepare and administer a $60,000,000 annual budget. Select drilling contractors and negotiate all contractual agreements with them.

- Successfully integrated the drilling operations of three small companies acquired by Pilot.
- Introduced measurement-while-drilling technology to the company, reducing drilling time for directional holes by an estimated 12%.
- Organized Pilot's first Drilling Technology Group, which was responsible for developing and disseminating innovative solutions to drilling problems commonly encountered in the Gulf Region.
- Created a drilling program for exploratory wells at a newly leased area in Asia where extensive and unexpected drilling problems were encounterd. (This was a special, one-year troubleshooting assignment.)

Lead Drilling Engineer   1982-1986

Working on the staff of the Regional Office, provided day-to-day technical advice and management supervision for up to 15 wells at a time in the Gulf of Mexico.

- Scheduled and budgeted all wells under my supervision.
- Prepared performance evaluations of all staff drilling engineers.
- Introduced the use of computers to monitor downhole conditions while drilling. When later adopted throughout the company, this technology increased the average life of drill bits by 17 hours.

Field Drilling Engineer   1977-1982

Supervised drilling operations and rig crews for both shallow and deep wells in Rocky Mountain and Alaskan oil fields.

FIGURE 2–5   Resume of a Person with Fourteen Years of Professional Experience (Prepared with Desktop Publishing)

Sharon Pollock

Previous Employment

Neptune Oil - Corpus Christi, Texas   1975-1976
  Engineering Summer Intern

Lawson Enterprises - Galveston,Texas   1974
  Roustabout during summer vacations from college

Education

University of Texas at Austin, B.S. in Petroleum Engineering   1977

Certification

Registered Professional Engineer in Alaska, Colorado, Louisiana, and Texas

Professional Honors and Service

Member, Industrial Advisory Council, University of Texas   1989-Present
Society of Petroleum Engineers Distinguished Lecturer   1988

Professional Memberships

Member, Society of Petroleum Engineers of AIME
Member, Society of Professional Well Log Analysts
Member, Pi Epsilon Tau (Petroleum Engineering Honorary)

Personal

Born: 8/24/54   Excellent Health   Married, three children

FIGURE 2–5   *(continued)*

However, you will miss the opportunity to fully display your qualifications if you merely name your former employers and give your job titles.

You can show your capabilities by mentioning any duties you performed that resemble duties you will have on the kind of job you are currently seeking. Because she has lots of professional experience, Sharon can do that throughout her resume. For instance, in the general description of her current job, she highlights her accomplishments in managing a multimillion-dollar budget, introducing new drilling technologies, and establishing a drilling program in an overseas oil field—all capabilities she would need in the type of job she seeks. Ramon similarly mentions that in his current job he has helped create a computer system that performs tasks previously done manually, something he will surely have to do as a systems analyst.

Even if your previous jobs seem unrelated to your career objectives, you may still be able to use them to show that you have skills and knowledge that qualify you for the job you seek. Consider, for instance, the way Jeannie describes her job as a sales intern. Realizing that her duties of stacking ice cream packages neatly in supermarket freezers might not seem relevant for a job in marketing, she includes a brief description of valuable knowledge she gained in that job: "Learned to see consumer marketing from the points of view of both the manufacturer and the retailer."

You can also use your work experience to show you are reliable, that others have entrusted you with significant responsibility. Ask yourself, "What did I do in that job that showed others relied on my sense of responsibility?" By asking himself that question, Ramon discovered that even his daily routine of posting checks and cash payments could be turned to advantage: "handled daily cash flow."

Your noteworthy achievements are also good to highlight. For experienced people like Sharon, that can be especially helpful. Notice how she uses a list format to draw attention to her impressive accomplishments in her current job. If you are just graduating from college, you may not have achievements of the sort Sharon can cite, but it will still be worthwhile for you to think back over your jobs to see whether you have any special accomplishments you could mention in your resume.

By the way, if you paid a large portion of your college expenses with your earnings, be sure to mention that fact.

The following paragraphs provide advice about two important decisions you will have to make when drafting your resume: deciding how to order your jobs (if you've had more than one) and deciding how to phrase your descriptions of your responsibilities and accomplishments.

- **Ordering your jobs.** When deciding on the order in which to present your jobs, remember that you want to enable your busy readers to see your most impressive qualifications as soon as possible. Most people can achieve that objective by listing their jobs in reverse chronological order because their most recent job is also their most impressive one. However, if your most recent job is *not* your most impressive, you may want to devise some other way of presenting your jobs that gives the most important one its proper emphasis. One way to do that is to display that job under a special heading, such as "Related Job Experience"—related, that is, to the job you are seeking. Then you might describe the less-impressive jobs under the heading "Other Job Experience."

● **Phrasing your statements.** Even the way you phrase your descriptions of your jobs can determine the success of your resume. Use verbs rather than nouns, because verbs portray you in action. For example, don't say you were responsible for the "*conversion* of manual accounting system to computer" but that you "*converted* manual accounting system to computer." Don't say you were responsible for the "*analysis* of test data" but that you "*analyzed* test data."

When choosing your verbs, choose specific, lively ones, not vague, lifeless ones. Avoid saying simply that you "worked with merchandise displays" but say you "designed" or "created" those displays. Don't tell your readers you "interacted with clients" but that you "responded to client concerns."

By the way, where you make parallel statements, be sure to use the grammatically correct parallel construction. Nonparallel constructions can slow reading and indicate a lack of writing skill. For example, when listing your major responsibilities in a job, don't say:

Not parallel

● Helped train new employees
● Correspondence with customers
● Prepared loan forms

Instead say:

Parallel

● Helped train new employees
● Corresponded with customers
● Prepared loan forms

Similarly, use parallel construction within sentences or sentence fragments. For example, revise the following

Not parallel

● My responsibilities included demonstrating the equipment, order preparation, and following up after delivery.

to read:

Parallel

- My responsibilities included demonstrating the equipment, preparing orders, and following up after delivery.

By using active, lively verbs and by using parallel constructions where appropriate, you can greatly increase the persuasiveness of your work-experience section.

Activities   At the very least, your participation in group activities indicates you are a pleasant person who gets along with others. Beyond that, your activities may also show you developed some specific abilities that are important in the job you want. Notice, for instance, how Jeannie Ryan describes her participation in one of her extracurricular activities:

```
Synchronized Swim Team.  As Vice President, I am
responsible for planning and directing two hour-long
public programs for this 50-member club.
```

This statement suggests Jeannie has been entrusted with considerable responsibility by people who know her well, and also that she has experience in an essentially managerial position.

In considering material to include in your resume, be sure not to overlook ways in which you have demonstrated unusual abilities of any sort, such as being selected for a high school choir that toured Europe or playing varsity sports at your college.

Interests   Although it is not necessary for you to tell about your interests in a resume, by doing so you can provide your readers with evidence that you are a pleasant person—one who can talk about something other than studies and jobs. Of course, the information you provide under the ''Activities'' heading may provide plenty of information about your interests, making a separate section unnecessary.

Personal Data   Federal law prohibits employers from discriminating on the basis of sex, religion, color, age, or national origin. It also makes it illegal for employers to inquire about matters unrelated to the job a person has applied for. For instance, employers cannot legally ask you if you are married or plan to be married. Many students welcome these restrictions on employers because they consider such questions to be personal or irrelevant to their qualifications. On the other hand, federal law does not prohibit you from giving employers information of this sort, if you wish. And many students do.

Guidelines for effective writing cannot help you decide whether to include personal information in a resume. Your understanding of your readers can help you predict whether information of this sort might help win the job you want, but you will have to decide for yourself whether you wish to share such information in your resume.

If you do include personal data, you should probably place it as the last item before your references, which by convention appear at the very end of a resume. Even if the data is helpful in making you seem qualified, it is almost certainly less impressive than the things you say in the other sections.

References   Your references are people who agree to tell employers about your qualifications. When choosing your references, select people who can speak about your *professional* qualifications, such as college instructors, employers, and advisers to campus organizations in which you had a leadership role. Avoid people like your parents and their friends, who (an employer might feel) are going to say nice things about you no matter what.

By convention, you should have at least three references. That number is so common that you may appear deficient if you have fewer. You may have *more* than three, of course. Remember also that your readers may draw inferences about you based on the mix of people you list as references. For example, if your list includes one or more instructors from your major, readers will be likely to assume that you made a good impression in your classes. In contrast, if you don't list any instructors from your major, they may assume that you didn't do very well in your courses. Of course, if it's been several years since you graduated, your readers will not expect you to list instructors.

You can list your references in your resume or on a separate sheet that has your name as its heading. If you use a separate sheet, you can either attach it to your resume or give it separately to an employer when asked for it. Although many students simply state on their resumes that the names of their references are "available on request," there are two good reasons for supplying your references on your resume or along with it. First, you make it easier for employers to contact them. Otherwise, employers will have to write to you and then await your reply before making contact. Second, the mere appearance of your references can strengthen your resume if you make them seem as important as possible. By showing that impressive people will speak on your behalf, you make yourself impressive in the eyes of your readers.

Although it is generally best to include references on (or with) a resume, some people find it difficult or impossible to do so. For example, many people with full-time jobs do not want their current employers to know that they are looking elsewhere until they have a serious expression of interest from someone else. If such a consideration prevents you from listing your references and if you think your readers will wonder why you haven't listed any, you might want to explain your situation briefly in your resume or letter of application.

If you do list your references, you should carefully consider the information you give about them. Too many resume writers write references like these:

John Douglas	Walter Williamson
Laws Hall	6050 Busch Boulevard
Miami University	Columbus, Ohio 43220
Oxford, Ohio 45056	(614) 376-2554
(513) 529-5265	

For all an employer will know, Laws Hall might be a dormitory and Mr. Douglas might be the applicant's roommate. Likewise, a reader might think that Mr. Williamson is the applicant's neighbor on Busch Boulevard. To prevent such misunderstandings, you should give the title and full business address of your references:

Dr. John Douglas, Chairperson	Mr. Walter Williamson, Director
Management Department	Corporate Services Division
Laws Hall	Orion Oil Company
Miami University	6050 Busch Boulevard
Oxford, Ohio 45056	Columbus, Ohio 43220
(513) 529-5265	(614) 376-2554

Be sure to list phone numbers (including area codes) because many employers would rather call references than write to them.

Before you include anyone as a reference in your resume, check with that person. This can benefit you in three ways. First, you can assure yourself that the person feels well enough acquainted with you to give a good, supportive account of you to employers. Second, it helps your reference person prepare for calls from employers: an instructor who had you in class a year ago or an employer who hired you two summers ago may have momentary trouble recalling you if they receive an unexpected phone call about you. Third, you can arrange to provide the person with your resume, so he or she will have specific information about you at hand when an employer calls.

Ramon includes his references within his resume (Figure 2–4). Jeannie presents hers on a separate sheet, shown in Figure 2–6.

Designing the Appearance

As you design the appearance of your resume, one of the most important things to remember is that you want to help your very busy readers *quickly* find the specific information they want.

You can do that by following one of the conventional layouts shown in Figures 2–2 through 2–5 or by creating a less conventional design if you think such layouts will be effective with a particular employer. Whether your layout is conventional or unconventional, use headings extensively to help readers

```
                          REFERENCES

                         Jeannie Ryan


        Tracey Sutton
        District Manager
        Ali Company
        200 West Wilson Bridge Road    Suite 240
        Wilke, Colorado   73402
        (303) 349-7773


        Professor Mark B. Niles
        Marketing Department
        University of Denver
        Denver, Colorado   76403
        (303) 459-7800


        Professor Samuel M. Chan
        Marketing Department
        University of Denver
        Denver, Colorado   76403
        (303) 459-7800


        Professor George M. Schwartz
        Computer Science Department
        University of Denver
        Denver, Colorado   76403
        (303) 459-2121
```

FIGURE 2–6   Reference List to Accompany the Resume Shown in Figure 2–2

locate the specific information that interests them. Use ample white space to separate blocks of information. And be sure your margins are large enough to avoid a cramped and crowded appearance.

Because resumes are usually sent to more than one company, they may be printed by either copy machine or offset press. Naturally, your resume should be neat and carefully proofread.

Evaluating

Throughout your work at drafting, you concentrate on developing strategies that you think will affect your readers in the way you desire. But will your strategies actually succeed? To find out, you need to step back from your communication, temporarily abandoning your role as writer. Instead try to see your communication as your readers will see it. By doing so, you will be able to determine which of your writing strategies will work and which won't.

Unfortunately, as writers we all have difficulty seeing our own work in just the way our readers will. When we look at our pages, we often see what we intended—which is not always the same as what we accomplished. Consequently, we all benefit from showing our drafts to other people, asking for their reactions and advice. When you are writing a resume, various people can help you. If the instructor in the course for which you are using this book asks you and your classmates to share drafts with one another, you will get (and give) helpful advice. Likewise, both your writing instructor and instructors in your major department can help you see where your draft works and where it might be strengthened.

Revising

The final activity of the writing process is revising. When you revise, you improve your communication by acting on the insights and advice you gained by evaluating it. To make these improvements, you will probably have to consider not only matters of right and wrong but also matters requiring your sensitivity, insight, and good judgment.

For example, the people who help you evaluate drafts of your resume may identify problems but not be able to tell you how to overcome them. In addition, some people may give you advice that contradicts the advice you get from others. That's partly because conventions about resume writing differ from field to field and partly because many people believe a ''correct'' way to create resumes exists but often disagree with one another about what that correct way is.

How can you choose between conflicting pieces of advice? You need to think back to your objectives, to your descriptions of your readers and purpose. Then, using your own good sense and creativity, revise in the way that you think best matches your communication to its particular readers and purpose.

In the preceding discussion of how to write your resume, the most important thing for you to notice is the great amount of help you get from thinking constantly about your readers. You define your purpose in terms of your readers' needs and attitudes so you will know how you want your resume to affect your readers as they read it. You create a mental portrait of your readers so you can imagine what they will want from your resume and how they are likely to react while reading it. Then, referring constantly to your understanding of your readers and to your reader-centered definition of objectives, you plan, draft, evaluate, and revise your resume.

The rest of this chapter shows how you can use this same approach when writing your letter of application.

WRITING YOUR LETTER OF APPLICATION

Your letter of application is a companion to your resume. It goes to the same readers and it has the same *general* objective: to get you the job you want. But the differences between the specific objectives of resumes and of letters of application, though slight, are great enough to make the two quite distinct from one another. Consequently, besides providing you with practical advice about writing letters of application, the following paragraphs show why you should think very carefully about your readers and the way you want to affect them *whenever* you communicate at work: even small differences in purpose or audience can greatly affect the way you should write or speak.

Defining Your Objectives

How do the objectives of your letter of application differ from the purpose of your resume? To begin with, your resume is a versatile document. You give it to employers whether you contact them first in person or through the mail. Furthermore, the people involved with hiring you will use it throughout the various steps of the hiring process, from their initial assessment of your qualifications through your visit to their facilities.

In contrast, you send a letter of application only when you first contact employers through the mail. Unlike your resume, your letter becomes less important to your readers after they have met you because your letter of application is, in essence, a written substitute for an initial interview.

Accordingly, one way to understand what readers look for in letters of application is to think about what they look for in interviews. The most important of these things is some sense of your personality. Your readers want to know what you are like as an individual, what you would be like as an employee, a co-worker. That is information you cannot readily provide in a resume, which is more like a table of data than an expression of your personal characteristics.

In addition, potential employers will want to learn from your letter how you feel your education and other experiences relate to the job for which you are applying. In the survey mentioned earlier of personnel directors of the 500 largest corporations in the United States, 88 percent ''agreed'' or ''strongly agreed'' that a letter should show how the applicant's education and experience fit the requirements of the particular job he or she has applied for. Also, employers will want your letter to tell them why you chose *their* organizations. Eighty-one percent of the personnel directors surveyed either ''agreed'' or ''strongly agreed'' that job applicants should provide that information in their letters of application.

Employers can't get that information from your resume because the convention is to present your qualifications in only a general way so you can send it to several employers. In your letter, however, readers will want you to demonstrate how your specific qualifications match the requirements of the particular job they offer.

Thus, the major objective of your letter is to answer your readers' questions about your personality, your interest in them, and your perception of the relationship between your qualifications and the particular job they offer. Of course, your letter also aims to answer all those questions in a way that persuades your readers you are a top candidate for their job.

Planning

That understanding of your letter's objectives, together with the facts about your readers discussed earlier in this chapter, forms a sound basis for planning the kinds of material you should include in your letter.

The best way to begin is to tell your readers why you are writing. This gives you an opportunity to pursue two of your objectives. First, if you begin by telling your readers what you like about their organization, you thereby answer your readers' question about why you are applying to their organization rather than to someone else's. Second, through your explanation you can show yourself to be likable. One of the most pleasant messages to include in a letter of application is ''I know about you and like you.'' In their first paragraphs, both Jeannie Ryan and Ramon Perez convey that message (see Figures 2–7 and 2–8). Sharon Pollock, whose letter is shown in Figure 2–9, was unable to do so because she was responding to a newspaper advertisement that did not give the employer's name.

Of course, in order to explain why you are applying to a particular organization you must first know something about it. If you aren't already acquainted with the companies to which you are applying, you can ask the staff of your campus placement center or library to help you locate information about them so you can say something specifically geared to each company as you write to it. That's what Jeannie Ryan did in order to construct the opening of her letter. You will be well rewarded for exerting a similar effort to learn about the companies to which you apply.

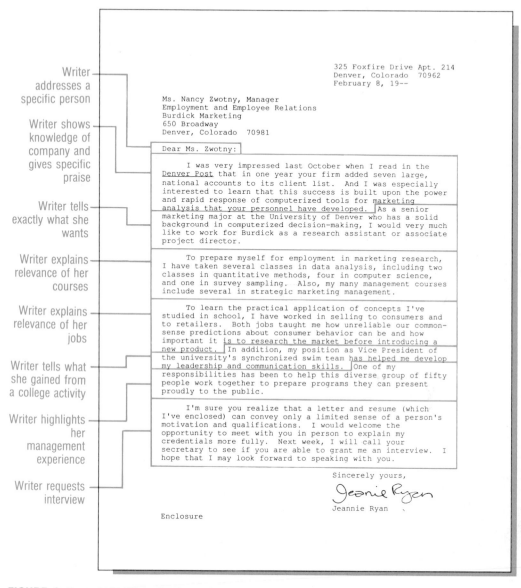

Writer addresses a specific person

Writer shows knowledge of company and gives specific praise

Writer tells exactly what she wants

Writer explains relevance of her courses

Writer explains relevance of her jobs

Writer tells what she gained from a college activity

Writer highlights her management experience

Writer requests interview

325 Foxfire Drive Apt. 214
Denver, Colorado 70962
February 8, 19--

Ms. Nancy Zwotny, Manager
Employment and Employee Relations
Burdick Marketing
650 Broadway
Denver, Colorado 70981

Dear Ms. Zwotny:

I was very impressed last October when I read in the Denver Post that in one year your firm added seven large, national accounts to its client list. And I was especially interested to learn that this success is built upon the power and rapid response of computerized tools for marketing analysis that your personnel have developed. As a senior marketing major at the University of Denver who has a solid background in computerized decision-making, I would very much like to work for Burdick as a research assistant or associate project director.

To prepare myself for employment in marketing research, I have taken several classes in data analysis, including two classes in quantitative methods, four in computer science, and one in survey sampling. Also, my many management courses include several in strategic marketing management.

To learn the practical application of concepts I've studied in school, I have worked in selling to consumers and to retailers. Both jobs taught me how unreliable our common-sense predictions about consumer behavior can be and how important it is to research the market before introducing a new product. In addition, my position as Vice President of the university's synchronized swim team has helped me develop my leadership and communication skills. One of my responsibilities has been to help this diverse group of fifty people work together to prepare programs they can present proudly to the public.

I'm sure you realize that a letter and resume (which I've enclosed) can convey only a limited sense of a person's motivation and qualifications. I would welcome the opportunity to meet with you in person to explain my credentials more fully. Next week, I will call your secretary to see if you are able to grant me an interview. I hope that I may look forward to speaking with you.

Sincerely yours,

Jeannie Ryan

Jeannie Ryan

Enclosure

FIGURE 2–7   Letter of Application to Accompany the Resume Shown in Figure 2–2

By the way, you should also take the trouble to learn the name of the person you are writing to. You may need to call the company itself (that is, its switchboard) to learn the name of an appropriate person to whom you can address your letter.

After explaining your interest in your readers' organization, you can discuss your qualifications. When doing that, you should do more than merely

16 Henry Street
Brooklyn, New York 11231
September 24, 19--

Estelle Ritter
Financial Systems Division
Medallion Software, Inc.
1655 Avenue of the Americas
New York, New York 11301

Dear Ms. Ritter:

About a year ago, I attended a reception held by **Medallion Software** for computer science majors at New York University. Not only did Medallion's representatives highlight your company's remarkable growth, but they also conveyed an excitement about their careers that I would like to share. Now that I am about to complete my degree, I ask you to accept this letter as my application for a position with your firm. I am especially interested in working with your division to create comprehensive computer systems for banks, insurance companies, and other financial institutions.

My preparation for this work includes not only a broad range of courses in systems analysis and design, but also several classes in business and finance. This term, for example, I am taking one course in computer security and another in international finance. I have also completed several classes in written and oral communication, which would help me represent Medallion effectively to clients. In addition, I have gained considerable insight into the structure and uses of two sophisticated financial systems while working as a data entry clerk for the Miller Health Spas chain and as a teller for the Meninger Bank. The enclosed resume contains additional details about my qualifications.

I feel that I am a well-disciplined, highly motivated person with a strong desire to succeed. I have taken the last half of my college courses while working full-time to support my wife and young son. After I become settled into the new job I obtain after completing my degree, I plan to begin work on a master's degree in computer science.

A job in Medallion's Financial Systems Division would offer me exactly the kind of challenge -- and opportunity -- that I have hoped my college degree would bring. Could we please meet to discuss my credentials? I can be reached at (212) 374-7631 during working hours.

Sincerely,

Ramon Perez

Ramon Perez

Enclosure

FIGURE 2–8   Letter of Application to Accompany the Resume Shown in Figure 2–4
(Prepared with Desktop Publishing)

Sharon Pollock
_____

8965 Portage Drive
Baton Rouge, LA 70811

July 17, 19—

Drawer 3765
Houston Chronicle
801 Texas Drive
Houston, Texas 77002

Dear Employer:

I am very interested in the position for a Corporate Manager of Drilling Operations that you advertised in the July 16 issue of the _Houston Chronicle_.

For the past fourteen years, I have worked for Pilot Petroleum Company in a series of positions of increasing responsibility. Currently, I am Regional Drilling Engineer for the Gulf of Mexico Region, a position I have held for six years. Through my various assignments in the states and in Asia, I have gained the broad technical and management experience needed by a person who directs the kind of extensive international drilling program you described in your ad.

In addition, throughout my career I have taken leadership in introducing new technologies and in encouraging others to do the same. Given the opportunity to direct your company's drilling programs, I would continue to emphasize technological innovation as one means of increasing competitiveness and profitability.

I would welcome an opportunity to discuss my professional experience and qualifications with you in person. I can be reached at 504/241-2324 during office hours and at 504/569-4230 evenings.

Sincerely,

Sharon Pollock

Sharon Pollock

FIGURE 2–9   Letter of Application to Accompany the Resume Shown in Figure 2–5
(Prepared with Desktop Publishing)

repeat what your resume says. You should explain precisely how your education and experiences together are relevant to the job you want—just as you would in an interview.

How do you answer your readers' questions about your personality? One way is through what you say about your qualifications and reasons for applying. When you tell what appeals to you about the organization, you reveal some of your personality. You do the same when you choose which qualifications to emphasize and when you explain their relevance to the employer's position.

Notice, for instance, how the first paragraph of Ramon's letter (Figure 2–8) shows him to be an enthusiastic person with a firm sense of direction. Similarly, notice how the first sentence of Jeannie's second paragraph (Figure 2–7) indicates she is an ambitious person who plans her work purposefully. Other passages throughout all three sample letters also reveal aspects of the writers' personalities.

Drafting

Even when you have made excellent and thorough plans, you may still find it a challenge to draft your letter of application. The most difficult thing for many people to achieve is an appropriate and persuasive tone. Yet in your letter nothing is more important than tone. In the survey mentioned earlier, 90 percent of the personnel directors "agreed" or "strongly agreed" that the tone of the letter of application is important.

Why is tone so important? As much as anything else in your letter or resume, it reveals your personality. Or it can hide your personality. If you try to control the tone of your letter by using standard phrases you have seen in other letters, your readers will know it. They've seen those same phrases before. Rather than being impressed, however, they will be uninterested.

Many people have difficulty achieving a tone appropriate to both the occasion and their personalities. This difficulty is especially great for people who haven't written such letters before. Here are some particularly troublesome areas for many people.

Achieving a Self-Confident Tone

Some people have difficulty indicating an appropriate level of self-confidence. In your letter you want to seem self-assured. However, you do not want to appear brash or overconfident. Thus you want to avoid statements like the following:

> I am sure you will agree that my excellent education qualifies me for a position in your Ball Bearing Department.

That sentence has two problems. The first is the offensive "I am sure," and the second is that the sentence pushes the readers out of consideration by asserting that the writer has already performed for them their job of evaluating the writer's qualifications. To improve the sample sentence, the writer might say:

> I hope you will find that my education qualifies me for a position in your Ball Bearing Department.

Achieving a Good Tone in the Conclusion

In addition, some students also have problems achieving the proper tone in the conclusions of their letters. You should avoid conclusions like this:

> I would like to meet with you at your earliest convenience. Please let me know when this is possible.

> I am available for interviews in the Columbus area on April 15. I will be expecting to hear from you soon.

To sound less demanding, you might revise the second sentence of the first sample to say, ''Please let me know *whether* this is possible.'' Likewise, you might revise the second sentence of the second sample to read, ''I *look forward* to hearing from you.''

Using a Conventional Format

Finally, when drafting your letter you should strive to follow a conventional letter format. Business letters usually follow one of several standard formats. You are not likely to impress your readers favorably by using any other. Standard formats are described in Appendix A and are used in the sample letters in this chapter.

Evaluating

You will undoubtedly find that writing your letter of application is a challenging undertaking. The chief reason is that no model can tell you how to do it. The personality you need to reveal is your own. If you merely imitate one of the sample letters out of this or any other book, you will be revealing someone else's personality. If you are like most college students, writing letters of application is also difficult because you have had very little experience in writing about yourself in a business situation.

Because of the difficulty and the novelty of writing these letters, it will probably take you many drafts to arrive at an effective letter. You would be wise to ask other people to read over your drafts. Ask them to play the role of the recruiter or the department head you hope to impress. Good choices for these role-playing readers would be the instructor of your writing course and instructors in your major department. They can tell you whether the letter you have written is likely to affect your readers in the way you intend.

Revising

When revising your drafts, you may encounter some of the same difficulties you encounter when revising your resume. The people who help you evaluate your letter may identify problems but not tell you how to solve them, or they may give you contradictory advice. To revise well you have to use your own creativity and good judgment. And you need to think constantly about the way your readers will react, moment by moment, to what you have written.

FINAL NOTE

In tracing the process of creating a resume and letter of application, this chapter has been organized around the five major activities of composing: defining objectives, planning, drafting, evaluating, and revising. As explained in Chapter 1, people generally move back and forth among these activities. Such movement is natural and helpful. Throughout your composing effort, you are likely to refine your sense of your purpose and audience and to see ways to make all kinds of improvements—from polishing the phrasing of a sentence to changing your fundamental strategy. To act on these insights means more work, but it can also make your writing much more effective. You will communicate most effectively if you remain open to what you learn while composing.

CONCLUSION

This chapter has had three major purposes:

- To show you how a reader-centered approach can help you make important decisions when writing a resume and a letter of application.
- To illustrate the writing process and show how its various activities are related to one another. In particular, this chapter has emphasized the importance of defining your objectives, which involves understanding your readers and deciding how you want to affect them. That early work will provide you with sound guidance throughout the later stages of the writing process.
- To show how completely the design of a communication depends on your particular purpose and readers. In particular, it has demonstrated how the differences between two similar communications, the resume and the letter of application, derive from small differences in their purposes.

Keep in mind that the reader-centered strategy described here for writing the resume and the letter of application is fundamentally the same as the strategy you should use when creating any other kind of communication related to your job. That strategy is explained in more detail in the rest of this book.

EXERCISES

1. **a.** Figures 2–2 and 2–3 contain marginal notes that point out the persuasive strategies Jeannie Ryan and George Shriver use in their resumes. Make similar notes explaining the strategies Ramon Perez (Figure 2–4) and Sharon Pollock (Figure 2–5) use.

    b. Make marginal notes explaining the persuasive strategies that Ramon Perez and Sharon Pollock use in their letters of application (Figures 2–8 and 2–9).

2. Find a sample resume in handouts from your college's job placement center or in books about resume writing available in your library. Evaluate that sample from the point of view of its intended reader.

3. Complete the assignment in Appendix C that involves writing a resume and letter of application.

| CASE | **ADVISING PATRICIA** |

This morning you stopped in the lobby of the library to talk with Patricia Norman, a friend who is majoring in marketing. As you exchanged news, she told you with a mixture of excitement and anxiety that she has finally decided to join the many other seniors who are busily looking for a job. She's even written a resume and begun writing letters to employers listed in a copy of the *College Placement Annual* she picked up at the Career Planning and Placement Center.

"And look," she exclaimed, "one of the department store chains I'm writing to is written up in this article that Professor Schraff asked us to read." She held out an article from *Retail Management*. "They've begun opening free-standing specialty shops in their stores. The managers order their own merchandise and run their own advertising campaigns. It's been a huge success. This sounds like such a great place to work. A big chain that welcomes innovators." Then her excitement turned to anxiety. "I'm so worried that they won't like my resume and application letter, though. I've gone over both again and again, and my roommate has, too. But I'm still worried."

As you tried to reassure her, she pulled out her drafts and put them into your hand. "But take a look at them and tell me what you *really* think. I need all the help I can get." You had to leave for an appointment, but agreed to look over her drafts and meet her back at the library in the evening.

Now it's afternoon and you've started to read through Patricia's resume and letter. As you do, you think back over some of the things you know about her. She's an active and energetic person, talkative, and fun to be with. Throughout her years in college, she has spent lots of time with a group called Angel Flight, a volunteer organization that sponsors lots of service activities on campus and off. In fact, this past year you've seen less of her because she has spent so much time serving as the organization's president. "As President, I'm responsible for *everything*,'' she once told you. ''Everything from running meetings, to getting volunteers, to seeing that the volunteers have done what they said they would.'' While a junior she held some other office, you recall— also a time-consuming one. But she's like that. In the Marketing Club, she edited the newsletter and handled lots of odd jobs, like putting up posters announcing speakers and meetings. She was also treasurer of the Fencing Club, another of her interests. Once when you marveled at how many things she was able to do, she responded, ''It's not so much, if you're organized.''

Despite all the time she has spent on such activities, Patricia has good grades, a 3.6 average she told you once. Although she's had to take lots of business courses, she's also squeezed in a few electives about one of her favorite subjects, art history. One Saturday last year, she even got you to travel two hundred miles with her to see an art exhibit—a ''major'' exhibit she had assured you.

But the trip you most enjoyed with her was to a shopping center, where she spent more time commenting on how the merchandise was organized and displayed than looking for things she might buy. She talked a lot about the way they did things at a Dallas department store she's worked in for the past three summers. She must have some interesting opportunities there, you note; after all, one of the people she lists as a reference is the store manager. Her other references are professors who've taught classes that you and Patricia have taken together. They were fun. Everything's fun with Patricia.

Your Assignment

Decide what you will tell Patricia about her letter (Figure 2–10) and resume (Figure 2–11). What strengths would you praise? What changes would you suggest? What questions would you ask to determine whether she might include additional things?

Box 88
Wells Hall
University of Washington
Seattle, Washington 98195
February 12, 19--

Kevin Mathews, Director
Corporate Recruiting
A.L. Lambert Department Stores, Inc.
Fifth and Noble Streets
San Diego, California 92103

Dear Mr. Mathews:

I saw A.L. Lambert's advertisement in the <u>College Placement Annual</u>. I was very impressed with your company. I hope that you will consider me for an opening in your Executive Developement Program.

In June, I will graduate from the University of Washington's retailing program, where I have focused my study on marketing management. I have learned a great deal about consumer behavior, advertising, and inovative sales techniques. Furthermore, I have gained a through overview of the retailing industry, and I have studied successful and unsuccessful retailing campaigns through the case-study method.

In addition to my educational qualifications, I have experience both in retail sales and in managing volunteer organizations. While working in a Dallas department store for the past four summers, I had many opprtunities to apply the knowledge and skills that I have learned in college. Likewise, in my extracurricular activities, I have gained experience working and communicating with people. For instance, I have been the President of Angel Flight, a volunteer service organization at the University of Washington. Like a manager, I supervised many of the organization's activities. Similarly, while holding offices in two campus organizations, I have developed my senses of organization and responsibility.

I would like to talk with you in person about my qualifications. Please tell me how that can be arranged.

Sincerely yours,

Patricia Norman

Patricia Norman

FIGURE 2–10   Letter of Application for Use with Case

PATRICIA NORMAN
Box 80, Wells Hall
University of Washington
Seattle, Washington 98195
(206) 529-5097

PERSONAL
Born: March 17, 19--
Health: Excellent
Willing to relocate

PROFESSIONAL OBJECTIVE
To work for an innovative and growing retailer.

EDUCATION
University of Washington, Seattle, Washington, B.S. in Retailing, May 19--.

Earned 23 credit hours in marketing managment, obtaining a working knowledge of the factors motivating today's consumer. Also learned how a product is marketed and distributed to the consumer. Took eight credit hours of study focused specifically on principles and problems of retail management

WORK EXPERIENCE
Danzig's Department Store, Dallas, Texas,
Summers 19-- through 19--.

Worked as a sales clerk. Helped customers choose their purchases and listened politely to their complaints. Cash register operation. Stocked shelves and racks. Provided assistence to several department managers.

ACTIVITIES
Fencing Club, served as treasurer
Angel Flight, President.
Marketing Club, member

REFERENCES

Derek Yoder, Store Manager
Danzig's Department Store
11134 Longhorn Drive
Dallas, Texas 75220

Gregory Yule
Pinehurst Hall
University of Washington 98195
(206) 579-9481

Lydia Zelasko
Putman Hall
University of Washington
Seattle, Washington 98195

FIGURE 2–11 Resume for Use with Case

II
DEFINING OBJECTIVES

CHAPTER 3
Defining Your Objectives

3

DEFINING YOUR OBJECTIVES

DEFINING
OBJECTIVES
PLANNING
DRAFTING
EVALUATING
REVISING

Y ou are now beginning a new portion of this book. In it you will find seventeen chapters of detailed, reader-centered advice about how to perform the five major activities of writing described in Chapter 1:

- Defining your objectives
- Planning your communication
- Drafting your prose and visual aids
- Evaluating your draft
- Revising

The chapter you are now reading tells how to perform the first of those activities: defining your objectives.

THE IMPORTANCE OF YOUR OBJECTIVES

All five writing activities are important. You must do each with skill and creativity in order to communicate effectively. However, as Chapter 2 explained, your work at defining objectives deserves special attention.

Consider, for instance, Todd's situation. A recent college graduate, Todd has been assigned to investigate ways of creating a more chip-resistant paint for his employer's products, which are stoves, refrigerators, and other large appliances. This morning, Todd's boss dropped by his office. After reassuring Todd that such requests are routine in this company, Todd's boss told him that two vice presidents have asked Todd for a report on his progress.

"How should I write this report?" Todd asks himself. "What should I tell the vice presidents and how should I say it?"

To answer those questions—and all the others he will face as he writes—Todd must decide what he wants to accomplish in his report: he must define his objectives. If Todd states his objectives clearly and precisely, they can guide him throughout all his work at writing. While he is planning and drafting, they can help him decide what to say and how to say it, what to present in prose and what in tables and illustrations, how to design his pages, and how to structure his sentences and paragraphs—how, in fact, to handle every aspect of his report. Similarly, when Todd is evaluating, his objectives can serve as the standard against which he assesses his draft to determine how it might be improved. And when Todd is revising, his objectives can describe the goal toward which all his changes are directed.

A READER-CENTERED APPROACH TO DEFINING OBJECTIVES

Like Todd, you will benefit in many ways from carefully and thoughtfully defining your objectives. This chapter presents a reader-centered approach to doing that. Its advice is organized around five guidelines. Guideline 1 suggests you begin by describing the final result you want from your communication. Guidelines 2 and 3 then urge you to tell precisely how your readers must react while reading your message if you are to achieve your goal. Finally, Guidelines 4 and 5 describe what you should learn about your readers so you can predict their likely reactions to your statements and write accordingly.

| GUIDELINE 1 | **DESCRIBE THE FINAL RESULT YOU DESIRE** |

When defining objectives, start by identifying the final result you want from your communication. Doing so will help you remember that when you write, you are performing an action: you are exerting your powers to create a certain outcome.

In many situations, you will be able to identify this final result with ease:

- You want your boss to approve the project you are describing in your proposal.
- You want the twenty-seven people in the department you manage to adhere to the new policy you will explain in the memo you are preparing.
- You want the employees in three departments to be able to perform the new procedure you have developed and are describing in an instruction manual.

In each of these sample situations, your overall objective is clear enough to you because you initiated the communication yourself. However, at work you will often write because someone else has asked you to, so your definition of the final result you want your communication to achieve must take into account the other person's reason for making the request. Even then, it can be easy to define the final outcome you desire—if the requester tells you exactly what he or she wants. But that won't always happen.

Imagine, for example, that Todd's boss left Todd's office without telling him the purpose of the progress report to the vice presidents. Without that information, Todd would not know which of several very different approaches he should take to the report. To see why, consider two possible uses the vice presidents might have for his report. First, the vice presidents might want to know what Todd has learned about the two or three most promising approaches to creating chip-resistant paint so they can select one for Todd to concentrate on in the weeks ahead. In order to achieve this objective, Todd would write a fairly detailed report telling what he has read and discovered in the laboratory about the two or three most promising approaches, focusing on information that would permit his readers to compare them. Alternatively, the vice presidents may want to know whether Todd will complete his project in time for them to use a superior paint on a new line of appliances the company will begin manufacturing in a few months. In that case, Todd would write a very brief report telling how much he has accomplished, how much he has left to do, and how much time he estimates it will take him to do it.

As you can see, the writing strategy that will achieve one of these possible purposes is not well suited to achieving the other. Of course, Todd could try to protect himself by writing a longer report that combines both strategies, but doing so would waste both Todd's and the vice presidents' time unless the vice presidents truly had both purposes in mind. Furthermore, for all Todd knows, the vice presidents might actually have in mind some third purpose that neither of these strategies would address effectively.

As the example of Todd's report illustrates, it is important that you describe quite specifically the final result you want from your communications. Whether you are writing at someone else's request or on your own initiative, that description will need to be specific if you are to get any practical help from it in writing your communication. A vague description like the one provided by Todd's boss is of little or no help.

What should you do if you are in Todd's situation and you aren't given complete information from the person who asks you to write something? Without doubt, you should ask. Your only alternative is to guess—and then you risk guessing incorrectly, which is likely to result in ineffective writing. Later in this chapter you will find advice about approaching your boss and others for help in defining your objectives. The key point for you to remember now is that whenever you write, you should develop a precise understanding of the final result you are seeking.

INTRODUCTION TO GUIDELINES 2 AND 3

Guidelines 2 and 3 shift your attention away from the final result you want from your communication and toward what must happen while your readers are reading if you are to succeed in achieving your goal. Specifically, they focus your attention on the tasks your readers will want to perform while reading and on the ways you want your readers to alter their attitudes as they read.

Why think about these two things? When people read, they respond in two ways simultaneously: they think and they feel. Imagine, for example, you are reading a sheet on which your instructor has written directions for an in-class writing exercise. You respond by performing the mental tasks required to understand and remember what he or she is saying. And you also respond emotionally: perhaps you feel eager because the assignment sounds interesting, or perhaps you feel indifferent because you don't see the point of the work. Consequently, to achieve his or her desired outcome, which is to have you complete the assignment correctly and thoughtfully, your instructor must explain the assignment in a way that achieves the following two objectives: to enable you to perform your mental task of accurately understanding the assignment and to encourage you to view it favorably.

Similarly, by identifying your readers' tasks (Guideline 2) and describing the way you want to change their attitudes (Guideline 3), you specify the ways you will need to affect your audience if your communication is to achieve the outcome you desire.

IDENTIFY THE TASKS YOU WILL HELP YOUR READERS PERFORM WHILE THEY READ

When you identify the tasks your readers will attempt to perform while reading, you are describing the *enabling element* of purpose.

At a general level, these tasks are the same for all readers. All seek to locate, understand, and use the information and ideas that are important or interesting to them. For example, if the vice presidents are reading Todd's report to determine which approach to creating chip-resistant paint they would like Todd to concentrate on, they will have to locate the information he provides about the two or three most promising approaches, they will have to determine the meaning of the statements and visual aids he uses, and they will have to evaluate that information in terms of their own knowledge and business objectives.

Readers perform the tasks of locating, understanding, and using information in many different ways, depending on the situation. For example, the vice presidents will probably locate the information they want in Todd's report by reading its entire contents sequentially from front to back. However, if they were at home trying to learn how to record a television show on their new VCRs, they probably would not read the entire owner's manual from start to finish, but rather skim the text to find the section that tells them how to program.

Not only do readers' specific tasks vary, but different readers' tasks are helped by different writing strategies. For example, the writers of the VCR instructions can probably help the vice presidents (and all other readers) by arranging the steps to be performed in a list. In contrast, Todd would probably make the reading of his report very difficult if he presented every statement as a separate item in a list.

When writing, part of your aim is to enable your readers to perform their reading tasks quickly and efficiently. That's why identifying those tasks is an essential step in defining your communication objectives. Once you have identified the tasks, you will have a solid basis for making many important writing decisions that might otherwise perplex you.

An Example

To see how thinking about your readers' tasks can help you write effectively, consider a report Lorraine must write. Lorraine works for a steel mill that has decided to build a new blast furnace. She has been asked to study the two types of furnaces the mill is considering. The final result of her communication is to help the mill's upper management decide which furnace to buy. In her report, she is to submit the results of her investigation together with her recommendation.

Having gathered all the relevant information about the two furnaces, she finds herself unable to make a basic decision about her report: how to organize the one hundred pages of material she has amassed. She knows to begin with

Divided Pattern	Alternating Pattern
Furnace A	**Cost**
Cost	Furnace A
Efficiency	Furnace B
Construction Time	**Efficiency**
Air Pollution	Furnace A
Et cetera	Furnace B
Furnace B	**Construction Time**
Cost	Furnace A
Efficiency	Furnace B
Construction Time	**Air Pollution**
Air Pollution	Furnace A
Et cetera	Furnace B
	Et cetera
	Furnace A
	Furnace B

FIGURE 3–1   Two Organizational Patterns Lorraine Can Use in Her Report

a summary of her findings and a statement of her recommendation. It's the material that will follow in the body of the report she is unsure about. She realizes she can use either of the organizational patterns shown in Figure 3–1, but she can't decide which is better.

Rules of logic won't help. From the point of view of logic, the *divided pattern* and the *alternating pattern* are equally correct. Consideration of length won't help. Even though the outline for the alternating pattern is longer than the outline for the divided pattern, both would yield reports of the same length. Her statement of the final outcome she hopes to achieve won't help either. Both patterns will provide the upper-level managers with the information they need to make their decision.

Perhaps Lorraine should choose the easiest pattern for her to write. For that the divided pattern is probably superior because when she conducted her investigation, Lorraine organized her notes into two big piles, one about blast furnace A and the other about blast furnace B. The divided pattern lets her transform those two piles of notes directly into the major sections of her report. But then ease of writing is a writer-centered consideration, not a reader-centered one.

By taking the reader-centered approach of considering the mental task her readers will perform while reading, Lorraine can see that the alternating pattern is best in this situation. She knows that regardless of which furnace she recommends, her readers will surely want to compare the two furnaces in detail in terms of the various criteria they consider most important. To make that detailed, point-by-point comparison in a report organized according to the divided pattern, Lorraine's readers would have to do a great deal of page-flipping. For example, to compare the two furnaces in terms of operating efficiency, they would have to search the first part of the report for pertinent information for furnace A and then search the second part for comparable information for furnace B. In contrast, if Lorraine uses the alternating pattern, her readers could make these comparisons much more easily because that pattern puts in one place the information for both furnaces about operating efficiency, construction time, air pollution, and so forth. For other purposes, the divided pattern—or some other pattern—may be a better way to organize a communication. But in this situation, where the readers have this particular task, the alternating pattern makes their work easier.

How to Identify Your Readers' Tasks

Lorraine's situation illustrates the general point that identifying the mental tasks your readers will perform while reading can help you make good decisions about how to compose a communication. But how can you identify your readers' tasks? There is no single correct way to do that. Sometimes you need merely call on memory, because in the past you have used communications similar to the one you are now preparing. At other times your direct knowledge of your readers lets you confidently predict their mental tasks. However, if you need some systematic technique to help you identify those tasks, try the following strategy, which involves thinking of your readers as people who ask questions they want answered in your communication.

First, identify the questions your readers will ask. Readers come to many communications with questions already formed. Some initial questions by Lorraine's readers will be: "Which furnace costs less?" "Which makes the higher grade steel?" Readers also ask questions in response to statements they read in the communication: "How did you find that out?" "What do you think is the consequence of that for us?" "What does that abbreviation mean?"

Once you've determined what your readers' major questions will be, identify their strategy for searching for the answers. They may search in many ways, including sequential reading (in which they start at the top of the first page and continue to the bottom of the last) and hunting to find particular facts (as when they are using a computer manual to learn how to change a printer ribbon).

Finally, identify the ways your readers will use the answers. Will they use them to make a point-by-point comparison, as Lorraine's readers will? Will they use the information as a step-by-step guide through some procedure, such as assembling a child's bicycle or conducting an electrical test?

All three steps produce insights that can help you enormously as you plan and prepare your communication. By identifying the questions your readers want answered, you learn what information you should include. Also, by learning how your readers will look for this information and what they will do with it, you gain important insights into ways to organize and present your message, just as Lorraine's knowledge of how her readers would use the information in her report helped her choose the alternating over the divided pattern of organization.

| GUIDELINE 3 | **TELL HOW YOU WANT TO CHANGE YOUR READERS' ATTITUDES** |

The second step in describing what you want to happen during your readers' reading is to consider how you want your communication to alter your readers' attitudes. Although every communication influences its readers' attitudes and is therefore persuasive, writers often set out with only a vague notion of their persuasive aims. By following the advice given here, you can identify the *persuasive element* of your communication's purpose, which you can use as a helpful guide to writing.

Begin by identifying the attitudes you want to alter. Usually you will be interested in shaping your readers' attitudes about your *subject matter*. For instance, when you suggest a certain improvement in departmental operations to your boss, you want that person to feel so good about the recommended course of action that he or she will approve it.

Even when you are not trying to influence people's attitudes toward your subject matter, you will want to affect your readers' attitudes toward *you*. As one recent college graduate observed, "As soon as I started my job, I realized that every time I communicate, people judge me based upon how well they think I write or speak." Thus, even when you are merely answering factual questions or reporting raw data, you will want your communication to positively influence your readers' attitudes toward you.

As you think about ways you want to change your readers' attitudes, keep in mind that you can seek to do so in more than one way. Sometimes you will want to *reinforce* an existing attitude. For example, your readers might already think you are a good employee who shows a lot of promise for a management position. You might aim to write your communication so it persuades your readers to feel even more favorably impressed by your qualifications. At other times, you might try to *reverse* an attitude you want your readers to abandon. For example, you might want to persuade them to like something they now dislike. Or you might need to persuade them to feel a problem exists where they currently feel everything is fine. Finally, at times you may want to *shape* your readers' attitudes on a subject that they haven't yet thought about seriously.

The distinction among these three kinds of attitude changes is important because the persuasive strategies that could succeed in achieving one kind could fail with another. To reinforce an attitude, you can build on the existing attitude and expect little resistance. Consequently, you might present only positive arguments and not bother addressing counterarguments to your position. But if you want your readers to reverse an attitude, you can expect lots of resistance: you must then not only cite positive points, but also address the counterarguments your readers will raise.

Because it is so important for you to identify the kind of attitude change you want to bring about, you should specify both your readers' present attitudes and those you want them to hold after reading.

Figure 3–2 shows some ways you might want to alter your readers' attitudes in several typical communications.

The Way Things Are Now	You Act (You Write)	The Way You Want Things to Be
Your reader is a manager who wants to decide whether or not to purchase a certain piece of equipment.	You write a memo evaluating the equipment in terms of the benefits it will bring the company.	The manager decides to buy the equipment and feels confident that he or she has made a good decision based on information you provided.
Your reader is the director of a plant that is using an outdated process. The director feels that the process is fine.	You write a report on problems with the current process and the ways they can be overcome by various new processes.	After reading your report, the plant director feels that the process being used now may be faulty and that one of the new ones is worth investigating further.
Your readers are bank clerks who will be using a new computer system. They are afraid of computers and are therefore reluctant to use the new system.	You write a procedures manual that shows how easy it is to use the new computers.	The bank clerks feel relaxed and self-confident after learning how to use the new system from your procedures manual.

FIGURE 3–2   Examples of Ways that Communications Prepared at Work Can Alter Attitudes

INTRODUCTION TO GUIDELINES 4 AND 5

The two preceding guidelines suggest that when defining objectives you tell what you want to happen while your readers are reading your communication. As you learned in Chapter 1, what actually happens during reading will depend as much on who your readers are as on what you say. Therefore, an essential step in defining objectives is to learn about your readers. Guidelines 4 and 5 provide advice about doing that.

GUIDELINE
4

LEARN YOUR READERS' IMPORTANT CHARACTERISTICS

What characteristics should you focus on as you think about your readers? That depends partly on the situation. For instance, if you are writing a progress report to a client, it probably doesn't matter how old your reader is. Although many factors might affect the types of questions your reader has about your project and the use he or she will make of the information you provide, age is not likely to be one of them. In contrast, if you are writing instructions for a science project to be read by elementary school children, the age of your audience becomes very important because you will probably need to take special measures to accommodate them.

Although the reader characteristics you should identify will vary, in most working situations the following four topics are among the most helpful ones you can consider: your readers' professional roles, their familiarity with the topic you discuss, the level of their knowledge of your specialty, and their preferred communication style. In addition, you should always remember to ask yourself whether you should consider any other special factors about your readers as you write.

Professional Roles

When a person is hired by an organization, he or she is expected to perform a certain role. By learning your readers' roles, you can often infer the kinds of information they will seek from your communication and the ways they will use it. For instance, when reading a report on the industrial emissions from a factory, an environmental engineer working for the factory might ask, ''How are these emissions produced and what ideas does this report give me about how to reduce them?'' whereas the corporate attorney might ask, ''Do these emissions exceed standards set by the Environmental Protection Agency and, if so, how can we limit our legal liability for these violations?''

The most obvious clues to your readers' roles are their job titles, for example, systems analyst, laboratory technician, bank cashier, or director of public relations. However, job titles often fail to indicate precisely the responsibilities of the people who hold the jobs. Furthermore, you need to know not only the general but also the specific roles of your audience. Why *exactly* will each person be reading your communication, and precisely what information will he or she be looking for?

As you try to answer those questions, you might find it helpful to think about the three general roles people play at work.

- **Decision-makers.** The decision-makers' role is to say how the organization, or some part of it, will act when confronted with a particular choice.
- **Advisers.** Advisers provide information and advice for decision-makers to consider when deciding what the organization should do.
- **Implementers.** Implementers carry out the decisions that have been made.

Of course, these general descriptions of the three basic roles are greatly simplified. For example, advisers usually are specialists who advise only on certain things: an accountant normally advises only on financial matters. Therefore, the specific questions any adviser—or decision-maker or implementer—will ask will be determined largely by the individual's particular responsibilities in the organization.

Also, an employee can play more than one of the three roles described above. Often, for example, those who are asked to advise about a particular decision also will be asked to carry it out. For instance, the shop foreman asked by the manager of a plastics factory about the wisdom of installing a new type of injection mold will probably be required to retrain his department in the use of that new mold if it is purchased.

Furthermore, although the roles of decision-maker, adviser, and implementer may seem to apply only to employees within large organizations, it is equally useful to consider these roles when thinking of people outside of organizations. For example, when deciding what brand of stereo to buy, a consumer may be considered a decision-maker, perhaps one who will consult the "advisers" who review these products for magazines and journals such as *Stereo Review* and *Consumer Reports*. Once the consumer has decided what to purchase, he or she becomes an implementer faced with the problems of transporting, unpacking, setting up, and using the equipment.

Familiarity with Your Topic

A second important characteristic to investigate is your readers' familiarity with your topic—company inventory levels, employee morale on the second shift, software problems, or whatever. Your readers' familiarity with your topic will determine the amount of background information you need to provide to make your communication understandable and useful to your audience. If you are writing about a subject or situation they know well, no background may be needed at all. However, if you are treating a subject they don't know much about, you may have to explain the general situation before you can proceed to the heart of your message. Remember, too, that background can be needed not only at the beginning of a communication but also throughout it. Keep in mind that people unfamiliar with your topic may also want you to explain how it relates to them so they can be reassured it is worth their time and effort to read your communication about it.

Level of Knowledge of Your Speciality

To use the information you provide, your readers will need to understand the terms and concepts you use. Consider, for example, the reader of an instruction manual that says the next step in operating a drill press is to "zero the tool along the Z-axis." If unfamiliar with this type of equipment, the reader might ask, "What is zeroing? What is the Z-axis?" If the instructions do not answer such questions, the reader would have to ask someone else for help, which would defeat the purpose of the instructions.

Of course, the writer of those instructions would not want to answer questions about zeroing and the Z-axis if all the readers already know the answers. Upon coming to such explanations, readers familiar with the terms would ask, "Why is this writer making me read about things I already know?"

Knowing the degree of your readers' familiarity with your specialty will help you anticipate and answer the questions your audience will ask of your communication and also help you avoid cluttering your message with unnecessary information.

Preferred Communication Style

Most people have preferences concerning the style of the communications they read. Some want brief messages and some want more detail. Some prefer reading prose and some like as much information as possible in tables, graphs, and other visual aids. Some prefer a formal writing style while others prefer informality.

To a certain extent, people's communication preferences are shaped by the customary practices in the organization that employs them. This is important to keep in mind as you begin to work for a new company, regardless of whether you just graduated from college or moved to the company from other employment. When people react to a communication by saying, "That's not the way we do things here," they are expressing communication preferences that probably apply to many or all the other people in the company.

The negative reactions people have to unfamiliar communication styles underscore the importance of learning and adjusting to these preferences on the job.

Special Factors

This category is a catchall. It reminds you that each reader is unique so you should always be on the lookout for important reader characteristics you would not normally need to consider.

For example, you might be addressing an individual who detests certain words, insists on particular ways of phrasing certain statements, or has an especially high or low reading level. Or you might be communicating to someone who is interested in certain kinds of information you would not have to supply to most people.

Also, you may be communicating on a subject about which your reader has especially strong feelings, perhaps because of current circumstances or past experiences. For example, imagine you are writing to request money to attend an important professional meeting. You will have to make an especially strong case for the trip if your department has just been reprimanded for excessive travel expenses. Or imagine you are going to announce the reorganization of a department that has just adjusted to another major organizational change. You will need to be especially skillful in presenting the new change in a positive light.

Sometimes you also may need to consider the setting in which your readers will be reading, especially if they will be reading outside the normal office environment. For example, suppose you are writing a repair manual for hydraulic pumps. You know your readers will use it around water, oil, and dirt. Therefore, you might consider printing your manual on paper coated to resist moisture and oil. Additionally, you realize your readers also may have difficulty finding a spot close to a pump where they can set your manual down and may have to read from a greater distance than they normally would. You therefore may want to use a large typeface for your printing.

Much of the time you may not need to consider special factors like those just described. However, each time you prepare a communication, you should ask yourself, ''Should I take into account any special factors when addressing this audience?''

How to Consolidate Information about Your Readers

Whatever information you find about your readers, you must consolidate it in some way that will let you keep the key facts about your audience in mind as you write. As suggested in Chapter 1, it is often helpful to create a mental portrait of your readers. Some writers even use real portraits. For instance, when writing an instruction manual telling small-business owners how to do their bookkeeping on a computer, one writer placed over his desk a photograph showing an elderly couple standing behind the counter of the neighborhood grocery they owned. ''Whenever I sat down to write,'' this person said, ''I looked at the photograph so that I would always begin by thinking of the people who would be using my manual.''

Whatever technique you use for keeping the key characteristics of your readers in mind, remember you should view them not merely as a list of facts but as a dynamic resource you can use to guide you as you perform the rest of your writing tasks.

LEARN WHO *ALL* YOUR READERS WILL BE

The discussion so far in this chapter has assumed you will know from the start who your readers will be. Often that will be the case. Without any thought and without making any inquiries you will know exactly who will be in your audience. In other cases, however, the composition of your audience will not be obvious. As explained in Chapter 1, what you say in a communication at work may be of use to a wide range of people in many parts of your organization. Numerous memos and reports prepared at work are seen routinely by one or two dozen people, and sometimes many more. Even brief communications you write to one person can be copied or shown to others. If your communication is to be effective, you must know all the people who will read it so you can write with those individuals in mind. This section will help you identify readers you might otherwise overlook.

Phantom Readers

In some situations, the most important readers for a communication may be hidden from you. That's because at work, written communications addressed to one person are often *used* by others. Those real but unnamed readers are called *phantom readers*.

You will most often encounter phantom readers when you write communications that require some sort of decision. One clue to the presence of phantom readers is that the person you are addressing is not high enough in the organizational hierarchy to make the decision your communication requires. Perhaps the decision will directly affect more parts of an organization than managed by the person addressed, or perhaps it involves more money than the person addressed is likely to control.

Much of what you write to your own boss may actually be used by phantom readers. For instance, your boss may be one of those managers who accomplishes the work assigned by superiors by asking assistants to perform parts of the necessary task. Thus when you turn in your report your boss may check it over and then pass it along to his or her superiors.

After people have worked at a job for a while, they usually learn which communications will be passed up the organizational hierarchy, but a new employee may be chagrined to discover that a hastily written memo has been read by people very high in the organizational structure. To avoid this kind of embarrassment, you need to be able to identify phantom readers. Also, this identifying will enable you to write in a way that meets their needs, not just the needs of the less influential person you are addressing. Thus, you, your named, and your unnamed readers will all benefit.

Future Readers

When identifying your readers, you should consider the possibility your communication may be used weeks, months, or even years from now. Lawyers say that the memos, reports, and other documents employees write today are evidence for court cases tomorrow. Most company documents can be subpoenaed for product-liability, patent-violation, breach-of-contract, and other lawsuits. If you are writing a communication that could have such use, keep lawyers and judges in mind as possible readers.

You should also consider whether other employees of your company might someday look back at your communication for information or ideas. By thinking of their needs, you may be able to save them considerable labor. In fact, you may even be able to save yourself a lot of work, as the chemists in an industrial research center found out.

These chemists worked in a department that analyzed plants from around the world to see whether the plants contained any compounds known to have medicinal value. Such analyses can be extremely complex. For instance, the chemists once searched through some plant samples for a compound whose presence was hidden by other compounds. The chemists finally isolated the compound through an ingenious procedure, and they dutifully placed a record of the results in the company's file. Three years later, they encountered a similar problem with another plant. Remembering their earlier adventure, they went to the files to see what they had done in the past.

To their dismay, they discovered they had recorded only the results of their ingenious analysis and not the procedures they had used. Consequently, they had to spend two weeks re-creating by a hit-and-miss method the procedure they could have imitated in a day if only they had considered themselves a potential audience for their original communication.

Complex Audiences

Besides overlooking phantom and future readers, writers sometimes neglect important members of their audience because they imagine their readers form a single large group of people sharing identical needs and concerns. Actually, audiences at work often consist of diverse groups of people with widely varying backgrounds and responsibilities.

That's partly because decisions and actions at work often have an impact on many people and departments scattered throughout an organization. Imagine, for instance, that you have proposed an improvement in one of your company's products. If your idea is approved, designers may have to redesign the product, the production department may have to alter its production process, the purchasing department may need to obtain different supplies, the marketing department may have to revise its sales materials and strategies, and the service department may have to learn new maintenance and repair procedures. Because so many departments would be affected by the implementation of your idea, each needs to be asked about the impact they foresee from their own point of view.

Even when only a few people are affected by a decision, many employers expect widespread consultation and advice on it. Each person consulted will have his or her own professional role and areas of expertise, and each will play that role and apply that special knowledge when studying your communication.

When you address a group of people who will be reading from many perspectives, you are addressing a *complex audience*. To do that effectively, you write in a way that will meet each person's needs without hampering clear and effective communication to the others. Sometimes you will have to make trade-offs to accomplish that, for instance by focusing on the needs and concerns of the most influential members of your audience. In any case, the first step in writing effectively to a complex audience is to identify each of its members or types of members.

How to Identify Your Readers

How can you identify all the readers in your audience—phantom, future, and otherwise? Sometimes you can simply ask another person, such as your boss or an experienced co-worker. Or you can brainstorm by asking yourself such questions as "Who in my department will read this? Who in departments that work closely with mine? And who in more remote parts of the organization?" Brainstorming will work especially well for you when you have been on the job long enough to know how your company works.

When you are new on the job, however, you may benefit from using a more systematic procedure that minimizes the chances of overlooking an important reader. One such procedure involves using your employer's organizational chart. Such a chart names all the groups within an organization. To find potential readers, check each block within the chart, asking yourself whether anyone in the group represented by the block would have reason to read your communication. As you look through the chart, however, remember you are trying to identify not groups but individuals so you can learn the characteristics of each. Therefore, once you have identified a group likely to contain one or more readers, you should try to find out who those people are.

An added complication occurs when you write to complex audiences outside your own organization: you may not have organizational charts for those other organizations. In such situations, you may nevertheless discover probable *types* of audience members by asking yourself who would read your communication if it were addressed to your employer's organization.

Example: Identifying Readers

To see how one person identified the readers of one of his communications, consider the efforts of Thomas McKay as he prepared to write the letter in Figure 3–3. In the letter, McKay addresses Robert Fulton, vice president for sales at the company that sold McKay's firm an air sampler that failed to work properly. Although the letter is addressed to a single reader, McKay actually

MIDLANDS RESEARCH INCORPORATED

2796 Buchanan Boulevard
Cincinnati, Ohio 45202

October 17, 19--

Mr. Robert Fulton
Vice-President for Sales
Aerotest Corporation
485 Connie Avenue
Sea View, California 94024

Dear Mr. Fulton:

In August, Midlands Research Incorporated purchased a
Model Bass 0070 sampling system from Aerotest. Our
Environmental Monitoring Group has been using--or trying to
use--that sampler to fulfill the conditions of a contract
that MRI has with the Environmental Protection Agency to test
for toxic substances in the effluent gases from thirteen
industrial smokestacks in the Cincinnati area. However, the
manager of our Environmental Monitoring Group reports that
her employees have had considerable trouble with the sampler
while trying to use it on the first two smokestacks.

These difficulties have prevented MRI from fulfilling
some of its contractual obligations on time. Thus, besides
frustrating our Environmental Monitoring Group, particularly
the field technicians, these problems have also troubled Mr.
Bernard Gordon, who is our EPA contracting officer, and the
EPA enforcement officials who have been awaiting data from
us.

I am enclosing two detailed accounts of the problems we
have had with the sampler. As you can see, these problems
arise from serious design and construction flaws in the
sampler itself. Because of the strict schedule contained in

FIGURE 3–3   Letter to a Complex Audience
(Enclosures Not Shown)

Robert Fulton   —2—   October 17, 19——

our contract with EPA, we have not had time to return our
sampler to you for repair. Therefore, we have had to correct
the flaws ourselves, using our Equipment Support Shop, at a
cost of approximately $1500. Since we are incurring the
additional expense only because of the poor engineering and
construction of your sampler, we hope that you will be
willing to reimburse us, at least in part, by supplying
without charge the replacement parts listed on the enclosed
page. We will be able to use those parts in future work.

Thank you for your consideration in this matter.

Sincerely,

Thomas McKay

Thomas McKay
Vice—President
Environmental Division

Enclosures: 2 Accounts of Problems
            1 Statement of Repair Expenses
            1 List of Replacement Parts

FIGURE 3–3   *(continued)*

has a wide and complex phantom readership in three different organizations: the company that employs McKay (Midlands Research), the company that supplied the materials (Aerotest), and the federal agency that paid for the work that Midlands Research could not complete on time (the Environmental Protection Agency).

When McKay set out to identify his readers, the only organizational chart he could obtain was for his own company, Midlands Research. Figure 3–4 shows how he marked that chart to designate the people within Midlands who would read his letter. He circled the legal department because he realized that Aerotest's failure to fulfill his request might lead to legal action. And he circled the field teams because the people who had encountered the problem would want to know what was being done about it.

To identify the likely readers within Aerotest, McKay used the same chart, thinking that every reader in Midlands Research would probably have a counterpart in Aerotest: lawyer for lawyer, vice president for vice president. To identify other probable readers in Aerotest McKay used his common sense. He thought the departments that engineered and manufactured the faulty equipment would read the letter because their work was being criticized. And he expected that some of Aerotest's sales force would read it also, especially the sales engineer who sold the product to Midland.

McKay also planned to send the Environmental Protection Agency (EPA) a copy of his letter as a way of explaining the delays in Midlands' work. He guessed that at the EPA the letter would not travel beyond the contracting officer and those most directly affected by the delay in obtaining the test results.

After he identified the many readers of his letter, McKay asked himself what questions each would want answered. He found that his readers would be playing a wide variety of roles and accordingly would have a wide variety of questions. For instance, the lawyer within McKay's own company would read the letter (before it was sent) to answer this question, "Has McKay said anything that will invalidate Midlands' right to sue Aerotest?" In contrast, the lawyer for Aerotest would ask the same question, but hope to find a different answer. Likewise, the people in the Aerotest engineering division would ask, "What evidence does McKay have that the problems encountered by Midlands were caused by poor work on our part?" To answer that question, McKay drew up the two detailed accounts of the causes of the problems. Similarly, the Aerotest repair shop would ask, "Did the repairs made by Midlands really cost $1500?" Anticipating that question, McKay prepared the detailed statement of repair expenses.

As you can see, the readership of a communication can be quite large and diverse. Yet all of these readers are important people with important questions the writer must answer. Thus, whether you use a systematic or an unsystematic method, you must be careful to find out exactly who will read what you write.

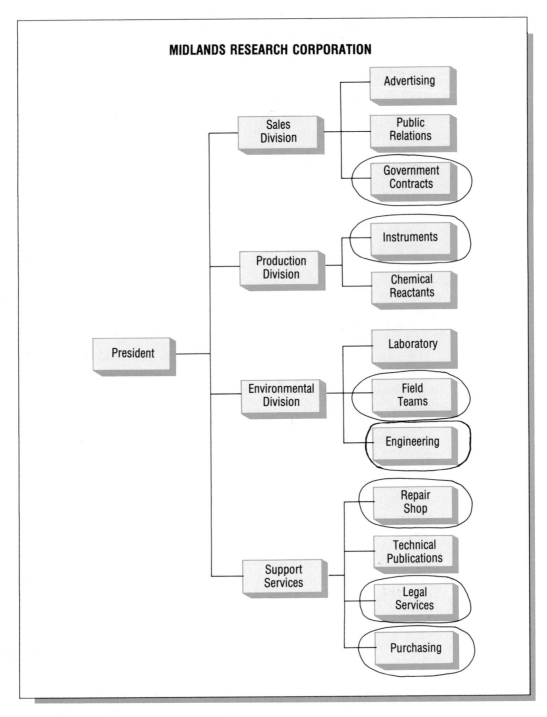

MIDLANDS RESEARCH CORPORATION

FIGURE 3–4 Organizational Chart Used to Identify Members of a Complex Audience

ASKING OTHERS ABOUT PURPOSES AND READERS

The five guidelines in this chapter have identified the kinds of information you should gather and develop to define your objectives. However, the discussion of these guidelines has largely left aside the question of *how* you can acquire this information.

In some cases, you will be able to describe your purposes and readers without conducting an extensive analysis. For instance, when writing on your own initiative to someone you work with daily, you will already know the final result you desire as well as the person's key characteristics and probable attitudes and reading tasks. On the other hand, at one time or another you almost surely will be asked to write or speak to people you do not know for reasons that are not fully explained to you. In such cases, you have two alternatives: to guess or to aggressively seek the information you need.

Check Your Assumptions

Given those two choices, many employees prefer to guess because they are reluctant to let anyone know they don't fully understand their assignment. But relying on guesses involves a great risk: you could guess incorrectly. If you write a communication that will achieve one final result excellently, but your boss has another result in mind, you are not likely to succeed with that communication.

Even experienced employees can make inaccurate assumptions about who their readers are. Consider, for example, the internal auditing department of an international corporation. This department's sole product is a continuous stream of reports. When an outside consultant once asked the two dozen auditors in this department who the primary readers of their reports were, every auditor thought he or she knew. However, only two of the twenty-four named the readers identified by their manager.

The manager was astounded. He thought his auditors knew—just as surely as they themselves thought they knew. It turned out that some of the problems with the department's reports were caused by the auditors' efforts to meet the needs of their assumed readers, not their actual readers. With an accurate identification of their readers, the auditors were able to make better decisions about many aspects of their reports, including what material to include, which conclusions to emphasize, the amount of background information needed, and even the tone to be used.

The moral of this situation is simple: if you make assumptions, be sure to check them.

Aggressively Seek Information

How can you check your assumptions so you avoid a similar error? Ask someone. One excellent source of information is the person who gives you an assignment. If your boss asks you to write a report, ask him or her about what the report should accomplish. Who will use it? How? In what ways does your boss want your report to affect the readers' attitudes? You may hesitate to ask such questions if your boss did not volunteer that information in the first place. Keep in mind, however, that your boss may assume if you need to know anything beyond what he or she has told you, you will ask.

You also can gain information about the purpose and audience for your communications by asking co-workers. These might be experienced people in your own department or co-workers employed in other departments. For example, imagine you are a technical specialist who helps design your company's products. To your surprise, you are asked by your boss to address a meeting of the company's major stockholders. To learn about the specific concerns and other characteristics of this audience, you might contact the company's public relations department.

Members of your intended audience are another important source of information. If you must prepare a report for people in another department or in a client's organization, visit or call the people who will use your report. Tell them you want to be sure your report meets their needs. Ask how they will use your communication, what kinds of questions they hope it will answer, what form they want the answer to be presented in, and so forth. By approaching your readers in this way, you will be using a technique employed by marketing and advertising personnel: you will be asking your target market (your readers) what would make them satisfied with your product (your communication).

How to Deal with Partial Disclosure

You should be aware that in some situations you will not be able to get complete information about the purposes of or even the audience for your communications. As explained in Chapter 1, bosses sometimes pass communications written by their subordinates along to their superiors or other phantom readers. Although many managers are quite open about this practice, you may encounter situations in which your manager will not disclose such information.

For example, your manager might ask you to write her a report evaluating some part of the department's operation. While the manager seems to be the intended audience, she actually might be planning to send the report to upper management to buttress her arguments for a budget increase. If your boss fears this strategy will be ruined if others find out about it, she may not tell you the real purpose of the report, or even that it will be shown to people outside the department.

Situations like this illustrate the complex purposes that communications serve on the job. You should not, however, let such complexities deter you from asking about the purpose and audience for the communications you prepare on the job. The more fully you understand your audience and the more clearly you can define your purpose, the more successful your communications will be.

WORKSHEET FOR DEFINING OBJECTIVES

Figure 3–5 is a "Worksheet for Defining Objectives" you can use to help you define the objectives of communications you prepare at work. Of special use to you will be the sections in which you describe the key characteristics of your readers, specify the way you want to alter your readers' attitudes, and identify your readers' mental tasks.

Example of a Completed Worksheet

Figure 3–6 shows how one person, Stephanie, filled out the worksheet for a recommendation report she planned to submit to the head of the organization for which she worked. The organization provides braille translations of textbooks and other reading materials in response to requests from blind people. The translations are assigned to volunteers, and Stephanie believes she knows how to improve the assignment procedures. At present, the assignments are made to each volunteer in turn, regardless of the volunteer's speed or reliability with translating. Stephanie wants to suggest instead that the most urgent and important translations be given to volunteers who have in the past demonstrated their speed and reliability. As you can see from the worksheet, however, she does not expect her boss to welcome the recommendation.

How to Use the Worksheet

The purpose of the worksheet, of course, is to guide you as you prepare your written or oral communication. The information you develop while filling it out can help you make many practical decisions about how to construct your message. Consider, for example, some of the insights Stephanie gained. In this case, the most important observations come from the section on readers' attitudes, which clearly indicates Mrs. Land will be defensive about the recommendation. Not only did she set up the present system, but also she feels it runs as well as it possible can. Furthermore, she is the type of person who naturally resists suggestions for change. That insight helped Stephanie realize she could not assume Land would readily agree a problem exists. Furthermore, the information about Land's insecurity and touchiness helped Stephanie see she would have to be very diplomatic in presenting the problem so she would not appear to be criticizing Land.

The worksheet also helped Stephanie focus on the fact that Land feels that Stephanie does not understand all the considerations that go into running the agency. Stephanie therefore decided she would have to take these considerations thoroughly into account so Land's attitude toward Stephanie and (consequently) her recommendation would be more favorable.

Similarly, the worksheet helped Stephanie see that Land would wish to compare the two systems in terms of cost, speed of producing translations, and

Worksheet

DEFINING OBJECTIVES

Overall Purpose

What are you writing?

What prompts you to write?

What outcome do you desire?

Reader Profile

Who is your primary reader?

What is your reader's relationship to you?

What are your reader's job title and responsibilities?

Who else might read your communication?

How familiar is your reader with your subject?

How familiar is your reader with your specialty?

Does your reader have any communication preferences you should take into account?

Should you take into account any other things about your reader when writing?

FIGURE 3–5   Worksheet for Defining Objectives

Reader's Informational Needs

What are the key questions your reader will ask while reading?

How will your reader search for the answers? (The reader may use more than one strategy.)
_____ Sequential reading from beginning to end
_____ Selective reading, as when using a reference book (what key terms will your
reader look for?)
_____ Other (explain)

How will your reader use the information you provide?

_____ Compare point by point (what will be the points of comparison?)
_____ Attempt to determine how the information you provide will affect him or her
_____ Attempt to determine how the information you provide will affect his or her
organization
_____ Follow instructions step by step
_____ Other (explain)

Reader's Attitudes

What is your reader's attitude toward your subject? Why? What do you want it to be?

What is your reader's attitude toward you? Why? What do you want it to be?

FIGURE 3–5   *(continued)*

Worksheet

DEFINING OBJECTIVES

Overall Purpose

What are you writing?   A proposal that the Kansas City office of the Society for the Blind adopt a new method of assigning Braille translations in which the most urgent requests go to the quickest and most reliable volunteers.

What prompts you to write?   I think that the method will enable us to respond more rapidly to requests for translations from people who need them quickly. These people would include students who need textbooks translated.

What outcome do you desire?   I would like the new method to be put into effect, at least on a trial basis.

Reader Profile

Who is your primary reader?   Mrs. Land.

What is your reader's relationship to you?
We see each other daily but still have a fairly formal relationship.

What are your reader's job title and responsibilities?   She is Director of the Braille Division, responsible for recruiting and maintaining a large group of volunteer Braille translators, advertising translation services to blind people, and responding to requests for translations by assigning volunteers the work of making them.

Who else might read your communication?
Rich Seybold and Mina Williams, Mrs. Land's chief assistants.

How familiar is your reader with your subject?   Mrs. Land knows the present system of assigning Braille translations as well as anyone because she set it up and has run it for the past twelve years. She does not know that I am thinking of proposing an alternative, but does know that systems similar to mine are used at some other offices of the Society for the Blind.

How familiar is your reader with your specialty?
Very familiar.

Does your reader have any communication preferences you should take into account?
She likes all communications to look "business-like." She does not like informality.

Should you take into account any other things about your reader when writing?   Mrs. Land gives the impression of being very sure of herself but feels threatened by suggestions for change.

FIGURE 3–6   Stephanie's Completed Worksheet for Defining Objectives for an Unrequested Recommendation

Reader's Informational Needs

What are the key questions your reader will ask while reading?
What makes you think anything is wrong with the present system?
What will happen if some volunteers learn that they are not in the top
  group for assignment of translations?
Won't your proposed system hurt the morale of our volunteers and cause them
  to quit working for us?
How, exactly, would your new system work?
What would I have to do differently?
How would the operations of this office be changed?
How would we determine which translations deserve highest priority?
How would we decide which translators are placed in our top group?
Would it cost anything?

How will your reader search for the answers? (The reader may use more than one strategy.)

__X__ Sequential reading from beginning to end
__X__ Selective reading, as when using a reference book (what key terms will your
        reader look for?)
_____ Other (explain)
        Mrs. Land will probably skip around through the report at first, but
        will later read it from front to back.

How will your reader use the information you provide?

__X__ Compare point by point (what will be the points of comparison?)
__X__ Attempt to determine how the information you provide will affect him or her
__X__ Attempt to determine how the information you provide will affect his or her
        organization
__X__ Follow instructions step by step
____ Other (explain)
        She will compare her system with mine in terms of cost, speed of
        producing translations, and effect on volunteer morale. Although
        she won't exactly look for instructions in my proposal, she will
        want to know in detail how the process would work.

Reader's Attitudes

What is your reader's attitude toward your subject? Why? What do you want it to be?
Mrs. Land thinks the present system runs as well as one possibly could.
She has said that she thinks a system like the one I am proposing would
create competition among volunteers that would destroy their morale. I
want her to see that a better system is possible and that morale problems
can be avoided.

What is your reader's attitude toward you? Why? What do you want it to be?
Although I have worked for her for three years, Mrs. Land still thinks of
me as a newcomer who knows little and has impractical ideas. I want her to
think that I am a helpful, knowledgeable, sensible person.

FIGURE 3–6   *(continued)*

so on. Stephanie resolved to devote one part of her discussion to those topics, each under its own heading.

When filling out the worksheet, Stephanie also realized how important volunteer morale was for Land. Stephanie decided that, to present her recommendation effectively, she would have to address Land's counterarguments on that point. In fact, Stephanie wrote to similar agencies employing the procedures she advocated to learn about the successes they had been enjoying.

As this example shows, filling out the Worksheet for Defining Objectives is much more than an academic exercise. The information you provide on the worksheet can give you important insights into ways to construct your communication so it gets the result you desire.

CONCLUSION

This chapter has described a reader-centered approach to defining objectives that focuses your attention on the people you are writing to, the ways you want them to respond, and the personal and professional characteristics that will shape their reactions as they read your message. As you plan, draft, evaluate, and revise your communication, use your objectives as your guide. Don't stray from them. At the same time, be flexible. If your continued work on your communication modifies or deepens your understanding of your situation, topic, or readers, be willing to refine your objectives accordingly. The rest of this book provides detailed advice about how you can make good use of your reader-centered objectives throughout all your other work on a communication.

EXERCISES

1. Study the completed Worksheet for Defining Objectives in Figure 3–7. Describe some of the writing strategies the author could use to achieve the purpose set out there. Try to think of at least one strategy to match each of the answers the writer provides.

2. Find an example of a communication you might write in your career. After studying that communication, fill out a Worksheet for Defining Objectives (Figure 3–5) in a way that describes the communication's objectives. Then explain the features of the communication by telling how they seem to be tailored to the objectives you have described. If you can think of ways the communication could be improved, make and explain your recommendations.

Worksheet

DEFINING OBJECTIVES

Overall Purpose

What are you writing? An instruction manual that will enable students to use the color graphics printer to create graphs for their courses. This printer requires some special procedures.

What prompts you to write? I am a lab assistant who has been assigned this project.

What outcome do you desire? I would like students to be able to run the printer using my manual only, so that they do not need to request assistance from the lab staff.

Reader Profile

Who is your primary reader? Juniors and seniors in advanced engineering courses.

What is your reader's relationship to you? I'm a junior engineering student myself.

What are your reader's job title and responsibilities? Not applicable.

Who else might read your communication? New members of the lab staff.

How familiar is your reader with your subject? They know lots about ordinary printers, but nothing about this one.

How familiar is your reader with your specialty? Because they come from many majors, the students will not necessarily know the terms used in my major (or in those of other students).

Does your reader have any communication preferences you should take into account? The truth is that they just don't like instructions. If they are going to read my manual, it will have to be direct and to the point—and attractive also.

Should you take into account any other things about your reader when writing? Not that I can think of.

FIGURE 3–7 Worksheet for Defining Objectives Completed by Student Preparing to Write Instructions

Reader's Informational Needs

What are the key questions your reader will ask while reading?
What can this printer do for me?
Is it easy to use?
Is there someone around I can ask for help so I don't have to read these
  instructions?
How do I do this particular thing (I don't want to learn about how to do
  other things)?
How long does it take to learn to use this printer?

How will your reader search for the answers? (The reader may use more than one strategy.)

__X__ Sequential reading from beginning to end
__X__ Selective reading, as when using a reference book (what key terms will your
        readers look for?)
____ Other (explain)
        Though some may read from front to back, most will search for just
        the information they think they need.

How will your reader use the information you provide?

_____ Compare point by point (what will be the points of comparison?)
_____ Attempt to determine how the information you provide will affect him or her
_____ Attempt to determine how the information you provide will affect his or her
          organization
__X__ Follow instructions step by step
____ Other (explain)

Reader's Attitudes

What is your reader's attitude toward your subject? Why? What do you want it to be?
They probably know that graphics printers can do wonderful things, but they
may fear that it will take a long time to learn to use this printer. Some
may also worry that they won't be able to learn enough about the printer to
do exactly what they want to do. I want them to feel that they can easily
master the printer by following my instructions.

What is your reader's attitude toward you? Why? What do you want it to be? The
students won't have any particular attitude toward me, except to see me as
another person who writes those confusing and boring instructions they hate
to read. I want them to feel that I have taken their needs and desires
into account and prepared an exceptionally good set of instructions for them.

FIGURE 3–7   *(continued)*

CASE	ANNOUNCING THE SMOKING BAN

As you sit in your office on the top floor of the four-story building owned by your employer, you look up from the memo that company president C. K. Mitchell dropped on your desk a few minutes ago. It's a draft and he's asked your opinion about it.

The memo announces a new no-smoking policy for the company, which employs about 250 people, all working in your building or the one next door. Developing that policy has been an arduous task. Recently, the state legislature passed a law requiring all employers to establish a smoking policy. The law doesn't say what the policy should be, only that every employer must have one.

To figure out what policy to establish, C.K. enlisted the help of a team of his top managers, who have hashed over the question for the past three months. One possibility would be to permit smoking anywhere. Although that is the current policy, an increasingly vocal group of nonsmokers clearly expects a different response to the new law. Because this group numbers nearly half the work force, including many influential people, that possibility had to be rejected.

Another idea was to allow smoking only in private offices. At first that seemed like a great idea because it meant smokers could smoke in places where they wouldn't bother nonsmokers. The policy sounded so good it was almost announced. At the last minute, however, Maryellen Rosenberg, Director of Personnel, pointed out that lots of employees don't have private offices. In fact, only the salaried employees do. The secretaries, mail clerks, janitors, and others without private offices who smoke would surely complain that the policy discriminated against them.

After considering and rejecting several other plans, C.K. finally decided the only thing to do would be to ban smoking altogether. "What could be more appropriate," he asked at yesterday's meeting, "for a company that tries to make people healthier?" Your company designs and markets exercise equipment for homes and fitness centers. After prompting from you and several others, C.K. decided to hire a consulting firm that offers a course to help employees stop smoking. The course will be free for the employees.

"C.K.'s memo on this one will sure need to be good," you thought to yourself when he left the draft with you. Although most employees realize that C.K. has been meeting with you and others to create a policy, the details of the ideas discussed have been kept secret to avoid inciting bad feelings among groups in the company. This caution has seemed justified because the publicity surrounding the new law has sparked some heated controversies among employees about smoking. Some have talked of quitting if they can't smoke at work, and others have talked of quitting if smoking isn't prohibited.

This controversy is harming morale, which is already low because C.K. recently reduced the company's contribution to the corporate profit-sharing plan even though revenues have been rising steadily. The money is being invested instead in purchasing the building next door, a move that will probably save the company money in the long run. However, C.K.'s action is widely viewed as yet another example of his heavy-handed and unfeeling management.

Your Assignment

First fill out a Worksheet for Defining Objectives that describes a reasonable set of objectives for C.K.'s memo. Remember that this one memo must successfully address all employees. Second, evaluate the draft of C.K.'s memo (Figure 3–8) in light of the objectives you have established for it and then revise the memo to make it more effective. Be prepared to explain your revisions to C.K.

MEMORANDUM

TO: All Employees
FROM: C. K. Mitchell
DATE: June 4, 19––
RE: Smoke–Free Environment

I hereby notify you that beginning September 1 of this year, Fitness Exercise Equipment, Inc. will institute a smoke–free environment. To wit, smoking will <u>not</u> be allowed anywhere inside the main building or the satellite building next door.

The delay between this announcement and the beginning date for this new policy will allow any employees who smoke the chance to enroll in courses which, I hope, will help them break or curtail their habit. In accordance with our concern for the wellness of all our employees, we will enhance our working environment by prohibiting all smoking.

All employees are thanked for their cooperation, understanding, and dedication to better health.

FIGURE 3–8   Memo for Use with Case

PLANNING

4

PLANNING TO MEET YOUR READERS' INFORMATIONAL NEEDS

I n the preceding chapter, you learned how to define objectives for the communications you write at work. You are now beginning the first of three chapters that will help you transform your definition of objectives into a successful plan of action. The chapters are as follows:

● Planning to Meet Your Readers' Informational Needs (Chapter 4)
● Planning Your Persuasive Strategies (Chapter 5)
● Using the Library (Chapter 6)

BENEFITS OF GOOD PLANNING

Planning is an inevitable part of writing. Whenever you start to write something to a friend, instructor, or co-worker, you begin with at least *some* idea of what you will say and how you will say it. However, there is a great difference between planning well and planning poorly. And good planning can benefit you in two ways.

First, good, thoughtful planning saves you time. Consider, for instance, some of the difficulties encountered by Charlotte because she drafted a report without first planning it carefully. She spent more than an hour writing and then polishing three long paragraphs she later realized she didn't need. She also spent half an hour revising two pages before she discovered they would be much more effective if reorganized and moved to another place. Had she planned more thoroughly, Charlotte could have saved herself much wasted time and effort.

Second, planning helps you improve the effectiveness of your writing. When Charlotte showed her report to her boss, he pointed out several places where she didn't present and support her major points as clearly and forcefully as she might have. How did that happen? Charlotte concentrated on the smaller parts of her report, focusing on each sentence and paragraph as she wrote it. She didn't think much about overall strategy. A more thorough planning effort would have drawn her attention to larger questions, such as how she could make all her points fit together and how she could most persuasively present her conclusions. With this broader focus, she could have devised a plan that would have avoided the weaknesses her boss pointed out.

In a way, Charlotte was lucky. Her boss gave her time to redo her report before she sent it to its intended readers. Consequently, she suffered only a loss of time and effort, together with some loss of pride. However, at work people often have no time to rewrite their communications. They must get them right the first time. Even if they don't, the communications are sent anyway. The costs of poor planning in such situations can be much greater than those Charlotte experienced.

ORGANIZATION OF THIS BOOK'S ADVICE ABOUT PLANNING

To help you plan effectively, this book presents two complementary sets of planning guidelines that correspond to the enabling and persuasive elements of purpose explained in Chapter 3. The four guidelines in this chapter will help you design plans for achieving the enabling element of purpose, which is to enable your readers to locate, understand, and use the information and ideas they need from you. Then, the five guidelines in Chapter 5 will help you create plans for achieving the persuasive element of purpose, which is to influence your readers' attitudes and actions in the ways you desire.

| GUIDELINE 1 | ANSWER YOUR READERS' QUESTIONS |

The first step in achieving the enabling element of purpose is to learn what information your readers need from you. Finding that out will help you determine what to say in your communication.

You can start learning about your readers' informational needs by taking the advice in Chapter 3 to think of your readers as people who ask questions. Using their questions as your guide, you can select from all you know about your topic the particular facts that belong in your communication, and you can identify items you need to investigate further because you don't yet have the information your readers need.

The amount of work you must do to identify your readers' questions can vary greatly from situation to situation. When writing to people with whom you've worked regularly in the past, you will probably have no trouble listing their questions. At other times you may have to be much more resourceful. For instance, you might use your imaginary portrait of your readers. If a person you've portrayed were to sit down with you now, what would he or she ask?

Another way to identify your readers' questions is to think about the readers in terms of three reader characteristics discussed in Chapter 3: professional role, knowledge of your subject or topic, and knowledge of your specialty. The following paragraphs describe how each of those characteristics can help you predict the readers' questions you must answer in your communication.

Questions Arising from Your Readers' Professional Roles

As Chapter 3 explained, readers at work typically play one of these three roles: decision-maker, adviser, and implementer. The following paragraphs discuss the kinds of questions people typically ask when playing each one.

Decision-makers   Decision-makers typically ask questions shaped by their need to choose between alternative courses of action. They will use the information and ideas in your communication to decide what the company should do in the future—next week, next month, next year. Consequently, as they read, decision-makers are likely to ask the following questions:

- **What are your conclusions?** Decision-makers are much more interested in the conclusions you draw than in the data you gathered or the procedures you used. Conclusions are the parts of your communication that most readily serve as the basis for a decision.
- **What do you recommend?** If you have any recommendations to make, decision-makers want to know what they are. Indeed, decision-makers often assign people the task of making written recommendations.
- **What will happen?** Decision-makers usually want to know what will happen if they follow your recommendations—and what will happen if they don't. How much money will be saved? How much will production increase? How will customers react to the change?

Decision-makers rarely ask for raw data or lengthy explanations of how you arrived at your conclusions, recommendations, and projections. Generally, the only use they would have for such details would be to check on the validity of your work, something they rarely want to do because they assume the people who report to them are experts in their own specialties. When decision-makers do want to check, for instance when they are reading a proposal submitted by another company, they often ask their advisers to look into the matter for them.

However, when answering a decision-maker's questions, be careful not to eliminate specificity and essential detail. If you are reporting sales and income projections for the next three years, your readers will probably *not* want to know all of your statistical techniques and technical assumptions. But they will want you to tell them your results as exactly as possible. For example, they will want to know not merely that both sales and income will increase, but also exactly how large those increases will be and which products will generate them.

Decision-makers also want brief answers. As the director of one government research office declared, "If it can't be held together with a paper clip, it is too long for me to read." That attitude is typical. All kinds of organizations have guidelines requiring reports to management to be no more than one or two pages long, even if the reports represent several months' work.

Many decision-makers prefer nontechnical answers. Often decision-makers know very little about the technical terms and concepts used in specialized fields. Even decision-makers who, at one time, did specialize in the same field as the writer may have lost touch with new developments in the field since being promoted to a managerial position. They depend on the writer to use terms they can understand and to provide background and explanatory information they need in order to grasp unavoidable technical content.

In sum, when writing to decision-makers, plan to emphasize your conclusions, recommendations, and predictions, which you should present with brevity and simplicity.

Advisers   Unlike decision-makers, advisers *are* interested in details. They need to analyze and evaluate the evidence supporting your general conclusions, recommendations, and projections. Perhaps they have been asked to do so by a decision-maker, or perhaps they will be affected by the decision being considered so they want to know its exact implications.

Consequently, advisers ask questions that touch on the thoroughness, reliability, and impact of your work:

- Did you use a reasonable method to obtain your results?
- Do your data really support your conclusions?
- Have you overlooked anything important?
- If your recommendation is followed, what will be the effect on other departments?
- What kinds of problems are likely to arise?

Though advisers need detailed answers to these questions—much more detailed than decision-makers want—they still are not likely to want every last particular of the procedures you used or every last piece of data you collected. They need to know just enough detail to assess the quality and consequences of your ideas and information.

Despite the differences between the kinds of answers decision-makers and advisers want, you will often find yourself writing a single communication that must meet the needs of both groups. (In that case, you would be addressing your communication to a complex audience, as discussed in Chapter 3.) Usually such reports fall into two distinct parts: (1) a very brief summary—called an *executive summary* or *abstract*—at the beginning of the report, designed for decision-makers, and (2) the body of the report, designed for advisers. Typically, the executive summary is only one page, or a few pages at most, while the body may exceed 100 pages.

In a study conducted by James W. Souther, a large group of managers was asked how often they read each of the major parts of a long report. As Figure 4–1 indicates, they reported that they read the summary 100 percent of the time, whereas they read the body only 15 percent of the time. When these decision-makers want more information than the summary provides, they usually go to the introduction (which provides background information) and to the recommendations (where the writers explain their suggestions more fully). The rest of the report is read by advisers.

Implementers   Decisions, once made, have to be carried out. People at all levels within an organization are entrusted with doing this. To aid them, you will write a wide variety of communications, including the following common types:

- **Step-by-step instructions.** For example, you might write instructions for using a new computer program or operating your employer's lathes.

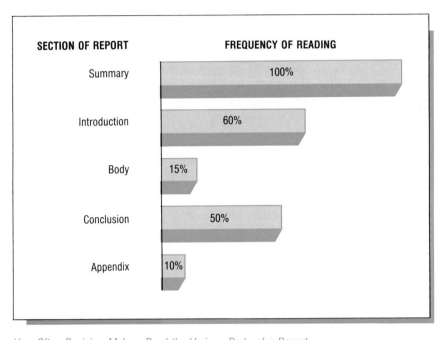

FIGURE 4–1 How Often Decision-Makers Read the Various Parts of a Report

From a study by James W. Souther described by Richard W. Dodge in ''What to Report,''
Westinghouse Engineer 22, nos. 4–5 (1962), 108–11.

- **General instructions.** General instructions include policy statements and general directives that readers must follow. You might, for instance, write up the company's new vacation policy that employees and managers will have to interpret and apply to their own situations.

- **Requests and orders.** You will write requests and orders when something needs to be done and your reader is the person who should do it. When you advance to a supervisory or managerial position, you will do a great deal of communicating of this sort. However, you also may do it when you are new on the job—for instance, when you request help from another department that is supposed to work with you and you must tell your readers exactly what you would like them to do.

The most important question implementers want answered is, ''What do you want me to do?'' Whether you are writing step-by-step instructions, requests, or orders, answer by providing clear, exact, easy-to-follow directions. Be sure to say explicitly what you want. All too often, people fail to get what they want for the simple reason that their readers cannot tell exactly what they are being asked to do.

You may find it particularly difficult to answer the question, ''What do you want me to do?'' when you are preparing general instructions. You must be sure each person in your audience can see how the policy or directive applies to his or her specific situation. One excellent way to do this is by using concrete examples to illustrate otherwise abstract instructions.

Another important question asked by implementers is, "What is the purpose of the actions you are asking me to perform?" Readers of policy statements and general directives, especially, often must know the intent of a policy if they are to produce satisfactory results. Imagine, for instance, the situation of the managers of a factory who have been directed to cut by 15 percent the amount of energy used in production. They will want to know whether they are to realize long-term energy savings or to compensate for a short-term energy shortage. If the latter, they may think of temporary actions, such as altering work hours and temporarily curtailing certain operations. However, if the reduction is to be long-term, they may think of long-term measures, such as purchasing new equipment and modifying the structure of the factory building. Similarly, the people to whom you send requests and orders can often help you best if they know *why* you are asking for their assistance.

In many situations, implementers also need the answer to the question, "How much freedom have I to choose the way I will perform this action?" People who are following instructions often devise shortcuts or alternative ways of doing things. Where any of several ways is sufficiently effective, efficient, and safe, you may not care to specify one over another. But where you have a preference, you should say so explicitly.

Finally, implementers often require the answer to the question, "When must I complete this task?" or "How long will this task take?" They can schedule their work only if they know when you want it done or how long it will take them to complete it.

The preceding discussion of decision-makers, advisers, and implementers has identified only the general questions typically asked by large groups of people. When listing the questions any particular reader will want your communication to answer, you will need to think also about that individual's specific job responsibilities. The key point for you to remember is that thinking about your readers' professional roles can provide valuable insights into the questions you will need to address when you write.

Questions Arising from Your Readers' Unfamiliarity with Your Subject

The kinds of questions your readers will want you to answer in your communication will depend not only on their roles but also on their familiarity with your subject. Imagine, for example, you are going to recommend to a group of executives at corporate headquarters some methods for improving the management structure in a branch office. If the executives already know how that office operates and why management there needs to be improved, you can probably proceed directly to the heart of your message. On the other hand, if the executives don't know much about that office or don't realize a problem exists there, they will have many background questions about the situation as they try to fill in the context for your recommendations.

When providing the background information, remember to supply only information your audience needs in order to grasp your message and its significance to them. Don't provide information that is overly detailed or that includes material unrelated to your readers' needs.

Questions Arising from Your Readers' Limited Knowledge of Your Specialty

When answering your readers' questions, you may use a word, concept, or symbol they don't understand. That gives rise to another sort of question, "What does that word (or concept or symbol) mean?"

The first step in answering this kind of question is to anticipate where your readers will ask it. As you decide what to include in a communication, you should look for places where you use terms that might be unfamiliar to your audience. Once you have identified them, ask yourself what kind of information you should provide to help your readers understand the terms. You might, for instance, provide a definition:

A *headgate* is a gate for controlling the amount of water flowing into an irrigation ditch.

A *rectifier* is an electrical device for converting alternating current to direct current.

Or you might explain by means of analogy:

Wood sorrel is a plant that resembles clover.

On the proposed extruders, we will use a feed system much like that found on a Banbury mixer.

You will find detailed advice about defining and explaining unfamiliar words and concepts in Chapter 12.

In sum, one very helpful way to determine what you should say in order to meet your readers' informational needs is to think about the questions your readers will ask while reading. As an aid to identifying their questions, you can think about your readers' professional roles, familiarity with your subject, and level of knowledge of your specialty.

<table>
<tr><td>GUIDELINE
2</td><td>**INCLUDE THE ADDITIONAL INFORMATION YOUR READERS NEED**</td></tr>
</table>

Although your efforts to decide what to say in a communication should certainly include an attempt to identify the questions your readers will want you to answer, you should remember that sometimes readers don't know all the questions they should ask. Through your expertise and experience you will see that some additional information is indispensable to them. Include it also.

For example, imagine Toni's situation. Toni has been asked to investigate three computer programs her employer might purchase to aid in designing products. While gathering information about the capabilities, performance, and cost of the programs, she has learned that the company that makes one is having serious financial difficulties. If the company goes out of business, Toni's employer would not be able to obtain the assistance and improved versions of the program it would normally expect to receive over the coming years. Thus, even though Toni's readers are not likely to ask "Are any of these programs made by companies that appear to be on the verge of bankruptcy?" Toni should include that information in her report.

As you consider information you should include even though your audience doesn't know enough to ask for it, be careful to avoid the temptation to incorporate facts that interest you but will not be important to your readers.

<table>
<tr><td>GUIDELINE
3</td><td>**ORGANIZE TO SUPPORT YOUR READERS' TASKS**</td></tr>
</table>

To fully meet your readers' informational needs, you must not only decide what to say in a communication, but also organize that material in a helpful way. This guideline suggests you do so by thinking once again about the enabling component of your communication's purpose: consider the tasks your readers will perform when searching for and using the information you will provide, and then organize in a way that will help them perform those tasks.

To follow this guideline's advice effectively, it's important for you to understand its relationship to typical advice about organization. When asked what method of organization a writer should use, most people reply, "A logical organization, of course." Unfortunately, a logical organization is not necessarily helpful to the reader.

Imagine, for example, that you work in the registrar's office of a small college. The school's computer has broken down, losing some files. Consequently, you have been asked to work manually to fulfill a request from the college's president, who wants a list of the home addresses, majors, and class standings (first-year, sophomore, junior, senior) of all students enrolled in a technical writing class. The instructors of these courses have circulated sign-up sheets to their students. Now, with the completed sheets in front of you, you are ready to combine all the information into a single list for the president.

How will you organize the list? Should you arrange the names by major? Class standing? Hometown? Some other principle? While any of these organizations would be *logical,* they will not all be equally *useful.* The most useful organization would be the one most compatible with the *use* the president will make of the list. If the president will use it to send out a series of letters, one to all the seniors, another to all the juniors, and so on, you could help your reader most by organizing the list according to class standings. Or if the president will use the list to determine which departments on campus send the most students to technical writing classes, you could arrange the list according to the students' majors. Each of these patterns is not only logical in itself but also succeeds in supporting the reader's task.

Of course, readers' tasks vary greatly from communication to communication. So, for each communication you write, you should think about what your readers will do with the information you provide. Nevertheless, some readers' tasks are common enough that it is possible to identify a few widely applicable strategies for organizing messages. The following paragraphs describe three: organize hierarchically, group together the items your readers will use together, and give the bottom line first. These strategies will not be appropriate for everything you write at work, but they will be very useful for many communications, especially long ones.

Organize Hierarchically

In any reading situation, people's most basic task is to determine the meaning of the message. They must combine the various bits of information derived from individual words and sentences into larger structures of meaning. In many instances, those structures are hierarchies in which the overall topic is broken down into subtopics and some or all of the subtopics are broken down into still smaller units. Figure 4–2 is a tree diagram showing the hierarchical organization of a report on parking problems in a small city; Figure 4–3 is an outline showing the hierarchy of the same report.

By building mental hierarchies, readers attempt to create a meaningful place for every piece of information in a communication. However, building hierarchies can be hard work. If the writer doesn't help, the reader must figure out how to gather the various bits of information into groups, then decide which groups go together into larger groups, and at each level identify some generalization that characterizes each group's contents or expresses its meaning. You can save your readers much of this work by organizing your writing hierarchically.

To plan a hierarchical organization, you can proceed in either of two directions. You can take your overall topic and subdivide it into progressively smaller parts, or you can begin with a list of items of information (such as you might generate through brainstorming) and gather them into progressively larger groups. People often work from both directions toward the middle.

Group Together the Items Your Readers Will *Use* Together

Whether or not you are organizing hierarchically, be careful to group together information your readers will *use* together. That will enable them to use your communication more efficiently.

To determine what information your readers will use together, think about the specific mental tasks they will perform while reading. In Chapter 3, you

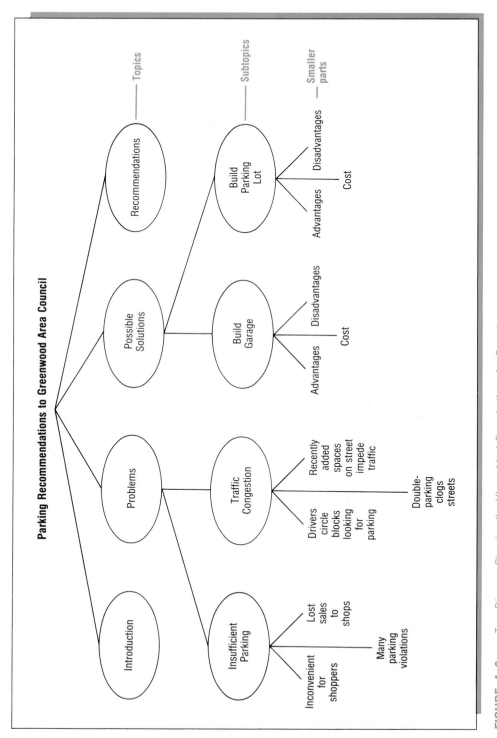

FIGURE 4–2   Tree Diagram Showing the Hierarchical Organization of a Report

```
               PARKING RECOMMENDATIONS TO GREENWOOD AREA COUNCIL
                 I. Introduction
                II. Problems
                    A. Insufficient parking
                       1. Shoppers complain about inconvenience
                       2. Shop owners complain about lost sales
                       3. Police report many parking violations
                    B. Traffic congestion
                       1. People circle through the area waiting for a
                          parking space to open up
                       2. Additional on-street parking impedes the free
                          flow of traffic
                       3. Though illegal, some double-parking occurs,
                          clogging streets
               III. Possible solutions
                    A. Build a parking garage in the center of the
                       shopping area
                       1. Advantages
                       2. Disadvantages
                       3. Cost
                    B. Build a parking lot on the edge of the shopping
                       area and run shuttle buses to major stores
                       1. Advantages
                       2. Disadvantages
                       3. Cost
                 IV. Recommendations
```

Topic
Topic
  Subtopic

    Smaller parts

  Subtopic

    Smaller parts

Topic
  Subtopic

    Smaller parts

  Subtopic

    Smaller parts

Topic

FIGURE 4–3   Outline Showing the Hierarchical Organization of a Report

have already read an example of how to do that. The example involved the report Lorraine was writing about the two blast furnaces her employer, a steel mill, was thinking of purchasing. Because Lorraine knew that her readers' primary task would be to make a point-by-point comparison, she grouped her information around the criteria her reader would use (cost, efficiency, etc.).

As another example, consider Daniel's efforts to organize a report to be used in a much different way than Lorraine's. Daniel works for a federal task force that has assigned him to write the final report on their efforts to determine the effects of sulfur dioxide (SO_2) emissions from automobiles and factories. The report will summarize the findings of hundreds of research projects conducted over many years.

Daniel might organize his report in many different ways. Suppose the purpose of his report is to teach readers about the history of research in SO_2 emissions. Then the readers would want to understand the relationships over time of the various studies he summarizes. To help them do that he could organize his report chronologically; its chapters might have titles such as "Research from 1960–1980" and "Research from 1980 to the Present."

However, suppose the purpose of Daniel's report is to help federal and state legislators evaluate the effects of SO_2 emissions so the legislators can pass appropriate laws regulating the emissions. Then his readers will want a summary of each kind of impact SO_2 emissions have, regardless of when the studies that evaluated the impact were made. To support that reading task, Daniel could organize his report in the following way:

I. Effects on Human Health
   A. Effects on skin
   B. Effects on internal organs
      1. Heart
      2. Lungs
      3. Etc.

II. Effects on the Environment
   A. Effects on man-made objects
      1. Buildings
      2. Cars
   B. Effects on Nature
      1. Lakes, rivers, and oceans
      2. Forests
      3. Wildlife
      4. Etc.

Of course, you can't plan a useful organization simply by imitating the ones Lorraine and Daniel created, but you can do it by following their procedure: think about how your readers will use the information and ideas you provide, and group your material accordingly.

Give the Bottom Line First

A task readers at work perform very energetically is to look for the writer's main point. To express their desire to find the main point as quickly as possible, people at work often urge writers to "put the bottom line first."

What Is the Bottom Line?   The *bottom line,* of course, is the last line of a financial statement. Literally, the writers are being told, "Before you immerse me in the details concerning individual expenditures and sources of income, tell me what I most want to know: whether we made a profit or took a loss."

However, this advice is applied figuratively to many kinds of communications prepared at work, not just financial reports. Before people at work read the details of a report, proposal, or other decision-making communication, they want to know what the main point is.

To say you should satisfy readers by giving the bottom line first doesn't mean you must give your main point in the very first sentence. It's still appropriate to provide necessary background information and transitional material. But don't keep your readers in suspense. Before giving the details, state the main point: Is the project on schedule, or must we take special action to meet the deadline? Will the proposed design for our product work or must it be modified?

Is It Illogical to Put the Bottom Line First?   Putting the most important information first seems illogical to some writers. They reason that the most important information is generally some conclusion they reached fairly late in their thinking about their subject matter. Before reaching it, they had to gather and analyze facts, and they therefore reason that they should give their readers those facts to consider before telling them what conclusions they reached.

Such a view is writer-centered. It assumes that because the writer gained information in a certain way, that is the way the information should be presented. While such an organization has logic, it is also *logical* to state a conclusion or recommendation and then to follow with explanations and supporting details. And when the reader is *first* interested in the conclusion or recommendation, then it is certainly more reader-centered to accommodate the reader.

Should You *Always* Put the Bottom Line First?   "Do I always need to give the bottom line first?" you may be asking. Not necessarily. The importance of doing so depends on your readers and the situation. Some readers are more patient than others when the main point is delayed, especially when the communication is brief. Furthermore, in some special situations, giving the bottom line first can undermine the success of your communication. You will find a discussion of them in the explanation of the fourth guideline of Chapter 5. As a general rule, however, you can increase your readers' satisfaction with your writing by giving the bottom line first.

GUIDELINE
4

PLAN YOUR VISUAL AIDS

At work your communications will often convey information not only in prose but also in visual aids. Tables, graphs, drawings, and similar devices are widely used with on-the-job writing because they can communicate some information and ideas much more succinctly and forcefully than words. Consequently, one very important decision you must make when planning is deciding what information you should present in prose and what in visual aids. Look for places where you can use visual aids to show how something looks (in drawings or photographs, for instance), to explain how something is organized (flow charts), to make detailed information readily accessible (tables), and to clarify the relationship among separate groups of data (graphs).

Chapter 13 provides detailed advice about where to use visual aids and how to construct them effectively. Chapter 14 explains how to create twelve types of visual aids that are very useful at work. However, don't wait until you read those chapters to begin incorporating visual aids into your communication plans.

USING MODEL COMMUNICATIONS AS PLANNING AIDS

This chapter's four guidelines provide general advice about how to plan the content and organization of communications to meet your readers' informational needs. All the guidelines suggest strategies that call on your skill and creativity in imagining your readers—their needs, preferences, and tasks while reading. At work, you also will have another valuable resource as you plan: model communications that show how other writers have attempted to meet their readers' needs in situations similar to your own. But using model communications also brings a danger: if you misuse the models, they can seriously impair your effectiveness as a writer.

The following paragraphs briefly describe two kinds of model communications—examples and superstructures—and then provide advice about how to use each effectively.

Example Communications

Example communications are simply samples of the very kind of message you are about to write. For instance, if you are going to write a feasibility report addressed to people inside your employer's organization, you might look for a feasibility report written by another employee so you can study it for ideas you can use. If the examples are well written, the ideas you gain from studying them can be quite valuable.

Superstructures

As explained in Chapter 1, a superstructure isn't an actual communication but a general pattern or framework for constructing a certain type of message— such as a proposal, progress report, or empirical research report. Like the blueprint for a home, these patterns tell writers how to build their communications: what topics to cover, what arrangement to give them, and what strategies to use to develop each topic and relate it to the others.

These superstructures have developed because they have proven successful with certain types of communication situations that occur repeatedly at work. Every day, in organizations all across the country, writers need to report on their research or tell people how to perform a procedure. Again and again, both writers and readers find that certain ways of framing those messages are effective. These successful patterns thus become widely known and used. To see some of the features of a superstructure, examine Figure 4–4, which describes a pattern writers often use when seeking approval to start a new project.

It's because superstructures work for *both* writers and readers that they are so valuable as planning aids. Consider, for instance, the ways the superstructure shown in Figure 4–4 will help your *readers* if you use it to propose a project. First, it will help them by prompting you to answer the questions they are most likely to ask in that situation—questions about what you want to do, why, and how. The superstructure will also help your readers by prompting you to *organize* your answers in a way they are likely to find useful. In fact, even the mere familiarity of this conventional pattern will help your readers. Because they know the pattern in advance, they will be able to use it to mentally organize the information you provide.

In addition to meeting your readers' informational needs, the superstructure can help you achieve your persuasive aims. That's because it suggests the persuasive points you should make in each part of your proposal; see the list in the right-hand column of Figure 4–4.

Because superstructures are such valuable aids in planning (as well as in drafting), Chapters 20 through 25 present detailed discussions of six that are widely used in the working world: four types of reports, proposals, and instructions.

Topic	Your Persuasive Point
Introduction	Briefly, here's the project I propose.
Problem	This proposed project addresses a problem that is important to you.
Objectives	This project has the following specific objectives.
Product or Outcome	Here's a detailed account of what I propose to produce and how it will achieve the project objectives.
Method	I will be able to carry out this project successfully because I have a sound plan for putting the product together: the necessary facilities, equipment, and other resources; a realistic schedule; appropriate qualifications; and an effective management plan for organizing and monitoring project activities and personnel.
Budget	The costs are reasonable.

FIGURE 4–4   General Superstructure for a Proposal

How to Use Example Communications and Superstructures Effectively

To use example communications and superstructures effectively, you must avoid the mistake of treating them as if they were cookbook recipes that require no thought or creativity on your part. Despite the overall similarities among many communication situations that arise at work, each situation is different, sometimes in small ways and sometimes in large ones. It makes sense for your communication to resemble an example or superstructure to the extent your audience, purpose, and circumstances are the same as those the model addressed. But to the extent your situation differs from the one for which the example or superstructure was intended, your communication also should differ.

Thus, to use an example communication or a superstructure effectively, you need to creatively adapt its strategies to your situation. Base your adaptation on a careful comparison of your purpose and audience with the purpose and audience of the model. If you aren't sure about the model's purpose or readers, study it carefully and make a guess. (The discussion of superstructures in Chapters 20 through 25 provides extensive information about the situations for which they are intended.)

CONCLUSION

This chapter has presented many suggestions for planning that will save you time and increase the effectiveness of your writing. How much time should you devote to following these suggestions when working on a particular communication? The answer depends upon many factors. If you are writing a brief communication on a familiar topic that is relatively easy to explain, you may need only a few moments to plan. For longer communications on complex and unfamiliar topics, you may need several hours—or days—along with the advice of other people. The forms you give your plans can also vary. Sometimes you can plan effectively by simply making some mental notes or jotting a few words on a sheet of paper. At other times, you may find it helpful to create an outline or other, more formal aid.

Note that when writing long communications, you plan in stages by beginning with top-level or overall plans, then developing progressively more detailed or lower-level plans as you prepare to draft the subparts of your message.

Once you begin drafting you are likely to discover ways to improve your plans. That's natural. It doesn't mean that you planned poorly or that your original planning was a waste of time. The more you think about a communication, the more good ideas you are likely to have about how to write it. Just as you would replace a confusing sentence with a clearer statement, replace good plans with the better ones that occur to you. Planning is an ongoing activity you should continue until you have completed your final draft.

SEVEN TECHNIQUES FOR GENERATING IDEAS ABOUT WHAT TO SAY

This special section describes seven ways of generating ideas about what to say in your communications.* They supplement the basic strategies for deciding what to say described in Guidelines 1 and 2: (1) anticipate your readers' questions, then answer them, and (2) figure out what else your reader needs to know, then include that information. In many situations those two strategies will be all you need in order to generate the complete contents for your communications. In other situations, however, you may feel the need to explore your topic more fully in order to ensure that you haven't overlooked something important. The seven techniques described in this section will help you do that:

Talk with Someone Else	Make an Idea Tree
Develop and Study a Table of Your Data	Draw a Flow Chart
Brainstorm	Make a Matrix
Write a Throw-Away Draft	

Note, by the way, that these techniques will assist you not only in identifying what you need to say in order to meet your readers' informational needs (the subject of this chapter) but also what persuasive points you could make in order to influence them to think or feel the way you want them to (the subject of Chapter 5).

On the other hand, keep in mind that all seven of these techniques are also capable of helping you identify items that are completely irrelevant to your purpose and readers. Therefore, it's best to think of them as techniques for generating lists of *possible* contents. To decide which items you should actually include in a communication, examine each one from the point of view of your readers.

Talk with Someone Else

Talking about your subject with someone else is one of the most productive—and enjoyable—ways to generate ideas about what to tell your readers. As you discuss your information and ideas with the other person, you not only get someone else's help in exploring your topic but also have an opportunity to figure out together how to make your message clear, complete, and compelling to your readers.

---

*Note to the Instructor. Several chapters in this book contain Special Topics. They are designed to give you added teaching flexibility; you can instruct students to skip them altogether, to read them with the chapters in which they appear, or to read them at some other appropriate time in your course.

A conversation about your communication can be especially valuable if you hold it with one of the readers you will be addressing. Asking your readers is certainly the simplest and most direct way of finding out what questions they want you to answer. You will find many readers eager to tell you exactly what they want to find in your communication.

Develop and Study a Table of Your Data

Often at work you will need to write communications about data, such as the results of a test you have run, costs you have calculated, or production figures you have gathered. In such cases, many people find it helpful to begin composing by making the tables they will include in their communication. Then they can begin to interpret the data arrayed before them, thinking about what the figures may mean and how they might be significant to their readers.

Brainstorm

When you brainstorm, you generate thoughts about your subject as rapidly as you can through the spontaneous association of ideas. You write down whatever thoughts occur to you.

The power of brainstorming arises from the way it unleashes your natural creativity. By freeing you from the confines imposed by outlines or other highly structured ways of organizing your ideas, brainstorming lets you follow your own fertile lines of thought.

Brainstorming is especially helpful when you begin with very little idea about what to tell your audience, or when you have thought a great deal over a long period of time about a subject and now want to recall all your thoughts. Brainstorming is helpful whether you are deciding what to say in a whole communication or in part of one. It also works well in group writing projects: all the group members can brainstorm aloud together. The ideas offered by one person often suggest additional ideas to other group members.

The usual procedure for brainstorming is as follows:

1. Review your purpose and understanding of your audience.
2. Ask yourself, "What do I know about my subject that might help me achieve my purpose?"
3. As the ideas come, write them down as quickly as you can, using single words or short phrases. As soon as you list one idea, move on to the next.
4. When your stream of ideas runs dry, read back through your list to see if your previous entries suggest any new ones.
5. When you no longer have any new ideas, gather related items in your list into groups to see if this activity prompts new thoughts.

The key to brainstorming is to write quickly and avoid evaluating or rejecting ideas that come to mind. Jot down everything. If you shift your task from generating ideas to evaluating them, you will disrupt the flow of free association that brainstorming relies on.

Figure 4–5 shows part of the list of ideas Nicole generated through brainstorming. She works for Meditech, a company that makes machines that keep surgery patients alive during organ transplants. Nicole is planning to recommend that Meditech use new procedures for ensuring that every one of its machines is in perfect working order when delivered to a hospital. After running dry of ideas while making the list in Figure 4–5, Nicole began to group items in the list, as suggested in step 5. Some of her results are shown in Figure 4–6.

Write a Throw-Away Draft

Writing a throw-away draft is very much like brainstorming. Here, too, you tap your natural creativity free from the confines of structured thought. You sit down and write out your ideas as they come to you. Only this time you write

Ideas for Quality-Control Recommendations

Present system is rushed, chaotic
Everyone is supposed to be responsible for quality
No one has specific responsibilities
Workers feel rushed, often ignore quality procedures
People's lives at stake—if our equipment fails during
    surgery
Nearly fatal failure last year in Tucson
Overall record of performance is excellent
Some problems with below-standard parts may not show up in
    present tests
Must ensure quality of component parts and of whole
    machine when assembled
Need to assign specific responsibilities for quality
New procedure will test critical components as they are
    delivered by suppliers
New procedure will require workers to use standard
    assembly procedures
Now people follow their own shortcuts and use personal
    assembly techniques
Don't realize the harm they can do

FIGURE 4–5   List of Ideas Produced by Brainstorming

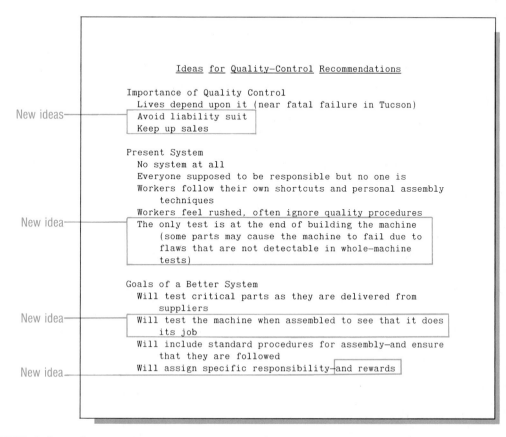

FIGURE 4–6   Grouping of Ideas Produced by Brainstorming
(Note the new ideas suggested to the writer by the grouping.)

prose rather than a list. As you write, you will often refine your sense of what to say. New ideas and new lines of thought will occur to you. You write them down without trying to make them fit into the material you have already written. The key is to keep your ideas flowing.

The resulting draft will be chaotic, but that is fine. You are trying to follow and develop your ideas as they occur to you, not produce a draft you can revise directly into a finished communication. When you finally do run out of ideas, you can read back throughout your material to select the ideas worth telling your readers. The rest you throw away.

Throw-away drafts can be especially helpful when you are trying to develop your main points. They are not as good at helping you identify minor points. You probably can use these drafts best when you are writing brief communications or the parts of a long communication where you have yet to decide how to interpret your data or state your recommendations. It would not be efficient for you to write a throw-away draft for an entire long communication.

Wind shear and microbursts have been blamed for several
recent crashes. (Find out which ones—Dallas??) People are
studying several technologies for detecting these
conditions. Then pilots can fly around them. There is also
much interest in using these or other technologies for
instruments to be used aboard planes. These would be much
more helpful to pilots. Four technologies are currently
being studied. The equipment would need approval of the
Federal Aviation Administration, which is eager for such
devices to be developed. Airlines would be eager to buy a
reliable system as soon as it becomes available. The key
to all devices is to allow pilots to detect in advance the
special conditions that give rise to wind shear and
microbursts, both very dangerous conditions that can
overwhelm a plane that is landing or taking off. A key
point is: we can make a lot of money if we can develop the
right instrument first. We need to pick the best one and
develop it. Criteria are: effectiveness, readiness for
commercial development, and likely competition.

FIGURE 4–7   Throw-Away Draft

Figure 4–7 is a throw-away draft written by Miguel, an employee of a company that makes precision instruments. Miguel has spent two weeks investigating technologies to be placed aboard airplanes for detecting microbursts and wind shear, two dangerous atmospheric conditions that have caused several crashes. Miguel wrote the throw-away draft when trying to decide what to say in the opening paragraph of his report.

The usual procedure for writing a throw-away draft is very much like that for brainstorming.

1. Review your purpose and understanding of your readers.
2. Ask yourself, "What do I know about my subject that will help me achieve my purpose?"
3. As new ideas come, write them down as sentences. Follow each line of thought until you come to the end of it, then immediately pick up the next line of thought that suggests itself.
4. Write without making corrections or refining your prose. If you think of a better way to say something, start the sentence anew.
5. Don't stop for gaps in your knowledge. If you find you need some information you don't have, note the place where you would use it, then keep on writing.

We have a substantial opportunity to develop and successfully market instruments that can be placed aboard airplanes to detect dangerous wind conditions called wind shear and microbursts. These conditions have been blamed for several recent air crashes, including one of a Lockheed L-1011 that killed 133 people. Because of the increasing awareness of the danger of these wind conditions, the Federal Aviation Administration is supporting considerable research into a variety of technologies for detecting them. Most of this research concerns systems placed at airports, but both the FAA and the airlines agree that systems that could be installed on airplanes would be even more helpful.

In this report, I will review the four major technologies now being studied. About each, I will answer four questions:

How does it work?

Will it result in a product that will perform well enough to attract substantial sales?

How long until the technology will be refined enough to make an effective instrument?

If we develop such an instrument, what will our competition be like?

FIGURE 4–8   Introduction Written from Ideas Generated in the Throw-Away Draft in Figure 4–7

Miguel's throw-away paragraph is, naturally, a jumble since he was working out his ideas freely. Figure 4–8 shows the first paragraph Miguel subsequently wrote. Note that it further develops some ideas from the throw-away draft and omits others. It is a fresh start—but one built on the thinking Miguel did while writing his throw-away draft.

Make an Idea Tree

An idea tree is a sketch of the various topics and subtopics you might discuss in your communication. It helps you explore your subject by logically identifying its parts. The following diagram shows what an idea tree might look like:

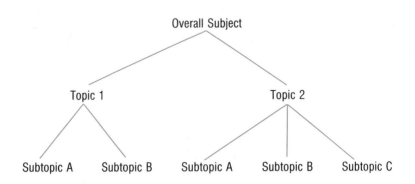

To create an idea tree, use the following procedure:

1. Review your purpose and understanding of your readers.
2. At the top of the sheet, write the overall subject of your communication. (Some people prefer to draw their trees from side to side, beginning at the left-hand margin.)
3. Ask yourself, "What are the major topics I might take up in my communication?" List those topics across the page, below the subject. Don't worry about their order.
4. Draw a line (or branch) from each topic to the main subject.
5. For each topic, identify the subtopics you might take up. Join them to their topics with lines.
6. Keep subdividing until you run out of ideas. It's fine if some of your branches have more levels than others do. Take each branch as far as you can, whether to the topic level, subtopic level, or some lower level.

Figure 4–9 shows an idea tree written by Carol, who works for a marketing manager for an engineering company that drills water wells. Carol leads a team that is helping a small city look for places it can drill new wells for its municipal water supply. Notice how easily Carol could add new topics and new levels of detail to the tree as she thinks of additional things to say.

Draw a Flow Chart

Flow charts can help you decide what you might say about a succession of events, such as when you are describing a process, explaining how to perform a series of steps, or proposing a sequence of actions. To generate these ideas, you can draw a flow chart by elaborating on the following basic form:

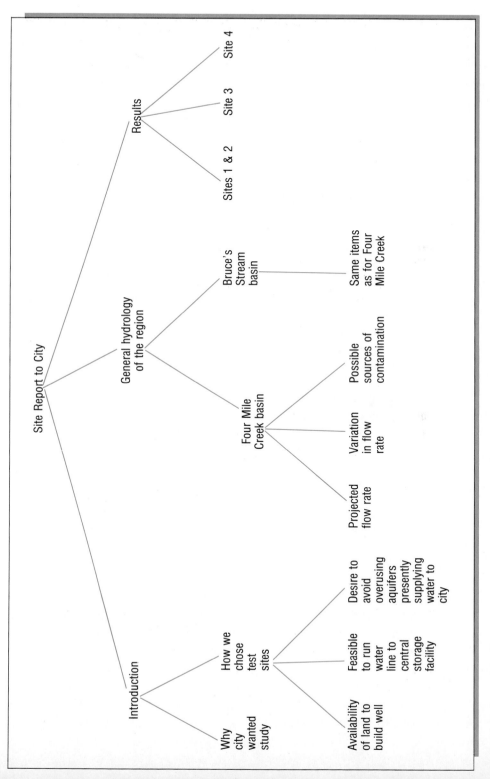

FIGURE 4–9 Idea Tree

Here are some suggestions for using flow charts to generate ideas about your subject:

1. Review your purpose and understanding of your readers.
2. List (perhaps mentally) the steps or activities in the process or procedure.
3. Draw the flow chart, leaving lots of blank space around each box, especially above and below.
4. Brainstorm about the things you might want to say about each activity or step. Write your ideas above or below the boxes.
5. When you no longer have additional ideas about a step, review what you have written about the others. Often a piece of information you might provide about one step or activity will suggest a parallel piece for others.

Figure 4–10 is a flow chart Nicole used in her recommendations to Medi-tech for improving their quality-control procedures during the manufacture and delivery of medical machinery.

Make a Matrix

A matrix is a special kind of table writers can use when comparing two or more things in a parallel fashion. Figure 4–11 shows the form of a matrix.

Here are some steps in using a matrix:

1. Review your purpose and understanding of your readers.
2. List (perhaps mentally) the items you will discuss and the general topics or issues you want to cover when discussing each item.
3. Draw a matrix, listing the items down the left-hand side and the topics across the top. Make as much room as possible in the matrix to write down your ideas.
4. Brainstorm to determine what you could say in each of the boxes of the matrix. Write your ideas in the appropriate box.
5. When you no longer have additional ideas for a box, review what you have written in the others. Often a comment you make in one box will suggest a parallel piece of information for the others.

Figure 4–12 shows a matrix Miguel used to determine the points he would make about each of the technologies he discussed in his report on airplane instruments.

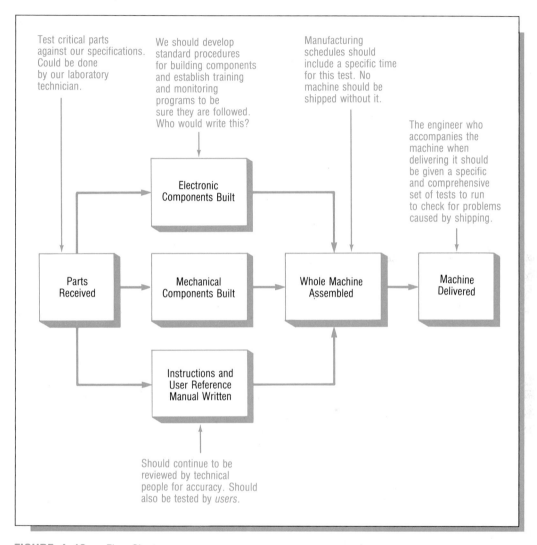

FIGURE 4–10   Flow Chart

Some Final Thoughts

In addition to the seven techniques you have just read about, there are many other methods for generating ideas about what to tell your audience. Earlier in this chapter, you read about one other very useful one: looking at model communications. Another technique, not yet described, is outlining. Outlining is discussed at the end of the next chapter, where it is treated both as a technique for generating ideas and as a tool for organizing your communication.

Whatever techniques you use to generate ideas, remember to take your readers' viewpoint when you begin to select from the list of possible contents

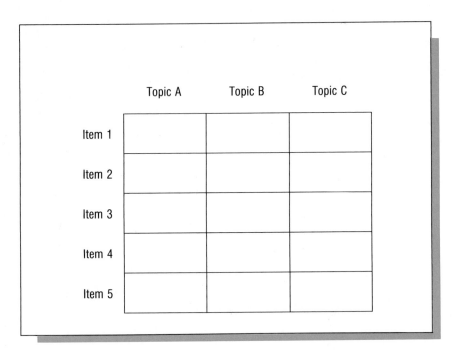

FIGURE 4–11 Basic Form of a Matrix

you have generated. Select the facts your readers will want or need to know, and choose the arguments they will find persuasive. While your efforts to generate ideas can be *writer-centered* (as when brainstorming or writing a throw-away draft) or *subject-centered* (as when studying a table of your data or drawing a flow chart), your selection of the final contents for your communication always should be *reader-centered*.

EXERCISES

1. Imagine you are now employed full-time. You have decided you would like to take a course at a local college. Begin this exercise by naming a course you might really like to take. Next, by following Guidelines 1 and 2, list the information you would include in a memo that asks your employer to pay your tuition and to permit you to leave work early two days a week to attend the class.

2. Imagine you have been hired by your major department to write a brochure that will present the department in a favorable light both to entering freshmen and to continuing students thinking of transferring out of other majors. Using two of the seven techniques described in the special topic sec-

TOPIC SYSTEM	How It Works	Technical Limitations	Readiness for Commercial Development	Existing or Anticipated Competition	Recommendation
Air Speed Sensor Systems	Sensors measure air speed at various places on the plane (nose, wing, tail, etc.). Large differences indicate wind shear.	Works only when plane is already in wind shear.	Now being used on some planes.	Sperry Delco Electronics.	Only if we can improve on currently marketed designs.
Infrared Detectors	Often wind shear raises temperature slightly. Infrared devices could detect rising temperatures.	Sometimes temperature may not rise.	Tests by Federal Aviation Administration have been suspended.	None expected.	Not worth pursuing.
Doppler Radar	New radar techniques can detect rapidly rotating masses of air, like those that accompany wind shear.	Technology still being researched.	Perhaps two more years of technical development needed.	None known.	Very promising. Should start preliminary design immediately.
Laser Sensors	Sudden wind shifts affect reflectivity of air. Lasers can detect that.	Limited to a 20-second warning for jet traveling at typical speed.	Ready.	Perhaps General Electric.	Reasonable to develop if we can do so quickly.

FIGURE 4–12 Matrix Used to Generate Ideas

tion of this chapter, generate ideas about what you might say in that brochure. How would you group and order your material?

3. Imagine you are head of the client-relations department of a company that prepares scientific studies for other companies and for government agencies. The U.S. Environmental Protection Agency (EPA) has refused to accept one of your company's reports because it is so poorly written. Because the company feels that the engineers who originally wrote the report can't do any better than they did the first time, you have been assigned to revise it.

Before you begin your revision, you might find it helpful to know some background information about the study on which it reports. Many of the plastic parts used in new cars give off a gas, called vinyl chloride monomer (VCM), which is hazardous to human health in large doses. The EPA commissioned your company to determine whether the level of VCM in new cars was high enough to constitute a health hazard. The passage you will work on describes an experiment in which the new cars were found to be safe. For the purpose of this exercise, suppose the readers of this passage are researchers in other laboratories who desire to repeat this experiment using other cars.

From the following list of tasks, perform those assigned by your instructor:

a. Write a topic outline for the passage.

b. Write a brief analysis of the passage's weaknesses your outline reveals. (*Note:* Your outline will have to include each of the various topics of the longer paragraphs to uncover all the major weaknesses.)

c. Compare your outline with the outline prepared by one or more of your classmates.

d. Following the advice given in the discussion of Guideline 3 (''Organize to support your readers' tasks''), write an outline that shows how the material in the passage could be reorganized to better serve the needs of a reader who wants to reproduce the experiment.

e. Compare your outline from *d* with the outline prepared by one or more of your classmates.

To collect vinyl chloride monomer (VCM) from the interior of each test automobile, we used glass collection-tubes purchased from SKC (Pittsburgh, PA). Each tube was filled with 100 milligrams of charcoal sorbent, through which we drew 8 liters of air at 50 milliliters per minute. The VCM present in the air was captured in the charcoal sorbent. We then capped the tubes, stored them in a freezer at minus 20 degrees centigrade, and shipped them to the laboratory, which performed a VCM analysis using an extraction/gas chromatographic technique.

For this procedure, we drew the air through the collection tubes using two Telmatic Air Samplers (Bendix Models 150 and C115), which we modified with remote controls so that we could turn them on and off from outside the test cars. We also placed the batteries used to power the pumps outside the cars. Each sampling pump was calibrated and adjusted for a constant flow rate of 50 milliliters per minute. We placed the collection tubes on the front seat next to the sampling pump. The collection tubes were attached to the air sampler by short connectors to minimize the exposure of the 8 liters of air to the connector materials before the air reached the carbon sorbent.

Each air sampler was equipped with a thermistor to measure the air temperature inside the test vehicle. An additional thermistor was used to measure the air temperature outside the car.

We placed the air sampler and thermistor on the front seat, either on the driver's side or on the passenger's side, wherever it seemed the sun would heat the seat the most. We reasoned that sampling the hottest seat during the summertime would establish the worst-case condition for concentrations of the pollutant, VCM. Then we inserted the collection tubes on the connector to the air sampler and ran the control wires through the car window, which we quickly rolled up to the top position. The seal at the top of the window was such that the window could be tightly closed and still allow the wires to pass through.

After placing the sampling package in the cars, we waited at least 30 minutes before beginning to sample so equilibrium could be reestablished inside the car. Because we were aware that disturbing the equilibrium would give us unreliable results, we had opened the car door carefully, slid the sampler in quickly, and closed the door with a minimum of air interchange. Furthermore, we always used the downwind door to minimize the circulation of outside air into the car.

We activated the pumps until approximately 8 liters of air had been sampled. Then we turned the pumps off and prepared collection tubes for storage as previously described.

CASE **FILLING THE COMPUTERIZED CLASSROOM**

The message took you completely by surprise. Dr. Baldwin shouted it over her shoulder as she rushed out of the room at the end of her class this morning: "The chairperson of the English department wants to see you. This afternoon if possible."

Before you could wedge through the crowd at the door to ask why, Professor Baldwin was down the hall and out of sight.

Now you stand before the English chairperson's office, still puzzled. As you prepare to rap on the door, you ask yourself for the hundredth time, "Have

I done something wrong? Something right? Why does the chairperson of the English department want to see me?''

"Come on in," he says, immediately guessing who you are. "I'm glad you could come. Professor Baldwin suggested you might be able to help me with a sort of marketing project."

What a relief. It seems that you aren't going to be thrown out of school after all.

"Have you heard about our new computerized classroom?" he asks. You tell him you know that a large room in the English building is being refurbished as a computerized classroom.

"That's right," the chairperson says. "In a month we will install a new computer system there. Twenty-five workstations."

"Twenty-five?" you marvel. "Why so many?"

"We're going to teach a few sections of First-Year College English in the room, and we want to have one workstation for each student in a class."

"I bet that will cost a lot."

"Yes, the project is being paid for by the Dean of the College of Arts and Science. Every year she is given more than $100,000 by the Alumni Office from what is called the Fund for Academic Excellence. The purpose of the fund is to provide our dean and the other deans with money they can use—in any way they see fit—to improve the academic program here. This year our dean has decided to invest much of her money in this computerized classroom."

"What made her decide to do that?" you ask.

"We sent her a proposal. In it we argued two things: first, that in a few years every incoming student will be familiar with computers, and second, that we may be able to teach First-Year English by using computers. As you might imagine, we had more difficulty making the second argument than the first. There is considerable debate among educators about whether computer technology can be used effectively to teach any writing skills except the most basic ones, like grammar and punctuation."

"Who's right?—in the debate, I mean."

"Several faculty in the department who have studied the matter think computers will help. They even cite reports of experiences at some other schools that have tried to teach First-Year English in this way. However, no one has tried to use computers in exactly the kind of course we teach or with exactly the sorts of students we have. One of the aims of the classes held in the computerized classroom in the fall will be to test out various ideas our faculty have about how to incorporate computers into First-Year English here at our school."

"That's interesting," you observe. "But what does it have to do with me?"

"What we'd like you to do, if you are willing, is to write a pamphlet we can send this summer to incoming students to entice some of them to sign up for the special sections of First-Year English that will use this new classroom."

"That should be easy," you say. "I bet we could find loads of incoming students who have already used computers for word processing and would welcome the chance to write on computers in college."

"Well," says the chairperson, "the faculty who will be teaching the course have something else in mind. This is an experiment of sorts. Except for using the computers as word processors, the computerized classes will be taught exactly like the other classes. Many tests will be run to determine the effects of using the computers. Consequently, the faculty members want the students in the computerized classes to represent a cross section of students in First-Year English. Therefore, they have asked the Admissions Office to generate a list of 250 randomly selected incoming students to whom the pamphlet will be sent. The aim of the pamphlet is to persuade at least half of the recipients to enroll in these special sections."

"In that case, writing the pamphlet will be a challenge," you agree.

"Will you accept the challenge?" he asks.

"Sure. Why not?"

Your Assignment

Fill out a Worksheet for Defining Objectives (Figure 3–5) for your pamphlet. Then, by following the guidelines in this chapter, list the things you would say in the letter to the incoming students. Identify the things you would need to investigate further before writing.

CHAPTER 5

PLANNING YOUR PERSUASIVE STRATEGIES

How Persuasion Works

GUIDELINE 1: Emphasize Benefits for Your Readers

Stress Organizational Objectives

Stress Growth Needs

GUIDELINE 2: Address Your Readers' Concerns and Counterarguments

GUIDELINE 3: Show That You Are Reasoning Soundly

How Reasoning Works

Present Sufficient and Reliable Evidence

Explicitly Justify Your Line of Reasoning Where Necessary

GUIDELINE 4: Organize to Create a Favorable Response

Choose Carefully Between Direct and Indirect Organizational Patterns

Create a Tight Fit among the Parts of Your Communication

GUIDELINE 5: Create an Effective Role for Yourself

Choose an Appropriate Voice

Present Yourself As a Credible Person

Present Yourself As Nonthreatening

DEFINING
OBJECTIVES
PLANNING
DRAFTING
EVALUATING
REVISING

I n Chapter 4, you learned how to plan communications that meet your reader's informational needs and thereby achieve the communication's enabling purpose. This chapter shifts your focus to the plans you make for achieving the persuasive purpose.

As explained in Chapter 3, persuasion is a crucial element in almost all on-the-job writing. In some communications, such as requests to management and proposals to clients, the persuasive aim is especially prominent. In other communications, such as instructions and technical reports, it is less obvious but still present: at the very least, writers want to create a favorable impression of themselves, their departments, and their companies.

Whether in the foreground or the background, your persuasive aims are almost always important enough for you to develop specific plans for achieving them. This chapter begins with a practical discussion of how persuasion works and then presents five guidelines for influencing your readers' thoughts, feelings, and actions.

HOW PERSUASION WORKS

According to researchers, when you want to influence the way people think, feel, or act, you should concentrate on shaping their *attitudes*.[1] At work, you will be concerned with your readers' attitudes about a wide variety of subjects, such as products, policies, actions, and other people. Sometimes you will use your persuasive powers to *reverse* an attitude you want your readers to abandon, sometimes to *reinforce* an attitude you want them to hold even more firmly, and sometimes to *shape* your readers' attitude on a subject about which they currently have no opinion.[2]

What determines a person's attitude about something? Researchers have found it isn't a single thought or argument, but the *sum* of the various thoughts the person associates with it.[3] This is a critical point to remember as you plan your persuasive strategies.

Consider, for example, Edward's attitude toward a photocopy machine he is thinking of purchasing for his department. As he thinks about the machine, his associated thoughts concern the amount of money in his department's budget, the machine's special features (such as automatic sorting), its appearance and repair record, the experience he has had with another product made by the same company, and his impression of the marketing representative who sells the machine. Each of these thoughts will make Edward's attitude more positive or more negative. He may think his budget has enough money (positive) or too little (negative); he may like the salesperson (positive) or detest the person (negative), and so forth.

Some of these thoughts will probably influence Edward's overall attitude more than others. For instance, Edward may like the machine's special features and color, but those two factors may weigh less in his mind than the machine's repair record, which he may have heard is very poor. If the sum of Edward's associations with the machine is positive, he will have a positive attitude toward it and may buy it. If the sum is negative, he will have an unfavorable attitude and probably will not buy it.

The following five guidelines suggest some strategies you can use to foster favorable attitudes toward actions you advocate and positions you support.

| GUIDELINE 1 | **EMPHASIZE BENEFITS FOR YOUR READERS** |

The most obvious way to prompt your readers to experience favorable thoughts is to tell how they will benefit from taking the position, performing the action, or purchasing the product you advocate. You are undoubtedly familiar with the way companies use this strategy in their advertisements. They tell about the whitener in their toothpaste, which will brighten the purchaser's teeth. They describe the computer-controlled braking system in their cars, which will protect the driver from skidding. They praise the nutritional content of their breakfast cereal, which will make the consumer healthier.

These ads stress *personal* benefits readers will enjoy in their private lives. Such a strategy works fine in some on-the-job situations, as when you are advertising a consumer product. However, in many situations you are much more likely to succeed if you focus instead on benefits related to your readers' *professional* responsibilities. You can do that in two ways: stress organizational objectives and stress your readers' growth needs.

Stress Organizational Objectives

At work you probably will write most often to co-workers in your company and to employees in other organizations. These people have been hired to advance their employer's interests. Consequently, you can often persuade them to look favorably at an action or decision if you can convince them it will help them pursue their employer's objectives.

At a general level, most business organizations share the same objectives: to operate efficiently, increase income, reduce costs, keep employee morale high, and so on. However, individual companies have distinctive variations on these general goals as well as their own unique goals. A company might seek to control 50 percent of the market for its type of product, to have the best safety record in its industry, or to expand into ten states in the next five years. In addition, different departments within a company have more specific objectives. Thus, the research department aims to pursue the company's overall objectives by developing new and improved products, the marketing department aims to identify new markets, and the accounting department aims to manage financial resources prudently.

When planning a communication, think about the organizational objectives—both general and specific—your readers feel responsible for pursuing. Then select the objectives you can tell your readers that your proposed action or decision will help them achieve.

Figure 5–1 shows a marketing brochure that uses organizational objectives to persuade its intended readers (purchasing managers for supermarkets and

FIGURE 5–1 Brochure that Stresses Organizational Objectives

supermarket chains) to carry a certain product. Notice the boldfaced statements that assert how the product will build sales volume, stimulate impulse purchases, and deliver profits.

As you think about which organizational goals you might stress, keep in mind that most companies and their employees have many goals, some of which involve social, ethical, and aesthetic concerns not directly related to profit and productivity. Some companies spell out those values in corporate credos; Figure 5–2 is an example. Even in companies that do not have an official credo, such broad human and social values can provide an effective foundation for persuasion, especially when you are advocating a course of

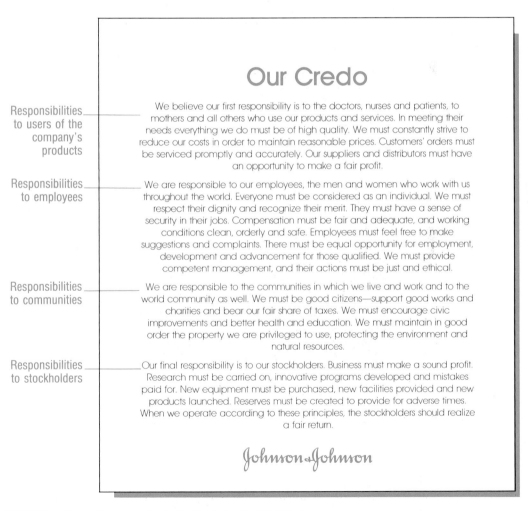

Our Credo

Responsibilities to users of the company's products

We believe our first responsibility is to the doctors, nurses and patients, to mothers and all others who use our products and services. In meeting their needs everything we do must be of high quality. We must constantly strive to reduce our costs in order to maintain reasonable prices. Customers' orders must be serviced promptly and accurately. Our suppliers and distributors must have an opportunity to make a fair profit.

Responsibilities to employees

We are responsible to our employees, the men and women who work with us throughout the world. Everyone must be considered as an individual. We must respect their dignity and recognize their merit. They must have a sense of security in their jobs. Compensation must be fair and adequate, and working conditions clean, orderly and safe. Employees must feel free to make suggestions and complaints. There must be equal opportunity for employment, development and advancement for those qualified. We must provide competent management, and their actions must be just and ethical.

Responsibilities to communities

We are responsible to the communities in which we live and work and to the world community as well. We must be good citizens—support good works and charities and bear our fair share of taxes. We must encourage civic improvements and better health and education. We must maintain in good order the property we are privileged to use, protecting the environment and natural resources.

Responsibilities to stockholders

Our final responsibility is to our stockholders. Business must make a sound profit. Research must be carried on, innovative programs developed and mistakes paid for. New equipment must be purchased, new facilities provided and new products launched. Reserves must be created to provide for adverse times. When we operate according to these principles, the stockholders should realize a fair return.

Johnson & Johnson

FIGURE 5–2   Corporate Credo that Includes Varied Objectives
Courtesy of Johnson & Johnson Company.

action that seems contrary to the organization's narrow business interests. For example, if you point to the good that will be done for a nearby community, you might be able to persuade your employer to spend money on additional water pollution controls even though the law does not require them.

Finally, remember that even within the same organization or department, a benefit that appeals to one individual may be completely unattractive to another. And a benefit that appeals strongly to you may be unattractive to your readers. Recognizing the importance of pitching their persuasive messages to the particular people they hope to influence, advertising agencies conduct extensive studies to determine what appeals to the target markets for the products they are trying to sell. You should imitate them when planning your persuasive strategies. Learn as much as possible about the organizational benefits that will be most appealing to your target readers.

Stress Growth Needs

Two informative and influential studies of employee motivation conducted by Abraham Maslow and Frederick Herzberg suggest another type of benefit you can use to appeal to people you will address on the job.[4] Both studies found that, as everyone knew, employees are motivated by such considerations as pay and safe working conditions. However, they also found that once most people feel they have an adequate income and safe working conditions, they cease to be so easily motivated by these factors. Consequently, factors like these are called *deficiency needs:* they motivate principally when they are absent.

Once their deficiency needs are met—and even before—most people are motivated primarily by another set of factors called *growth needs*. These include the desire for recognition, good relationships at work, a sense of achievement, personal development, and the enjoyment of work itself.

Many other studies have confirmed Maslow and Herzberg's findings. For instance, one study showed that for supervisors, recognition, achievement, and personal relationships are much more potent motivators than money.[5] Another found that computer professionals are not motivated by salary or status symbols nearly as much as by the opportunity to gain new skills and knowledge in their field.[6]

These findings mean that a powerful persuasive strategy is to show how the decisions or actions you advocate will help your readers satisfy their growth needs. When you request information or cooperation from a co-worker, point out how much you value his or her assistance. When you evaluate a subordinate's performance, be sure to recognize that person's accomplishments even as you also point out shortcomings. When you ask someone to undertake additional duties, emphasize the challenge and opportunities for achievement that lie ahead for the person.

Figure 5–3 shows a recruiting brochure by a company (Procter & Gamble) trying to attract potential employees. Notice that under ''Key Considerations,'' the company devotes much more space to growth needs (''Early and Meaning-

Procter & Gamble

Procter & Gamble is known primarily for its preeminent position in the consumer packaged goods field. Familiar brands such as Tide laundry detergent, Crest toothpaste, Pampers disposable diapers, Duncan Hines cake mixes, Ivory soap, Head & Shoulders shampoo, Folgers coffee, Crisco shortening, Cascade dishwashing detergent and other market leaders including Oil of Olay, Pepto-Bismol and Vicks cough and cold medicines. More recently, P&G's consumer product base has been expanded to include soft drinks (Crush and Hires), orange juice (Citrus Hill) and other citrus products and pharmaceuticals. In addition; institutional, industrial and cellulose pulp products have contributed importantly to the business for many years.

Sales has consistently placed us among the top 25 U.S. industrial corporations. More than 75% of this business is in products developed and introduced within your lifetime.

KEY CONSIDERATIONS

Growth and Diversification

The continuing development of our traditional cleaning, food, paper, and health and beauty care business and our entry into soft drinks, citrus products and pharmaceuticals, plus the recent acquisition of Richardson-Vicks ensure new growth opportunities.

Promotion From Within

P&G recruits for entry level positions and promotes from within. This means that new managers start their careers in entry level positions and earn promotions based on a proven track record within the P&G organization.

Early and Meaningful Responsibility

In order to "learn by doing," managers are given significant responsibilities very early. This provides the opportunity for an immediate impact on the business and frequently leads to key positions being held by very young people.

Training and Development

Training and development focus on building both managerial and technical skills. There are no set timetables for advancement. The responsibility for training and development lies with operating managers supported by staff groups. Since managers in our business are aware that the ability to develop others is important to their own prospects for promotion, this responsibility is taken seriously.

A Challenging Work Environment

We operate in a fast-paced, competitive market. We take an analytical approach to the business that requires thoroughness and self-discipline in an environment where facts and logic carry authority. We have a healthy dissatisfaction with the status quo. We seek deliberate change for the better in every area of the business. We're always willing to test new ideas and to question current practices. Above all, there is the challenge and stimulation of working with other top caliber people.

Top Salaries and Benefits

Our pay and benefits make us a leader among those companies with which we compete for people. Tower, Perrin, Forster & Crosby, in a new study showing how P&G benefits stack up against 20 other U.S. companies that are leaders in their field (10 business competitors for top people), ranked P&G first in the total dollar value of employee benefits. In fact, the value of our benefits is shown to be nearly one-third greater than the average for the other 20 companies.

FIGURE 5–3   First Page from a Recruiting Brochure that Stresses Growth Needs

ful Responsibility,'' ''Training and Development,'' and ''A Challenging Work Environment'') than to deficiency needs (''Top Salaries and Benefits''). Furthermore, to emphasize growth needs, the brochure discusses them *before* deficiency needs.

GUIDELINE
2

ADDRESS YOUR READERS' CONCERNS AND COUNTERARGUMENTS

When people read, they focus not only on the writer's statements but also generate thoughts on their own. For example, while reading a brochure about a copy machine he might purchase, Edward may respond to a description of some special feature by thinking, ''Hey, that sounds useful'' (positive) or ''I don't see what good that will do me'' (negative). Some of a reader's thoughts even arise independently of any specific statement the writer makes. For instance, while reading the brochure, Edward may remember a positive or negative comment about the copy machine made by another manager whose department already owns one.

Whether a reader's self-generated thoughts are favorable or unfavorable, research shows that they can have a greater influence on the reader's overall balance of positive and negative associations with the writer's subject than any point made in the communication.[7] Consequently, when persuading you should not only write positively about your subject, but also carefully avoid prompting negative thoughts. And you should also try to counteract any negative thoughts that seem likely to arise in your readers' minds despite your best efforts.

One way to avoid arousing negative thoughts is to answer all the important questions your readers are likely to ask. At work, people will ask many questions while reading your communications. Some questions will simply reflect their efforts to understand your position or proposal thoroughly. What are the costs of doing as you suggest? What do other people who have looked into this matter think? Do we have employees with the education, experience, or talent to do what you recommend? If you answer these questions satisfactorily, they create no problem whatever. Such questions can generate negative thoughts, however, if you ignore them. Readers may think you have overlooked some important consideration or have decided to ignore evidence that counters your position. The remedy, quite simply, is to anticipate what these questions will be and answer them satisfactorily.

In addition to asking questions, readers may generate arguments against your position. For example, if you say there is a serious risk of injury to workers in a building used by your company, your readers may think to themselves that the building has been used for three years without an accident. If you say a new procedure will increase productivity by at least 10 percent, your readers may think to themselves that some of the data on which you based your estimate are inaccurate. Readers are especially likely to generate counterarguments when you attempt to *reverse* their attitudes. Research shows that people resist efforts to persuade them that their present attitudes are incorrect.[8]

To deal effectively with counterarguments, you must offer some reason for placing greater reliance on your position than on the opposing position. For example, imagine you are proposing the purchase of a certain piece of equipment, and that you predict your readers will complain it is more expensive than a competitor's product your readers believe to be its equal. You might explain that despite its lower purchase price the competitor's product really costs more because it is more expensive to operate and maintain. Or you might argue that the competitor's product really isn't equal to the product you advocate because the latter possesses some capabilities that make it well worth the additional cost.

To identify the questions and counterarguments you must address, follow Chapter 3's advice for identifying readers' questions, but focus specifically on questions a *skeptical* reader would pose. Another way of anticipating readers' counterarguments is to learn their reasons for holding their present attitudes. Why do they do things the way they do? What arguments do they feel most strongly support their present position? It's important for you to understand these reasons because you will be able to change your readers' minds only if you show that their reasons for supporting their present position aren't as strong as the reasons you advance in support of your position. In many situations, you won't be able to do that unless you show not only the strengths of your reasons but also the weaknesses of your readers'. Showing those weaknesses requires considerable diplomacy because you must uncover the flaws in your readers' reasoning without arousing the defensiveness of the persons whose mistakes you are disclosing.

In your efforts to address your audience's concerns and counterarguments, be careful not to mention objections your audience is unlikely to raise spontaneously. Otherwise you run the risk that your audience will find an objection you raise more persuasive than the rebuttal you offer.

Figure 5–4 is a letter in which the marketing director of a radio station attempts to persuade an advertising agency to switch a portion of its ads to her station. In the letter, the writer anticipates and addresses some possible counterarguments.

GUIDELINE 3	SHOW THAT YOU ARE REASONING SOUNDLY

A third strategy for writing persuasively is to show that you are reasoning soundly. In most business situations, one of the most favorable thoughts your readers could possibly have is, ''Yeah, that makes sense''—and one of the most unfavorable is, ''Hey, there's a flaw in the reasoning here.''

Sound reasoning is especially important when you attempt to influence your readers' decisions and actions. In such situations, it's not enough merely to identify the benefits to your readers of doing what you recommend. You also must persuade them that the decision or action you advocate will actually bring about the benefit you name—that the new equipment you suggest really will reduce costs enough to pay for itself in just eighteen months or that the modification you recommend in a product really will boost sales 10 percent in the first year.

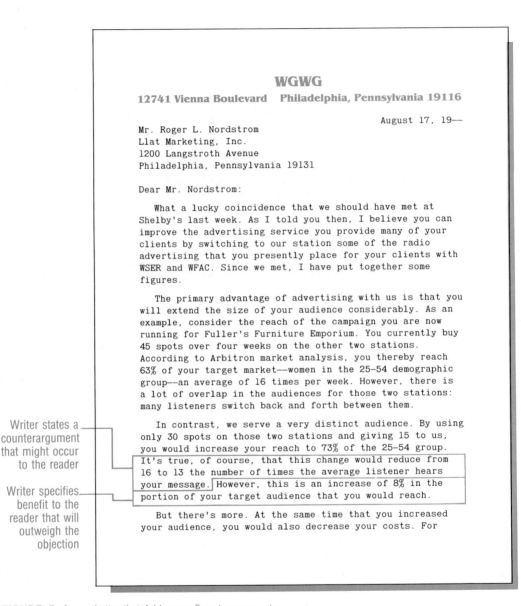

WGWG

12741 Vienna Boulevard   Philadelphia, Pennsylvania 19116

August 17, 19--

Mr. Roger L. Nordstrom
Llat Marketing, Inc.
1200 Langstroth Avenue
Philadelphia, Pennsylvania 19131

Dear Mr. Nordstrom:

What a lucky coincidence that we should have met at Shelby's last week. As I told you then, I believe you can improve the advertising service you provide many of your clients by switching to our station some of the radio advertising that you presently place for your clients with WSER and WFAC. Since we met, I have put together some figures.

The primary advantage of advertising with us is that you will extend the size of your audience considerably. As an example, consider the reach of the campaign you are now running for Fuller's Furniture Emporium. You currently buy 45 spots over four weeks on the other two stations. According to Arbitron market analysis, you thereby reach 63% of your target market—women in the 25–54 demographic group—an average of 16 times per week. However, there is a lot of overlap in the audiences for those two stations: many listeners switch back and forth between them.

In contrast, we serve a very distinct audience. By using only 30 spots on those two stations and giving 15 to us, you would increase your reach to 73% of the 25–54 group. It's true, of course, that this change would reduce from 16 to 13 the number of times the average listener hears your message. However, this is an increase of 8% in the portion of your target audience that you would reach.

But there's more. At the same time that you increased your audience, you would also decrease your costs. For

Writer states a counterargument that might occur to the reader

Writer specifies benefit to the reader that will outweigh the objection

FIGURE 5–4   Letter that Addresses Counterarguments

Sound reasoning is also essential in many other situations. Time and again you will be asked to draw conclusions from a group of facts. Maybe the facts will be the results of a laboratory experiment you ran, a consumer survey you conducted, or a comparison you made of alternative manufacturing methods. When writing in such circumstances, you must persuade your readers that the conclusions you draw are justified in light of your evidence.

Roger L. Nordstrom   – 2 –   August 17, 19—

example, if you switched the Fuller's Furniture Emporium
spots in the way described, your costs would drop from
$6,079 to $4,850. Our audience is smaller than the
audiences of those stations, but our rates are only about
half of theirs.

As you consider these figures and others that might be
given to you by other radio stations, be careful to
evaluate them carefully. As we are all aware, anyone can
manipulate statistics to their advantage. We recently
discovered that one of our competitors was circulating
data regarding the portion of the audiences of
Philadelphia stations that are "unemployed." In that
analysis, our station had a significantly higher portion
of "unemployed" listeners than did either WSER or WFAC.

However, after some investigation, we discovered that in
those statistics, "unemployed" meant "unemployed outside
the home." That means that people "unemployed" are not
necessarily breadwinners who are unable to provide an
income for their families. They could be housewives with
considerable power over family decisions about spending.
In fact, surveys indicate that housewives are often the
primary decision-makers about such things as household
furnishings—a point that is quite important with respect
to your account with Fuller's Furniture Emporium.

I suggest that we get together soon so that we can
sketch out the details of some radio advertising schedules
that would include WGWG. I'll call you next week.

Cordially,

Ruth Anne Peterson
Marketing Manager

Writer states a counterargument the reader is likely to hear from competitors

Writer exposes faulty reasoning underlying the counterargument

FIGURE 5–4   *(continued)*

Notice that in each of the situations just mentioned (as well as in any other you might encounter on the job), it's not enough merely to use sound reasoning. You also must *convince your readers* that your reasoning is sound. The ability to do so is one of the most valuable writing skills you can develop. The following discussion, based largely on the work of Stephen Toulmin, will help you master this skill.[9]

How Reasoning Works

To feel confident about your reasoning, your readers generally must feel that they understand all three of the following elements:

- **Your claim.** The position you want your reader to agree with.
- **Your evidence.** The facts, observations, or other information you offer in support of your claim.
- **Your line of reasoning.** The connection linking your claim and your evidence; the reasons your readers should agree that your evidence proves your claim.

The following diagram illustrates the relationship among a claim, the evidence supporting it, and the line of reasoning that links the two.

Evidence ⟶ Claim

The following facts, observations, and other evidence support the claim.

Such and such should be done.

Line of Reasoning

The evidence should be accepted as adequate support for the claim for the following reasons.

Imagine, for instance, that you work for a company that manufactures cloth. You have found out that one of your employer's competitors recently increased its profits by installing computers to run some of its textile mills. If you were to recommend that your employer buy similar computers, your basic argument could be diagrammed as follows.

Evidence ⟶ Claim

Our competitor has increased profits by using computers.

By using computers we will increase our profits.

Line of Reasoning

Experience has shown that actions that increase the profits of one company in an industry will usually increase the profits of other companies in the same industry.

To accept your position, readers must be willing to place their faith in *both* your evidence and your line of reasoning. The next two sections provide advice about how to persuade them to do so.

Present Sufficient and Reliable Evidence

First you must convince your readers your evidence is both sufficient and reliable.

To provide *sufficient* evidence, you need to provide the details you expect your readers to request. For instance, in the example of the textile mill, you could predict your readers would feel your evidence was too skimpy if you produced only a vague report that the other company had somehow used computers and saved some money. Your readers would want to know how the company had used computers, how much money they had saved, whether the savings had justified the cost of the equipment, and so forth.

To provide *reliable* evidence, you need to produce the kinds of evidence your readers are willing to accept. The type of evidence varies greatly from field to field. For instance, in science and engineering fields, certain experimental procedures are widely accepted as reliable, whereas common wisdom and ordinary, unsystematic observation usually are not. In contrast, in many business situations, personal observations and anecdotes provided by knowledgeable people often are accepted as reliable evidence. As a writer you need only apply your common sense to determine what kind of evidence is needed. For instance, in the example of the textile mill, you can be reasonably certain your readers would probably reject your evidence if it were based on rumors, but would probably accept it if it were based on an article published in a trade journal.

If you realize your readers might question your evidence, you can do several things. If you think they will object that your evidence is skimpy, try to provide more. If you expect them to object that you have overlooked important data, add a discussion of that evidence. Alternatively, you might narrow your claim to correspond more closely with the evidence you already have: instead of contending that your claim applies to all people or all situations, show why it applies especially well to the particular people or situations you treat in your communication. Finally, you can acknowledge the limitations and assumptions of your argument. For instance, if you argue that a certain savings will be enjoyed, acknowledge your assumption that prices will remain the same, or that loans will be granted at certain interest rates, or that demand will continue to increase.

Explicitly Justify Your Line of Reasoning Where Necessary

In addition to presenting sufficient and reliable evidence, you can also persuade your readers you are reasoning soundly by ensuring they will accept as valid the line of reasoning that links your evidence and your claim. In writing at work, as in everyday conversation, people often deliberately omit explicit justification of their line of reasoning. They believe their reasoning will be perfectly evident to their readers or listeners, who will therefore accept it without question. For instance, in the construction industry, people generally agree that if you use the appropriate formulas to analyze the size and shape of a bridge you are designing, the formulas will accurately predict whether the bridge will be strong enough to support the loads it must carry. You don't need to justify a line of reasoning your readers will accept without question.

However, when writing persuasively, you must look carefully for situations where your readers won't automatically accept your reasoning. For example, readers at work often aggressively look for places where a writer has made a logical mistake by drawing a broad conclusion based on too few specific instances or a conclusion based on an untypical example. Similarly, they are sensitive to places where a writer has missed some crucial difference between two situations the writer has assumed to be the same. For example, if your readers believe your textile mill has some crucial difference from the one that increased its profits after installing computers, then they would suspect you are incorrect in concluding that your mill would enjoy a similar growth in profits if it computerized.

If you identify a place where your readers might question your reasoning, explicitly state your line of thought there and argue on behalf of it. For example, where your audience might suspect you are drawing a hasty generalization, explain why you believe that you do have enough examples or why the examples you have chosen are in fact typical. If you think your readers might object that your reasoning doesn't apply in this case, tell why you think it does.

GUIDELINE 4   ORGANIZE TO CREATE A FAVORABLE RESPONSE

Guidelines 1, 2, and 3 have focused on the kinds of things you should—and should not—say when trying to shape your readers' attitudes. In contrast, this guideline focuses on the way you organize your communication.

How you organize your message sometimes can have almost as great an effect on your persuasiveness as what you say. That point was demonstrated by researchers Sternthal, Dholakia, and Leavitt, who presented two groups of people with different versions of a talk on behalf of establishing a federal consumer protection agency.[10] One version *began* by saying that the speaker was a highly credible source (a lawyer who graduated from Harvard and had extensive experience with consumer issues); the other *ended* with the same information.

Among people initially opposed to the speaker's position, those who learned about his credentials at the beginning responded more favorably to his arguments than those who learned about his credentials at the end. Why? This outcome can be explained in terms of two principles you learned in Chapter 1. First, people react to persuasive messages moment by moment as they read them. Second, (and here's the key point), their reactions in one moment will affect their reactions in subsequent moments. In the experiment, those who heard the speaker's credentials before hearing his arguments were relatively open to what he had to say. They listened without creating numerous counterarguments as they heard each of his statements. In contrast, those who heard the speaker's credentials at the end worked more vigorously at creating counterarguments as they heard each of his points. Once they had created their counterarguments and recorded them in memory, the counterarguments could not be erased simply by adding information about the speaker's credibility. The self-generated counterarguments remained influential and affected each listener's final assessment of the speaker's position.

Thus, it's not merely the array of information in a communication that is critical in persuasion, but also the way the readers process the information. The following sections suggest two strategies for organizing to elicit a favorable response: choose carefully between direct and indirect organizational patterns, and create a tight fit among the parts of your communication.

Choose Carefully between Direct and Indirect Organizational Patterns

As you learned in Chapter 4, by far the most common organizational pattern at work is one that begins by telling the bottom line—the writer's main point. Communications organized this way are said to use a *direct* pattern of organization because they go directly to the main point and only afterward present the evidence and other information related to it. For example, in a memo recommending the purchase of a new computer program, you would begin with the recommendation and then present your arguments on behalf of the purchase, perhaps by explaining the situation that makes you think the new program is desirable and then telling how the new program will improve it.

The alternative is to postpone your bottom-line statement until you have presented much or all of your evidence or related information. This is called the *indirect* pattern. For example, in a memo recommending purchase of a new computer program, you would first discuss the situation that makes you think a new program is desirable, withholding your purchasing recommendation until later. Figure 5–5 compares direct and indirect organizational patterns.

To choose between a direct and an indirect organizational pattern, focus your attention on your readers' all-important initial response to your message. The direct pattern will get your readers off on the right foot when you have good news to convey: "You're hired," "I've figured out a solution to your problem," or something similar. By starting with that good news, you put your readers in a favorable frame of mind as they read the rest of your message.

The direct pattern also works well when you are offering an analysis or recommending a course of action that you expect your readers to view favorably—or at least objectively—from the start. Leah is about to write such a memo, in which she will recommend a new system for managing the warehouses for her employer, a company that manufactures hundreds of parts used to dig oil and gas wells. Leah has chosen to use the direct pattern shown in the left-hand column of Figure 5–6. This is an appropriate choice because her

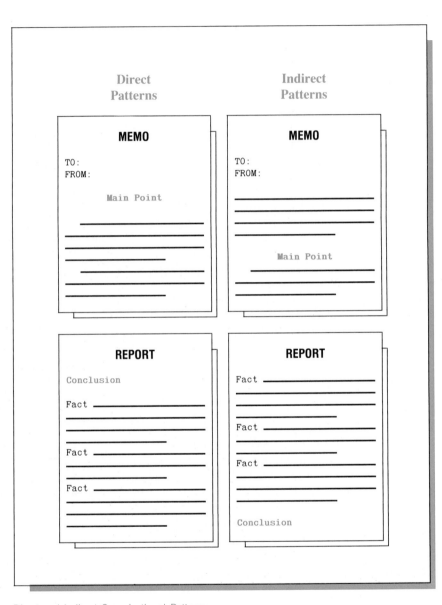

FIGURE 5–5   Direct and Indirect Organizational Patterns

readers (upper management) have expressed dissatisfaction with the present warehousing system. Consequently, she can expect a favorable reaction to her initial announcement that she designed a better system.

Although the direct pattern works well when you are delivering welcome information to open-minded readers, it doesn't work so well when you are conveying information your reader might view as bad, alarming, or threaten-

Direct Patterns	Indirect Patterns

REQUEST

Request Made

Reason _____
_____
_____

Reason _____
_____
_____

Reason _____
_____
_____

REQUEST

Reason _____
_____
_____

Reason _____
_____
_____

Reason _____
_____
_____

Request Made

RECOMMENDATION

Recommendation Stated

Argument _____
_____
_____

Argument _____
_____
_____

Argument _____
_____
_____

RECOMMENDATION

Argument _____
_____
_____

Argument _____
_____
_____

Argument _____
_____
_____

Recommendation Stated

FIGURE 5–5   *(continued)*

ing. Imagine, for example, that Leah's readers are the same people who set up the present warehousing system and they believe it is working well. In that case, if Leah begins her memo by recommending a new system, she might put her readers immediately on the defensive because they might feel she is criticizing their competence. Then she would have little hope of receiving an open and objective reading of the points she makes later on behalf of her recommen-

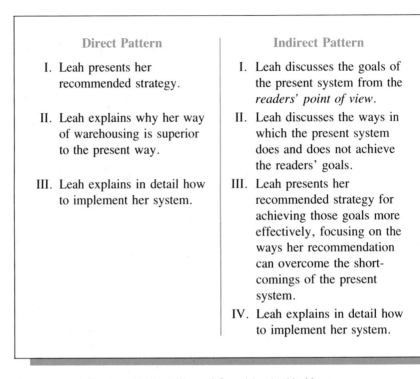

Direct Pattern	Indirect Pattern
I. Leah presents her recommended strategy.	I. Leah discusses the goals of the present system from the *readers' point of view.*
II. Leah explains why her way of warehousing is superior to the present way.	II. Leah discusses the ways in which the present system does and does not achieve the readers' goals.
III. Leah explains in detail how to implement her system.	III. Leah presents her recommended strategy for achieving those goals more effectively, focusing on the ways her recommendation can overcome the short-comings of the present system.
	IV. Leah explains in detail how to implement her system.

FIGURE 5–6   Comparison of Direct and Indirect Ways of Organizing Leah's Memo

dation. By using an indirect pattern, however, Leah can attempt to *prepare* her readers for her recommendation by first getting them to agree that it might be possible to improve on the present system. The right-hand column of Figure 5–6 shows an indirect pattern she might use.

Of course, in addition to choosing the best organization, Leah will have to write skillfully to suggest a problem exists without arousing defensiveness. In sections I and II of the indirect version of her memo, she uses one effective strategy for doing that: she discusses the present system in terms of her *readers' goals,* showing its shortcomings in meeting their goals.

You may wonder why you shouldn't just use the indirect pattern all the time. It presents the same information as the direct organization (plus some more), and it avoids the risk of inciting an initial negative reaction. The trouble is that it frustrates the readers' desire to know the bottom line first. In certain situations, if you don't present your main point first, your readers can become impatient and irritated, and may skip ahead to find your main point anyway.

In sum, the choice between direct and indirect patterns can greatly affect the persuasiveness of your communications. To choose, you need to follow the basic strategy suggested throughout this book: think about your readers' moment-by-moment reactions to your message.

Create a Tight Fit among the Parts of Your Communication

In addition to making a well-calculated choice between a direct and indirect organization, you can also increase the persuasiveness of your writing by planning your communication so the parts fit together tightly. This advice applies particularly to longer communications, where the overall argument often consists of two or more subordinate arguments. That happens, for example, in long proposals. In an early section, writers often describe a problem their proposed project will solve. The writers' persuasive points in this section are that a problem exists and that their readers should view it as a serious one. In a later section, the writers often describe their proposed project in detail.

It's critical that these two sections of the proposal work harmoniously together because the writers' overall argument has to be that their proposed project will effectively solve the particular problem they have just described. If their discussion of the proposed project doesn't address the issues raised in the discussion of the problem, the readers will decide that, however important the problem might be, the proposed project isn't a reasonable way to solve it. Similarly, the writers' discussion of their budget should include only expenses that are clearly related to the project they describe, and their schedule should show when they will carry out each activity necessary to complete the project successfully. In short, every part of the proposal must bear a clear and strong relationship to the other parts.

What do you do if you discover you do not have a tight fit among the parts of a persuasive communication you are writing? You can make adjustments on either end. For example, to make your descriptions of a problem and its solution fit together more tightly, you can adjust your discussion of the problem *or* you can adjust your discussion of the solution, or both.

Of course, the need for a tight fit applies not only to proposals but to any communication whose various parts work together to affect your readers' attitudes. Whenever you write, think about ways to make the parts work harmoniously together in mutual support of your overall position.

GUIDELINE
5

CREATE AN EFFECTIVE ROLE FOR YOURSELF

The four guidelines you have just read concern the points you make to persuade. You have learned how to choose those points and present them as persuasively as possible. However, the points you make for your position are not the sole determinants of your success as a persuasive communicator. Many other factors come into play.

One of the most important is how your audience feels about you. You will remember from Chapter 1 that you should think of the communications you prepare at work as interpersonal interactions. If your readers feel well disposed toward you, they are likely to consider your points openly and without bias. If they feel irritated, angry, or otherwise unfriendly toward you, they may immediately raise counterarguments to every point you present, making it extremely difficult for you to elicit a favorable reaction, even if all your points are clear, valid, and substantiated. Good points rarely win the day in the face of bad feelings.

The following sections describe three ways you can present yourself to obtain a fair—or even a favorable—hearing from your readers at work: choose an appropriate voice, present yourself as a credible person, and present yourself as nonthreatening.

Choose an Appropriate Voice

Your most important tool in establishing a good personal relationship with your readers is the *voice* with which you write. Through the voice you choose, you indicate the particular role you see yourself playing with respect to your audience. For example, when writing to workers in the department you manage, you might assume the voice of a stern authority or an open-minded leader. When instructing a new employee, you might assume the voice of a demanding instructor or a helpful guide.

The voice you choose is such an important element of your persuasive strategy because it indicates not only the role you assign yourself, but also the one you assign your readers. If you are a manager who assumes the voice of an equal writing to respected peers, then the people in your department will probably accept their implied role as your equals. But if you assume the voice of a superior, unerring authority, they may resent their implied role as your error-prone inferiors. Although they may not speak the words aloud, they may think, ''You have no right to talk to me like that.'' If your audience responds to your voice in that way, they are not likely to receive your message openly.

The voice that will work for a particular communication depends on many factors:

● **Your professional relationship with your audience.** For example, whether you are a customer, a supervisor, or a subordinate.
● **Your purpose.** For example, whether you are asking for something, apologizing for an oversight, providing advice, ordering your readers to do something, granting a request, or selling a product.
● **Your subject.** For example, whether you are writing about a routine matter or an urgent problem.
● **Your personality.**
● **Your audience's personality.**

Many things contribute to the creation of voice in communications, including choice of words, sentence rhythm, and content. You are already familiar enough with human relationships to know which tones generally upset other people and which they generally find agreeable. To check your voice, you might read your communication aloud, listening to yourself. As you read, ask yourself this question, "If I heard someone say these things to me, how would I feel that person was viewing me and treating me?"

Present Yourself as a Credible Person

Your credibility is your readers' beliefs about whether or not you are a good source for information and ideas. If people believe you are credible, they will be relatively open to the things you say. In some situations, they may even accept your judgments and recommendations without inquiring deeply into your reasons. If people do not find you credible, they may refuse to give you a fair hearing no matter how soundly you state your case.

To see how much your audience's perception of you can affect the way they respond to your message, consider the results of the following experiment. Researchers H. C. Kelman and C. I. Hovland devised a tape-recorded speech advocating lenient treatment of juvenile delinquents. Before one group of people heard the tape, the speaker was identified as an ex-delinquent out on bail. Another group was told that the speaker was a judge. Only 27 percent of those who heard the "delinquent's" talk responded favorably to it, while 73 percent of those who heard the "judge's" talk responded favorably.[11]

Researchers have conducted many studies into the factors that affect an audience's impression of a person's credibility. In summarizing this research, Robert Bostrom has identified five key factors.[12]

- **Expertise.** Expertise is knowledge, education, and experience that the readers perceive to be relevant to the topic or situation under discussion.
- **Trustworthiness.** This factor mainly concerns the readers' perceptions of the writer's motives. If that person seems to be acting out of self-interest, his or her credibility is low; if the person is acting objectively or for goals shared by the audience, the credibility is high.
- **Group membership.** People who are members of groups admired by the audience have more credibility than those who aren't. Members of the readers' own group have high credibility.
- **Dynamic appeal.** A person who is aggressive, forceful, and frank has much higher appeal than someone who is passive, forceless, and closed.
- **Power.** Readers assign high credibility to those who control what the readers want. Thus, simply by virtue of his or her position, a boss acquires some credibility with subordinates.

In many situations at work, your readers will have an initial impression of you even before they begin to read to your message. For instance, if you are a manager writing to people who work in your department, you will already enjoy some credibility as a result of your power and your membership in their group. Of course, initial impressions can also work against you. You may be writing to a group who has had a bad experience with your employer, or you may speak before a group who holds positions contrary to those they know your employer holds.

Whether your audience has a positive or negative initial impression of you, or no initial impression at all, you can compose your message to raise your credibility. One or more of the following strategies may help you in a particular situation.

- **Expertise.** Mention your credentials. Show you have a command of the facts by demonstrating a detailed knowledge of the situation or subject. Avoid oversimplifying. Mention or quote people who are experts in the field so their expertise supports your position.
- **Trustworthiness.** To avoid appearing biased, demonstrate a knowledge of all the factors in the situation. Avoid drawing attention to personal advantages to you. Argue in terms of values and objectives important to your readers.
- **Group membership.** If you are associated with a group admired by your readers, allude to that relationship. If you are addressing others in your own organization, affirm that relationship by showing that you share the group's objectives, methods, and values. Use common terms in your organization.
- **Dynamic appeal.** State your message simply and directly. Use an energetic style. Show your enthusiasm for your ideas and subject.
- **Power.** If you are in a position of power, identify your position if your readers don't know it. If you are not in a position of power, associate yourself with a powerful person by quoting the person or by saying that you consulted with him or her or were assigned the job by that individual.

Figure 5–7 shows a letter in which an advertising company attempts to persuade a bank to let it make a sales presentation. The writer's chief persuasive strategy is to build the advertising company's credibility.

Present Yourself as Nonthreatening

Psychologist Carl Rogers has identified another important strategy in fostering open, unbiased communication: reduce the sense of threat people often feel when others are presenting ideas to them.[13] According to Rogers, people are likely to feel threatened even when you make *helpful* suggestions. As a result, your readers may see you as an adversary even though you don't intend at all to be one.

Here are four methods, based on the work of Rogers, you can use to present yourself as nonthreatening to your readers.

- **Praise them.** In many working situations, you can induce your readers to feel more favorably disposed to you if you let them know you think they

3645 WEST LAKE ROAD · ERIE, PENNSYLVANIA 16505 · 814/838-4505

30 September 19--

Mr. Jeff Frances
Vice President, Marketing
First Federal Savings and Loan
724 Boardman-Poland Road
Youngstown, OH 44512

Dear Mr. Frances:

Our firm is very interested in presenting its financial
advertising and marketing credentials to you for consideration.

While you may never have heard about Tal, Inc., chances are
you're familiar with our work. Here in Youngstown, Tal, Inc.
has been responsible for recent marketing communications
programs of the Western Reserve Care System that includes
Northside and Southside Medical Centers, Tod Children's
Hospital, and Beeghly Medical Park.

But more important to you as Vice President of Marketing at
First Federal Savings and Loan is Tal, Inc.'s experience as a
full-service <u>financial</u> advertising agency. That experience
includes work for such clients as Pennbank; The First
National Bank of Pennsylvania; Permanent Savings Bank,
Buffalo, NY; North Star Bank, Buffalo, NY; and Citizens
Fidelity Bank, Louisville, KY.

Our growth mission is to expand our financial client base to
the Youngstown market. Toward that end, our firm is
interested in presenting our award-winning financial
portfolio and case histories for your review. The portfolio
features a broad range of financial capabilities. In addition
to TV, radio, newspaper, and outdoor advertising, you'll find
samples of P.O.S., Collateral Brochures and Pamphlets, Annual
Reports, Direct Mail, and Sales Support Materials. We're
confident you'll like what you see and hear. But we're also
confident you'll recognize the possibilities a Tal, Inc./
First Federal Savings and Loan relationship would afford.

ADVERTISING

BUFFALO · ERIE · LANCASTER

FIGURE 5–7   Sales Letter that Relies Heavily on Building Credibility

Mr. Jeff Frances
30 September 19--
Page 2

Our goal is to apply our capability to the marketing
challenges in the Youngstwon market during the coming year.
Please let me know how I might proceed in presenting our
qualifications to you. I can be contacted during regular
business hours at 814/838-4505.

Thank you for your consideration of Tal, Inc.

Sincerely,

Peter M. Sitter

Peter M. Sitter
Financial Services
Account Supervisor

FIGURE 5-7   *(continued)*

are doing a good job. If you are writing to an individual, allude to some accomplishment of his or hers. If you are writing to a company, mention something it prides itself on. Of course, general praise will sound insincere if it doesn't seem to be based on a real knowledge of the audience and isn't related to your topic. Therefore, mention specifics.

● **Present yourself as their partner.** Identify some personal or organizational goal of *theirs* that you will help your readers attain. If you are already your readers' partner, allude to that fact and emphasize the goals you share.

● **Show that you understand them.** Even where you disagree with your audience, state their case fairly. Focus on areas on which you agree.

● **Maintain a positive and helpful stance.** Present your suggestions as ways to help your readers do an even better job. Avoid criticism or blaming.

To see how to apply this advice, consider the situation of Martin Lakwurtz, who is a regional manager for a company that employs nearly 1,000 students in many cities during the summer months to paint houses and other large structures. Each student is required at the beginning of the summer to make a $100 deposit, which is returned at the end of the summer. Students complain that the equipment they work with isn't worth anywhere near $100 so the deposit is unjustified. Furthermore, the students believe that the company is simply requiring the deposit so as to have money to invest during the summer. The company keeps all the interest on the deposits, not giving any to the students when it returns their deposits at the end of the summer.

Martin agrees that the students have a good case, so he wants to suggest a change. He could do that in a negative way by telling the company that it should be ashamed of its greedy plot to profit by investing other people's money. Or he could take a positive approach, presenting his suggested change as a way the company can better achieve its own goals. In the memo shown in Figure 5–8 he takes the latter course. You can see how one writer skillfully presented himself as his audience's partner.

A Caution against Evoking Negative Thoughts

One caution is in order here. When you are trying to show how much you sympathize with your readers or understand their position, be careful not to call up negative thoughts accidentally, or you will be working against yourself. For example, imagine you are writing a memo notifying people in your company that the head office will be moving to a new city. You will want the employees to feel favorably inclined to the company and the move. For that reason, you would want to avoid saying things like, "I know you are made anxious by the move," which would bring their anxiety to the forefront of their thoughts, or "I know you have heard rumors about the move for weeks," which may make them resentful for those weeks they had to live on rumors rather than hard facts from upper management.

PREMIUM PAINTING COMPANY

February 12, 19—

TO: Marjorie Sneed

FROM: Martin Lakwurtz $\mathcal{ML}$

RE: Improving Worker Morale

Now that I have completed my first year with the company, I have been thinking over my experiences here. I have enjoyed the challenges that this unique company offers, and I've been very favorably impressed with our ability to find nearly 1000 temporary workers who are willing to work so energetically and diligently for us.

In fact, it's occurred to me that the good attitude of the students who work for us is one of the indispensable ingredients in our success. If they slack off or become careless, our profits could drop precipitously. If they become careless, splattering and spilling paint for instance, we would lose much of our profits in cleaning up the mess.

Because student morale is so crucial to us, I would like to suggest a way of raising it even further—and of ensuring that it won't droop. From many different students in several cities, I have heard complaints about the $100 security deposit we require them to pay before they begin work. They feel that the equipment they work with is worth much less than that, so that the amount of this security deposit is very steep. Furthermore, there is a widespread belief that the company is actually cheating them by taking their money, investing it for the summer, and keeping the profits for itself.

Because the complaints are so widespread and because they can directly affect the students' sense of obligation to the company, I think we should try to do something about the security deposit. First, we could reduce the deposit. If you think that would be unwise, we could give the students the interest earned on their money when we return the deposits to them at the end of the summer.

I realize that it would be difficult to calculate the precise amount of interest earned by each student. They begin work at different times in May and June, and they end at different times in August and September. However, we could establish a flat amount to be paid each student, perhaps basing it on the average interest earned by the deposits over a three-month period.

FIGURE 5–8   Memo in Which the Writer Establishes a Partnership with the Readers

Marjorie Sneed
February 12, 19--
Page 2

I'm sure that it has been somewhat beneficial for the company to have
the extra income produced by the deposits. But the amount earned is
still rather modest. The additional productivity we might enjoy from
our students by removing this irritant to them is likely to increase
our profits by much more than the interest paid.

If you would like to talk to me about this idea or about the feelings
expressed by the students, I would be happy to meet with you.

FIGURE 5—8   *(continued)*

This caution doesn't mean you should ignore your audience's counterproductive feelings. You should, for example, do what you can to reduce their anxiety by telling them favorable and reassuring things about the move. But try to avoid making direct reference to negative thoughts or to topics that will evoke negative thoughts.

Roles and Integrity

Perhaps you are hesitating right now as you think about all the advice you have just read about how to create an effective role for yourself. You may be wondering whether you would be acting insincerely or artificially if you were to take so many factors into account as you decide how you will present yourself each time you write on the job. Wouldn't it be more honest simply to say what's on your mind without all this concern for how your readers will perceive you?

To answer that question, imagine you are walking across campus. On the way, you talk with each of three people: a close friend, one of your instructors, and an interviewer from a company you would like to work for. The subjects you discuss, the words you choose, even the way you stand are likely to be different in each of these three conversations. In fact, each of these individuals would think it strange if you didn't adjust your conversation to him or her. What would the interviewer think if you spoke to her in the same casual way you speak to your close friend? And what would your friend think if you were to address him in the way you would speak to your instructor?

These adjustments in the way you present yourself are perfectly natural. You have different relationships with each person and adjust your behavior accordingly. On the job you will make similar adjustments. You will talk one way with the person working next to you and another with your boss, one way with someone in the department down the hall and another way with a potential customer. In each case you will strive to present yourself in a way that will prompt the other person to respond to you receptively. The discussion of this guideline has been aimed at helping you do that.

TAKING EXPECTATIONS AND REGULATIONS INTO ACCOUNT

So far, all the advice about planning you have read in Chapters 4 and 5 has focused on the ways your understanding of your purpose and audience can guide you as you plan the content and organization of something you write on the job. You also may need to take into account various expectations—and even regulations—that limit what you say and how you can say it.

These expectations and restrictions can affect *any* aspect of a communication. Constraints on what the writer should and should not say are very common. Some employers and readers also want to control organization, tone, use of abbreviations, layout of tables, size of margins, and length (usually specifying a *maximum* length, not a minimum).

These various limitations on writers can arise from a myriad of sources. Some come directly from the employer, reflecting such motives as the company's desire to cultivate a particular corporate image through its writing, to protect its legal interests (since any written document can be subpoenaed in a lawsuit), and to preserve its competitive edge (for example, by preventing employees from accidentally tipping off competitors about new technological breakthroughs). Writing limitations also can originate from outside the company, for instance from the traditional writing practices of certain professions (''This is how engineers write''; ''This is how accountants write'') or from government regulations that specify how patent applications, environmental impact reports, and many other types of documents are to be prepared. In addition, most organizations develop writing customs—''the way we write things here.'' Finally, personal preferences of bosses and influential readers also can restrict the range of writing strategies an employee can use.

One way to learn about the expectations and restrictions that apply to you is to ask other people, including your boss and co-workers. You also can ask your intended audience, who will probably be delighted to tell you how they would like you to write. Many companies publish *style guides* that tell some of their regulations about writing. You also can learn a lot about what is expected by reading communications similar to your own.

In many organizations, the established restrictions and expectations about writing help people write effectively, but in others some or all of the required ways of writing actually hinder effective communication. If you encounter constraints that hamper good writing, you might see how receptive people are to changes. Your diplomatic championing of improved writing practices is as important a contribution to your employer as any other recommendation you make to improve operations.

CONCLUSION

This chapter has focused on advice for writing persuasively. As you can see, nearly every aspect of your communication affects your ability to influence your readers' attitudes and actions. While the guidelines in this chapter will help you write persuasively, the most important persuasive strategy of all is to keep in mind your particular readers' needs, concerns, values, and preferences whenever you write.

SPECIAL TOPIC   OUTLINING

In this chapter and the preceding one, you have learned several important principles for planning the content and organization of your communications. This special section discusses outlining, a powerful tool that can—in certain situations—help you immensely in your planning effort.

Many students are skeptical of outlining because some instructor has asked—or required—them to outline in situations where it did not help them and only added an unnecessary burden to their writing effort. In fact, outlining can be a waste of time in some situations, whether at work or at school. On the other hand, outlining can be very helpful in the right situations. The following discussion will help you decide when to outline and it will help you use your outlines effectively when you do.

Outlining versus Organizing

As mentioned at the beginning of Chapter 4, outlining and organizing are closely related but not the same thing. *Organizing* is the activity through which you structure your writing. *Outlining* is but one *technique* you can use to develop and describe structure. There are many others. You can, for instance, draw idea trees or flow charts. Furthermore, none of these techniques—not even outlining—can guarantee that the structure you create will be effective. To make it so, you must follow the guidelines you have already read in this and the preceding chapter.

What, then, can outlining do for you? It can help you look at your communication solely from the point of view of structure. When you create an outline, you draw a sketch of your communication, displaying its contents in two dimensions: hierarchy and sequence. This sketch might consist merely of an indented list of topics you will treat (see Figure 5–9), or it might employ a special notational system, such as those shown in Figure 5–10. In either case,

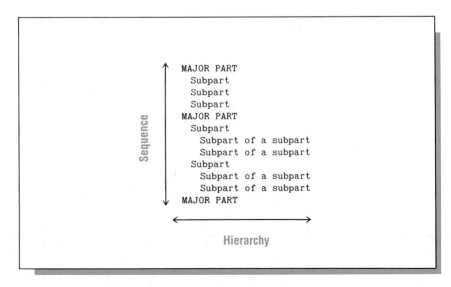

FIGURE 5–9   Two Dimensions of an Outline

Alphanumeric System	Decimal System
I. . . .	1.0 . . .
A. . . .	1.1 . . .
B. . . .	1.2 . . .
C. . . .	1.3 . . .
II. . . .	2.0 . . .
A. . . .	2.1 . . .
1. . . .	2.1.1 . . .
2. . . .	2.1.2 . . .
B. . . .	2.2 . . .
1. . . .	2.2.1 . . .
2. . . .	2.2.2 . . .
III. . . .	3.0 . . .

FIGURE 5–10   Two Notational Systems for Outlines.

as you develop and refine this sketch, you can concentrate solely on deciding what topics you will cover and how you will organize your discussions of them, without simultaneously needing to consider phrasing, grammar, and other details.

When Outlining Can Help

The concentrated look at structure an outline provides can help you in several ways. While you are planning your communication, outlining enables you to play with and settle on its overall design before you invest any time in creating drafts you might otherwise want to change. Also, when you want to talk your plans over with someone else, such as a co-worker or your boss, an outline can help you tell the other person quickly and clearly what you are thinking about doing. And if you are working with a team of people to write something, creating an outline can help you obtain the advance agreement about content and structure so important for writing teams to achieve before individual members begin investing their time and creativity in drafting.

Besides helping you plan your communications before you begin to draft them, outlining also can help you discover and solve problems you encounter after you draft. By outlining problematic passages or communications, you gain the advantage of stepping back from the details of your draft to look at its general structure. From that perspective you can detect problems rooted in weak organization or poor selection of contents, such as repetition and digression. Outlining a problematic draft also can help you decide on the points you really want to emphasize so you can revise your draft to make those points more prominent.

Outlining can bring you some additional benefits if you create a sentence outline rather than a topic outline. To see how helpful a sentence outline can be, compare the topic outline in Figure 5–11 with the sentence outline in Fig-

```
              PROPOSAL FOR FUNDS TO MAKE AN INSTRUCTIONAL
                 TELEVISION SERIES ABOUT ENERGY AND THE
                   ENVIRONMENT FOR HIGH SCHOOL STUDENTS

        I. Need
           A. Students' need
           B. Teachers' need

       II. Proposed project
           A. Television shows
              1. Contents
              2. Technical specifications
           B. Teacher's Guide

      III. Plan of action
           A. Planning and scripting
           B. Field testing
              1. Location
              2. Evaluation
           C. Production
           D. Promotion and distribution

       IV. Qualifications
           A. Institutional
           B. Personnel

        V. Budget
```

FIGURE 5–11   Topic Outline

ure 5–12; both are for the sample proposal written by a television production company that sought a government grant to create a series on energy and the environment. First, notice how the sentence outline has helped the writers sharpen their focus by inviting them to go beyond merely identifying their topics to deciding the main point they want to make about each topic. Second, notice how it has helped the writers focus on the links and transitions they can use to weave their sections and subsections together. For some communications, the special help sentence outlining can give is irrelevant. However, with longer, more complex communications, sentence outlining can provide substantial help focusing your message and tying it together.

When Outlining Won't Help

Although outlining can be beneficial in the situations just mentioned, in many circumstances it won't be. For example:

- When you are writing a short communication that does not require much thought about organization
- When you are writing a kind of communication you have written several times before, so you have already worked out the organizational issues
- When you are confident you can hit on the best organization intuitively
- When you are confident you can organize effectively by imitating the organization of a similar communication someone else has written

Keep in mind also that outlining may not help you if you are still unsure about the general nature of what you want to say. You may be a writer who needs to start drafting in order to figure out what you have to say—or even to figure out what your purpose is. If that's the case, go ahead and draft first. *Then* outline if it seems helpful. On the other hand, for some writers, outlining helps even when they are unsure of their general message because sitting down to outline forces them to think it through.

A Caution

Whenever you outline, keep in mind the distinction made earlier in Chapter 4 between a logical organization and an effective one. It is possible to create an outline that looks very reasonable but won't work for your readers. And even a promising outline must be translated into words, sentences, and paragraphs that create an effective interchange between you and your audience. Consequently, even when outlining can help you most, you still need to consider your readers as you plan your writing strategies.

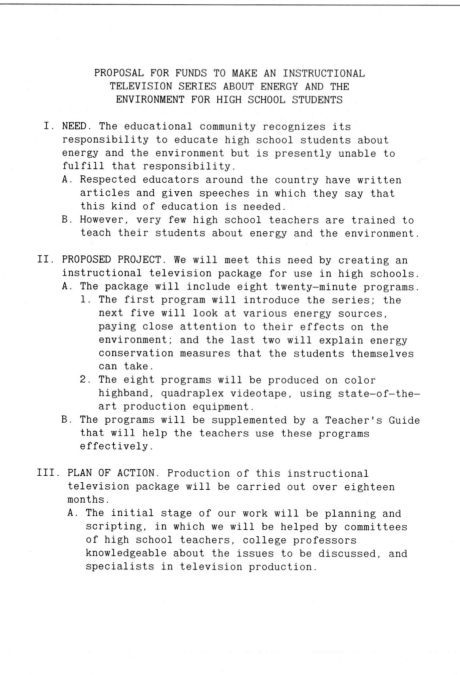

PROPOSAL FOR FUNDS TO MAKE AN INSTRUCTIONAL
TELEVISION SERIES ABOUT ENERGY AND THE
ENVIRONMENT FOR HIGH SCHOOL STUDENTS

I. NEED. The educational community recognizes its
responsibility to educate high school students about
energy and the environment but is presently unable to
fulfill that responsibility.
   A. Respected educators around the country have written
   articles and given speeches in which they say that
   this kind of education is needed.
   B. However, very few high school teachers are trained to
   teach their students about energy and the environment.

II. PROPOSED PROJECT. We will meet this need by creating an
instructional television package for use in high schools.
   A. The package will include eight twenty-minute programs.
      1. The first program will introduce the series; the
      next five will look at various energy sources,
      paying close attention to their effects on the
      environment; and the last two will explain energy
      conservation measures that the students themselves
      can take.
      2. The eight programs will be produced on color
      highband, quadraplex videotape, using state-of-the-
      art production equipment.
   B. The programs will be supplemented by a Teacher's Guide
   that will help the teachers use these programs
   effectively.

III. PLAN OF ACTION. Production of this instructional
television package will be carried out over eighteen
months.
   A. The initial stage of our work will be planning and
   scripting, in which we will be helped by committees
   of high school teachers, college professors
   knowledgeable about the issues to be discussed, and
   specialists in television production.

FIGURE 5–12   Sentence Outline

B. When scripting is completed, we will construct a slide-tape mock-up of the series, which we will field test.
   1. The field testing will occur in urban, suburban, and rural schools to involve all the kinds of locations in which the series might be used.
   2. The test results will be evaluated with the assistance of an expert in educational research.
C. The entire package will be produced at our own facilities.
D. We will use established channels for distributing instructional television shows.

. . . And so on.

FIGURE 5–12   *(continued)*

EXERCISES

1. Figures 5–13 and 5–14 show two versions of the same memo. Which one is more likely to work? Explain your choice in terms of the guidelines in this chapter.

2. Analyze the memo shown in Figure 5–8 to identify the ways the writer, Martin Lakwurtz, has applied each of the general guidelines presented in this chapter.

3. **a.** Using your common sense about how people react to one another, rewrite the memo by Donald Pryzblo (Figure 1–6) so it is more likely than his to persuade the personnel manager to follow his recommendation. You may assume that Pryzblo knows the manager's clerks are miscopying because he examined the time sheets, time tickets, and computer files associated with 37 incorrect payroll checks; in 35 cases the clerks made the errors. Remember to take into account the initial reaction you expect the personnel manager to have at finding a memo from Pryzblo in his mail; make sure the first sentence of your memo addresses a person in that frame of mind and that your sentences lead effectively from there. Leave the last sentence unchanged.

    b. In the margin of your memo or on a separate sheet of paper, tell what you expect the personnel manager's reaction to be after reading each of the individual sentences.

4. Playing the role of the personnel manager, write two replies to Donald Pryzblo. In the first, write in a manner likely to increase the antagonism between the two of you. In the second, write in a manner that could possibly end the antagonism and encourage Pryzblo to cooperate with you to solve the problem of the incorrect payroll checks.

5. Find two persuasive messages, one with at least twenty-five words of prose. These messages may be advertisements, marketing letters, memos from school or work, or similar business-type communications. Examine the persuasive strategies used in each. What are they? Do they work well? What other strategies might have been used? Present your responses in a memo addressed to your instructor and attach copies of the messages. For a brief assignment, write 200 to 500 words. For a long assignment, write 800 to 1200 words.

6. Write a memo to your instructor about a situation in which you tried to persuade someone to make a particular decision or to take a certain action. If possible, choose a business situation. If not, choose one involving a college instructor, administrator, friend, or parent. Describe the person, the situation, the person's attitude before you made your persuasive effort, and the outcome you desired. Tell whether the attitude change you wanted to bring about involved reinforcement, reversal, or shaping. Then describe your persuasive strategies in terms of the guidelines and concepts used in this chapter. For instance, if you anticipated counterarguments, identify them and tell how you dealt with each. Did your persuasive effort work?

MEMORANDUM

TO: Freida Van Husen, Vice President, Marketing

FROM: Terry Prindle, Marketing Specialist

RE: PROPOSED NEW OFFICE ARRANGEMENT

Current use of our outer office space does not meet our needs.
Specifically:

1. A private work area for me.
2. A clearly designated reception area.
3. A videotape storage and production room.
4. A conference room.

I recommend the following changes to better meet those needs:

1. Place partitions around my work area to separate it from the
    receptionist's part of the room.
2. Furnish the vacant office adjacent to my area with cabinets and
    other things that will enable me to use it for a conference room
    and to store the videotape equipment.

With the current office arrangement, I am constantly interrupted by
visitors who should be going to the receptionist's desk instead. Also,
I am frequently embarrassed when I see the puzzlement of visitors who
look at the pile of videotape equipment near my desk. I am further
embarrassed when I have to drag my chairs and a card table into the
empty office in order to review advertising layouts with our agency's
representatives.

FIGURE 5–13   First Version of Memo for Exercise 1

MEMORANDUM

TO: Freida Van Husen, Vice President, Marketing

FROM: Terry Prindle, Marketing Specialist

RE: IMPROVEMENT OF YOUR OFFICE AREA

The large area outside your office serves two purposes. The most important is to create a favorable impression on visitors who wish to do business with our marketing department. In addition, the room provides a work area for your receptionist and me.

Currently, the area is not meeting those needs.

1. The area creates an unfavorable impression on visitors because it is cluttered by the videotape equipment. In addition, when our agency's representatives want to meet with me to work out the details of layouts you have requested, they must accompany me into the vacant office next to my area, dragging chairs behind them while I fetch a card table.

2. Visitors frequently come to my desk instead of the receptionist's because they are unsure which of us is your receptionist. The numerous interruptions they cause make it difficult for me to complete in a timely fashion the assignments you have given me.

Therefore, I recommend that you ask the maintenance department to (1) place a partition around my area and (2) convert the vacant office into a conference room and storage area for the videotape equipment by supplying it with the necessary cabinets and other furniture.

FIGURE 5–14   Second Version of Memo for Exercise 1

What might you have done differently if you had read this chapter before making that persuasive effort? For a brief assignment, write 200 to 500 words. For a long assignment, write 800 to 1200 words.

7. Analyze the letter in Figure 5–15 to identify its strengths and weaknesses. Relate your points to the guidelines in this chapter. The following paragraphs describe the situation in which the writer, Scott Houck, is communicating.

Before going to college, Scott worked for a few years at Thompson Textiles. In his letter he addresses Thompson's vice president, Georgiana Stroh. He is writing because in college he learned many things that made him think Thompson would benefit if its managers were better educated in modern management techniques. Thompson Textiles could enjoy this benefit, Scott believes, if it offered courses in management to its employees and if it filled job openings at the managerial level with college graduates. However, if Thompson were to follow Scott's recommendations, it would be changing its current practices considerably. Thompson has never offered courses for its employees and has long sought to keep payroll expenses low by employing people without a college education, even in management positions. (In a rare exception to this practice, the company has guaranteed Scott a position after he graduates.)

To attempt to change the company's policies, Scott decided to write a letter to one of the most influential people on its staff, Mrs. Stroh. Unfortunately for Scott, throughout the three decades that Stroh has served as an executive officer at Thompson, she has consistently opposed company-sponsored education and the hiring of college graduates. Consequently, she has an especially strong motive for rejecting Scott's advice: she is likely to feel that, if she were to agree that Thompson's educational and hiring policies should be changed, she would be publicly admitting that she had been wrong all those years to support the current policies.

8. After you have decided what to include in a project you are preparing for your technical writing class, make two outlines that show alternative ways of arranging this material. Write a paragraph comparing the two outlines from the point of view of the intended reader of the project.

CASE   **DEBATING A COMPANY DRUG-TESTING PROGRAM**

"Have you opened your mail yet this afternoon?" Hal asks excitedly on the telephone. The two of you started working at Life Systems, Inc., in the same month and have become good friends even though you work in different departments. By now you've both risen to management positions.

"My mail isn't here yet. What's up?"

"Wait 'til you hear this," Hal replies. "Tonti's sent us all a memo announcing that she is thinking about starting a company drug-testing program." The person Hal calls Tonti is Maria Tonti, the company president, who has

616 S. College #84
Oxford, Ohio 45056

April 28, 19—

Georgiana Stroh
Executive Vice President
Thompson Textiles Incorporated
1010 Note Ave.
Cincinnati, Ohio 45014

Dear Mrs. Stroh:

As my junior year draws to a close, I am more and more eager to return to our company where I can apply my new knowledge and skills. Since our recent talk about the increasingly stiff competition in the textile industry, I have thought quite a bit about what I can do to help Thompson continue to prosper. I have been going over some notes I have made on the subject, and I am struck by how many of the ideas stemmed directly from the courses I have taken here at Miami University.

Almost all of the notes featured suggestions or thoughts I simply didn't have the knowledge to consider before I went to college! Before I enrolled, I, like many people, presumed that operating a business required only a certain measure of commonsense ability—that almost anyone could learn to guide a business down the right path with a little experience. However, I have come to realize that this belief is far from the truth. It is true that many decisions are common sense, but decisions often only appear to be simple because the entire scope of the problem or the full ramifications of a particular alternative are not well understood. A path is always chosen, but how often is it the BEST path for the company as a whole?

In retrospect, I appreciate the year I spent supervising the Eaton Avenue Plant because the experience has been an impetus to actually learn from my classes instead of just receiving grades. But I look back in embarrassment upon some of the decisions I made and the methods I used then. I now see that my previous work in our factories and my military

FIGURE 5–15 Letter for Use in Exercise 7

From Paul V. Anderson, "Unrequested Recommendation," in *What Makes Writing Good: A Multiperspective*, ed. William E. Coles and James Vopat (Lexington, Mass.: D. C. Heath, 1985), 221–23. Reprinted by permission.

Mrs. Stroh   -2-   April 28, 19--

experience did not prepare me as well for that position as I
thought they did. My mistakes were not so often a poor
selection among known alternatives, but were usually sins of
omission. For example, you may remember that we were
constantly running low on packing cartons, and we sometimes
ran completely out, causing the entire line to shut down. Now
I know that instead of haphazardly placing orders for a
different amount every time, we should have used a
forecasting model to determine demand and establish a reorder
point and a reorder quantity. But I was simply unaware of
many of the sophisticated techniques available to me as a
manager.

   I respectfully submit that many of our supervisory
personnel are in a similar situation. This is not to downplay
the many contributions they have made to the company.
Thompson can directly attribute its prominent position in the
industry to the devotion and hard work of these people. But
very few of them have more than a high school education or
have read even a single text on management skills. We have
always counted on our supervisors to pick up their management
skills on the job without any additional training. While I
recognize that I owe my own opportunities to this approach,
this comes too close to the commonsense theory I mentioned
earlier.

   The success of Thompson depends on the abilities of our
managers relative to the abilities of our competitors. In the
past, EVERY company used this commonsense approach and
Thompson prospered because of the natural talent of people
like you. But in the last decade many new managerial
techniques have been developed that are too complex for the
average employee to just "figure out" on his own. For
example, mankind had been doing business for several thousand
years before developing the Linear Programming Model for
transportation and resource allocation problem-solving. It is
not reasonable to expect a high school graduate to recognize
that his or her particular distribution problem could be
solved by a mathematical model and then to develop the LP

FIGURE 5–15   *(continued)*

Mrs. Stroh   -3-   April 28, 19--

model from scratch. But as our world grows more complex,
competition will stiffen as others take advantage of these
innovations. I fear that what has worked in the past will not
necessarily work in the future: we may find that what our
managers DON'T know CAN hurt us. Our managers must be made
aware of advances in computer technology, management theory,
and operations innovations, and must be able to use them to
transform our business as changing market conditions demand.

    I would like to suggest that you consider the value of
investing in an in-house training program dealing with
relevant topics to augment the practical experience our
employees are gaining. In addition, when management or other
fast-track administrative positions must be filled, it may be
worth the investment to hire college graduates whose
coursework has prepared them to use state-of-the-art
techniques to help us remain competitive. Of course, these
programs will initially show up on the bottom line as
increased expenses, but it is reasonable to expect that, in
the not-so-long run, profits will be boosted by new-found
efficiencies. Most important, we must recognize the danger of
adopting a wait-and-see attitude. Our competitors are now
making this same decision; hesitation on our part may leave
us playing catch-up.

    In conclusion, I believe I will be a valuable asset to the
company, in large part because of the education I am now
receiving. I hope you agree that a higher education level in
our employees is a cause worthy of our most sincere efforts.
I will contact your office next week to find out if you are
interested in meeting to discuss questions you may have or to
review possible implementation strategies.

                              Sincerely,

                              Scott Houck

                              Scott Houck

FIGURE 5–15   *(continued)*

increased the company's profitability greatly in the past five years by spurring employees to higher productivity while still giving them a feeling they were being treated fairly.

"A what?"

"Drug testing. Mandatory urine samples taken at unannounced times. She says that if she decides to go ahead with the program, any one of our employees could be tested, from the newest stockroom clerk to Tonti herself."

"You're kidding!" you exclaim as you swivel your chair around and look out of your window and across the shady lawn that extends from the Life Systems Building to the highway.

"Nope. And she wants our opinions about whether she should start a program or not. In the memo she says she's asking all of her 'key managers' to submit their views in writing. She wants each of us to pick one side or the other and argue for it, so she and her Executive Committee will know all the angles when they sit down to discuss the possibilities next week."

Hal pauses, waiting for your reaction. After a moment, you say, "Given the kinds of products we make, I suppose it was inevitable that Tonti would consider drug testing sooner or later." Life Support, Inc., manufactures highly specialized, very expensive medical machines, such as ones that keep patients alive during heart and lung transplants. The company has done extremely well in the past few years because of its advanced technology and—more important—the unsurpassed reputation for reliability of its machines.

You add, "Does she say in her memo what made her start thinking about drug testing? Has there been some problem?"

"I guess not. In fact, in her memo Tonti says she hasn't had a single report or received any other indication that any employee has used drugs. I wonder what kinds of responses she'll get."

You speculate, "I bet she gets the full range of opinions—all the way from total support for the idea to complete rejection of it."

Just then someone knocks at your door. Looking over your shoulder, you see it's Scotty with some papers you asked him to get.

"I've got to go, Hal. Talk to you later."

"Before you hang up," Hal says, "tell me what you're going to say in your memo to Tonti."

"I don't know. I guess I'll have to think it over."

Your Assignment

Write Maria Tonti the memo she has requested. Begin by deciding whether you want to argue for or against the drug program. Then think of all the arguments on your side you can. Next think of the strongest counterarguments that might be raised against your position by people who disagree with you. In your memo, argue persuasively on behalf of your position, following the guidelines in Chapters 4 and 5. Remember to deal effectively with the arguments on the other side.

6

USING THE LIBRARY

DEFINING
OBJECTIVES
PLANNING
DRAFTING
EVALUATING
REVISING

An important step in planning many communications is to conduct library research. In some situations, you may need only a single fact; in others you may need to gain a thorough understanding of a subject that's entirely new to you. Whatever your need, this chapter will help. First, it describes specific library tools and techniques you can use to find books, magazine and journal articles, government documents, and other sources on your subject. Then it presents four guidelines for organizing and carrying out your work in the library.

BOOKS

Libraries vary in size from a few hundred books to millions of them. Whether the libraries you use are small or large, the techniques for locating books that will help you are the same.

Using the Card Catalog

The best known technique for locating books is to look up your subject in the card catalog. However, beginning your library research by opening a drawer and thumbing through the cards can be inefficient—and it can cause you to overlook useful books. First, the card catalog uses a special set of words for filing cards, and the words it uses to describe a subject may not be the same as the ones you would use. What you might call *gene splicing,* the card catalog calls *genetic engineering;* what you might call *solar power,* the card catalog calls *solar energy;* and what you might call *artificial heart,* it calls *heart, artificial.* Also, for many topics, the number of cards in a card catalog can be quite large. For instance, in an average-sized college library, the cards that begin with the word *cancer* fill more than two drawers. To help you, the cards on many topics are grouped under subheadings, such as Cancer—Diagnosis, Cancer—Prevention, and Cancer—Treatment. Even so, you can waste a lot of time looking through the drawer unless you know in advance what those subgroups are. For cancer, there are 73 subtopics! Finally, cards related to many general topics are scattered around the card catalog under various, more specific headings. For instance, cards for books related to genetic engineering appear under genetic recombination, biotechnology, protein engineering, molecular cloning, and eight other subject headings.

Fortunately, there is a guide to the card catalog that can greatly simplify your search. Called the *Library of Congress Subject Headings List,* it does four very helpful things:

- Lists all the subject headings used in the card catalog.
- Provides cross-references from the terms you might use to the ones the card catalog uses.

● Lists related terms under which books of interest to you may be cataloged.

● Identifies all the subheadings used under larger headings, so that you don't have to spend a lot of time flipping through drawers of cards to discover these subheadings.

Figure 6–1 shows the entry in the *Subject Headings List* for artificial intelligence. Once you've used the *Subject Headings List* to identify the headings that relate to your topic, you can find sources in the card catalog very efficiently.

In the card catalog itself, you will find three kinds of cards: subject, author, and title. Figure 6–2 shows examples. These cards contain much information that can help you decide which books are worth examining. For example, they give the year of publication (so you can identify the books that have the most up-to-date information); they list the number of pages (so you can guess whether the book treats your subject at the level of detail that you require); and they tell if the book contains illustrations (which might suggest visual aids you can use in your communication) or a bibliography (which could help you locate additional sources for information about your subject).

As you search through the card catalog, remember that the cards are alphabetized according to the word-by-word method, which differs substantially from the letter-by-letter method used in dictionaries. Notice that in the following lists, only the first word keeps the same place.

Word-by-Word Alphabetizing	Letter-by-Letter Alphabetizing
San Antonio	San Antonio
San Diego	sanctuary
San Pedro	sandals
sanctuary	sand dollars
sand dollars	San Diego
sandals	sandstone
sandstone	San Pedro

Some libraries have computerized their card catalogs. Instead of walking from place to place and thumbing through long drawers of cards, you can sit down at a computer terminal and type in the author, title, or subject you are looking for. Almost instantly, the computer screen will show full bibliographic information and the call numbers for the book or books you want. Even when you are using a computer, however, the *Library of Congress Subject Headings List* can still save you time and reduce the risk that you will overlook important books.

Heading used in the card catalog——

Terms sometimes used for this subject——
that do *not* appear as headings in the
card catalog. Each appears in the
*Library of Congress Subject Headings
List,* but the entry tells users to look
under *Artificial intelligence* instead.

Card catalog headings for *broader*——
subjects within which artificial
intelligence fits

Card catalog headings for *related*——
subjects

Card catalog headings for *narrower* or——
more specialized subjects that are part
of the general topic of *artificial
intelligence*

Card catalog *subheadings* under the
general heading of *Artificial intelligence*

Artificial intelligence	
UF	AI (Artificial intelligence)
	Artificial thinking
	Electronic brains
	Intellectronics
	Intelligence, Artificial
	Intelligent machines
	Machine intelligence
	Thinking, Artificial
BT	Bionics
	Digital computer simulation
	Electronic data processing
	Logic machines
	Machine theory
	Self-organizing systems
	Simulation methods
RT	Fifth generation computers
	Neural computers
NT	Adaptive control systems
	Automatic hypothesis formation
	Automatic theorem proving
	Computer vision
	Distributed artificial intelligence
	Error-correcting codes (Information theory)
	Expert systems (Computer science)
	GPS (Computer program)
	Heuristic programming
	Machine learning
	Machine translating
	Natural language processing (Computer science)
	Perceptrons
	PURR-PUSS (Computer program)
	Question-answering systems
	VL1 system
	VL21 system
—**Computer programs**	
NT	Blackboard systems (Computer programs)
	SOLOMON (Computer program)
—**Data processing**	
NT	ROSIE (computer system)

FIGURE 6–1   Entry from the *Library of Congress Subject Headings List*

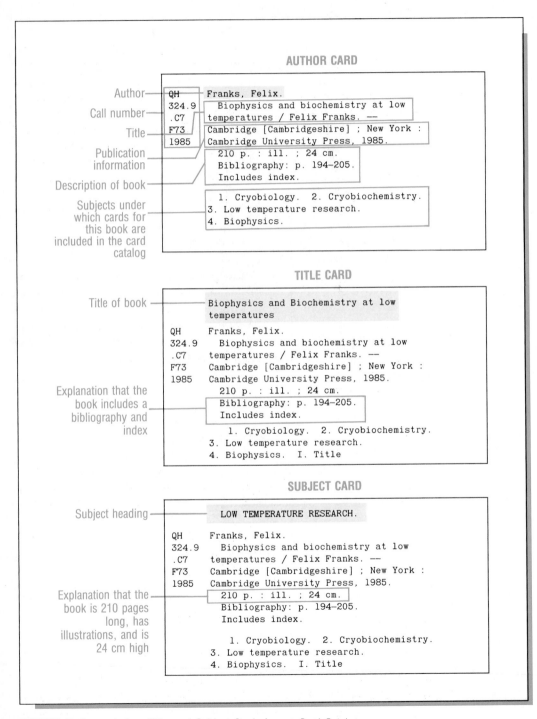

AUTHOR CARD

Author —
Call number —
Title —
Publication information —
Description of book —
Subjects under which cards for this book are included in the card catalog —

QH
324.9
.C7
F73
1985

Franks, Felix.
   Biophysics and biochemistry at low
temperatures / Felix Franks. —
Cambridge [Cambridgeshire] ; New York :
Cambridge University Press, 1985.
   210 p. : ill. ; 24 cm.
   Bibliography: p. 194–205.
   Includes index.

   1. Cryobiology. 2. Cryobiochemistry.
3. Low temperature research.
4. Biophysics.

TITLE CARD

Title of book —

Explanation that the book includes a bibliography and index —

Biophysics and Biochemistry at low
temperatures

QH   Franks, Felix.
324.9   Biophysics and biochemistry at low
.C7   temperatures / Felix Franks. —
F73   Cambridge [Cambridgeshire] ; New York :
1985   Cambridge University Press, 1985.
    210 p. : ill. ; 24 cm.
    Bibliography: p. 194–205.
    Includes index.

    1. Cryobiology. 2. Cryobiochemistry.
3. Low temperature research.
4. Biophysics. I. Title

SUBJECT CARD

Subject heading —

Explanation that the book is 210 pages long, has illustrations, and is 24 cm high —

LOW TEMPERATURE RESEARCH.

QH   Franks, Felix.
324.9   Biophysics and biochemistry at low
.C7   temperatures / Felix Franks. —
F73   Cambridge [Cambridgeshire] ; New York :
1985   Cambridge University Press, 1985.
    210 p. : ill. ; 24 cm.
    Bibliography: p. 194–205.
    Includes index.

    1. Cryobiology. 2. Cryobiochemistry.
3. Low temperature research.
4. Biophysics. I. Title

FIGURE 6–2   Author, Title, and Subject Cards from a Card Catalog

Using Bibliographies

A real shortcut to research would be to find a list prepared by somebody else that identifies books on your subject. A bibliography is just such a list. You can find bibliographies in many places. The books you locate through the card catalog may contain them. So, too, may the journal articles you find. Another good place to look for bibliographies is at the end of encyclopedia articles.

Moreover, bibliographies are sometimes published by themselves. Such bibliographies can be especially helpful because they often list more items than are included in the bibliographies at the ends of books and articles. To look for bibliographies on your subject, you can use the following three resources:

- *Bibliographic Index.* Issued three times a year, it lists bibliographies that have appeared in more than 2500 periodicals, as well as bibliographies that appear in newly published books and ones that have been published recently as separate books.
- Card catalog. The card catalog lists bibliographies under the subtopic "Bibliographies" or "Information Services" (for example, *Forestry—Bibliographies*).
- Sheehy's *Guide to Reference Works.* Sheehy's *Guide* lists major bibliographies in these broad areas: science, technology, and medicine; social and behavioral sciences; history and area studies; and humanities.

Following Leads

Once you have located some books on your subject, you can use them to lead you to others. First, note the author of each book you find, then look under that author's name in the card catalog to see if the person has written other books related to your subject. Also, if you are using a library that lets you get your books from the shelves yourself (some libraries don't), you can look up and down the shelves. Because books on a topic are generally shelved together, you may find additional sources.

ARTICLES IN MAGAZINES, JOURNALS, AND NEWSPAPERS

To locate articles in magazines, journals, and newspapers, you can use both periodical indexes and computerized data bases.

Using Periodical Indexes

Periodical indexes list articles that have appeared in selected magazines and journals on hundreds or thousands of topics. The *Readers' Guide to Periodical Literature* is a typical and familiar example. Every two weeks, a new issue appears that tells what has been published in more than one hundred and sev-

Animal Behavior Abstracts	Geographical Abstracts
Applied Science and Technology Index	Index to Legal Periodicals
Aquatic Sciences and Fisheries Abstracts	Index Medicus
	International Index to Film Periodicals
Art Index	Land Use Planning Abstracts
Astronomy and Astrophysics Abstracts	Metals Abstracts
Bibliography and Index of Geology	Microbiology Abstracts
Biological Abstracts	Mineralogical Abstracts
Biological and Agricultural Index	Music Index
Business Index	Nuclear Science Abstracts
Ceramic Abstracts	Nursing Studies Abstract
Chemical Abstracts	Nutrition Abstracts and Reviews
Computer Literature Index	Oceanic Abstracts
Consumers Index to Product Evaluations and Information Sources	Physical Education Index
	Physical Fitness/Sports Medicine
Criminology Index	Physics Abstracts
Current Technology Index	Plant Breeding Abstract
Ecology Abstracts	Psychological Abstracts
Education Index	Science Citation Index
Electrical and Electronics Abstracts	Social Sciences Index
Energy Research Abstracts	Sport Fishery Abstracts
Engineering Index	Urban Mass Transportation Abstracts
Environment Abstracts	Wildlife Review
Forestry Abstracts	World Textile Abstracts
General Science Index	Zoological Record

FIGURE 6–3   Some of the Many Specialized Indexes

enty popular magazines (such as *Time, Newsweek, Popular Mechanics,* and *Scientific American*). It also indexes selected scholarly journals. Every three months, the citations from these semimonthly issues are integrated into bound volumes, and at the end of the year the citations are integrated again.

Readers' Guide is just one among hundreds of periodical indexes. What distinguishes these indexes from one another is the group of journals upon which each reports. For example, the *Biological and Agricultural Index* covers more than two hundred journals such as *Advances in Agronomy, Journal of Dairy Science,* and *Weed Research,* while the *Applied Science and Technology Index* reports on more than three hundred journals that include *Adhesives Age,*

Monthly Weather Review, and *Plastics World.* To illustrate the great diversity of fields covered by indexes, Figure 6–3 lists some of the indexes available in many libraries.

Some indexes include the word *abstracts* in their titles. These indexes not only list articles but also provide an abstract (or summary) of each one so that you will know what the article covers—and what it does not. Figure 6–4 shows an index entry that includes an abstract. Sometimes abstracts are located right next to the listing, and sometimes in a separate volume. Wherever they are placed, abstracts can save you much time because they help you determine which articles are worth finding and which you should ignore.

Because there are so many specialized indexes, you will probably be able to find one or more that covers your specific field. To search for them, you can use these three sources:

● Katz's *Magazines for Librarians.* This not only lists 6,500 journals according to their field but also tells where each is indexed.

● Sheehy's *Guide to Reference Books.* This discusses the major indexes in many fields.

● Ulrich's *International Periodical Directory.* This provides the same type of information as Katz's but covers 65,000 periodicals.

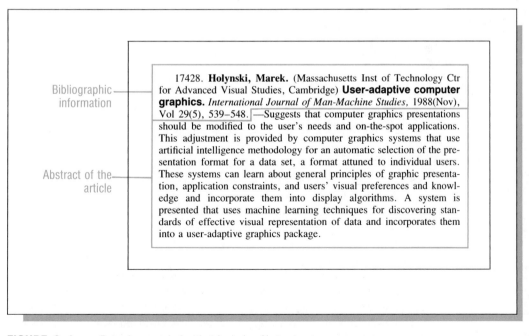

Bibliographic information

Abstract of the article

17428. **Holynski, Marek.** (Massachusetts Inst of Technology Ctr for Advanced Visual Studies, Cambridge) **User-adaptive computer graphics.** *International Journal of Man-Machine Studies,* 1988(Nov), Vol 29(5), 539–548. —Suggests that computer graphics presentations should be modified to the user's needs and on-the-spot applications. This adjustment is provided by computer graphics systems that use artificial intelligence methodology for an automatic selection of the presentation format for a data set, a format attuned to individual users. These systems can learn about general principles of graphic presentation, application constraints, and users' visual preferences and knowledge and incorporate them into display algorithms. A system is presented that uses machine learning techniques for discovering standards of effective visual representation of data and incorporates them into a user-adaptive graphics package.

FIGURE 6–4   Entry from an Index that Includes Abstracts

From *Psychological Abstracts,* 76 (1989): 1651. Reprinted with permission of the American Psychological Association.

If you think that you might be able to find useful material in newspapers, you can turn to the indexes for them. Among the many newspapers for which indexes are published are *The Atlanta Constitution, Chicago Tribune, Houston Post, London Times, Los Angeles Times, The New York Times, Pravda,* and *Wall Street Journal.*

Indexes arrange their listings in either of two ways. Some, such as the *Readers' Guide,* arrange their articles under subject headings, just as is done in the card catalog. Others, such as the *Science Citation Index,* rely on the words used in the articles' titles and (sometimes) their abstracts. For instance, an article entitled "Song Learning and Mate Choice in Robins" would be listed under *song, learning, mate, choice,* and *robins.* Indexes that list articles in this way are called *key word indexes.* Whether an index arranges its articles by subject or by key words, you must determine what words to use to locate articles on your topic. To help you convert your terms into the special vocabulary used in the indexes, many print at the front of each volume a list of the subject headings they use, cross-references, and even thesauruses that list synonyms for indexing terms. Other indexes require you to rely solely on your creativity and, perhaps, the aid of a reference librarian.

Once you find an entry for an article on your subject, you will need to decipher the abbreviations used in the index to save space. Figure 6–5 shows a sample section from the *Biological & Agricultural Index.* This form of entry is used in many other indexes—but certainly not all—because the publisher of this index (H. W. Wilson Company) also issues the *Readers' Guide* and many other useful indexes. For detailed instructions about how to use the indexes you find most valuable, look at the explanations that the publishers place at the beginning of each issue or volume.

Using Computer Searches

Many periodical indexes are now available on computer, which greatly simplifies looking through them. Consider for instance, the time you can save by using *Biosis Previews,* which contains computerized abstracts for more than four million articles that are also indexed in *Biological Abstracts* and *Biological Abstracts/RRM.* To look through printed copies of those indexes for articles on a subject could take many hours; as you found each entry you would have to copy it down, being sure that you have every detail of the citation correct. In contrast, by using *Biosis Previews,* you can have the computer search through the same abstracts in a matter of minutes; when the search is over, you can ask the computer to print out the abstracts for you. Figure 6–6 lists a few of the more than one thousand indexes now available on computer.

Some computerized indexes are available only by telephone. Your library probably has one or more computer terminals that are used to call up these indexes; the library may charge you a small fee for using them. Other computerized indexes are available by subscription. At regular intervals, the company that produces the index sends libraries compact disks (called CD-ROM disks) containing updated information that is loaded into computers at the libraries.

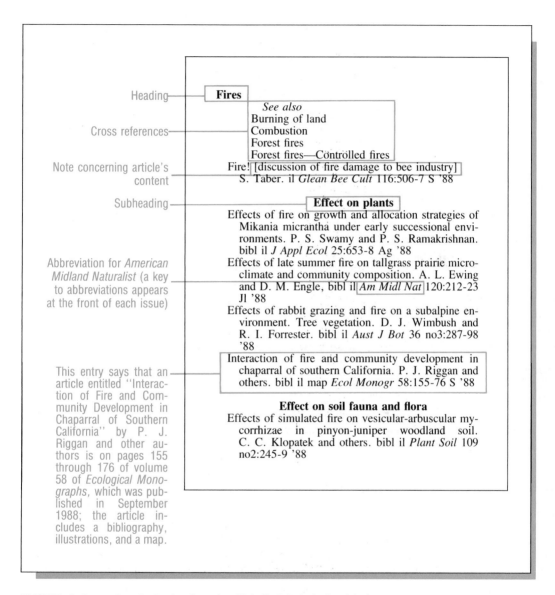

Heading

Cross references

Note concerning article's content

Subheading

Abbreviation for *American Midland Naturalist* (a key to abbreviations appears at the front of each issue)

This entry says that an article entitled "Interaction of Fire and Community Development in Chaparral of Southern California" by P. J. Riggan and other authors is on pages 155 through 176 of volume 58 of *Ecological Monographs*, which was published in September 1988; the article includes a bibliography, illustrations, and a map.

Fires

See also
Burning of land
Combustion
Forest fires
Forest fires—Controlled fires
Fire! [discussion of fire damage to bee industry] S. Taber. il *Glean Bee Cult* 116:506-7 S '88

Effect on plants
Effects of fire on growth and allocation strategies of Mikania micrantha under early successional environments. P. S. Swamy and P. S. Ramakrishnan. bibl il *J Appl Ecol* 25:653-8 Ag '88
Effects of late summer fire on tallgrass prairie microclimate and community composition. A. L. Ewing and D. M. Engle, bibl il *Am Midl Nat* 120:212-23 Jl '88
Effects of rabbit grazing and fire on a subalpine environment. Tree vegetation. D. J. Wimbush and R. I. Forrester. bibl il *Aust J Bot* 36 no3:287-98 '88
Interaction of fire and community development in chaparral of southern California. P. J. Riggan and others. bibl il map *Ecol Monogr* 58:155-76 S '88

Effect on soil fauna and flora
Effects of simulated fire on vesicular-arbuscular mycorrhizae in pinyon-juniper woodland soil. C. C. Klopatek and others. bibl il *Plant Soil* 109 no2:245-9 '88

FIGURE 6–5   Sample Section from the *Biological & Agricultural Index*
From *Biological & Agricultural Index*, 74:5 (January 1989): 263. Used with permission.

The *General Science Index* is an example of a bibliographic index available on CD-ROM at many libraries. You can probably use your library's CD-ROM indexes for free.

When you use computerized indexes, be sure to see what years they cover. Most contain only items published since the indexes were computerized, even though the printed indexes with which they are associated began much earlier.

Agricola

Arts and Humanities Citation Index

Biological & Agricultural Index

Business Periodicals Index

Computer Data Base

Environmental Bibliography

Food Service and Technology Index

International Pharmaceutical Abstracts

Magazine Index

Microcomputer Index

National Newspaper Index

Philosopher's Index

PsycINFO

Sociological Abstracts

Trade and Industry Index

FIGURE 6–6   Some of the Many Computerized Periodical Indexes

For example, *Chemical Abstracts,* which contains information not only for chemistry but also biochemistry and physics, was established in 1907. *Chemical Abstracts* was computerized in 1967, but only for things published that year or later. To find items from the first sixty years, you must turn to the printed volumes.

To instruct a computerized index to perform a search for you, you must type into the computer a word that describes your subject. Typically, the computer will then search through its records for titles or abstracts that contain the words you specify. In many indexes, you can give two or more terms, and the computer will give you only items in which those terms appear together. This greatly simplifies your work. For instance, if you were looking for articles on computers that play chess, you could enter both "computers" and "chess." If you were particularly interested in the design of the computer programs, you might also enter the term "design" or "program." With some indexes, you can also exclude articles that contain certain terms. For instance, if you wanted to exclude programs that use a certain brand of computer or a certain computer language, you could tell the computer that you do not want articles discussing that brand or language.

Typically, after you enter your request, the computer will tell you how many entries it has found using your key words. If the list is too small, you may need to ask for another search using a more general term for one or more of your key words. If it is too large, you may need to substitute more specific terms or add an additional term that you want to appear along with the others that you specified. Once you have refined your key words so that the index produces a list with an appropriate number of items, you can tell it to show you those entries on the computer screen or print them out for you.

REFERENCE WORKS

When you hear the term *reference works,* you probably think immediately (and quite correctly) of encyclopedias, dictionaries, and similar storehouses of knowledge designed to help you obtain specific facts and information quickly. What you may not realize is that there are thousands of reference works, many of which will probably be of great assistance to you in your career.

Take encyclopedias, for example. In addition to the familiar general encyclopedias, such as the *Americana, Britannica,* and *World Book,* there are hundreds of other, more specialized encyclopedias. They provide in-depth coverage of subjects that are too narrow to be treated in general encyclopedias, and they provide more detailed coverage of topics that general encyclopedias treat only briefly. Some examples follow:

> *Encyclopedia of Atmospheric Sciences and Astrogeology*
> *Encyclopedia of the Biological Sciences*
> *Encyclopedia of Psychoanalysis*
> *Encyclopedia of Science Fiction and Fantasy*
> *Encyclopedia of Social Work*
> *McGraw-Hill Encyclopedia of Science and Technology* (20 volumes)
> *Universal Encyclopedia of Mathematics*
> *World Encyclopedia of Food*

Whatever the subject of your research, there is probably a specialized encyclopedia that covers it.

Dictionaries, too, come in specialized versions. For example, there are the *Dictionary of Behavioral Science, Elsevier's Medical Dictionary, Encyclopedic Dictionary of Physics, Harper's Dictionary of Music, McGraw-Hill Dictionary of Science,* and *Petroleum Dictionary.*

Yearbooks are another helpful reference source. Some are annual supplements to encyclopedias, designed to bring the basic volumes up to date, primarily by describing the major events in the year that is covered. Among the more widely useful are the following:

> *Americana Annual*
> *Britannica Book of the Year*
> *McGraw-Hill Yearbook of Science and Technology*
> *World Book Yearbook*

In addition, there are also yearbooks in many specialized subjects, such as astronomy, drug therapy, education, nutritional medicine, dance, and archeology. Quite useful in many situations are statistical yearbooks. For example, the annually published *Statistical Abstract of the United States* presents numerous facts about many facets of American life, including education, business, transportation, health care, and the like.

In addition to encyclopedias, dictionaries, and yearbooks, there are many other helpful kinds of reference works. These include almanacs, directories of corporations, maps, movies, transcripts of network news broadcasts, and more. With such a wealth of reference sources available to you, how can you find the ones that will help you with a specific research project? First, consult Sheehy's *Guide to Reference Books*. Second, ask a reference librarian.

GOVERNMENT DOCUMENTS

Every year, the United States Government issues more than twenty-five million publications, ranging from pamphlets and brochures to periodicals, reports and books. Some are addressed to the general public, or special subgroups of it (such as children or homeowners), while others are addressed to specialists in various fields, such as disease control, space exploration, and finance. The following list illustrates the variety of topics covered by government publications:

Acid Rain

Building a Solar Home

Camper's First Aid

Chinese Herbal Medicine

Defining Death: A report on the Medical, Legal, and Ethical Issues in the Determination of Death

Diplomatic Hebrew

The Educational System in Switzerland

Laser Technology—Development and Applications

Poisonous Snakes of the World

A Report on the U.S. Semiconductor Industry

Severe Storms Research

Government publications that may be especially useful to you are reports on research projects undertaken by government agencies or supported by government grants and contracts. Each year there are more than seventy thousand such reports on topics ranging everywhere from nuclear physics to the sociology of Peruvian squatter settlements. The chances are great that some of these reports relate to your subject.

Many libraries house government documents separately from the rest of their collections, and many do not catalog government documents in the general card catalog. However, these documents are indexed by a wide variety of guides published by the government and private companies. Two general and very helpful sources are the following:

- *Government Reports Announcements and Index (GRAI).* This lists all the technical and research reports handled by the National Technical Information Service.
- *Monthly Catalog of U.S. Government Publications.* This lists all government documents not included in the Government Reports Announcements and Index.

To use these and the other indexes to government publications can require special skills. Consequently, the best way to begin searching for government documents is to ask for help from a reference librarian. Many libraries have one or more librarians who specialize in working with these publications.

Because the government publishes so many documents each year, no library (except for the Library of Congress) has them all. Therefore, once you have identified the government documents you want, you may have to ask your librarian to order them for you.

MICROFORM PUBLICATIONS

The library resources discussed so far in this chapter are all printed on paper. Most libraries also provide resources that are printed on microfilm or microfiche. Generally, libraries catalog these microform resources outside the general card catalog, so you should ask a reference librarian what your library has available. Three very useful microform resources are:

- *Envirofiche.* This indexes a vast array of scientific literature related to the environment, as well as a wide variety of publications dealing with the educational, legal, management, socioeconomic, and other aspects of environmental issues. (Similar microform resources exist for many other scientific and technical topics.)
- *ERIC.* This is a very large collection of conference papers and other documents pertaining to education. (ERIC stands for Educational Resources Information Center.)
- *Disclosure System.* This includes the annual reports that companies issue for their stockholders and the 10K reports that they submit to the Securities and Exchange Commission.

Like government documents, microform resources may be housed in their own part of the library, and the catalogs and indexes for them may be there also.

DEFINE YOUR RESEARCH OBJECTIVES

The first part of this chapter has provided an overview of many kinds of re-sources available to you in the library, along with a general sense of how to find the information you want in each of them. The rest of the chapter is de-voted to four guidelines that will help you plan and carry out your library research.

The first of these guidelines suggests that you begin your library research by defining your research objectives. In a library, there is so much information on most subjects, and that information is presented from so many points of view, that you can streamline your research considerably by deciding in ad-vance exactly what you are looking for. Your surest guide is the statement of overall objectives that you created for the communication in which you will use the results of your research. Specifically, remind yourself of who your readers are, what information they need, and how they will use the information you gather. Doing so will help you decide such things as whether you should look for general information or specific details, whether you can use only the most up-to-date sources or also older ones, and whether you need to find sources that consider your information from the point of view of the engineer or the accountant, the consumer or the producer.

SEARCH SYSTEMATICALLY

Many people conduct library research haphazardly. They go to the card cata-log, where they look under a subject heading or two, stop at the *Readers' Guide to Periodical Literature,* which they use in much the same way, and then head off to the stacks, hoping that a few of the items they found listed will be on the shelves. Such a strategy can waste time and lead you to overlook very useful resources, including (perhaps) the ones that include exactly the information you need.

You will be a much more efficient and effective researcher if you search systematically. Of course, there is no single procedure that works for all situ-ations. Nevertheless, the following pieces of advice are widely useful:

- **Survey and select from the research aids available to you.** Instead of going directly to the card catalog, *Readers' Guide,* or the ready-reference shelf, consult a guide to reference sources (such as Sheehy's *Guide to Reference Books*) or a reference librarian. Maybe there's a bibliography or periodical index that is especially suited to your needs.

- **Use a variety of sources.** Instead of using only books or only articles, combine the two, if appropriate. Consider which encyclopedias, govern-ment documents, or other publications might help you.

- **Develop a list of key words.** Your path to the information you seek is through the words you use to describe your subject as you consult a peri-odical index, conduct a computerized search, or hunt through the card catalog. Before going to these sources, develop a list of key words. If you need help in developing this list, you can read general articles on your subject or talk with people who are knowledgeable about it (these people may also be able to direct you to useful sources).

Once you have developed your list of key words, remember to check it against the terms used in the research aid you are employing (for example, the *Library of Congress Subject Headings List* if you are using the card catalog, or the thesaurus of terms adopted by the computerized periodical index you are employing).

● **Move from general to particular.** When you are trying to learn about a subject (not just obtain isolated facts about it), plan to look at general sources first, then proceed to more specialized ones. This procedure will give you the overall understanding necessary to comprehend and relate the more particular facts and ideas when you encounter them. To find general discussions, look in encyclopedias (including specialized ones), at articles in popular magazines, and at articles in specialized journals that summarize knowledge about your subject. You can usually identify the latter articles, which are often called *review articles,* by their titles or abstracts.

● **Assess each source before you begin to study it.** Some of the books, articles, and other sources you find will be more useful than others. Don't waste time by studying or taking notes on every source you locate. First, size up each source. If you find that one treats your subject at the wrong level or from the wrong perspective, put it back on the shelf and move on to something more useful.

● **Plan when to stop.** When researching, some people are tempted to keep going until they have located and read everything available to them. Others quit after they've looked at a few sources, even if they haven't yet discovered precisely what they need. Instead, you should match your research results against your objectives. If you haven't got what you need, keep going (if time permits). If you have found what you need, quit, even if time remains.

GUIDELINE 3   CONSULT REFERENCE LIBRARIANS EFFECTIVELY

Reference librarians can help you in two very important ways. First, they can assist you in finding bibliographies, indexes, and other aids that are most likely to help you. Second, they can teach you how to use those aids. Some sources are fairly difficult for novices, even after reading the instructions that accompany them. If you find that you are having trouble using a source to which a librarian has guided you, do not hesitate to ask for additional assistance. Similarly, if your search through the card catalog, a bibliography, or index turns up too few key words, tell a reference librarian what ones you have used. The librarian may be able to draw upon his or her experience with that particular source to help you think of other words that will produce better results.

When you approach reference librarians for assistance, the more specific you are about what you want, the more helpful they can be. Accordingly, you should give them all the information you generated when defining the objectives of your research: exactly what your topic and purpose are and who your readers will be. If you have devised a list of key words you plan to use in your search, you should show that to them also.

In situations where you need to begin by gaining a general understanding of a topic, you might plan to enlist the help of a reference librarian at two stages: in finding a source that will give you a general overview of your subject and in finding the more specific information that you subsequently identify as important to you.

<table>
<tr><td>GUIDELINE
4</td><td>**TAKE GOOD NOTES**</td></tr>
</table>

The product of your research will generally be a set of notes. The more care you take with these notes, the easier your job will be later on. Generally, you will take two types of notes: those that contain the information you obtain from your sources and those that record bibliographic information about your sources.

Informational Notes

There are many ways to record your informational notes. Some people find it convenient to use large note cards (usually 4″ x 6″), while others prefer to use looseleaf paper or bound notebooks. Of course, you can also take notes by photocopying the relevant pages from your source and then highlighting and making marginal notations on them.

In your informational notes, be sure to distinguish between material you are quoting directly and your paraphrases of things you find in your sources so that you can properly identify quotations in your communcation. Also, clearly distinguish ideas you get from your sources and your own ideas in response to what you find there. If you are using books that you can check out of the library, you may have no need to copy out extensive notes from them; your notes can simply tell the pages on which you will find various kinds of information when you begin to draft your communication.

Bibliographic Notes

When you are recording bibliographic details, be sure to take down all the facts you will need for the particular citation system you will use (see Appendix B, ''Formats for Reference Lists, Footnotes, and Bibliographies''). For books, this typically includes the author's or editor's full name, exact book title, city of publication, publisher, year of publication, edition, and page numbers. For a magazine article, it typically includes the author's full name, exact article title, full name of the journal, volume number, date of publication, and page number. Inexperienced researchers sometimes forget one or more of these facts, such as a book's city of publication or the number of the volume in which the article appears. As a result, they end up having to make an unnecessary trip to the library.

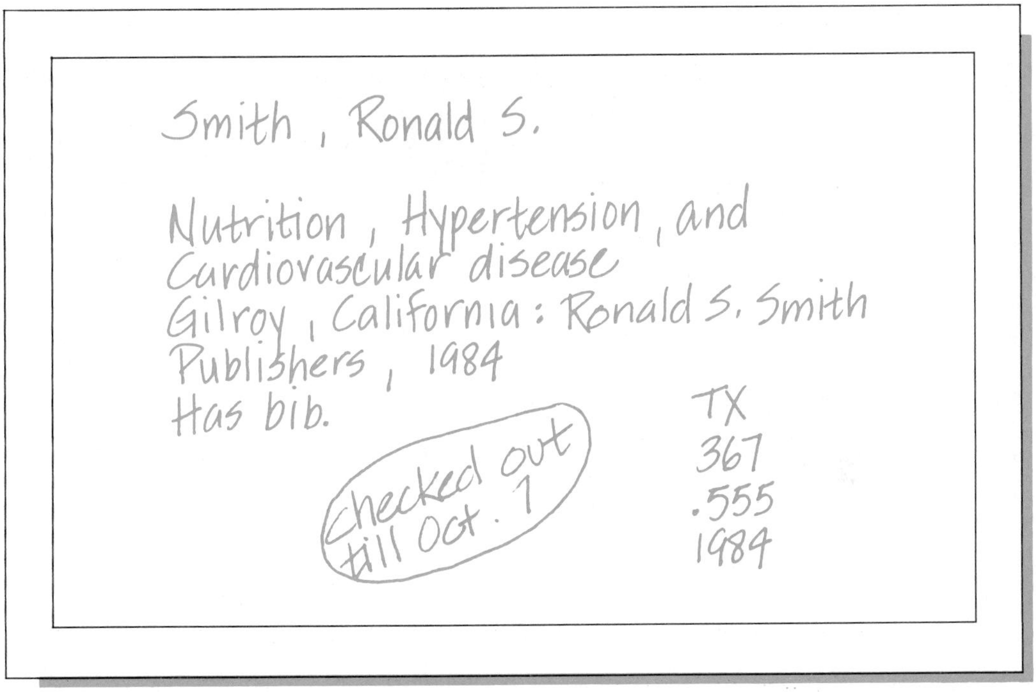

Smith, Ronald S.

Nutrition, Hypertension, and
Cardiovascular disease
Gilroy, California: Ronald S. Smith
Publishers, 1984
Has bib.

Checked out
till Oct. 1

TX
367
.555
1984

FIGURE 6–7   A Typical Bibliographic Note card

Using 3″ x 5″ note cards can be a convenient way to record bibliographic information. You can fill out one card for each source as you use a bibliography, periodical index, or similar research aid. Once you go to the card catalog or other list of library holdings, you can enter the call number or other location information on the card, then sort through the cards so you can easily look for all the sources that are in one place. You can also use the cards to keep records of your efforts to locate items (for example, "Feb. 10, asked the librarian to recall this book from the person who checked it out"). Figure 6–7 shows a sample bibliographic note card.

As you proceed with your search, be sure to record sources you checked but found useless. Otherwise, you may later find another reference to the same sources but be unable to remember whether you have already looked at them.

CONCLUSION

In addition to books and periodicals, the library has many other kinds of resources that can be very helpful to you. It also has many tools you can use to locate the information you are seeking. To use these resources and tools effectively, you should begin by defining your research objectives, then search systematically, ask reference librarians for help when you need it, and take careful, thorough notes.[1]

IV
DRAFTING

7

DRAFTING PARAGRAPHS, SECTIONS, AND CHAPTERS

DEFINING
OBJECTIVES

PLANNING

DRAFTING

EVALUATING

REVISING

T his chapter marks a major transition in your study of on-the-job writing. In the preceding four chapters, you have learned how to build a solid foundation for successful writing by defining your communication objectives and then creating plans for achieving them. When you have completed these activities, you will have created written or mental notes that tell what you want your communication to do and describe your strategies for creating a message that will do just that. You are now going to begin studying advice you can use on the job as you transform your plans into action by drafting the communication you will deliver to your readers.

At work, drafting can be a very major undertaking. Of course, sometimes you will draft very rapidly. You will simply sit down, write your draft, and send it off without making any changes at all. However, this will usually happen only when you are writing very short communications on routine matters. People at work generally devote considerable effort to writing longer, less routine messages. For example, in response to a survey, 122 professional and managerial personnel said that when they prepare reports, they spend an average of three hours on *each page*.[1]

Why does writing at work require so much effort? First, the standards for writing quality are high enough that even experienced writers cannot always meet them in a single draft. Also, the messages people write on the job are often complex enough that the writers need several drafts to figure out how to convey their messages clearly and persuasively.

This chapter and the next eight will help you draft clear, persuasive messages—and do so as efficiently as possible. The chapters provide practical suggestions about drafting paragraphs, sections, and chapters (Chapters 7 and 8); creating the beginning and end of a communication (Chapters 9 and 10); composing sentences (Chapter 11); and choosing words (Chapter 12). They also provide helpful advice about using visual aids (Chapters 13 and 14) and designing the layout of your pages (Chapter 15).

APPLICATION OF THIS CHAPTER'S ADVICE

This chapter focuses on helping you draft paragraphs and the groups of related paragraphs that make up the sections and chapters of longer communications. You can also apply its advice to the small groups of related sentences found within many longer paragraphs. You can even apply the advice to the very largest group of paragraphs in a communication—the whole communication itself.

For convenience's sake, this chapter uses the single word *segments* to designate all of these various prose units to which its advice applies.

How can this chapter's advice apply with equal validity to segments that range in size from a few sentences to an entire communication? First, all segments have the same basic structure.

You may have heard a paragraph defined as a group of sentences about the same thing. With only slight variation, that definition applies equally well to segments of other sizes: a subsection is a group of paragraphs on the same subject, a section is a group of subsections on the same topic, and so on.

Second, all segments make the same basic demand upon readers: to understand one, readers must determine what the segment's topic is and discern how the various parts (sentences, paragraphs, etc.) of the segment fit together into a coherent discussion of that topic. Also, in many segments, readers need to know which of the various points discussed is most important—which of the problems identified is most serious, which of the actions is most urgent. Because all segments have the same basic structure and present the same challenges to readers and writers, the same strategies can help you draft them all effectively.

Of course, you already know many of these strategies and can use them almost automatically. Even without thinking specifically about how you are doing it, you can often put sentences together into a successful paragraph and put many paragraphs together into a successful longer passage. At work, however, you may sometimes encounter difficulty writing some segments. That's especially likely to happen when you are writing long, formal communications or ones addressed to unfamiliar readers. The advice provided in this chapter can help you in situations where your natural ability to write segments isn't enough to see you through.

Because the suggestions in this chapter are especially useful with longer communications that might have a hierarchical structure (see Chapter 4), you may find it helpful to look at Figure 7–1, which shows how one writer translated a hierarchical organizational plan into the chapters, sections, and subsections of a finished report. This report was written by Jonathan, an employee of an engineering consulting firm that advises amusement parks, museums, and similar businesses. In the report, Jonathan recommends a series of construction projects that will help a zoo that is losing money to attract enough visitors to regain financial stability. Figure 7–2 presents a short passage from Jonathan's report so you can see the final result of Jonathan's efforts to convert his plan into prose.

This chapter's suggestions for writing segments are presented in four guidelines. The first three provide advice about arranging the contents of your segments, and the fourth suggests ways you can make the organization evident to your readers.

GUIDELINE 1   BEGIN BY ANNOUNCING YOUR TOPIC

You have undoubtedly heard that you should begin your paragraphs with topic sentences. This guideline extends that advice to all your segments, large and small: begin segments of every size by telling your readers what they are about.

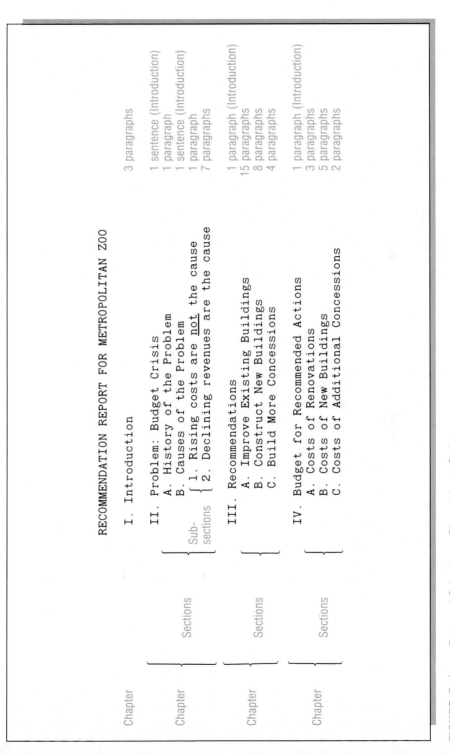

RECOMMENDATION REPORT FOR METROPOLITAN ZOO

I. Introduction   3 paragraphs

II. Problem: Budget Crisis   1 sentence (Introduction)
    A. History of the Problem   1 paragraph
    B. Causes of the Problem   1 sentence (Introduction)
        1. Rising costs are not the cause   1 paragraph
        2. Declining revenues are the cause   7 paragraphs

III. Recommendations   1 paragraph (Introduction)
    A. Improve Existing Buildings   15 paragraphs
    B. Construct New Buildings   8 paragraphs
    C. Build More Concessions   4 paragraphs

IV. Budget for Recommended Actions   1 paragraph (Introduction)
    A. Costs of Renovations   3 paragraphs
    B. Costs of New Buildings   5 paragraphs
    C. Costs of Additional Concessions   2 paragraphs

Chapter

Chapter   Sub-
           sections
           Sections

Chapter   Sections

Chapter   Sections

FIGURE 7–1   Parts and Subparts as Planned in an Outline

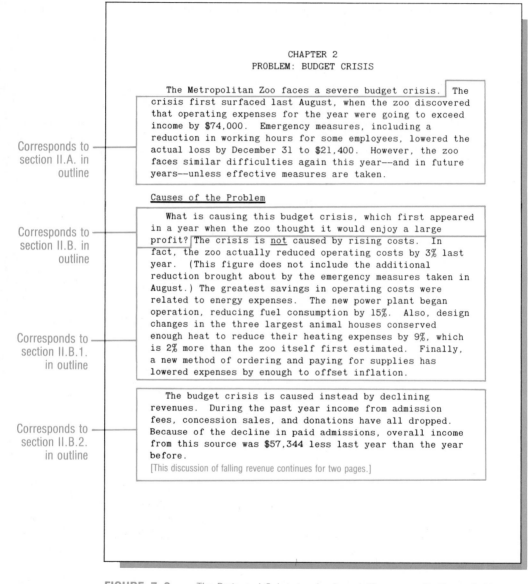

CHAPTER 2
PROBLEM: BUDGET CRISIS

The Metropolitan Zoo faces a severe budget crisis. The crisis first surfaced last August, when the zoo discovered that operating expenses for the year were going to exceed income by $74,000. Emergency measures, including a reduction in working hours for some employees, lowered the actual loss by December 31 to $21,400. However, the zoo faces similar difficulties again this year—and in future years—unless effective measures are taken.

Corresponds to section II.A. in outline

Causes of the Problem

What is causing this budget crisis, which first appeared in a year when the zoo thought it would enjoy a large profit? The crisis is _not_ caused by rising costs. In fact, the zoo actually reduced operating costs by 3% last year. (This figure does not include the additional reduction brought about by the emergency measures taken in August.) The greatest savings in operating costs were related to energy expenses. The new power plant began operation, reducing fuel consumption by 15%. Also, design changes in the three largest animal houses conserved enough heat to reduce their heating expenses by 9%, which is 2% more than the zoo itself first estimated. Finally, a new method of ordering and paying for supplies has lowered expenses by enough to offset inflation.

Corresponds to section II.B. in outline

Corresponds to section II.B.1. in outline

The budget crisis is caused instead by declining revenues. During the past year income from admission fees, concession sales, and donations have all dropped. Because of the decline in paid admissions, overall income from this source was $57,344 less last year than the year before.
[This discussion of falling revenue continues for two pages.]

Corresponds to section II.B.2. in outline

FIGURE 7–2 The Parts and Subparts of a Report (Compare with Figure 7–1)

How Topic Statements Help Readers

Why are topic statements so important? In order to understand a segment, readers must establish in their own minds a meaningful pattern to the various pieces of information conveyed within it. Unless the writer helps, this task of meaning construction can be very difficult. Consider, for example, how hard it is to understand the following sentences from a technical report written by an engineer at a coal-fired electric plant.

> Companies that make cement and wallboard cannot use wet gypsum cakes. The cakes must be transformed into dry pellets, using a process called agglomeration. We could enter the agglomeration business ourselves or hire another company to agglomerate the gypsum for us. Also, the chloride content of the cakes is too high for use in wallboard. Our engineers can probably devise inexpensive cake-washing equipment that will reduce the chlorides to an acceptable level.

Although you could probably figure out what the paragraph is about if you studied it long enough, your job as a reader would be much easier if the writer had added a statement telling you what it is about. Try reading the paragraph again, this time with a topic statement.

Topic statement
> Before we can sell the gypsum produced by our stack scrubbers, we will have to process the wet gypsum cakes they produce.
> Companies that make cement and wallboard cannot use wet gypsum cakes. The cakes must be transformed into dry pellets, using a process called agglomeration. We could enter the agglomeration business ourselves or hire another company to agglomerate the gypsum for us. Also, the chloride content of the cakes is too high for use in wallboard. Our engineers can probably device inexpensive cake-washing equipment that will reduce the chlorides to an acceptable level.

Why Topic Statements Help Most at the Beginning

While topic statements aid readers no matter where the statements are placed in the segment, they are especially helpful when placed at the beginning. To understand why, you may find it useful to know something about the two thinking strategies people use to create meaningful mental structures when they are reading. Researchers call these strategies bottom-up and top-down processing.

In *bottom-up* processing, readers proceed in much the same way as people who are working a jigsaw puzzle without having seen a picture that tells them what the finished puzzle will look like. They try to understand the individual sentences and then guess how the small pieces of information fit together to form the general meaning of the overall paragraph, section, or chapter.

In *top-down* processing, readers work in the opposite direction. Like people who have seen a picture of the finished jigsaw puzzle, they begin with a sense of the segment's overall structure. Then, as they read each detail in the segment, they place it immediately with that mental structure.

Although readers engage continuously in both kinds of processes, the more top-down processing they can perform while reading, the more easily they can understand and remember the message. To demonstrate that to yourself, consider the results of an experiment in which researchers asked two groups of people to listen to the following passage.[2]

> The procedure is actually quite simple. First you arrange things
> into different groups. Of course, one pile may be sufficient
> depending on how much there is to do. If you have to go somewhere
> else due to lack of facilities, that is the next step; otherwise you are
> pretty well set. It is important not to overdo things. That is, it is
> better to do too few things at once than too many. In the short run
> this may not seem important but complications can easily arise. A
> mistake can be expensive as well. At first the whole procedure will
> seem complicated. Soon, however, it will become just another facet
> of life. . . . After the procedure is completed, one arranges the
> materials into different groups again. Then they can be put into their
> appropriate places. Eventually they will be used once more and the
> whole cycle will then have to be repeated. However, that is part
> of life.

The researchers told one group the topic of this passage before they heard it; they told the other group afterwards. To see how well each group understood the passage, the researchers asked the people to write down everything they remembered from it. People who were told the topic (washing clothes) before hearing the passage remembered much more than did those who were told afterwards.

In addition to helping your readers understand and remember your message, topic statements placed at the beginning of your segments also help readers who are skimming through your communication for particular facts or who are scanning to determine your main points as quickly as possible.

Because topic statements are such powerful aids to efficient communication, you should generally use them at the beginnings of all your segments, small and large. When you do this, your communications will have hierarchies of topic statements. Figure 7–3 shows how Jonathan provided topic statements for three levels of segments in a passage from his report to the zoo. Notice how the topic statements interlock to form a hierarchy that corresponds to the hierarchy he planned in Section II of his outline (Figure 7–1).

How to Indicate Your Topic

There are many ways to indicate the topic of a segment. For example, you can devote an entire sentence to announcing the topic. The first sentence of the paragraph you are now reading is such a sentence. Sometimes you can indicate the topic through a single word. For example, the first word ("First") of the second sentence in the passage about washing clothes tells the reader, "You

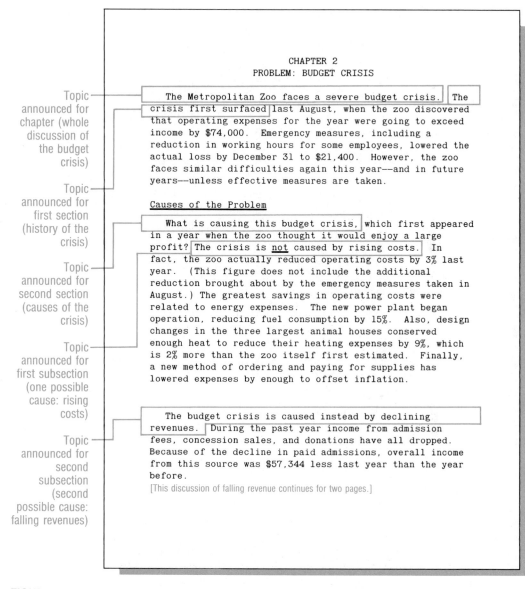

FIGURE 7–3   Topic Sentences Used in Three Levels of the Report Outlined in Figure 7–1

are now going to read a segment that explains the steps in the 'simple procedure' just mentioned.'' You can even indicate the topic of a segment by asking a question. For example, the question (''Why are topic statements so important?'') that begins the second paragraph in the discussion of this guideline told you that you were about to read a segment explaining the importance of topic statements.

Whether you use a sentence, a word, a question, or some other device, remember to help your readers understand your segments by telling them at the beginning what the segments are about.

GUIDELINE
2

PRESENT YOUR GENERALIZATIONS BEFORE YOUR DETAILS

In addition to announcing the topic at the beginning of a segment, it is often helpful to state the general point you want to make about the topic. For instance, instead of saying only, "The topic of this section is the relative costs of shipping our company's products by truck and by train," say "We can save 15 percent of transportation costs by shipping certain products by train rather than truck." Of course, you could provide your generalizations at the ends of your segments rather than at beginnings. When presented at the beginnings, however, your generalizations make your writing more understandable and useful to your readers, and they make your writing more persuasive.

How Initial Generalizations Make Your Writing More Understandable and Useful

Initial generalizations make your writing more understandable and useful by saving your readers the work of figuring out what your main point is. As they read, they can then concentrate on other tasks. Imagine, for instance, that you are a manager who finds the following sentences in a report:

> Using the sampling technique described above, we passed a gas sample containing 500 micrograms of VCM through the tube in a test chamber set at 25°C. Afterwards, we divided the charcoal in the sampling tube into two equal parts. The front half of the tube contained approximately ⅔ of the charcoal while the back half contained the rest. Analysis of the back half of the tube revealed no VCM; the front half contained the entire 500 micrograms.

As you read these facts, you might find yourself repeatedly asking the question, "So what?" Without any good answer to that question and without any general principle around which to organize the facts contained in the passage, you might not remember much from it. However, you would read the facts in a much different way if the writer had begun the segment with the following statement:

> We have conducted a test that demonstrates the ability of our sampling tube to absorb the necessary amount of VCM under the conditions specified.

Knowing from the outset what conclusion the writer draws allows you, as a reader, to focus in a different, more productive way on the details that follow: you can use them to evaluate whether or not the conclusion is valid. Of course, you may ultimately reject the writer's conclusion, but at least the writer has helped you read more efficiently by placing his or her general statement at the beginning of the passage.

How Initial Generalizations Make Your Writing More Persuasive

Furthermore, by providing your generalizations at the beginnings of your segments you actually increase your chances of persuading your readers to accept the conclusions, adopt the attitudes, and take the actions you advocate. This is because people can derive a wide variety of generalizations from any passage, and you will want to influence which of those generalizations they draw. Consider the following sentences:

Richard moved the gas chromatograph to the adjacent lab.

He also moved the electronic balance.

And he moved the XT computers.

As uninvolved observers, we might note that everything Richard moved is a piece of office equipment, and then we might generalize that "Richard moved some *office equipment* from the third floor to the fifth." Or, we might observe that everything he moved is heavy, so we might generalize that "Richard is strong." Similarly, someone more closely related to Richard's company could draw a variety of generalizations from those sentences. For example, a member of the labor union in Richard's organization might generalize that "Richard got plenty of exercise," or that "Richard persisted in performing jobs that should have been performed by union employees, not by a manager." If she made the first generalization, the union member might forget the incident; if she made the second generalization, she might file a grievance. Different generalizations will lead to different outcomes.

Because some generalizations will affect your readers in the way you desire and others will not, you should do what you can to influence the generalizations that your audience draws from your passages. Otherwise, they might not draw the ones you want them to draw. The first step in influencing your readers' generalizations, of course, is to suggest generalizations yourself rather than leaving your readers to draw their own conclusions. However, to enjoy the full benefits of such generalizing statements, you must usually place them at the beginnings of your passages. If you wait until later in a passage, your readers may already have formed their own generalizations before they encounter yours. Once your readers have formed their own generalizations, they will be less open to accepting yours.

When *Not* to Present Your Generalizations First

One caution is in order. Although it is usually most effective to present your generalizations first, there are situations when it's better to withhold them.

You've already read about them in Chapter 5's discussion of the direct and indirect organizational patterns. To review briefly, you would probably not want to state your generalization first if you expect it to provoke strongly negative reactions, as when your generalization is a conclusion that your readers will view as unwelcome or critical of them. Review Chapter 5 for advice about what to do under such circumstances.

How Guideline 2 Relates to Guideline 1

Finally, note the close relationship between this guideline's advice to begin your segments with your generalizations and the preceding guideline's advice to begin with your topic sentence: your generalization will be about your topic. Consequently, you can often follow both guidelines in a single sentence. Here, for instance, is the first sentence of a two-page discussion of a series of experimental results:

> Our tests showed three shortcomings in plastic resins that make it undesirable for us to use them to replace metal in the manufacture of the CV-200 housing.

After reading this sentence, people would know that the segment it introduces is about the shortcomings of plastic resins (topic) and that the writers believe that the resins will not make good substitutes for metal (the generalization the writers draw from the facts that follow). Of course, you won't always be able to present both your topic and your generalization in a single sentence this way. The key point for you to remember is that your readers will find both elements helpful at the beginning of each of your segments.

GUIDELINE
3

MOVE FROM MOST IMPORTANT TO LEAST IMPORTANT

Guidelines 1 and 2 concerned the way you begin your segments. This guideline focuses instead on the way you order the material that follows your opening sentence or sentences. It applies particularly to passages in which you present parallel pieces of information, such as a list of five recommendations you have developed, or an explanation of three causes of a problem that your employer faces, or a discussion of four reasons for using a part design you have created. Whether you present each of the items in a single sentence or in several paragraphs, you should usually put the most important item first, then proceed in descending order of importance.

Putting the most important information first has several advantages. First, by beginning with the most important information, you emphasize it. Second, you help readers (such as decision-makers) who are scanning your communication to find the key points. Finally, by placing your most important information first, you increase the likelihood that it will be read. At work, people frequently suffer interruptions when reading. Sometimes they never resume reading, so what follows the place where they were interrupted never gets seen. Also, if you delay presentation of your major points, you increase the chances that your audience will simply quit paying attention before they reach those points.

To decide what information is most important, consider your segment from your readers' point of view. What information will they want most or find most persuasive? For example, in the segment in which the writers will describe three shortcomings of plastic resins, the readers will certainly be more interested in the major shortcoming than in the more minor ones. Similarly, if they have to be persuaded to accept the writers' generalization that the plastic resins aren't a good substitute for metal, they are sure to find the major shortcoming to be more compelling than the others.

In some situations, of course, you should *not* put the most important information first. For example, sometimes you will have to withhold it in order to present your overall message clearly and economically. This might be necessary, for instance, if you are enumerating the causes of something and the historical relationship of the causes is significant. Perhaps the second cause is the more important; however, because it must be understood as an outgrowth of an earlier, less significant development, you must present the lesser cause first. In general, though, you will be the most persuasive and helpful to your readers if you organize the material within your paragraphs and groups of paragraphs so that you present the most important information first.

GUIDELINE 4	**REVEAL YOUR ORGANIZATION**

Guidelines 1, 2, and 3 provided you with advice about how to organize your segments. This one suggests ways you "reveal" that organization by drawing a "map" of it for your readers.

"But," you may ask, "if I've organized my communication carefully, why will my readers need a map of it? Won't a good, sensible organization be obvious to them?" Unfortunately, it may not be obvious. Communications contain two distinct kinds of information:

- Information about the subject matter
- Information about how the writer has organized the discussion of that subject matter.

The relationship between these two kinds of information is illustrated in Figure 7–4.

When you read something you wrote, you see clearly both kinds of information, even if you included only subject-matter information and no organizational information. That's because you created the organization and you remember what you planned.

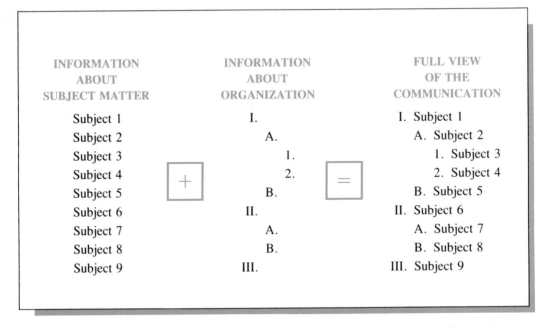

FIGURE 7–4 Relationship of Information about Subject Matter and Information about Organization

Your readers, however, must construct a mental map of your organization as they read. If you neglect to supplement your information about your subject matter with information about your organization, your readers' map may look very different from yours. As Figure 7–5 implies, if they know the contents but not the hierarchical organization of what they have already read, they will probably be unable to foresee either the content or the organization of the information that lies ahead.

You can demonstrate to yourself how difficult it is for people to puzzle out the organization of a communication when the writer doesn't provide a map of it: try to revise the table of contents shown in Figure 7–6 by converting it from a mere list into an outline. You will find that you must work hard to make the conversion. (The topics are listed in the correct order; your job is simply to reveal the hierarchical relationships among them.)

In contrast, when a writer provides a good map of his or her message, readers can discern the organization without having any knowledge at all of the contents. Look, for example, at Figure 7–7, which shows a page of a Chinese textbook on geology. Even if you cannot read the language, the writers have provided enough organizational information for you to grasp the structure of the presentation.

The following sections tell how you can reveal your communication's organization by using forecasting statements, transitions, headings, and the visual layout of your page.

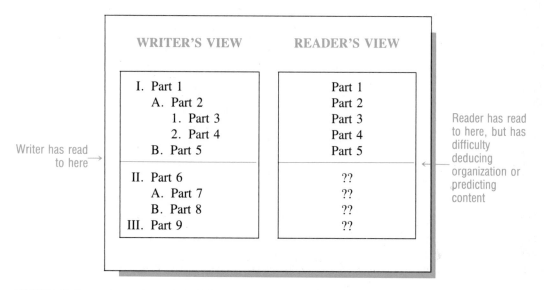

Writer has read to here →

Reader has read to here, but has difficulty deducing organization or predicting content

FIGURE 7–5   Comparison of the Writer's View and the Readers' View of a Communication if the Writer Neglects to Include Organizational Information

	Pegasus Automatic Balance	
1.	Introduction	1
2.	Right-Hand Application Knob	2
3.	Left-Hand Application Knob	2
4.	Micrometer Knob	3
5.	Zero-Point Adjustment Knob	4
6.	Right-Hand and Left-Hand Weight Dials	5
7.	Projection Weight Dial	5
8.	General Features of the Weighing Components	6
9.	Special Features of the Weighing Pan	6
10.	Special Features of the Weighing Stirrup	7
11.	Adjusting the Sensitivity	8
12.	Adjusting the Zero Point	9
13.	Determining Weight	9
14.	Weighing to a Preselected Weight	10

FIGURE 7–6   Table of Contents from an Instruction Manual Whose Writer Has Failed to Provide Organizational Information

第十章 鲸鱼铬铁矿床

一 岩体地质概况

鲸鱼岩体位于新疆西部华力西褶皱带的一个复向斜轴部，与萨尔岩体毗邻。

岩体的围岩为上泥盆统石英砂岩、粉砂岩及中基性火山岩。岩体与围岩为侵入接触，外接触带具弱石墨化，内接触带蚀变为硅化碳酸盐化超镁铁岩。接触面产状同围岩产状斜交。

全部岩体被厚约 20 米的第四系所覆盖。经地面磁测及钻探圈定，岩体呈北东 60° 方向延伸，与区域构造线走向一致。岩体长约 2.4 公里，最宽处 840 米，面积 1 平方公里余。南侧与北侧接触面相向倾斜，成一漏斗状（图 77），北侧较南侧缓。岩体最大延深部位据物探推断在岩体东南部，深度大于 565 米，可能为岩浆通道。

二、岩相带划分

组成岩体的岩石主要是斜辉辉橄岩（占全岩体的 85%）和斜辉橄榄岩（占 13%）及少量纯橄岩（约占 2%）。

岩相中基性程度略高的岩相——斜辉辉橄岩（斜方辉石和绢石含量为 5—15%）杂有少量纯橄岩、斜辉橄榄岩异离体组合分布于岩体下部和边缘及中部的狭长地带；酸性程度略高的岩相——斜辉橄榄岩（绢石含量 15—25%）和斜辉橄榄岩组合分布于岩体的上部和中部。物探表明，边缘岩相为低磁带，上部及中部岩相为高磁带，矿群主要分布于具低磁高重力特点的中部狭长地带。

三、岩石学特征

1. 纯橄岩

以黑绿色为主。镜下观察已全部蛇纹石化，具网格结构，无橄榄石残晶。副矿物铬尖晶石含量一般可达 1—3% 或更多，半自形—他形，粒径 0.1—0.5 毫米，多蚀变为不透明。偶含少量绢石，成为含绢纯橄岩。

2. 斜辉辉橄岩和斜辉橄榄岩

呈黄绿色。镜下观察，岩石亦遭强烈的蛇纹石化，斜方辉石变为绢石，呈他形，偶见残晶。 $2V = (+)77°—80°$，$N_g = 1.675±0.02$，$N_p = 1.664±0.02$，Fs = 7—8，为顽火辉石。橄榄石利蛇纹石化，具网格结构，可见少量残晶，$2V = (+)85°—88°$，$N_m = 1.662—1.665$，Fa = 7—8，属镁橄榄石。偶见少量单斜辉石，$2V = (+)58°—59°$，$c \wedge N_g = 35°—44°$，为透辉石。

· 167 ·

FIGURE 7–7 Page from a Chinese Book Whose Writers Have Provided a Good Map for the Reader
From Heng-sheng Wang, Wen-ji Bai, Bix-xi Wang, and Yao-chu Chai, *China's Chromium Iron Deposit and the Cause of Formation* (Beijing: The Science Press, 1983), 167.

Forecasting Statements

The first place to provide your readers with organizational information is in your prose. You have already read about one way of doing this: when you provide topic sentences, you notify your readers of the topic of the paragraph or section immediately following. This announcement can help readers immensely. However, it does not provide them with another kind of information that can also help them understand your message—information about how you have *organized* your discussion of the topic.

Consider the following sentence, which opens one section of a brochure published by a large chain of garden nurseries:

> Our first topic is the trees found in the American Southwest.

The sentence certainly adheres to the advice given earlier in this chapter to begin each passage with a topic statement. However, a section beginning with this sentence could be organized in a wide variety of ways. It might first take up evergreen trees and then deciduous ones. Or it might first discuss healthy trees and then diseased ones. Or it might be arranged in any of various other patterns, and the readers would have no way of knowing which pattern they were about to encounter.

A very effective way to tell your audience in advance about the organization of a passage is to begin the passage with a *forecasting statement* that previews what lies ahead.

You can forecast the organization of a passage in a full sentence that supplements your topic statement:

Forecasting statement
> Our first topic is the trees found in the American Southwest.
> Some of the trees are native, some imported.

Also, you can both announce the topic and forecast the organization in a single sentence:

Forecasting statement
> Our first topic is the trees—both native and imported—found in the American Southwest.

Your forecasting statements may vary greatly in the level of detail they provide. The sample sentences you have just read provide specific detail since the readers know both the number and the names of the categories of trees to be discussed. A much more general preview is this one:

> To solve this problem, the department must take the following actions.

This statement tells its readers to expect a list of actions but doesn't tell what they are or even how many there are.

When deciding how much detail to include in a forecasting statement, consider the following pieces of advice:

- **Provide enough detail so that your audience knows something specific about the arrangement of the section that follows.** Usually, the more complex the relationship among the parts, the greater the amount of detail needed.

- **Do not provide more detail than your readers can easily remember.** The purpose of a forecasting statement is to help your audience understand what is to come, not to test anyone's memory. If you are introducing the three steps of a solution, you might want to name the three steps before explaining them. However, if the solution contains eight steps, you will probably be better off stating the number of steps without naming them in the forecasting statement.

- **Do not forecast more than one level at a time.** People sometimes are tempted to explain all the contents of a document at its beginning. But this burdens their readers with a confusing amount of detail. To work effectively, a forecasting statement tells only the major divisions of a particular section. If those divisions are themselves divided, then each of the divisions will have its own forecasting statement.

Like topic statements, forecasting statements give readers an initial sense of a passage's meaning and organization. When you provide this help, your readers can more easily read and remember your message.

Transitions

As readers proceed from one part of a communication to another—from sentence to sentence, paragraph to paragraph, and section to section—they must figure out how these parts relate to one another. In Chapter 11 (Writing Sentences), you will find considerable advice about leading your readers from one sentence to the next. The discussion that follows focuses on transitions between larger units of prose—from one paragraph or group of paragraphs to the next.

The importance of transitions between parts of a communication cannot be emphasized too heavily. As people move from one part to another, they can become confused about the relationships among the parts. If this happens, they will have difficulty figuring out how the part they are just beginning to read or hear fits into the overall communication. You can help your readers avoid such confusion by providing *transitional statements* between your segments.

Transitional statements indicate two things: what the upcoming part is about and how that part relates to the one that just ended. Transitional statements often serve as topic sentences, because, like topic sentences, they tell what is coming. To make true transitions, however, they must also indicate the relationship between what is coming and what has just concluded.

How to Write a Transition   You can write transitions in a variety of ways. For example, you can say almost literally, ''We are now making a transition from one topic to the next.'' You might even take two full sentences to do that.

> In our guidelines for handling Exban, we now have completed our suggestions for guarding against accidental spills during transportation. We turn now to our guidelines for storing the product.

In more subtle transitions, you can devote only a word or phrase rather than an entire sentence to naming the discussion that has just been completed.

> After we developed our hypothesis, we were ready to design our experiment.

> Having described the proposed accounting procedure, we will now discuss its advantages over the present one.

In these examples, the introductory phrase reminds readers of what was just described in the previous section (the development of the hypothesis and the proposed accounting procedure, respectively). The rest of the sentence tells what will be discussed in the next passage (the design of the experiment and the advantages of the proposed procedure over the present one).

Some transitions achieve even greater subtlety by including the reference to the section just completed in the main clause itself (not in an introductory phrase):

> We then designed an experiment to test our hypothesis.

> The proposed accounting procedure has several advantages over the present one.

These sample transitions would all appear at the opening of the passage that is just beginning. However, though rare, some transitions come at the end of the passage that is just ending. Consider the last sentence of a section in a report about the sales performance of Big Q soft drinks during the past year:

> Against this background information concerning the factors affecting the soft-drink market overall, the performance of the Big Q line can be interpreted more precisely.

The section that this sentence concludes deals with general background information about the entire soft-drink market, and the section coming up will discuss the performance of the Big Q brand in light of that background. Of course, even though that section does not begin with a transition, it does begin with a topic sentence:

> During the past twelve months, Big Q soft drinks performed extremely well in the parts of the market that were generally weak, but it performed below average in those that were strong.

Other Means of Making Transitions   Not all transitions are made in sentences. Sometimes headings and other visual devices are used. For example, in a report that presents three brief recommendations, a writer might arrange the recommendations in a numbered list. The transition from the first recommendation to the second would simply be the appearance of the number *2* on the page.

Similarly, in a memo that takes up a number of separate topics, the transition from one to the next might be simply the appearance of the heading for that next topic.

Such nonprose transitions are generally used only with lists of parallel items. You must be careful to guard against relying solely upon such devices when your readers will need a fuller explanation of the relationships between the parts of your communication.

Will You Bore Your Audience with Transitions?   Transitions of one kind or another are needed at all levels in a communication. You need them to alert your audience when you are moving from one topic to another, whether from general to specific, from specific to general, or between topics at the same level in the organizational hierarchy of your message. However, some writers hesitate to provide transitions at all the points where readers need them. They fear they will bore their audience by saying what is obvious.

Of course, it is possible to include more transitional material than is required. However, relationships that are obvious to you as a writer may not be obvious to your readers. In general, it is wiser to risk boring your readers by including too much transitional material than to risk confusing them by leaving out something they may need.

Headings

A third technique for revealing organization is to use headings. Usually, headings are simply inserted into the text, without requiring any change in the prose. At work, writers use headings often, not only in long documents, such as reports and manuals, but also in short ones, such as letters and memos. To see how effectively the insertion of headings can reveal organization, look at Figures 7–8 and 7–9 (pp. 224–25), which display two versions of the same memo, one without headings and one with them.

Headings help readers because they are signposts that tell readers what the parts of a communication are about. They are especially useful for helping readers find information quickly. To see how much help they can be, use the headings in Figure 7–10 (p. 226) to learn why you should ride your bicycle on a street with parked cars rather than on a street without parked cars. Could you find the answer quickly? Imagine how much longer it would have taken you to find that same information if you had not had the help of headings.

Headings also help readers see the organizational hierarchy of a communication. They do this by indicating the relationships of coordination and subordination among its parts. You can see how clearly headings reveal those relationships to readers by looking at Figure 7–11 (pp. 227–28), which is from a research proposal submitted to a federal agency. The headings on this page tell you immediately that the subsections on " 'Wandering' Defined" and "Causes" are at the same level in the organizational hierarchy. The headings also clearly indicate that the subsections on "Organic Factors" and "Psychosocial Factors" are at the same level, and that both are subparts of the discussion of causes.

How to Write Headings   Of course, some headings are more helpful than others. The most important way to make your headings helpful is to write them so that they tell clearly and specifically what kind of information is included in the passages they label. Vague headings are not useful. Consider three strategies you can use to write headings:

- **Pose the question that the segment will answer for your readers.** Headings that ask questions such as, "What happens if I miss a payment of my loan?" or "Can I pay off my loan early?" are especially useful in communications that will help readers decide what to do. Figure 7–12 (p. 229) shows an informational brochure about donating organs (such as eyes and hearts) that uses questions for its headings.
- **State the main idea of the segment.** This strategy was used by the person who wrote the brochure on bicycling safety shown in Figure 7–10. Its headings say such things as "Ride with the Traffic" and "Use Streets with Parked Cars." Headings of this sort are particularly effective in focusing the readers' attention on the key point of a passage.
- **Use a key word or phrase.** This type of heading can be especially effective when you are writing a communication in which a full question or statement would be unnecessarily wordy. Imagine, for instance, that you are writing a feasibility report on purchasing a desktop publishing system for the sales department you manage. You could label the section that discusses prices with this title: "How Much Will This Equipment Cost?" However, in the context of the proposal, the simple heading "Cost" would provide the same information much more succinctly.

Should you restrict yourself to one type of heading in a particular communication, so that they are all questions, all statements, or all key words? Not necessarily. Each heading has its own individual function to perform—to announce the topic of its section as clearly and usefully as possible. In many

Garibaldi Corporation
INTEROFFICE MEMORANDUM

June 15, 19—

TO: Vice Presidents and Department Managers

FROM: Davis M. Pritchard, President

RE: PURCHASES OF MICROCOMPUTER AND WORD–PROCESSING
         EQUIPMENT

Three months ago I appointed a task force to develop corporate–wide policies for
the purchase of microcomputers and word–processing equipment. Based on the
advice of the task force, I am establishing the following policies.

The task force was to balance two possibly conflicting objectives: (1) to ensure
that each department purchase the equipment that best serves its special needs,
and (2) to ensure compatibility among the equipment purchased so the company can
create an electronic network for all our computer equipment.

I am designating one "preferred" vendor of microcomputers and two "secondary"
vendors.

The preferred vendor, YYY, is the vendor from which all purchases should be made
unless there is a compelling reason for selecting other equipment. To encourage
purchases from the preferred vendor, a special corporate fund will cover 30% of
the purchase price so that individual departments need fund only 70%.

Two other vendors, AAA and MMM, offer computers already widely used in Garibaldi;
both computers are compatible with our plans to establish a computer network.
Therefore, the special corporate fund will support 10% of the purchase price of
these machines.

We will select one preferred vendor and no secondary vendor for word–processing
equipment. The Committee will choose between two candidates: FFF and TTT. I
will notify you when the choice is made early next month.

FIGURE 7–8   Memo Without Headings (Compare with Figure 7–9)

Garibaldi Corporation
INTEROFFICE MEMORANDUM

June 15, 19---

TO: Vice Presidents and Department Managers

FROM: Davis M. Pritchard, President

RE: PURCHASES OF MICROCOMPUTER AND WORD-PROCESSING
         EQUIPMENT

Three months ago I appointed a task force to develop corporate-wide policies for the purchase of microcomputers and word-processing equipment. Based on the advice of the task force, I am establishing the following policies.

Objectives of Policies

The task force was to balance two possibly conflicting objectives: (1) to ensure that each department purchase the equipment that best serves its special needs, and (2) to ensure compatibility among the equipment purchased so the company can create an electronic network for all our computer equipment.

Microcomputer Purchases

I am designating one "preferred" vendor of microcomputers and two "secondary" vendors.

   Preferred Vendor.   The preferred vendor, YYY, is the vendor from which all purchases should be made unless there is a compelling reason for selecting other equipment.   To encourage purchases from the preferred vendor, a special corporate fund will cover 30% of the purchase price so that individual departments need fund only 70%.

   Secondary Vendors.   Two other vendors, AAA and MMM, offer computers already widely used in Garibaldi; both computers are compatible with our plans to establish a computer network.   Therefore, the special corporate fund will support 10% of the purchase price of these machines.

Word-Processing Purchases

We will select one preferred vendor and no secondary vendor for word-processing equipment.   The Committee will choose between two candidates: FFF and TTT.   I will notify you when the choice is made early next month.

FIGURE 7–9   Memo with Headings (Compare with Figure 7–8)

WELCOME TO THE WORLD OF CYCLING

Welcome to the world of bicycling! As a bike rider you're one of a fast-growing group of vigorous outdoors-loving people.

Bicycling in America has been growing at an amazing rate. Bicycles used to be sold to parents for their children. Now those same parents are buying them for themselves, as well as for their children. And grandma and grandpa and college and high school students are cycling too.

Many young executives ride bikes to work as an alternative to adding to the smog of smoky cities, and to fighting traffic jams. Young mothers are finding a way to do their shopping, complete with child carrier and shopping baskets, without having to drive hubby to work and fight for a parking place at the shopping center. College and high school students find bikes an economical alternative to cars or buses.

Unfortunately, the rise of ownership and riding of bicycles has seen a corresponding rise in traffic accidents involving bicyclists.

According to the National Safety Council, 1,000 U.S. bicyclists were killed and 60,000 injured in bicycle-motor vehicle accidents during a recent year. Over the same period, Ohio recorded 36 cyclist fatalities and 2,757 injuries.

This brochure is presented in an attempt to show how more people can happily enjoy the sport of biking without suffering needless tragedy.

GUIDES FOR SAFE RIDING

In traffic situations, the bicyclist must remain as alert as the driver of a car or other motor vehicle. For adult cyclists, this means applying defensive driving techniques to cycling as well as auto driving. The adult cyclist may even find he is becoming a better driver because his experience on a more vulnerable bicycle makes him more aware of traffic hazards when he is in a car. For the child rider, safe cycling mastery will lead to safer adult driving.

Ride with the Traffic

Riding in the same direction as the traffic on your side of the road minimizes the speed at which oncoming cars approach you. It unnerves drivers to see a cyclist heading toward them, yet this seems to be the most common mistake of young riders. One should always ride in the right hand lane, as close to the curb as possible. If you have to walk your bike, however, you become a pedestrian, and you should walk facing traffic, the same as any other pedestrian.

Use Hand Signals

Signals should be given with your left hand, well before you get to the place where you want to stop or turn. Keep your hand out until you start to make your move, or until you need to put on hand brakes, if you have them. A straight-out arm indicates a left turn, a right turn is signaled by holding your arm up bent at the elbow, with your thumb pointing to the right. (Don't use your right hand to signal right, because drivers can't see it as well.) A stop is signaled by holding your left hand straight down and out from your side.

LEFT

STOP or SLOW

RIGHT

Ride One on a Bike

You may "look sweet upon the seat of a bicycle built for two" but unless you have a tandem, only one will do. Carrying passengers on handlebars or luggage carrier makes the bike unwieldy, hard to handle in traffic, and easy to spill. The same rule should apply to large packages. Put packages on the luggage carrier and don't tie up your hands with bulky objects.

A child can be carried safely in a child seat attached to the bicycle. Expert riders recommend rear child carriers because front seats make it difficult to steer and offer less protection for the child.

Ride Single File

Always ride single file in congested areas. Double-file riding is only safe on deserted country roads or special bike paths where cars aren't allowed.

Use Streets with Parked Cars

Expert cyclists say city cycling is safest on streets with parked cars. A lane about three-feet wide is left between the parked cars and the moving traffic, so if you can ride in a straight line, you should be able to maneuver easily between the parked cars and the traffic.

Watch for car doors opening, however. Always use extra caution when you see a person sitting in a parked car; he might just open his door and jump out without warning, throwing you into moving traffic.

Streets which do not allow parking are usually more heavily traveled and not safe for cyclists, say the experts. Naturally, you should never ride on freeways, toll roads or heavily traveled state and federal highways.

Exercise Care at Intersections

The safest way to cross an intersection is to walk your bike across with the pedestrians. Left turns can be made from the center lane, as a car does, or by walking across one street and then the other. The latter procedure is safer for busy intersections and less experienced riders.

FIGURE 7–10   Brochure with Headings
Courtesy of State of Ohio Department of Highway Safety.

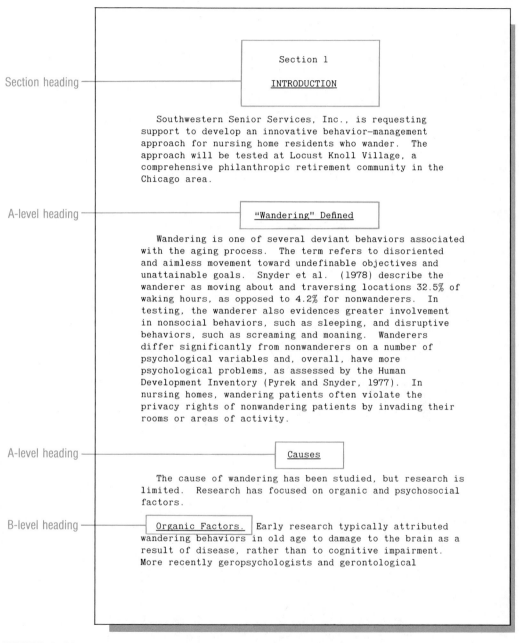

Section heading

Section 1

INTRODUCTION

Southwestern Senior Services, Inc., is requesting support to develop an innovative behavior—management approach for nursing home residents who wander. The approach will be tested at Locust Knoll Village, a comprehensive philanthropic retirement community in the Chicago area.

A-level heading

"Wandering" Defined

Wandering is one of several deviant behaviors associated with the aging process. The term refers to disoriented and aimless movement toward undefinable objectives and unattainable goals. Snyder et al. (1978) describe the wanderer as moving about and traversing locations 32.5% of waking hours, as opposed to 4.2% for nonwanderers. In testing, the wanderer also evidences greater involvement in nonsocial behaviors, such as sleeping, and disruptive behaviors, such as screaming and moaning. Wanderers differ significantly from nonwanderers on a number of psychological variables and, overall, have more psychological problems, as assessed by the Human Development Inventory (Pyrek and Snyder, 1977). In nursing homes, wandering patients often violate the privacy rights of nonwandering patients by invading their rooms or areas of activity.

A-level heading

Causes

The cause of wandering has been studied, but research is limited. Research has focused on organic and psychosocial factors.

B-level heading

Organic Factors. Early research typically attributed wandering behaviors in old age to damage to the brain as a result of disease, rather than to cognitive impairment. More recently geropsychologists and gerontological

FIGURE 7–11 Typewritten Headings that Indicate Organizational Hierarchy

practitioners are beginning to believe that organic
pathologies do not account for displays of functional
disorders or explain the causes of disruptive behavior
(Monsour and Robb, 1982). The main reason for this shift
in thinking is the research by Busse (1977) and Verwoerdt
(1976), who found no consistent relationship between
impairment and organic changes in brain vessels.

B-level heading ——— <u>Psychosocial Factors</u>. Monsour and Robb (1982) and
Snyder et al. (1978) suggest that the following three
psychosocial factors are associated with wandering:

1. <u>Lifelong patterns of coping with stress.</u> Halmes and
   Rahne's (1967) Social Readjustment Rating Scale shows
   wanderers, as opposed to nonwanderers, typically
   responding to dramatic changes in routine or
   environment through motoric responses. Methodical
   releases from tension developed early in life may have
   been brisk walks or long strolls.

2. <u>Previous work roles.</u> High rates of physical activity
   required by jobs held early in life may permanently
   influence behaviors, so that, for example, a mail
   carrier who delivered on foot may experience a
   compulsion to walk long after retirement.

3. <u>Search for security.</u> Some wanderers call out to dead
   parents or spouses, indicating that they are searching
   for security.

FIGURE 7–11 *(continued)*

DONATION OF TISSUES AND ORGANS

Florida law allows you to indicate on your driver license or identification card if you want to donate your organs or tissues after your death.

Reasons for making tissue and/or organ donations:

1. Organs and tissues are given so others may live, or see or walk.
2. It is a humanitarian thing to do — it is a priceless gift.
3. The beneficiaries are unknown, but they could be anybody — a friend, a relative, a neighbor or a stranger.
4. Whoever receives your gift will be grateful for a gift of life.

If you needed a cornea so you could see, or a kidney so you could live, or a bone so you could walk, or skin if you were burned — would you get it?

You would if people are generous in giving permission to remove tissues and organs for transplantation after death.

Medical advances now make it possible to restore sight, to keep burn victims from dying, to restore normal life of many patients with kidney disease and to preserve limbs by transplanting bone and joints. Progress

is also being make with vital organs such as the liver, pancreas and heart. It requires the right organ — from people like you who care enough to give this priceless gift.

Is there a need for tissue and organ donors?

Yes there is! In Florida there are now more people who need organs and tissues than there are donors.

How are tissues and organs obtained?

They are donated by people like you.

Will tissue and organ donation affect funeral arrangements?

No. Removal of tissues or organs will not interfere with the customary funeral arrangements.

Are there limits on who can donate tissues and organs?

Yes. Tissues and organs can't be taken if there is a chance of transmitting disease to the person receiving the transplant. Vital organs such as kidneys are taken only when medical experts have determined that there is no brain activity, but the body is being kept functioning by artificial means. Corneas, skin, bone, etc. can be taken from most donors.

Can you limit the tissues or organs you donate?

Yes, by saying so on the registration form. You can say eyes only or kidneys only. If you like, you can also specify who you want to receive the organ or tissue.

How can you make sure that doctors know you want to donate tissues or organs?

Simply fill out the registration form on the back of this pamphlet and bring it with you to the Driver License Examination Station when you go to apply for a driver license or identification card. The examiner will arrange for your driver license or identification card to have the words "Organ Donor" appear on the license or card, and send the completed will to Tallahassee for filing.

FIGURE 7–12   Headings that Ask Questions

Courtesy of the Statewide Tissue and Organ Donor Program, Department of Health and Rehabilitative Services, Tallahassee, Florida.

communications, different sections require different types of headings to achieve this goal.

On the other hand, you can often increase the clarity and usefulness of headings by writing them in a parallel fashion. This occurs especially when the headings label parallel sections, such as ones that describe parallel parts of a process (turning on the machine, calibrating the instruments, and so on) or parallel items in a list (fruits, vegetables, grains, and so on).

Figure 7–13 shows the table of contents from a booklet entitled *Getting the Bugs Out*. Like most tables of contents, it is constructed by listing the headings used inside the document. Notice that all three types of headings are used. Notice also where parallel headings are used—and where they are not.

How to Design Headings Visually   As mentioned above, headings can tell readers not only what your sections are about but also how the various parts of your communication fit into its overall organizational hierarchy. To use headings to indicate hierarchical relationships, do the following:

- Make headings for major segments look *more prominent* than those for minor ones.
- Make the headings at the same level in the hierarchy look *the same*.

The simplest way to make major headings look prominent is to make them large. This is easy to do if you are using a desktop publishing system that lets you choose different print sizes or, at least, make fatter letters by printing them in boldface. If you aren't using such equipment, you can use a cut-and-paste technique of creating large letters on another sheet and pasting them into the appropriate places in your typed text. The most inexpensive way to make the large headings is to type all the major headings on a separate sheet of paper, then enlarge them using a photocopier. Alternatively, you could use the large, rub-on letters available at many bookstores, or a Kroy lettering machine, available at many copy shops, to produce photographic strips with your headings in large letters.

Sticking strictly with what a typewriter can do, you could type major headings in all capital letters; for minor headings capitalize only the first letter of each major word. Consider underlining your major headings but not the minor ones.

You can also make major headings prominent by adjusting their location on the page. Headings that are centered look more prominent than those that are tucked against the margin. Likewise, a heading on a line of its own will be more prominent than one that is on the same line as the first sentence of the section it labels.

You can see how two writers adjusted the size and location of their headings by looking at Figures 7–11 and 7–14. The writer whose page is in Figure 7–11 used a typewriter only, and the writer whose page is in Figure 7–14 used press-on letters.

As those sample pages suggest, you can use the variables of size and location to supplement each other. For another example, look at Figure 7–10,

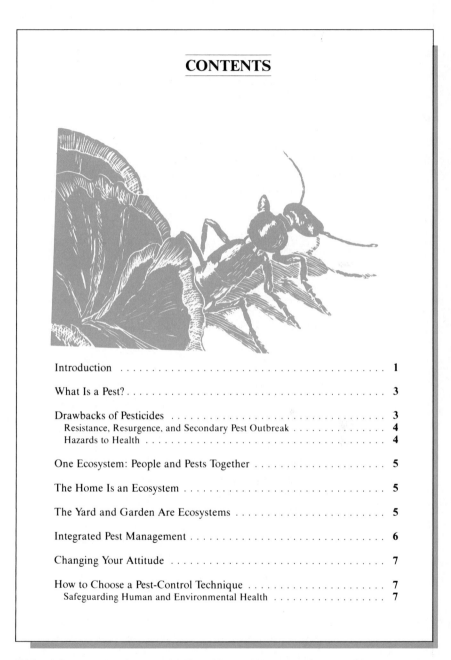

CONTENTS

FIGURE 7–13 Table of Contents that Contains All Three Kinds of Headings (Question, Statement, Key Word or Phrase)

Copyright © National Audubon Society. Used with permission.

FIGURE 7–13   *(continued)*

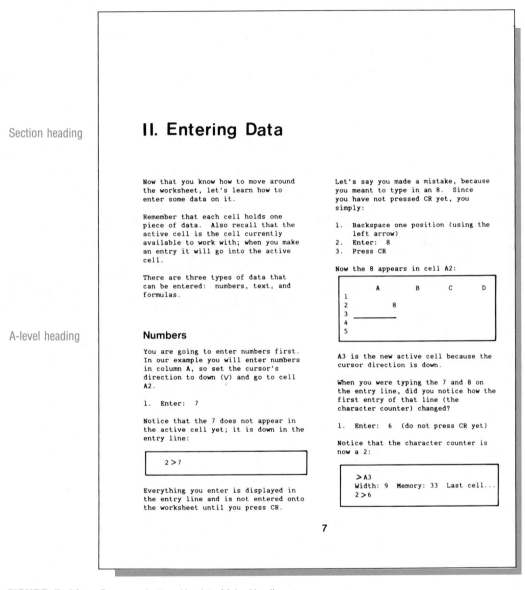

Section heading

II. Entering Data

A-level heading

Now that you know how to move around the worksheet, let's learn how to enter some data on it.

Remember that each cell holds one piece of data. Also recall that the active cell is the cell currently available to work with; when you make an entry it will go into the active cell.

There are three types of data that can be entered: numbers, text, and formulas.

Numbers

You are going to enter numbers first. In our example you will enter numbers in column A, so set the cursor's direction to down (V) and go to cell A2.

1. Enter: 7

Notice that the 7 does not appear in the active cell yet; it is down in the entry line:

```
    2 > 7
```

Everything you enter is displayed in the entry line and is not entered onto the worksheet until you press CR.

Let's say you made a mistake, because you meant to type in an 8. Since you have not pressed CR yet, you simply:

1. Backspace one position (using the left arrow)
2. Enter: 8
3. Press CR

Now the 8 appears in cell A2:

```
        A       B       C       D
  1
  2       8
  3  _____
  4
  5
```

A3 is the new active cell because the cursor direction is down.

When you were typing the 7 and 8 on the entry line, did you notice how the first entry of that line (the character counter) changed?

1. Enter: 6 (do not press CR yet)

Notice that the character counter is now a 2:

```
    > A3
    Width: 9  Memory: 33  Last cell...
    2 > 6
```

7

FIGURE 7–14   Press-on Letters Used to Make Headings
From an instructional manual written by a student.

where the major headings (for example, ''GUIDES FOR SAFE RIDING'') are in all-capital letters *and* are centered, while subheadings (for example, ''Ride with the Traffic'') have only the initial letters of major words capitalized *and* are aligned against the left-hand margin.

How Many Levels of Headings Should You Use? Theoretically, you could establish very elaborate visual hierarchies for your headings by manipulating their size and location. However, as the number of levels in a hierarchy increases, so too does the difficulty that readers have in comprehending the hierarchy. For most communications, two or three levels are about as many as readers find useful.

If additional levels are desirable, you can create them in either of two ways. The first is to make the most important sections into separate chapters. Because each chapter usually begins its own page and because the titles of chapters often use another, still larger size of type than do headings, chapters can add one level to the hierarchy of signposts in your communications. The second method is to reinforce the headings with the letters and numbers (or decimals) of an outlining system, which will enable readers to keep track of the many levels of headings more easily than if they have to rely only on the cues of size and location. See Figures 7–15 and 7–16 for examples.

Once you have designed a hierarchy that works, you can use it for many different documents that require the same number of levels of subordination. In fact, some employers require that one particular hierarchy be used in all long communications prepared by their employees.

How Many Headings Are Helpful? As you can see from the examples shown in this chapter, some pages of on-the-job writing have a great many headings. The exact number that's appropriate for any given communication depends on the readers' needs. In communications to be read as connected prose, you can usually meet your readers' needs by using headings every few paragraphs. Major shifts in topic occur frequently in on-the-job writing, and headings help to alert readers to those shifts. Also, frequent headings help hurried readers find specific pieces of information and review parts of a discussion that they have already read.

On the other hand, in connected prose, writers rarely use headings for every paragraph. If they did, their headings would emphasize not the connectedness of the prose but its disjointedness; every paragraph would begin a major new topic.

An exception involves communications that catalog information. There, you may need to use headings for every paragraph in order to meet your readers' needs. In fact, writers sometimes use headings to label sentence fragments or brief bits of data. The effect, often, is to turn such a communication into something very much like a table of facts. By looking at the advertising brochure shown in Figure 7–17, you can see how such frequent headings help readers locate specific pieces of information. (Look especially at the right-hand column, which describes the camera's ''specifications.'')

Headings and Topic Sentences Because headings and topic sentences both announce the subject of the section that follows them, you may be wondering whether it is redundant to include a topic sentence where a heading is used. In connected prose, the two are used to reinforce each other, with the topic sentence repeating one or more key words from the heading. For example, the

IEEE TRANSACTIONS ON SYSTEMS, MAN, AND CYBERNETICS, VOL. SMC-14, NO. 2, MARCH/APRIL 1984

257

Anatomical and Physiological Correlates of Visual Computation from Striate to Infero-Temporal Cortex

ERIC L. SCHWARTZ, MEMBER, IEEE

Abstract—A review of the calculus of two-dimensional mappings is provided, and it is shown that cortical mappings are in general locally equivalent to the sum of a conformal mapping and a shear mapping. Currently available data suggests that a conformal (isotropic) mapping is sufficient to model the global topography of primate striate cortex, although the addition of a shear component to this mapping is outlined. A similar analysis is applied to the local map structure of striate cortex. Models based on a local reiteration of the complex logarithm and on Radon (and backprojection) mappings are presented. A variety of computational functions associated with the novel architectures for image processing suggested by striate cortex neuroanatomy are discussed. Applications to segmentation, perceptual invariances, shape analysis, and visual data compression are included. Finally, recent experimental results on shape analysis by neurons of infero-temporal cortex are presented.

I. Introduction

RECENT ADVANCES in understanding the anatomy and physiology of visual cortex have focused attention on the problem of visual data formatting in the central nervous system (CNS). It appears that the primate visual system may utilize a variety of novel architectures to represent visual data in the CNS. This possibility is reinforced by the use of the term "functional architecture" that Hubel and Wiesel [1] have used to describe the spatial patterns of columnar architecture in striate cortex. Here the term functional architecture will be expanded to include the global topographic structure of striate cortex, as well as the local columnar architecture represented by the orientation and ocular dominance column systems described by Hubel and Wiesel [1].

It will be shown that a conformal mapping provides a good model for global cortical topographic mapping, as well as a possible model for the local hypercolumn pattern of striate cortex. Furthermore, this mapping suggests that several integral transforms (Radon transform, backprojection) may provide workable models of the local architecture of striate cortex that are in agreement with current phenomenological descriptions.

Finally, a variety of computational applications of these anatomical models will be reviewed. These include applications to perceptual invariance, binocular segmentation, data compression, and shape analysis. Recent experimental work

Manuscript received August 8, 1982; revised June 1983. This work was supported in part by the AFOSR Image Understanding Program no. F49620-83-C-0108 and in part by the Systems Development Foundation.
The author is with Brain Research Laboratories, New York University Medical Center, 550 First Avenue, New York, NY 10016

(in monkey infero-temporal cortex), which is related to this analysis, will be reviewed. It will be suggested that the novel architectures of image representation suggested by primate visual cortex neuroanatomy may provide an example of "computational anatomy," which has relevance both to contemporary machine vision applications, as well as to current attempts at understanding the biological basis of visual computation.

II. Mappings of Two-Dimensional Surfaces into Two-Dimensional Surfaces

The concept of "topographic" mapping of the surface of the retina to the surface of striate cortex naturally invites an attempt at mathematical modeling. Which mapping fits the available data? In the following a complete characterization of planar mapping will be provided in an attempt to answer this question.

A. Whitney Mapping Theorem

A broad classification of two-dimensional mappings is provided by the Whitney mapping theorem, which states that topologically stable planar mappings are locally equivalent to either a "fold," a "cusp," or a "regular" mapping [2]. In the following, we will ignore the singularities represented by folds and cusps. Regular maps are characterized by having a Jacobian matrix of partial derivatives whose determinant is finite and nonzero. In the following section we will regroup the terms of this Jacobian in order to better understand the biological significance of the isotropic (ie., conformal) and nonisotropic maps that might occur in the nervous system.

B. Jacobian Matrix and Magnification Tensor

The (retinal) point (x, y) is taken to a cortical point (f, g) by the functions $f(x, y)$ and $g(x, y)$. The Jacobian of this map [2] is

$$ J = \begin{pmatrix} \dfrac{\partial f}{\partial x} & \dfrac{\partial f}{\partial y} \\ \dfrac{\partial g}{\partial x} & \dfrac{\partial g}{\partial y} \end{pmatrix} = \begin{pmatrix} f_x & f_y \\ g_x & g_y \end{pmatrix}. \tag{1} $$

A common manipulation from the literature of continuum mechanics (and fluid mechanics) will be useful: the Jacobian is rewritten in terms of its symmetrical and

0018-9472/84/0300-0257$01.00 ©1984 IEEE

FIGURE 7–15   Headings that Use the Roman Outlining System

From *IEEE Transactions on Systems, Man, and Cybernetics*, Vol. SMC-14, No. 2, © 1984 IEEE. Used with permission.

FIGURE 7–16 Table of Contents from a Research Report that Uses the Decimal Outlining System in Its Headings

From General Electric, *Aerodynamic/Acoustic Performance of YJ101/Double Bypass VCE with Coannular Plug Nozzle* (Cincinnati, Ohio: General Electric, 1981), NASA CR-159869.

RCA "Small Wonder"

CLC020 Color Video Camera

- **Solid state MOS image sensor**
- **Infrared auto focus system**
- **Electronic viewfinder with adjustable diopter**
- **Illuminated viewfinder indicators**
- **Remote pause control with quick review**
- **f1.2—6:1 lens with power zoom**
- **Optional titler**
- **Macro focus**
- **Constant automatic white balance**
- **Automatic iris**
- **Built-in microphone**
- **VHS compatibility switch**
- **Standby switch**
- **Built-in tripod mount**

Solid state MOS image sensor

Metal oxide semiconductor (MOS) pickup device delivers an excellent color picture with none of the image "lag" associated with ordinary camera tubes. This miniature circuit also permits a more compact, rugged and efficient camera. Operates at light levels as low as 10 lux.

Infrared auto focus system

Sophisticated infrared triangulation system keeps moving objects in sharp focus. Automatically maintains a sharp image—even during zooms. Includes override for manual control of focus.

Electronic viewfinder

Electronic B&W viewfinder (EVF) displays exactly what will be recorded on the tape and doubles as a B&W monitor for viewing "instant replays" after taping. Adjustable diopter allows EVF focus to be changed to permit users who wear eyeglasses to remove them when operating camera.

Illuminated viewfinder indicators

Camera constantly monitors taping conditions and relays information to user via lights in the viewfinder. Indicators signal when: 1) VCR is recording; 2) insufficient lighting is available; and 3) VCR battery strength is low.

Remote VCR function controls

With VCR in record mode. *Pause* function is controlled with a single switch on camera's hand grip. Press to start a recording scene; press again to stop. For a quick look at the previously taped scene, press *Review* button during camera pause mode. *Review* feature available with RCA convertible VCRs.

f1.2 lens with 6:1 power zoom

High-speed f1.2 lens includes 6-to-1 zoom ratio and motor-driven zoom. Hand grip control lets operator zoom in for exciting close-ups or zoom out for panoramic shots. Lens will stop at desired perspective when control is released. Zoom ratios can also be manually adjusted by rotating lens ring on body of camera.

Optional titler

All RCA cameras may be used with RCA's optional character generator (Part No. CGA010), which makes titling and special effects easy. Compose up to 20 different titles and store them in title memory for later insertion in tapes. Other features include: choice of four colors and four sizes for characters; 20-page title memory; scroll up or down (variable speed); user-programmable 40-word permanent storage capacity; memory back-up; curtain (image is "wiped" off the screen); window (edges of the screen converge on the image); and calendar/stopwatch.

Macro focus

Obtain sharp images as close as ⅞" from the subject. Useful for shooting small objects without loss of detail.

Constant automatic white balance

Built-in circuitry continuously adjusts for proper color balance indoors or out. A manual color balance control is included for creating special visual effects, or for unusual lighting conditions.

Automatic iris

To assure correct exposure, camera automatically responds to available light conditions and adjusts aperture accordingly. Manual override provided for unusual lighting conditions.

Built-in microphone

Records sound without need for a separate microphone. Accessory jack included for optional stereo microphone (RCA Part No. SM002), providing two-track sound when recording with a stereo VCR.

VHS compatibility switch

Adapts camera's record/pause circuitry for use with virtually all VHS-format video recorders.

Standby switch

Puts camera into power-saving standby mode during extended breaks in recording. Conserves valuable battery recording time during portable VCR use.

Built-in tripod mount

Built-in fitting accommodates standard camera tripod to provide a steady picture during recording.

SPECIFICATIONS

Signal System:
EIA standard: NTSC color

Video Output Level:
1.0V p-p composite/75 ohms

Pickup Device:
2/3" MOS solid state circuit

Lens:
f1.2—6:1 motorized zoom and macro focus to within ⅞"

Minimum Required Illumination:
1.0 foot candles (10 lux at 25 IRE)

Power Source:
12VDC

Power Consumption:
5.4 watts nominal

Dimensions:
H-5¹¹⁄₁₆", W-5¹⁄₁₆", D-7½"

Weight:
2.2 lbs. (approx.)

Specifications subject to change without notice.

This product is backed by RCA Authorized Servicenters and the RCA Service Company.

For provisions of the limited warranty on this product, see separate warranty sheet form VR4856.

Form VR8642 Printed in U.S.A. 2/85
Trademark(s)® Registered® Marca(s) Registrada(s)

RCA
©1985 RCA Corporation

FIGURE 7–17 Advertising Brochure with Headings for Every Paragraph

Used with permission of Thomson Consumer Electronics.

heading ''Research'' might be followed immediately by a topic sentence that says, ''To test the three hypotheses described in the preceding section, we used a *method* that has proven reliable in similar situations.''

In contrast, writers sometimes use headings alone—without topic sentences—in communications that catalog information (that is, where readers treat each topic of a communication in isolation from the rest of the communication). Documents of this sort include warranties, contracts, troubleshooting guides, computer reference manuals, and fact sheets. Here, the headings are markers that tell readers where to find the answers to particular questions: How long does my warranty last? How do I get reimbursed for the car I rented on my business trip? In such communications, topic sentences can become unnecessary barriers between headings and the information they mark. Topic sentences are also sometimes eliminated where the information following the heading is presented in a list, where the heading is a sentence that serves as a topic sentence, or where the heading poses a question.

In the end, of course, conventional practices form only one source of advice about how to handle the relationship between headings and topic sentences. Your key consideration should be the specific needs of the particular readers you are addressing.

Visual Design

You can reveal organization not only through statements and headings in your prose but also through the visual arrangement of your text on the page. Two techniques for doing that are described next.

● **Adjust the location of blocks of prose.** Figure 7–18 shows a page on which the writer has used indentation (and headings) to indicate his hierarchy. Notice how the indented paragraphs appear subordinate to those that extend farther to the left.

Sometimes writers help their readers perceive their communication's hierarchy by leaving more blank lines at the end of a major section than at the end of a minor section. When writing long reports, they often signal the largest divisions by beginning each of the major sections (such as chapters) on its own page, regardless of where the preceding section ended on the previous page.

● **Use lists.** You can sometimes use a list format to present parallel pieces of information in an especially readable form. Figure 7–19 shows a variety of formats you can use for lists.

When constructing lists, note that the entries should have parallel grammatical construction: all the items should be nouns, or all should be full sentences, or all should be questions, and so on. Mixing grammatical constructions distracts readers and sometimes indicates a shift in point of view that breaks the tight relationship that should exist among the items in the list.

B. PRODUCT HANDLING

In our guidelines on processing, we now have completed our suggestions for guarding against microbial contamination. We turn now to our guidelines on how to handle the product. Here, we consider the receiving of the new raw material and the processing of it.

1. Receiving Raw Materials

By raw materials, we mean both the fish and any other raw materials used in processing.

a. Fish.—We consider first the fresh fish and then the frozen fishery products.

(1) Fresh fish.

Check fresh fish for sign of spoilage, off odors, and damage upon their arrival at your plant. Discard any spoiled fish.

Immediately move fresh fish under cover to prevent contamination by insects, sea gulls, other birds, and rodents. If the fish are to be scaled, scale them before you wash them.

Unload the fish immediately into a washing tank. Use potable, nonrecirculated water containing 20 parts per million of available chlorine and chill to 40°F or lower. Spray wash the fish with chlorinated water after taking them from the wash tank (Fig. 11).

If incoming fresh fish cannot be processed immediately, inspect them, cull out the spoiled fish, and re-ice the acceptable fish in clean boxes; then store them preferably in a cold room at 32° to 40°F or, at least, in an area protected from the sun and weather and from insects and vermin. Wash, rinse, and steam-clean carts, boxes, barrels, and trucks used to transport the fresh fish to the plant if any of these are to be used again. Reusable containers should be rinsed again with chlorinated or potable water just before use. *Note:* wooden boxes and barrels should not be reused. It is virtualy impossible to satisfactorily sanitize used wooden containers such as fish boxes and barrels. If disposable-type containers are used, rinse them off and store them in a screened area until you remove them from the premises.

(2) Frozen fishery products.

Use a loading zone that provides direct access to a refrigerated room.

32

FIGURE 7–18   Indentation of Text Used to Indicate Organizational Hierarchy (This page also shows a new section beginning at the top of a page.)
From Perry Lane, ''Sanitation Recommendations for Fresh and Frozen Fish Plants,'' *Fishery Facts* 8 (November 1974): 32. National Oceanic and Atmospheric Administration, Department of Commerce.

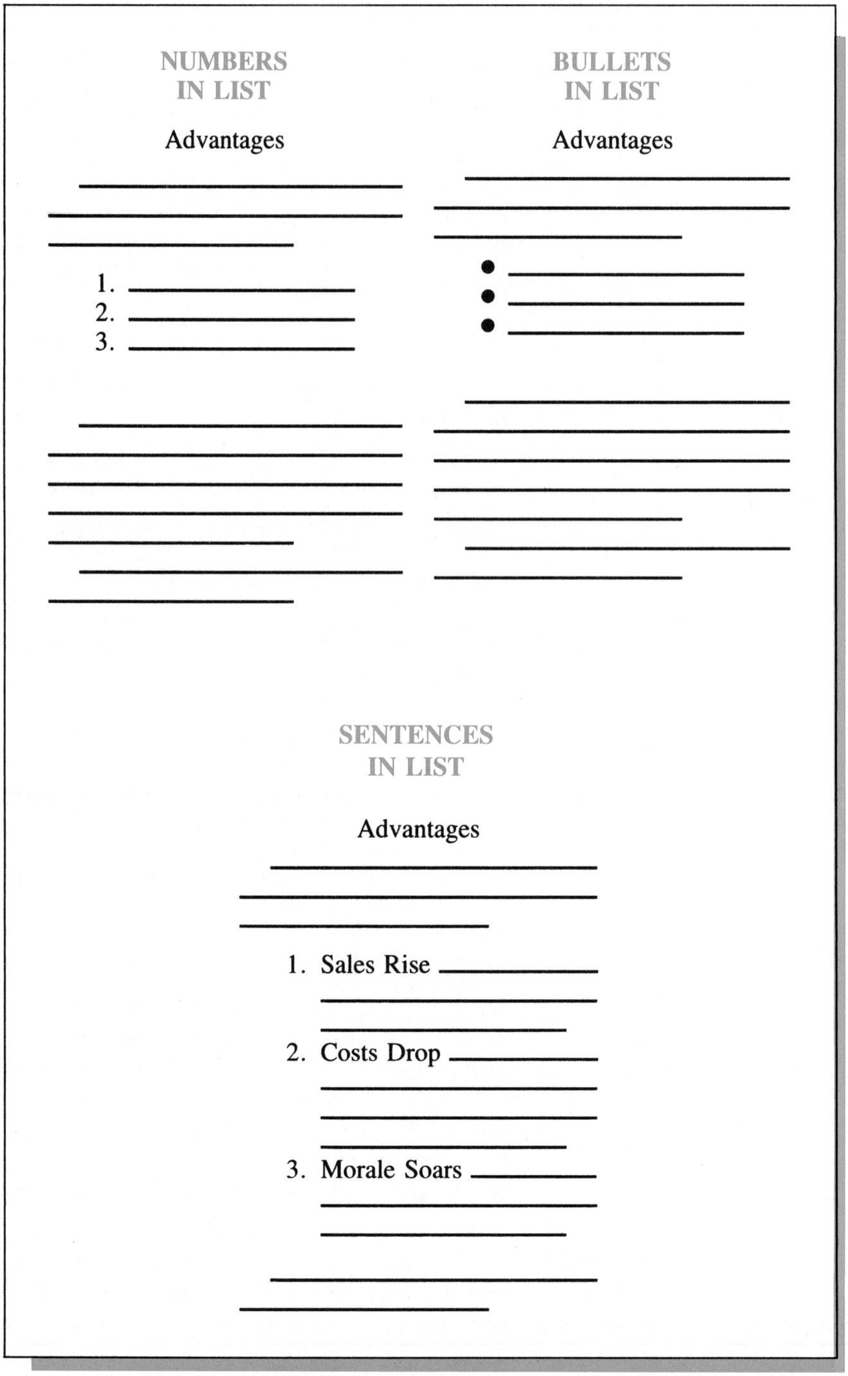

NUMBERS
IN LIST

Advantages

_____
_____
_____

1. _____
2. _____
3. _____

_____
_____
_____
_____
_____
_____

BULLETS
IN LIST

Advantages

_____
_____
_____

● _____
● _____
● _____

_____
_____
_____
_____
_____

SENTENCES
IN LIST

Advantages

_____
_____
_____

1. Sales Rise _____
   _____

2. Costs Drop _____
   _____
   _____

3. Morale Soars _____
   _____

   _____
   _____

FIGURE 7–19   Several Kinds of Lists

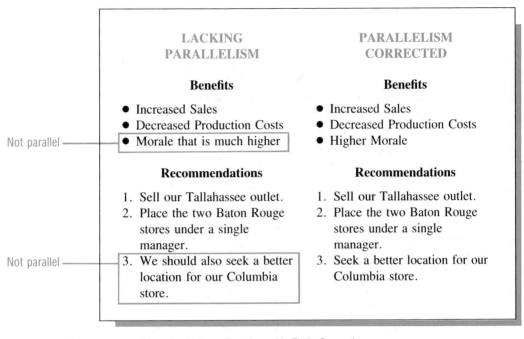

LACKING PARALLELISM	PARALLELISM CORRECTED
Benefits	**Benefits**
• Increased Sales	• Increased Sales
• Decreased Production Costs	• Decreased Production Costs
• Morale that is much higher	• Higher Morale
Recommendations	**Recommendations**
1. Sell our Tallahassee outlet.	1. Sell our Tallahassee outlet.
2. Place the two Baton Rouge stores under a single manager.	2. Place the two Baton Rouge stores under a single manager.
3. We should also seek a better location for our Columbia store.	3. Seek a better location for our Columbia store.

Not parallel ── Morale that is much higher

Not parallel ── 3. We should also seek a better location for our Columbia store.

FIGURE 7–20   Lists Lacking Parallelism, Together with Their Corrections

Figure 7–20 gives two examples of lists that lacked parallelism and shows how they were corrected.

CONCLUSION

This chapter suggested that in many situations you can make your writing more easily understandable and more persuasive if you begin your segments with topic sentences, present your generalizations before your details, organize from most important to least important, and use headings, forecasting statements, and similar devices to reveal your communication's organization to your readers.

As you seek to apply this chapter's advice, remember that its guidelines are suggestions, not rules. The only ''rule'' for writing segments is to be sure your readers know what you are talking about and how your various points relate to one another. Sometimes you may be able to do this without thinking consciously about your techniques for doing so. In numerous other situations, however, you will be able to increase the clarity and persuasiveness of your paragraphs, sections, and chapters by thinking of ways to employ creatively the strategies presented here.

EXERCISES

1. Circle the various parts and subparts of the passage given in Figure 7–21 to show how the smaller segments are contained within larger ones.

2. Identify the topic sentences in Figure 7–21 by putting an asterisk before the first word of each sentence that indicates the topic of a segment.

3. To see how one author provided statements that preview the organization of a passage, circle all of the forecasting statements in Figure 7–18. For those segments that do not have explicit forecasting statements, explain how the reader comes to know how they are organized.

4. Below are two versions of a passage from a proposal asking the federal government to sponsor a research project. In both versions, the writer uses explicit generalizing statements. However, only in the second version is the generalizing reader-centered. Write a brief paragraph describing the ways that reader-centered generalization improves the passage. Why is the second version shorter? (Incidentally, this passage is *intended* to be on a subject completely unfamiliar to you, so that you can see all the more clearly how much reader-centered generalizations assist readers.)

First Version

There are two types of magnetic susceptibility. Substances on which the molecules are arranged essentially at random (gases and liquids) are magnetically isotropic; that is, the magnetic susceptibility is the same in all directions. Solids, too, may be effectively isotropic if their components are randomly oriented. The magnetic susceptibility of these types of material is referred to as the average susceptibility. The magnetic susceptibilities along each axis (often unequal) are referred to as principal susceptibilities. Because the magnetic susceptibility of test dusts would be measured on bulk randomly oriented quantities, average susceptibility measurements would be appropriate.

There are many methods available for measuring magnetic susceptibility. Choice of method depends on the problem to be solved. There is no single method of universal applicability. One method applicable to the testing of finely divided powders is the Guoy method. The principle by which this method operates is shown in Figure 13 [not shown]. A cylindrical sample is suspended with one end in a uniform magnetic field and the other end in a region where the field is negligible. The force exerted on the sample by the magnetic field is measured by suspending the cylinder from the arm of a sensitive balance and finding the change in weight due to the field.

Importing Insects

By importing insects from other parts of the world, nations can sometimes increase the productivity of their agricultural sector, but they also risk hurting themselves. By importing insects, Australia controlled the infestation of its continent by the prickly pear, a cactus native to North and South America. The problem began when this plant, which has an edible fruit, was brought to Australia by early explorers. Because the explorers did not also bring its natural enemies, the prickly pear grew uncontrolled, eventually rendering large areas useless as grazing land, thereby harming the nation's farm economy. The problem was solved when scientists in Argentina found a small moth, <u>Cactoblastis cactorum</u>, whose larvae feed upon the prickly pear. The moth was imported to Australia, where its larvae, by eating the cactus, reopened thousands of acres of land.

In contrast, the importation of another insect, the Africanized bee, could threaten the well-being of the United States. It once appeared that the importation of this insect might bring tremendous benefits to North and South America. The Africanized bee produces about twice as much honey as do the bees native to the Americas.

However, the U.S. Department of Agriculture now speculates that the introduction of the Africanized honey bee into the U.S. would create serious problems. The problems arise from the peculiar way the Africanized honey bee swarms. When the honey bees native to the United States swarm, about half the bees leave the hive with the queen, moving a small distance. The rest remain, choosing a new queen. In contrast, when the Africanized honey bee swarms, the entire colony moves, sometimes up to fifty miles. If Africanized bees intermix with the domestic bee population, they might introduce these swarming traits. Beekeepers could be abandoned by their bees, and large areas of cropland could be left without the services of this pollinating insect. Unfortunately, the Africanized honey bee is moving slowly northward to the United States from Sao Paulo, Brazil, where several years ago a researcher accidentally released 27 swarms of the bee from an experiment.

Thus, while the importation of insects can sometimes benefit a nation, imported insects can also alter the nation's ecological system, thereby harming its agricultural business.

FIGURE 7–21   Passage with Several Levels of Segments

Second Version

Average magnetic susceptibility is the appropriate magnetic property for the characterization of test dusts. Average susceptibility describes the magnetic behavior of substances that are magnetically isotropic; that is, it describes substances whose magnetic susceptibility is the same in all directions. This behavior is exhibited by gases, liquids, and randomly oriented solids such as test dusts.

One of the many suitable approaches for characterizing the average magnetic susceptibility of finely divided powders is the Guoy method. This method is illustrated in Figure 13 [not shown]. A cylindrical sample is vertically suspended with one end in a uniform magnetic field and the other end in a region where the field is negligible. To calculate average magnetic susceptibility, the force exerted on the sample by the magnetic field is measured by suspending the cylinder from the arm of a sensitive balance and finding the change in weight due to the field.

5. The following paragraphs form one segment from a long proposal submitted by an architectural firm to a company that owns a chain of fast-food restaurants. After reading the paragraphs, imagine what advantage restaurant owners could enjoy as a result of the design features discussed in them. Then write a persuasive opening sentence that emphasizes that advantage.

In conventional restaurants, energy (in the form of heat) is pushed out of the building in several ways. For example, the heat generated by grills and ovens for cooking is captured in hoods and forced outside through vents and flues. Hot water is poured down drains. And the heat generated by refrigerators and air conditioners is also discharged outside the building.

The design we propose, however, will capture this heat so that you can use it in the normal operation of your restaurants. For instance, heat pumps will capture the waste heat generated from each restaurant's ovens, grills, refrigeration units, and dishwashing machines. This heat will be used to raise the temperature of cold water coming into the restaurant. Once warmed, the water will be stored in two 3,000-gallon tanks, from which it can be drawn when needed for cleaning, dishwashing, and other purposes.

6. Figure 7–22 shows a memorandum whose contents have been scrambled. Each statement in it has been assigned a number.
   a. Write the numbers of the statements in the order in which the statements would appear if the memorandum were written in accordance with the guidelines in this chapter. Place an asterisk before each statement that would begin a segment. (When ordering the statements, ignore the particular phrasing of the sentences. Order them according to the information they provide the reader.)
   b. Using the list you have just made, rewrite the memorandum by rephrasing the sentences so that the finished memorandum conforms with all the guidelines in this chapter.

TO: Jimmie Ru, Plant Engineer
FROM: Chip Bachmaier, Polymer Production
SUBJECT: NEW STUFFERS

1. When materials are stuffed into the extruder by hand,
   they cannot be stuffed in exactly the same way each time.

2. We have not been able to find a commercial source for an
   automatic stuffer.

3. A continuing problem in the utilization of our extruders
   in Building 10 is our inability to feed materials
   efficiently into the extruders.

4. An alternative to stuffing the materials by hand is to
   have them fed by an automatic stuffer.

5. If the materials are not stuffed into the extruder in
   exactly the same way each time, the filaments produced
   will vary from one to another.

6. I recommend that you approve the money to have the
   company shop design, build, and install automatic
   stuffers.

7. The company shop will charge $4,500 for the stuffers.

8. An automatic stuffer would feed material under constant
   pressure into the opening of the machine.

9. Currently, we are stuffing materials into the extruders
   by hand.

10. The shop has estimated the cost of designing, building,
    and installing automatic stuffers on our $\frac{3}{4}$-inch, 1-inch,
    and 1$\frac{1}{2}$-inch extruders.

11. It takes many worker-hours to stuff extruders by hand.

12. No automatic stuffers are available commercially because
    other companies, which use different processes, do not
    need them.

FIGURE 7–22   Memo for Exercise 6

7. Refer to Figure 7–6, which shows a table of contents from an instruction manual for an electronic balance, a device used to weigh samples in a laboratory. The author has done a good job of telling what the communication includes, but has done a poor job of telling about the relationships of coordination and subordination among those contents. Your assignment is to use the techniques described in Guideline 4 to create an improved table of contents that reveals the manual's organization more effectively. Note that after the "Introduction," the manual falls into two major parts, each with subparts. You will need to add some headings to these groupings.

USING SIX PATTERNS OF DEVELOPMENT

DEFINING
OBJECTIVES
PLANNING
DRAFTING
EVALUATING
REVISING

In Chapter 7, you read four guidelines for writing paragraphs, sections, and chapters. You can use those versatile guidelines regardless of the purpose, size, or subject of your passages. This chapter also looks at paragraphs, sections, and chapters, but in a different way. It concerns passages that have special purposes, such to describe a physical object, explain a process, or make a comparison. For many of these commonly used types of passages, conventional patterns for writing exist.

This chapter describes six of these conventional patterns, including the three just mentioned (description of an object, description of a process, and comparison) and three others (classification, cause and effect, and problem and solution). These six were chosen because of their great helpfulness in the kinds of writing you will do on the job.

Studying these conventional patterns can be very beneficial to you for many of the same reasons that it is helpful to study the conventional superstructures for reports, proposals, instructions, and other types of communications. Like superstructures, these patterns have become conventional because they work. Most importantly, they work for readers. That's because they organize information and ideas in ways that readers find understandable and useful. The patterns also help writers by providing frameworks they can use with confidence to achieve various commonly encountered communication objectives.

FOUR IMPORTANT POINTS BEFORE YOU READ FARTHER

Before studying any of the six patterns of development described in this chapter, you should note four important points.

First, you must remember to employ these conventional patterns only as guides. To make them work in your communications, you must adapt them to your purpose, readers, and situation.

Second, although the six types of passages described here are very common in writing at work, they aren't the only types of passages you will need to write. When you aren't using one of these six patterns, you can often create effective passages by following the guidelines in Chapter 7. Further, there are many other helpful conventional patterns that you can learn by reading other people's writing, especially the writing of people in your own company or field.

Third, although this chapter treats its six patterns separately, in actual communications the types are often mixed with one another and with passages using other patterns. This will happen, for instance, in a report that Gene is writing to tell decision-makers in his organization about a new technique for applying coatings to the insides of television screens. Like many technical reports, his will employ a problem-and-solution pattern for its overall structure. First, he will describe the problem that makes the old technique undesirable for his employer, then he will describe the technique he has developed as a solution to that problem. Within this overall pattern, Gene will use many others, including the ones for describing an object (the equipment used in his technique), explaining a process (the way his technique works), and making a comparison (the performance of his new technique versus the performance of the current one). Similarly, when you write at work, you will often weave together various patterns, each one suited to a special aim of one particular part of your message.

The fourth point for you to remember as you read the rest of this chapter is that when using the patterns described here, you should keep in mind the guidelines given in Chapter 7. In fact, much of the information given below is really advice about how to apply Chapter 7's guidelines to the special types of passages described here.

THE ORGANIZATION OF THIS CHAPTER

In order to describe the six patterns presented here as usefully as possible, this chapter is organized differently than are most of the others in this book. Instead of being built around a single set of guidelines, it contains a separate discussion of each pattern. This organization will enable you to use the chapter as a reference source in which you can rapidly find complete information about the particular pattern you want to study or use.

CLASSIFICATION

On the job, you will often need to write communications in which you must provide a sense of order to what is initially a miscellaneous list of facts and ideas. Maybe you'll work for a chemical company and be asked to write an advertising brochure that discusses in an organized manner the various adhesives your employer produces. Or maybe you'll work for a hospital and be asked to write a brochure telling patients in an organized manner about various kinds of cancer.

Whenever you want to present an organized discussion of a list of parallel items, you can *classify* your information by arranging it into groups that share common characteristics. For instance, in the brochure about adhesives, you might group those made for bonding metal, for bonding wood, and for bonding plastic. Or in the booklet about cancer, you might group those that affect internal organs, those that affect the skin, and so on. You can also subdivide these initial groups to create a hierarchy. For example, you can divide the adhesives used for wood into those that are waterproof and those that are not.

Two Kinds of Classification: Formal and Informal

The following paragraphs provide guidelines for two classification patterns, one formal and the other informal. In *formal* classification, you group the items in your list according to some objective, observable characteristic they possess. For example, when classifying the adhesives into those used for wood, metal, and plastic, you are using the objective characteristic of their intended application, something that all of your company's adhesives possess.

In contrast, you use *informal* classification when it is impossible or undesirable to classify according to observable characteristics. Suppose, for instance, that you want to classify the persuasive strategies used in advertisements that appear in trade journals read by people in your field. Different people might group the advertisements in different ways because the very creation of the categories requires interpretation and judgment.

Despite their differences, the formal and informal classification procedures share the same overall objectives:

- To arrange information usefully for your readers.
- To create groups (and subgroups) that include a place for every item you are trying to classify.
- To create a classification that has *only one* place for every item.

Guidelines for Formal Classification

1. **Choose principles of classification suited to your readers and purpose.** A principle of classification is the characteristic you look at when you create your groups and assign items to them. In the example of the adhesives, you might use any number of characteristics as your principle of classification, including color, the date first manufactured by your company, and so on. When writing, you should always choose a principle of classification that will make your information as useful as possible to your readers. Since the readers of your advertising booklet will, presumably, be looking for adhesives they can use for specific bonding needs, intended use makes a better principle of classification than the others mentioned above (color, date of first manufacture).

2. **Use only one principle of classification at a time.** As mentioned above, one aim of classification is to provide one and only one place for each item at each level of the resulting hierarchy. To accomplish this, you should use only one principle of classification at a time. For example, you might classify in the following way the cars owned by a large corporation:

   Cars built in the United States

   Cars built in other countries

In this classification, you use only one principle of classification—the country in which the car was manufactured. Because each car was built in only one country, each would fit into only one group. On the other hand, suppose you grouped the cars in this way:

Cars built in the United States

Cars built in other countries

Cars that are expensive

In this faulty classification, two principles are used at the same time—country of manufacture and cost. An expensive car built in the United States would fit into two categories.

Of course, you can use different principles at different levels in your hierarchy:

Cars built in the United States
   Expensive ones
   Inexpensive ones

Cars built in other countries
   Expensive ones
   Inexpensive ones

In this case, an expensive car built in the United States would have only one place at each level of the hierarchy.

Guidelines for Informal Classification

1. **Group your materials in ways suited to your readers and purpose.** Even when you are not able to use formal classification, you are still attempting to achieve a particular purpose with respect to specific readers. For example, if your employer wants you to discuss the various persuasive strategies used by trade journal advertising in your field, you might decide to group the strategies according to their focus, such as the feature of the product, the general image of the product, or the image of the manufacturer. Your readers could then judge which strategy might be most effective in ads for your employer's products.

2. **Use parallel groups at the same level.** For instance, if you are classifying advertisements, don't create a list of categories like this:

    Focus on price

    Focus on established reputation

Focus on advantages over a competitor's product

Focus on one of the product's key features

Focus on several of the product's key features

The last two categories are at a lower level of detail than are the other three. To make the categories parallel, you could combine them in the following way:

Focus on price

Focus on established reputation

Focus on advantages over a competitor's product

Focus on the product's key features
  Focus on one key feature
  Focus on several key features

3. **Avoid overlap among groups.** Even when you cannot use strict logic to classify your material, strive to provide one and only one place for each item. To do this, you must avoid overlap among categories that would give an item two places. For example, in the following list, the last item overlaps the others because photographs can be used in any of the other types of advertisements listed.

Focus on price

Focus on established reputation

Focus on advantages over a competitor's product

Focus on the product's key features

Use photographs

Sample Classification Segments

Figure 8–1 shows the outline of a research report that the writers organized using informal classification. In the report, they describe the ways that investigators should go about identifying water plants that might be cultivated, harvested, and dried to make fuel for power generators.

Figure 8–2 shows a group of paragraphs that the writers organized using formal classification. The paragraphs talk about two kinds of spruce forests. Both kinds are scattered throughout a large area of Alaska that was being studied for an environmental impact statement.

IDENTIFYING EMERGENT AQUATIC PLANTS THAT MIGHT BE USED AS
FUEL FOR BIOMASS ENERGY SYSTEMS

INTRODUCTION

BOTANICAL CONSIDERATIONS
 Growth Habitat
 Morphology
 Genetics

PHYSIOLOGICAL CONSIDERATIONS
 Carbon Utilization
 Water Utilization
 Nutrient Absorption
 Environmental Factors Influencing Growth

CHEMICAL CONSIDERATIONS
 Carbohydrate Composition
 Crude Protein Content
 Crude Lipid Content
 Inorganic Content

AGRONOMIC CONSIDERATIONS
 Current Emergent Aquatic Systems
  Eleocharis dulcis
  Ipomoea aquatica
  Zizania palustris
  Oryza sativa
 Mechanized Harvesting, Collection, Densification, and Transportation of
  Biomass
 Crop Improvement
 Propagule Availability

ECOLOGICAL CONSIDERATIONS
 Water Quality
 Habitat Disruption and Development
 Coastal Wetlands

ECONOMIC CONSIDERATIONS
 Prior Research Efforts
  Phragmites communis
  Arundo donax
  Other Research
 Production Costs for Candidate Species
  Planting and Crop Management
  Harvesting
  Drying and Densification
  Total Costs
 End Products and Potential Competition

SELECTION OF CANDIDATE SPECIES

FIGURE 8–1 Outline of a Report Organized Using Informal Classification

From S. Kresovich et al., *The Utilization of Emergent Aquatic Plants for Biomass Energy Systems Development,* (Golden, Colo.: Solar Energy Research Institute, 1982).

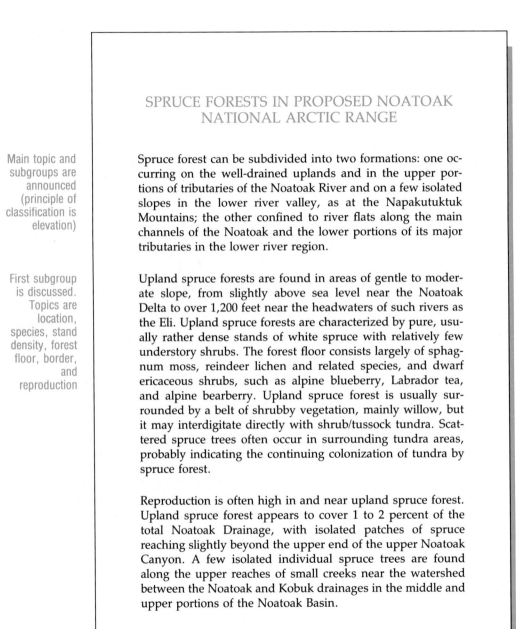

SPRUCE FORESTS IN PROPOSED NOATOAK
NATIONAL ARCTIC RANGE

Main topic and
subgroups are
announced
(principle of
classification is
elevation)

Spruce forest can be subdivided into two formations: one occurring on the well-drained uplands and in the upper portions of tributaries of the Noatoak River and on a few isolated
slopes in the lower river valley, as at the Napakutuktuk
Mountains; the other confined to river flats along the main
channels of the Noatoak and the lower portions of its major
tributaries in the lower river region.

First subgroup
is discussed.
Topics are
location,
species, stand
density, forest
floor, border,
and
reproduction

Upland spruce forests are found in areas of gentle to moderate slope, from slightly above sea level near the Noatoak
Delta to over 1,200 feet near the headwaters of such rivers as
the Eli. Upland spruce forests are characterized by pure, usually rather dense stands of white spruce with relatively few
understory shrubs. The forest floor consists largely of sphagnum moss, reindeer lichen and related species, and dwarf
ericaceous shrubs, such as alpine blueberry, Labrador tea,
and alpine bearberry. Upland spruce forest is usually sur-
rounded by a belt of shrubby vegetation, mainly willow, but
it may interdigitate directly with shrub/tussock tundra. Scattered spruce trees often occur in surrounding tundra areas,
probably indicating the continuing colonization of tundra by
spruce forest.

Reproduction is often high in and near upland spruce forest.
Upland spruce forest appears to cover 1 to 2 percent of the
total Noatoak Drainage, with isolated patches of spruce
reaching slightly beyond the upper end of the upper Noatoak
Canyon. A few isolated individual spruce trees are found
along the upper reaches of small creeks near the watershed
between the Noatoak and Kobuk drainages in the middle and
upper portions of the Noatoak Basin.

FIGURE 8–2   Passage Organized Using Formal Classification

From U.S. Department of the Interior, *Proposed Noatoak National Arctic Range, Alaska: Final Environmental Impact Statement* (Washington, D.C.: U.S. Department of the Interior, 1974), 51–52.

2

Second
subgroup is
discussed.
Topics are same
as for first
subgroup

Lowland spruce forests are found on islands and low-lying river banks in and near the main channel of the Noatoak River, mainly between the lower canyon and the lower reaches of the upper canyon. The dominant species here is also white spruce, but the understory and other associated vegetation differ markedly from the upland white spruce forest. The lowland forest is usually rather open and consists mainly of mature trees, usually 15 to 20 meters tall and with a diameter at breast height of 25 to 40 cm. Ring counts on cut logs from these forests indicated that the average age of the mature spruce trees was 150 to 300 years. Reproduction is limited or nonexistent in undisturbed stands, yet mortality from damage by porcupines and bark beetles of several species is considerable.

Much of the understory of lowland spruce forest is tall willow brush. These species occasionally form pure stands within the forest, particularly along the shores of oxbow lakes and in abandoned river channels. Forest floor vegetation consists largely of typical subarctic forest species, such as one-sided wintergreen, and small northern bog orchis, as well as a number of forest-tundra species such as large-flowered wintergreen. Sphagnum moss is rare, but other mosses are common, and often form a significant portion of the forest floor vegetation. Lowland spruce is normally bordered by wet tundra, riparian willow thickets, and shoreline vegetation.

FIGURE 8–2   *(continued)*

PARTITION: DESCRIPTION OF AN OBJECT

At work you will often need to describe a physical object to your readers. If you write about an experiment, for example, you may need to describe your equipment. If you write instructions, you may need to describe the machines your readers will use. If you propose a new purchase, you may need to describe the thing you want to buy.

To organize these descriptions, you can use a procedure called *partitioning*. In partitioning, you divide your subject into its major components and (if appropriate) divide those into their subcomponents. Partitioning is really a special kind of classification you use when you want to describe an object.

How Partitioning Works

When partitioning, you follow the same basic procedure used in classification. You think of the object as a collection of parts and you use some principle to identify groups of related parts. Most often, that principle is either function or location. Consider, for instance, the ways that these two principles might be used to organize a discussion of the parts of a car.

If you organize your discussion of the car by location, you might talk about the interior, the exterior, and the underside. The interior would be those areas under the hood, in the passenger compartment, and in the trunk. The exterior would be the front, back, sides, and top, while the underside would include the parts under the body, such as the wheels and transmission.

However, if you were to organize your discussion in terms of function, you might focus on parts that provide power and parts that guide the car. The parts that provide power are in several places (the gas pedal and gear shift are in the passenger compartment, the engine is under the hood, and the transmission, axle, and wheels are on the underside of the body). Nevertheless, you would discuss them together because they are related by function.

Often, these two commonly used principles of classification—location and function—coincide. For instance, if you were to partition the parts of a stereo system, you could identify the following major components: the compact disk player (whose function is to create electrical signals from the impressions made on the surface of a disk), the amplifier (whose function is to increase the amplitude of those signals), and the speakers (whose function is to convert those signals into sound waves).

Of course, other principles of classification are possible. You could organize your description of the parts of a car according to the materials they are made from, a classification that could be useful to someone looking for ways to decrease the cost of materials in a car. You could also partition the parts of a car according to the countries in which they were manufactured, a classification that could be useful to an economist studying the effects of import tariffs or quotas on American manufacturers.

The Aims of Partitioning

Like classification, partitioning can yield a hierarchy that might have one level or many levels. In either case, the aims are the same:

- To arrange your description in a way useful to your readers.
- To create groups (and perhaps subgroups) that include a place for every part you want to discuss. A rigorously thorough partitioning would identify every part of the object being described. In practice, however, partitioning is usually restricted to those parts relevant to the purpose and readers of the communication.
- To create an organization that has only one place for every item.

Guidelines for Describing an Object

1. **Choose a basis for partitioning suited to your readers and purpose.** For instance, if you are trying to describe a car for new owners who want to learn about what they have purchased, you might organize according to location, so they will learn about the comforts and conveniences they enjoy when sitting in the passenger compartment, when using the trunk, and so on. In contrast, when describing a car for mechanics who will have to diagnose and correct problems, it would make the most sense to organize your material around functions. Thus, when the mechanics must correct problems with the steering, they will understand the entire set of parts that make up the steering system.

2. **Use only one basis for partitioning at a time.** To assure that you have one place and only one place for each part you describe, you must use only one basis for partitioning at a time, just as you use only one basis for classifying at a time.

   In some communications, you may need to describe the same object from more than one point of view. For instance, in the introduction to an advertising brochure for a new car, you might need to describe the parts of a car from the point of view of their comfort and visual appeal to the owner. There you could organize by location. Later in the same brochure, you might describe the car from the point of view of performance, in which case you could organize by function. Each part of the car would appear in both sections of the brochure but would have only one place in each section.

3. **Arrange the parts of your discussion in a way your readers will find useful.** For instance, if you are partitioning by location, you might move systematically from left to right, front to back, or outside to inside. If you are describing things functionally, you might treat the parts in the order in

which they are involved in an activity of interest to your readers. For instance, if you are describing the parts that provide power to a car, you might discuss them in the order in which the power flows through them: from engine to transmission, to drive shaft, to axle, and so on. If you are describing the sections of a food processing factory, you might describe the sections in the order in which the food flows through them: from the railroad car in which the raw materials arrive to the loading dock where the finished product is shipped to customers. At times you can arrange material from most important to least important. For instance, if you are writing a repair manual, you might treat first the parts that most often cause problems.

Sample Description of an Object

Figure 8–3 shows an outline of part of a manual for owners of a word-processing system. In the manual, the writers have organized their description by systematically partitioning the system into its components and subcomponents.

SEGMENTATION: DESCRIPTION OF A PROCESS

A description of a process explains the relationship of events over time. You may have either of two purposes for describing a process to your readers. First, you may want to describe a process so your readers can *perform* it. For example, you may be writing instructions so that your readers can analyze the chemicals present in a sample of liver tissue, make a photovoltaic cell, apply for a loan, or run a computer program.

Second, you may describe a process so your readers can *understand* it. For example, you might want your readers to understand any of the following kinds of things:

● How something is done. For instance, how coal is transformed into synthetic diamonds.
● How something works. For instance, how the lungs provide oxygen to a person's bloodstream.
● How something happened. For instance, how the United States developed the space programs that eventually landed a person on the moon.

Whether your description is intended to enable your readers to perform a process or to understand it, your aim is the same: to show your readers the overall structure of the process. To do that, you segment the process. Like partitioning, segmenting is similar to classification. You begin with the long list of steps or events involved with the process, then you separate those steps or events into related groups. If the process is long enough, you also divide those groups of steps into subgroups, in which case you create a hierarchy.

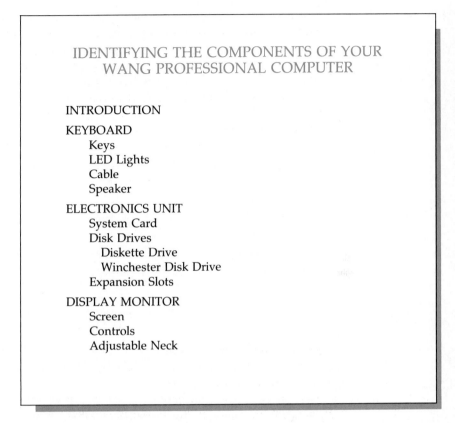

IDENTIFYING THE COMPONENTS OF YOUR
WANG PROFESSIONAL COMPUTER

INTRODUCTION

KEYBOARD
     Keys
     LED Lights
     Cable
     Speaker

ELECTRONICS UNIT
     System Card
     Disk Drives
         Diskette Drive
         Winchester Disk Drive
     Expansion Slots

DISPLAY MONITOR
     Screen
     Controls
     Adjustable Neck

FIGURE 8–3   Chapter Organized by Partitioning an Object

Outline of nine pages from *The Wang Professional Computer: Introductory Guide,* (Lowell, Mass.:
Wang Laboratories, 1983), 1–2 through 1–10.

Throughout, you treat the steps and events in sequence. The result might look
like this outline of instructions for building a cabinet.

    Obtaining Your Materials

    Building the Cabinet
       Cutting the wood
       Routing the wood
       Assembling the parts
          Making the frame
          Mounting the doors

    Finishing the Cabinet
       Sanding
       Applying the stain
       Applying the protective sealant

As with partitioning, you need a principle of classification to guide you as you group the steps or events. Commonly used principles include the time when the steps are performed (first day, second day; spring, summer, fall), the purpose of the steps (to prepare the equipment, to examine the results), and the tools that are used to perform the steps (for example, steps that use the function keys on a word processor, steps that use the numeric pad, and so on).

Guidelines for Segmenting

1. **Choose a principle for segmenting suited to your readers and purpose.** If you are writing instructions, this usually means grouping them in ways that reflect an efficient or comfortable rhythm of work. If you are trying to help your readers understand a process, this means organizing around concerns that are of interest or use to your readers. For example, imagine that you are writing a history of the process by which the United States placed a man on the moon. If your readers are legislators or government historians, you might segment your account according to the passage of various laws and appropriation bills, or according to the terms of the various administrators who headed the National Aeronautics and Space Administration (NASA). In contrast, if your readers are scientists and engineers, your segments might focus on the efforts to overcome various technical problems.

2. **Make your smallest groupings of a manageable size.** One of the most important things for you to do when describing a process is to make your smallest groupings manageable. If you include too many or too few steps in them, your readers will lose the benefit of your segmenting. They will not see the process as a structured hierarchy of activities or events, but merely as a long, unstructured list of steps—first one, then the next, and so on. If you are writing instructions, the lack of structure will make it harder for your readers to learn the task, and if you are describing a process you want your readers to understand, the lack of structure will make the process more difficult for them to comprehend.

3. **Make clear the relationships among the steps.** Finally, keep in mind that your segments should help your readers understand and remember the entire process. To do this, your readers will probably need to understand the relationships among the events and steps that make up the process. Sometimes, the relationship can be stated as simply as, ''Here are the steps you perform to make your cake.'' In other instances, the relationships are much more complex.

   There are many ways you can make clear how the various steps and events are related to one another. Where the relationships are obvious, you may be able to do this simply by providing informative headings. At other times, you will need to explain the relationships in an overview at the beginning of the segment, weave additional explanations into your discussions of the steps themselves, and explain the relationships again in a summary at the end. The important thing is that your readers understand the relationships among the groups and subgroups in the hierarchy of your description.

Sample Process Descriptions

Figure 8–4 shows a section of a research report. The report, which describes a special method for making solar cells, is organized by segmenting.

Figure 8–5 shows a group of paragraphs in which the writer has described the general process of the way wood burns. Notice how the writer emphasizes how each phase of the process is significant to his particular readers, people who are concerned with the amount of energy that can be derived from burning wood.

COMPARISON

At work, you will have many occasions to write segments that compare two or more things. Overall, these occasions fall into two categories.

- **When you want to help your readers make a decision.** The workplace is a world of choices. People are constantly choosing among available courses of action, competing products, alternative strategies. For this reason, you will often write communications in which you must compare and contrast two or more things.
- **When you want to help your readers understand something by means of an analogy.** One of the most effective ways to tell your readers about something new to them is to tell how it is like—and unlike—something they are familiar with.

In some ways, a comparison is like a classification. You begin with a large set of facts about the items that you will compare and then you create related groups of facts. In a comparison, you group your facts around points about the items that enable your readers to see how the items are like and unlike one another.

How Comparisons Work

When you want to compare two things, you begin by choosing the points upon which you want to base the comparison. For instance, to compare two pieces of equipment, you might select criteria such as cost and performance.

Section 2

Section 2
EXPERIMENTAL METHOD

General
background is
explained

The wafers for this study were processed simultaneously under identical conditions, and they lacked only a front contact. The 1-to-3 cm, boron-doped Czochralski-grown, single-crystal silicon wafers were texture-etched in NaOH. Then phosphorus, by means of phosphine gas, was diffused in to form a junction at a depth of 0.3 mm. The cells also were etched in nitric and hydrofluoric acid (HF) mixture, and a 2% aluminum, silver-based ink was screen-printed and dried on their backs to form the electrical contact. The ink processing sequence included five major steps: printing, drying, firing, HF etching, and solder dipping.

Overview of the
procedure is
presented

2.1 PRINTING

First step is
discussed.
Topics are:
procedure
used,
parameters
adjusted, and
data collected

A Presco (Model 435) automatic screen printer was used to print the thick-film inks. A grid pattern with lines 250 μm (10 mil) wide, a 2.5-mm square for adhesion testing, and two short 125-μm-wide lines were printed on each cell (Fig. 2-A). The adjustable parameters in the printing step included the

A

Figure 2–A.   Top View of Solar Cell with Screen-Printed Front Contact Showing 250-μm Grid Pattern with (A) Adhesion Test Pad, and (B) 125-μm Line.

FIGURE 8–4   Section Organized by Segmenting a Process

From Steve Hogan and Kay Firor, *Evaluation of Thick-Film Inks for Solar Cell Grid Metallization* (Golden, Colo.: Solar Energy Research Institute, 1981), 3–5.

2

screen mesh size, the snapback (distance of the screen above the substrate), and the squeegee pressure. After printing, data were obtained by observing how well the 125-μm line printed and how easily the ink cleaned off the screen after spraying it with trichloroethane solvent.

2.2 DRYING

Second step is discussed. Topics are: procedure used and parameters adjusted

After printing, the ink was allowed several minutes to settle. Visual inspection after this settling period determined how well the ink flow had smoothed out the screen impressions left on the printed surface of the film. The ink was then air-dried under an infrared (IR) lamp. Adjustable parameters during the drying stage included the time allowed for settling and the exposure time and temperature under the IR lamp.

2.3 FIRING

Third step is discussed. Topics are: procedures used, parameters adjusted, and data collected.

The cells were placed in a four-temperature-zone, Watkins-Johnson belt furnace for firing immediately after the ink dried. Figure 2-B shows a typical furnace profile, or plot, of

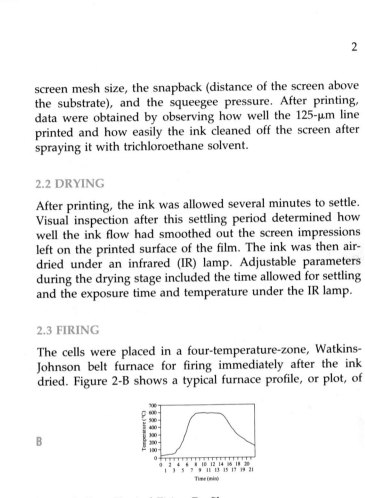

Figure 2–B.   Typical Firing Profile.

FIGURE 8–4   *(continued)*

3

temperature versus time. Firing took place in air, as recommended by all ink manufacturers. On several occasions, however, a nitrogen firing atmosphere was used to test for improved cell performance. Air flow rates, zone temperatures, and firing time were the adjustable parameters in this step. Immediately after the firing process, the first I-V curve was measured for each cell.

2.4 HF ETCH

Fourth step is discussed. Topics are: procedure used and parameters adjusted

The next processing step for most Ag-based inks was an HF etch. The cells were dipped in aqueous solution of either 2% or 5% (by weight) HF for 5 to 12 seconds. The cells then were rinsed in deionized water for 5 minutes, dried with air, and tested with another I-V curve. HF concentration and the etching time were the parameters adjusted during this step.

2.5 SOLDERING

Fifth step is discussed. Topics are: procedure used, data collected, and parameters adjusted

Each cell was solder-dipped to coat and somewhat protect the cell metallization from air moisture, which causes degradation of film adhesion. The cell metallization was fluxed with either Kester 1544 or Kester 1589 flux. The entire cell then was dipped into a bath of tin—lead solder (with 2% silver), typically heated to 210°C. A visual inspection then determined how well the solder coated the grid. An I-V curve was taken after soldering to determine the effects of this step on cell performance. Dip time, solder bath temperature, and the type of soldering flux were the adjustable parameters.

FIGURE 8–4   *(continued)*

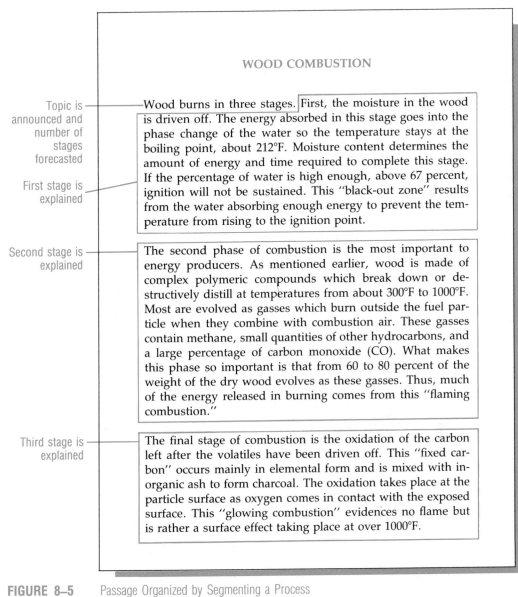

WOOD COMBUSTION

Topic is announced and number of stages forecasted

First stage is explained

Wood burns in three stages. First, the moisture in the wood is driven off. The energy absorbed in this stage goes into the phase change of the water so the temperature stays at the boiling point, about 212°F. Moisture content determines the amount of energy and time required to complete this stage. If the percentage of water is high enough, above 67 percent, ignition will not be sustained. This "black-out zone" results from the water absorbing enough energy to prevent the temperature from rising to the ignition point.

Second stage is explained

The second phase of combustion is the most important to energy producers. As mentioned earlier, wood is made of complex polymeric compounds which break down or destructively distill at temperatures from about 300°F to 1000°F. Most are evolved as gasses which burn outside the fuel particle when they combine with combustion air. These gasses contain methane, small quantities of other hydrocarbons, and a large percentage of carbon monoxide (CO). What makes this phase so important is that from 60 to 80 percent of the weight of the dry wood evolves as these gasses. Thus, much of the energy released in burning comes from this "flaming combustion."

Third stage is explained

The final stage of combustion is the oxidation of the carbon left after the volatiles have been driven off. This "fixed carbon" occurs mainly in elemental form and is mixed with inorganic ash to form charcoal. The oxidation takes place at the particle surface as oxygen comes in contact with the exposed surface. This "glowing combustion" evidences no flame but is rather a surface effect taking place at over 1000°F.

FIGURE 8–5   Passage Organized by Segmenting a Process
From N. Elliot, "Wood Combustion," in *Decision-Maker's Guide to Wood Fuel for Small Energy Users,* ed. Michael P. Levi and Michael J. O'Grady (Golden, Colo.: Solar Energy Research Institute, 1980), 32.

Then you gather relevant facts about the items being compared. If the comparison is intended to help the readers make a decision, you can also draw general conclusions from the comparison.

If you have read Chapter 3, you already know that there are two patterns for presenting comparison. In the first, called the alternating pattern, you organize around your criteria:

Statement of Criteria

Overview of the Two Alternatives

Evaluation of the Alternatives
   Criterion 1
      Alternative A
      Alternative B
   Criterion 2
      Alternative A
      Alternative B
   Criterion 3
      Alternative A
      Alternative B

When you use this pattern, you enable your readers to make a point-by-point comparison without flipping back and forth through the communication. The pertinent information about both alternatives is presented together. As the outline given above suggests, when this pattern is used it is often necessary to precede the evaluation with an overview of the alternatives. Otherwise, readers may have to piece together an overall understanding of each alternative from the specific details you provide when discussing the alternative point by point.

In the second pattern, called the divided pattern, you organize not around the criteria but around the alternatives themselves:

Statement of Criteria

Description of Alternatives
   Alternative A
      Overview
      Criterion 1
      Criterion 2
      Criterion 3
   Alternative B
      Overview
      Criterion 1
      Criterion 2
      Criterion 3

Comparison of Alternatives

The divided pattern is well suited to situations where both the general nature and the details of each alternative can be described in a short space, say one page or so. This pattern was used, for instance, by an employee of a restaurant who was asked to investigate the feasibility of buying a new turntable and amplifier for the restaurant's music system. He described each of three turntables and each of three amplifiers in about one page. In the final section, his comparison of alternatives, he briefly provided an analysis that enabled his readers to choose the best purchase.

Guidelines for Writing Comparisons

1. **Choose points of comparison suited to your purpose and readers.** When you are preparing comparisons for decision-makers, be sure to include not only the criteria that those people consider important but also any additional criteria that you—through your expert knowledge—realize are significant.

   When you are writing comparisons to create analogies, be sure to compare and contrast the things under consideration in terms of features that will help the readers understand the points you are trying to make. Avoid comparing extraneous details.

2. **If you are making complex comparisons, arrange the parts of your comparison hierarchically.** For example, group information on all kinds of cost (purchase price, operating cost, maintenance cost, and so on) in one place, information on all the aspects of performance in another place, and so on.

3. **Arrange your groups in an order your readers will find helpful.** In comparisons made for the purpose of helping readers make decisions, you should usually discuss at the outset the criteria that show the most telling difference between the things being compared. In comparisons designed to aid understanding, you should usually discuss points of likeness first. In this way, you can begin with what your readers will find familiar, and then lead them to the less familiar.

Sample Comparisons

Figure 8–6 shows a segment that compares two methods of studying the ways that mothers might affect their children if they take drugs during pregnancy.

CAUSE AND EFFECT

You may use the cause-and-effect pattern in two very different ways. In one, your aim is *descriptive*. You help your readers understand the cause or consequences of some action or event, as when you answer a child's question, "What makes a rainbow?"

In other circumstances, your aim is *persuasive*. You try to persuade your readers that some event or action had the specific causes you believe it had or that some event or action will have the specific consequences you predict. For

CLINICAL AND EPIDEMIOLOGICAL STUDIES OF THE EFFECTS OF PRENATAL EXPOSURE TO DRUGS

The need for epidemiological studies is explained

Since experimental administration of drugs to pregnant women is unethical, studies of the effects on their children of their using drugs is limited to clinical observation and epidemiological investigations. Although clinical reports can be of considerable importance in alerting physicians and health care providers to possible agents causing abnormal development, they are difficult to evaluate. For example, two early clinical reports of malformations in children born to marijuana users (Hecht et al. 1968; Carakushansky et al., 1969) were inconclusive since the mothers of these children used other drugs as well.

Comparison of two types of epidemiological studies is announced

When clinical reports are followed by epidemiological studies involving larger numbers of patients, a better appreciation of incidence and causation is possible. Such epidemiological studies can be divided into two types, retrospective and prospective, each of which has its own strengths and shortcomings.

First type is discussed: strength, then weaknesses

In most retrospective studies, information from large numbers of cases is obtained from hospital records. However, such records are often inadequate or incomplete, thoroughness of reporting varies widely, and criteria for assessment of anomalies may also vary. In contrast to retrospective

Second type is discussed: strength matched to weakness of first type, its weaknesses

studies, prospective studies carefully establish criteria and protocols for maternal histories and examination of infants in prenatal health clinics. However, women who are usually most seriously at risk for giving birth to infants with drug-related anomalies may not attend prenatal health care facilities and, therefore, do not participate in prospective studies, resulting in underestimation of whatever problem is being investigated. Because prospective studies are so rigid in their design, they also are less flexible in allowing for changes to be incorporated as new information is obtained. Also, prospective studies cannot anticipate knowledge. For example, in the U. S. Collaborative Perinatal Project (Heinonen et al. 1977) which prospectively evaluated 55,000 consecutive births, no information was obtained with respect to maternal marijuana consumption because, at the time of the original protocol, marijuana was not a suspected teratogen.

FIGURE 8–6   Passage Organized by Comparison

From Ernest L. Abel, "Effects of Prenatal Exposure to Cannabinoids," *Current Research on the Consequences of Maternal Drug Abuse,* ed. Theodore M. Pinkert (Washington, D.C.: U.S. Department of Health and Human Services, 1985), 20–21.

example, you might try to persuade your readers that the damage to a large turbine generator (effect) resulted from metal fatigue in a key part (cause) rather than from a failure to provide proper lubrication. Or you might try to persuade your readers that cutting the selling price of a product (cause) will increase sales enough to bring about a greater profit than the old price did (effect).

Because discussions of cause and effect deal with processes, they resemble discussions organized by segmenting a process. The key difference is that in segmenting a process you focus on the steps of the process, and in discussing cause and effect you focus on the links between the steps.

Guidelines for Describing Causes and Their Effects

1. **Begin by identifying the cause or effect you are going to describe.** Your readers will want to know from the beginning of your segment exactly what you are trying to explain so that they will know what they should be trying to understand as they read it. Sometimes you can describe the cause or effect in a single sentence. At other times a full description of it will take several sentences or even longer.

2. **Carefully explain the links in the chain of cause and effect that you are describing.** Remember that when you are describing cause and effect, you are not simply listing the steps in a process. You want your readers to understand how each step leads to the next one or is caused by the preceding one.

3. **If you are dealing with several causes or effects, group them into categories.** When you create categories, you build a hierarchy that your readers will find helpful as they try to understand a complex chain of events.

Guidelines for Persuading about Causes and Their Effects

1. **State your claim at the beginning of your passage.** Your claim will be that some particular effect was created by some particular cause, or that some particular cause will lead to a particular effect.

2. **Present your evidence and lines of reasoning.** Where possible, focus on undisputed evidence, because your readers' willingness to agree with your claim depends upon their willingness to accept your evidence. Use lines of reasoning your readers will agree are logically sound and appropriate to the situation you are discussing.

3. **Anticipate and respond to objections.** In cause and effect passages, as in any persuasive passage, your readers may object to your evidence or to your line of reasoning. The kind of reasoning that most often spurs objections by readers of cause and effect segments is one that contains the *post*

hoc ergo propter hoc fallacy. In this form of faulty reasoning, a writer argues that because something happened after another event, it was caused by that event. You can see the fault with this type of reasoning if you consider this example. In an attempt to persuade his employer, a furniture company, that the company should use computerized machinery for some of its manufacturing operations, Samuel argued that a competitor's profits rose substantially after that company invested in similar computerization. Samuel's boss pointed out that the other company's increased sales might have been caused by other changes the company had made in the same period, such as new designs and a reconfiguration of its sales districts. To persuade that the computerization *caused* the increase, Samuel would have to do *more* than show that computerization preceded it.

When writing persuasive cause-and-effect passages, you may benefit from reviewing the discussion of Guideline 3 in Chapter 5, which discusses ways to show your readers that you are reasoning soundly.

Sample Cause-and-Effect Segment

Figure 8–7 shows a segment in which the author explains one theory about the cause of the extinction of the dinosaurs.

PROBLEM AND SOLUTION

Like cause-and-effect patterns, problem-and-solution patterns fall into two categories: descriptive and persuasive. You can use a problem-and-solution pattern *descriptively* to explain things that have happened, such as how you and your co-workers solved the problem of the leaking reservoir, how you solved the problem of the large number of service calls on your employer's products, and so on. You can use a problem-and-solution pattern *persuasively* when you want your readers to agree that the particular actions you recommend will solve problems that they would like to overcome.

Guidelines for Describing Problems and Their Solutions

1. **Begin by identifying the problem that was solved.** Remember to make the problem seem significant to your readers. Emphasize the parts of the problem that were most directly addressed by the solution.

2. **Carefully explain the links between the problem and the solution.** Remember that you want your readers to understand *how* the solution overcame the problem.

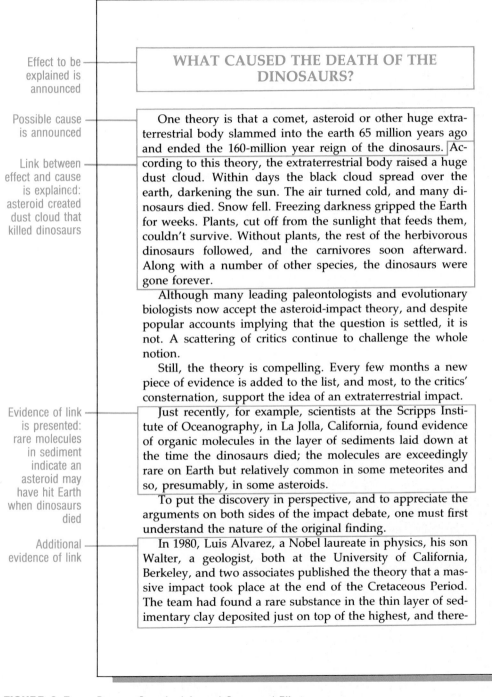

Effect to be explained is announced

WHAT CAUSED THE DEATH OF THE DINOSAURS?

Possible cause is announced

One theory is that a comet, asteroid or other huge extraterrestrial body slammed into the earth 65 million years ago and ended the 160-million year reign of the dinosaurs. According to this theory, the extraterrestrial body raised a huge dust cloud. Within days the black cloud spread over the earth, darkening the sun. The air turned cold, and many dinosaurs died. Snow fell. Freezing darkness gripped the Earth for weeks. Plants, cut off from the sunlight that feeds them, couldn't survive. Without plants, the rest of the herbivorous dinosaurs followed, and the carnivores soon afterward. Along with a number of other species, the dinosaurs were gone forever.

Link between effect and cause is explained: asteroid created dust cloud that killed dinosaurs

Although many leading paleontologists and evolutionary biologists now accept the asteroid-impact theory, and despite popular accounts implying that the question is settled, it is not. A scattering of critics continue to challenge the whole notion.

Still, the theory is compelling. Every few months a new piece of evidence is added to the list, and most, to the critics' consternation, support the idea of an extraterrestrial impact.

Evidence of link is presented: rare molecules in sediment indicate an asteroid may have hit Earth when dinosaurs died

Just recently, for example, scientists at the Scripps Institute of Oceanography, in La Jolla, California, found evidence of organic molecules in the layer of sediments laid down at the time the dinosaurs died; the molecules are exceedingly rare on Earth but relatively common in some meteorites and so, presumably, in some asteroids.

To put the discovery in perspective, and to appreciate the arguments on both sides of the impact debate, one must first understand the nature of the original finding.

Additional evidence of link

In 1980, Luis Alvarez, a Nobel laureate in physics, his son Walter, a geologist, both at the University of California, Berkeley, and two associates published the theory that a massive impact took place at the end of the Cretaceous Period. The team had found a rare substance in the thin layer of sedimentary clay deposited just on top of the highest, and there-

FIGURE 8–7   Passage Organized Around Cause and Effect

From Boyce Rensberger, "Death of Dinosaurs: A True Story?" *Science Digest* 94:5 (May 1986): 28–32.

2

Additional evidence continued

fore the most recent, stratum of rock contemporary with those bearing dinosaur fossils. It was the element iridium, which is almost nonexistent in the Earth's crust but 10,000 times more abundant in extraterrestrial rocks such as meteorites and asteroids. Deposits above and below the clay, which is the boundary layer separating the Cretaceous layer from the succeeding Tertiary, have very little iridium.

Because the same iridium anomaly appeared in two other parts of the world, in clay of exactly the same age, the Alvarez team proposed that the element had come from an asteroid that hit the Earth with enough force to vaporize, scattering iridium atoms in the atmosphere worldwide. When the iridium settled to the ground, it was incorporated in sediments laid down at the time.

Link is restated

More startling was the team's proposal that the impact blasted so much dust into the atmosphere that it blocked the sunlight and prevented photosynthesis (others suggested that a global freeze would also have resulted). They calculated that the object would have had to be about six miles in diameter.

Additional evidence of link

Since 1980, iridium anomalies have been found in more than 80 places around the world, including deep-sea cores, all in layers of sediment that formed at the same time.

Challenge to link is explained: molecules perhaps from volcano, not asteroid

One of the most serious challenges to the extraterrestrial theory came up very quickly. Critics said that the iridium could have come from volcanic eruptions, which are known to bring up iridium from deep within the Earth and feed it into the atmosphere. Traces of iridium have been detected in gases escaping from Hawaii's Kilauea volcano, for example.

Challenge is refuted by new evidence: other molecules couldn't have come from volcano

The new finding from Scripps appears to rule out that explanation, though, as a source for iridium in the Cretaceous–Tertiary (K–T) boundary layer. Chemists Jeffrey Bada and Nancy Lee have found that the same layer also contains a form of amino acid that is virtually nonexistent on Earth—certainly entirely absent from volcanoes—but abundant, along with many other organic compounds, in a type of meteor called a carbonaceous chrondrite.

FIGURE 8–7   *(continued)*

3. **If the solution involves several actions, group the actions into categories.** When you create categories, you build a hierarchy that your readers will find helpful as they try to understand a complex solution.

Guidelines for Persuading about Problems and Their Solutions

1. **Describe the problem in a way that makes it significant to your readers.** Remember that your overall aim is to persuade your readers to take the action you recommend because the action will solve some problem. Your readers will not be very interested in an action that solves a problem they don't care about.

2. **Present your evidence and lines of reasoning.** Be sure to use evidence that your readers will find sufficient and reliable, and lines of reasoning your readers will agree are logically sound and appropriate to the situation at hand.

3. **Anticipate and respond to objections.** In problem and solution passages, as in any persuasive passages, your readers may object to your evidence or your lines of reasoning. In responding to objections you expect your readers to raise, you can use any of the strategies discussed in the section on persuasive segments.

When writing persuasive problem-and-solution passages, you may wish to refer to Guideline 3 in Chapter 5, which discusses ways to show your readers that you are reasoning soundly.

Sample Problem-and-Solution Segment

Figure 8–8 shows a memo organized into a discussion of a problem and its possible solution.

EXERCISES

1. Some principles of classification you might use to group students include major and year in school. Think of five other principles you might use to classify students. Then identify a possible reader and purpose that would be served well by each. Are there any that seem to be useful to no one?

2. To choose the appropriate principle of classification for organizing a group of items, you need to consider your readers and purpose. Below you will find four topics for classification, each listed with two possible readers. First, identify a purpose that each reader might have for using a communication on that topic. Then identify a principle of classification that would be appropriate for each reader and purpose.

Problem is identified

Problem is explained

Solution is announced

Solution is explained

Link between problem and solution is explained

MANUFACTURING PROCESSES INSTITUTE
Interoffice Memorandum   June 21, 19—

To: Cliff Leibowitz

From: Candace Olin

RE: Suggestion to Investigate Kohle Reduktion Process for Steelmaking

As we have often discussed, it may be worthwhile to set up a project investigating steelmaking processes that could help the American industry compete more effectively with the more modern foreign mills. I suggest we begin with an investigation of the Kohle Reduktion method, which I learned about in the April 1986 issue of High Technology.

A major problem for American steelmakers is the process they use to make the molten iron ("hot metal") that is processed into steel. Relying on a technique developed on a commercial scale over 100 years ago by Sir Henry Bessemer, they make the hot metal by mixing iron ore, limestone, and coke in blast furnaces. To make the coke, they pyrolize coal in huge ovens in plants that cost over $100 million and create enormous amounts of air pollution.

In the Kohle Reduktion method, developed by Korf Engineering in West Germany, the hot metal is made without coke. Coal, limestone, and oxygen are mixed in a gasification unit at 2500°. The gas rises in a shaft furnace above the gasification unit, chemically reducing the iron ore to "sponge iron." The sponge iron then drops into the gasification unit, where it is melted and the contaminants are removed by reaction of the limestone. Finally, the hot metal drains out of the bottom of the gasifier.

The Kohle method, if developed satisfactorily, will have several advantages. It will eliminate the air pollution problem of coke plants, it can be built (according to Korf estimates) for 25% less than conventional furnaces, and it may cut the cost of producing hot metal by 15%.

This technology appears to offer a dramatic solution to the problems with our nation's steel industry: I recommend that we investigate it further. If the method proves feasible and if we develop an expertise in it, we will surely attract many clients for our consulting services.

FIGURE 8–8   Segment Organized around a Problem and Its Solution

Types of instruments or equipment used in your field
Student majoring in your field
Director of purchasing at your future employer's organization
Types of communications you might write on the job
You
Your future typist
Intramural sports
Director of intramural sports at your college
Student
Flowers
Florist
Owner of a greenhouse that sells garden plants

3. Use a classification to create a hierarchy having at least two levels. Some suggested topics are listed below. After you have selected a topic, identify a reader and purpose for your classification. Depending upon your instructor's request, show your hierarchy in an outline or use it to write a brief discussion of your topic. In either case, state your principle or basis of classification. Have you created a hierarchy that, at each level, has one and only one place for every item?

Boats

Cameras

Pets

Computers

Footwear

Physicians

The skills you will need on the job

Tools, instruments, or equipment you will use on the job

Some groups of items used in your field (for example, rocks if you are a geologist, or power sources if you are an electrical engineer)

4. Partition an object in a way that would be helpful to someone who wants to use it. Some topics are selected below. Whichever item you choose, describe one specific instance of it. For example, describe a certain brand and model of food processor, not a generic food processor. Be sure that your hierarchy has at least two levels, and state the basis of partitioning you use at each one. Depending upon your instructor's request, show your hierarchy in an outline or use it to write a brief discussion of your topic.

Food processor

Microwave oven

Theater

Bicycle

Some instrument or piece of equipment used in your field that has at least a dozen parts

5. Segment a procedure to create a hierarchy that you could use in a set of instructions. Give it at least two levels. Some topics are listed below. Show the resulting hierarchy in an outline. Be sure to identify your readers and purpose. If your instructor requests, use the outline to write a set of instructions.

Changing an automobile tire   Developing a roll of film
Making a pizza from scratch   Some procedure used in your field
Making homemade yogurt   that involves at least a dozen
Starting an aquarium   steps
Rigging a sailboat   Some other procedure of interest to
Planting a garden   you that includes at least a dozen
                                         steps

6. Segment a procedure to create a hierarchy that you could use in a general description of a process. Give it at least two levels. Some suggested topics are listed below. Show the resulting hierarchy in an outline. Be sure to identify your readers and purpose. If your instructor requests, use the outline to write a general description of the process addressed to someone who wants to understand it.

How the human body takes oxygen from the air and delivers it to the parts of the body where it is used

How television signals from a program originating in New York or Los Angeles reach television sets in other parts of the country

How aluminum is made

Some process used in your field that involves at least a dozen steps

Some other process of interest to you that includes at least a dozen steps

7. In courses in your major field, you may be learning about alternative ways of doing things. For instance you may be learning how to keep patient records for a doctor's office on paper and on a computer. Write an outline comparing a pair of alternatives from the point of view of someone who has a practical need to choose between them. Use the alternating pattern of organization. If your instructor requests, use that outline to write a memo to the person you had in mind.

8. Imagine that one of your friends is thinking of making a major purchase. Some possible items are listed below. Help your friend by creating an outline that has at least two levels and compares two or more good alternatives. If your instructor requests, use that outline to write your friend a letter.

Stereo

Binoculars

VCR

Personal computer

Bicycle

Some other type of product for which you can make a meaningful comparison on at least three important points

9. Think of some way that things might be done better in a club, business, or other organization you know of. Imagine that you are going to write a letter to the person or people who actually have the power to bring about the change you desire. Write an outline that has at least two levels and compares the way you think things should be done and the way they are done now.

10. Imagine that a friend of yours has asked you to explain the causes of some occurrence or event. Some suggested topics are listed below. Write that person a brief letter explaining those causes.

Sunspots

Static on radios and televisions

Freezer burns in foods

Twin births

Yellowing of paper

Earthquakes

11. Think of a problem that you have some ideas about solving. The problem might be noise in your college's library, shoplifting from a particular store, or parking problems on campus. Then, briefly describe the problem and list actions you would take to solve it. Next, explain how each action will contribute to solving the problem. If your instructor requests, use your outline to write a brief memo explaining the problem and your proposed solution to a person who could actually take the actions you suggest.

9

BEGINNING A COMMUNICATION

DEFINING
OBJECTIVES
PLANNING
DRAFTING
EVALUATING
REVISING

I magine that you are attending the first day of a new course. You don't know any of the other students. Afterwards, the person sitting next to you starts a conversation. How will you respond?

Your response will probably depend upon the other person's opening words. The subject this person talks about, the person's tone of voice, even the way the person phrases his or her comments may all determine whether you decide to chat or rush off to your next destination. If you stay, you will probably shape your own comments in light of the first things the other person says.

The opening words of the memos, reports, and other communications you write at work will be very much like the opening words of a conversation. They may influence greatly the way your readers react to your overall message. They may even determine whether your readers decide to keep on reading.

This chapter presents five guidelines that will help you begin your communications in ways that will get the responses you want from your readers.

THE TWO FUNCTIONS OF A BEGINNING

Before reading these guidelines, however, you may find it helpful to think about the two distinct functions that the beginning of a communication performs. First, the beginning introduces your audience to your message. You use it to persuade your readers to pay close attention and to influence their responses to what follows. The guidelines in this chapter will help you achieve those objectives.

Second, the beginning of a communication serves as the beginning of a group of paragraphs; in this case, the group consists of *all* the paragraphs in the message. For that reason, several of the things Chapter 7 suggested you do at the beginning of a paragraph, section, or chapter are things you should usually do when writing the beginning of a whole communication:

● Announce the topic.
● Begin with your main point.
● Provide a forecasting statement.

Those important pieces of advice from Chapter 7 are incorporated in Guidelines 1, 2, and 4 of this chapter so that you can see how to apply them when writing the beginning of a communication.

GUIDELINE 1   GIVE YOUR READERS A REASON TO PAY ATTENTION

The most important function of a beginning is to attract your readers' attention.

Unfortunately, attracting attention can be a very difficult task. At work, people frequently complain that they are sent too much mail. They routinely select which items will get a great deal of attention, which ones a little attention, and which ones no attention. As they pick up a memo or report from you, the first question they will ask is, "Why should I read this?" If they don't find a persuasive answer quickly, they may file or throw away your communication without examining it thoroughly.

In the face of these difficult circumstances, you must strive not merely for *some* attention from your audience, but for their *close* attention. Doing so will be especially important when your communication's overall purpose is primarily persuasive. Research has shown that the more deeply people think about a message while reading or listening to it, the longer they are likely to adopt the attitudes advocated by the message, the more likely they are to resist subsequent attempts to reverse those attitudes, and the more likely they are to act upon those attitudes.[1]

To attract readers' serious attention, you must usually do two things at the very beginning of your communications:

● Announce your topic.
● Tell your readers how they will benefit from the information your communication provides.

Be sure to do *both* things: guard against assuming that your readers will automatically see the benefit your communication offers if you simply state its topic. The benefit that appears obvious to you may not be evident to your audience. Further, even if they can deduce the benefit after a moment's thought, you still increase your readers' efficiency and satisfaction with your communication by indicating the benefit clearly and early. Compare the following sets of statements:

STATEMENTS OF TOPIC ONLY (AVOID THEM)	STATEMENTS OF TOPIC AND BENEFIT (USE THEM)
This memo tells about polymer coatings.	This memo answers each of the five questions you asked me last week about polymer coatings.
This report discusses step-up pumps.	Step-up pumps can save us money and increase our productivity.
This manual concerns the Cadmore Industrial Robot 2000.	This manual tells how to prepare the Cadmore Industrial Robot 2000 for routine welding tasks.

The following sections describe two strategies for persuading people that they will benefit from reading your communication: refer to your readers' request and offer to help your readers solve a problem.

Refer to the Readers' Request

At work, you will often write in response to a request from the person you are addressing. You can explain the benefit of these communications simply by referring to the request.

> Here are the test results you asked for.

> As you requested, I am enclosing a list of the steps that we have taken in the past year to tighten security in the Data Processing Department.

> Thank you for your inquiry about the capabilities of our Model 1770 Color Photocopier.

Offer to Help Your Readers Solve a Problem

A second way to persuade your readers to pay attention to your communication is to explain that it will help them solve a problem that is important to them. This sort of beginning can be especially effective at work because most employees see themselves as problem-solvers. Sometimes they are concerned with solving technical problems, such as finding out how to preserve the freshness of their company's meat products without using harmful chemicals or how to detect tiny flaws in the metal used to build airplane wings. At other times, they are concerned with solving organizational problems, such as finding ways to improve morale or reorganize the quality assurance department to increase its efficiency. Although specialists in different fields work on different kinds of problems, all usually respond favorably to offers to help them solve problems with which they are concerned.

Communication experts J. C. Mathes and Dwight W. Stevenson have suggested an especially powerful approach to writing beginnings that builds on readers' concerns with problem-solving.[2] First, think of the problems that are important to the people you are going to address in your communication. Then, identify from that list a problem that you will help your readers solve by presenting them with the information and ideas to be included in your communication. When you've done that, you have begun to think of yourself and your readers as partners in a joint problem-solving effort in which your communication plays a critical role. Figure 9–1 illustrates this relationship between you, your readers, and your communication.

Once you have determined how to describe a problem-solving partnership between you and your readers, you are ready to draft the beginning of your communication. In it, you would typically explain the following three things:

- **What the problem is.** Be careful to identify a problem that your readers will think important and to discuss it from your readers' point of view.

- **What you have done toward solving the problem.** These will be activities you have performed as a specialist in your own field, such as developing a new feature for one of your employer's products, investigating products offered by competitors, writing a new policy for handling pur-

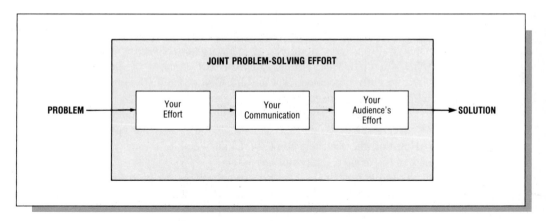

FIGURE 9–1   The Writer and the Readers Are Problem-Solving Partners

Based on J. C. Mathes and Dwight W. Stevenson, *Designing Technical Reports* (Indianapolis: Bobbs-Merrill, 1976).

chase orders, or whatever. Remember to describe only the activities that are significant to your readers, rather than everything you have done.

● **How your communication will help them.** Let your readers know what your communication will do for them as they perform their own part of your joint problem-solving effort: enable them to compare competitive products with your company's product, assist them in developing a new policy for making purchase orders, guide them as they create a new marketing plan, and so on.

Figure 9–2 shows the relationships among the three elements of a beginning.

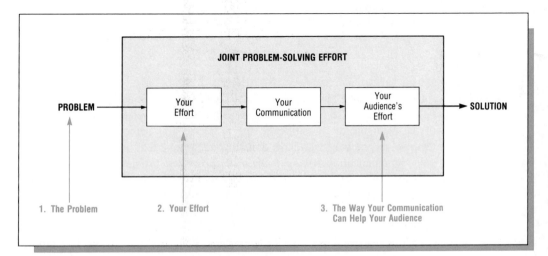

FIGURE 9–2   Three Elements of a Beginning that Describe a Problem-Solving Situation

An Example   To see how you can describe these three elements of a problem-solving situation, consider the beginning of a report by Carla, a systems analyst employed by a large hotel chain. Carla's report concerns a trip to Houston, where she studied the billing system used by one of the hotels that her employer recently purchased. She has already drafted the body of her report; now she must write the beginning.

Carla starts by describing the problem that her communication will help her readers solve. To do this, she simply states why her boss asked her to go to Houston: the Houston hotel is making very little money, perhaps because the hotel is using a faulty billing system.

To explain what she has done to help solve the problem, Carla tells what she did in Houston and afterwards to evaluate the hotel's billing system and to recommend improvements in it.

To tell how her report will help her readers perform their own problem-solving activities, Carla explains that her report provides the results of her analysis, together with her recommendations. Using this information, her readers can decide what action to take.

Here is the beginning Carla wrote:

Overall problem Carla will help her readers solve	Over the past two years, our Houston hotel has shown a profit of only 4 percent, even though it is almost always 78 percent filled. A preliminary examination of the hotel's operations suggests that its billing system may be inadequate: it may be too slow in billing customers, and it may be inefficient and needlessly unsuccessful in
Carla's work toward solving that problem	collecting overdue payments. Therefore, I have thoroughly examined the hotel's billing cycle and its collection procedures, and I have considered ways to improve them.
How Carla's report will help her readers solve the problem	In this report, I present the results of my analysis, together with my recommendations. To aid in the evaluation of my recommendations, I have included an analysis of the costs and benefits of each.

Must You Always Provide All This Information?   Descriptions of problem-solving situations can be much shorter than Carla's. Sometimes they need to be much longer. The crucial point is to be sure your readers *understand* each of the elements of the situation: the problem your communication will help them solve, the work you have done toward solving that problem, and the way your communication will help them perform their problem-solving activities. In many situations, your audience will understand some of the elements of the situation without being told explicitly. For example, if you are responding to your readers' request for information, you may not need to remind them what the problem is. It may be sufficient to tell them that you are providing the requested material. In such a situation, your activities toward solving the problem may likewise be obvious to them.

Consider, for example, a brief explanation of relevance that Carla might have used had she been sure all her readers were familiar with the problem with the Houston hotel and with her activities there:

> In this report, I evaluate the billing system in our Houston hotel and recommend ways of improving it.

Before using such an abbreviated description, however, you should be sure that all the likely members of your audience will understand immediately your communication's relevance to them even if you don't state the relevance explicitly.

Some Signs That You Should Provide a Full Description   Here are some signs that you probably should provide a full description of the problem-solving situation.

- **Your communication will be read or heard by people outside your immediate working group.** The larger your audience, the less likely it is that all audience members will be very familiar with the context of your message.
- **Your communication will have a binding and a cover.** Documents prepared in this way are usually prepared for large groups of readers. They may also be filed so that people years in the future can use them. Unless you tell them, many readers of bound reports will have no idea of the organizational situation in which you wrote.
- **Your communication will be used to make a decision involving a significant amount of money.** Such decisions are often made at upper levels of management. These managers usually need to be told of the organizational context of the reports they read. For example, although Carla's boss asked her for the report, the people who will make a decision based upon her recommendations are high-level managers responsible for the overall operation of 247 hotels. Carla cannot assume that they know in advance about the situation in Houston.

Defining the Problem in Unsolicited Communications   There is one situation in which you will have to devote a special effort to determining the problem that your communication will help solve: when you write unsolicited communications (communications you prepare on your own initiative, without having been asked to do so). Such situations often arise at work. According to a survey of college alumni, the majority write on their own initiative at least as often as they write on assignment.[3]

When you write unsolicited communications, you can't ask someone else what the problem is. You must decide what it is on your own. Furthermore, you can't pick just any problem. You must identify one that your readers will think is important.

The key step in identifying a problem relevant to your readers is to forget (at least for the time being) your own reasons for writing. Instead, consider the situation you are writing about from the point of view of your audience.

Suppose, for instance, that you want your boss to purchase a personal computer for you to use at work. You may want this computer primarily so that your job will be easier. In particular, the computer will enable you to complete rapidly a lengthy report that you must prepare at the end of each month. To complete that report on time each month, you currently work many overtime hours. However, you know that your boss also regularly works overtime and that reducing the amount of overtime you work is not a top priority to her.

Accordingly, you must find another problem that the computer will solve—a problem that will seem important to your boss. For example, you may know that she very much wants you to work on several projects that the two of you agree you do not have time to address. You might make this the problem you identify in your statement of relevance: with a computer, you could finish your monthly report more rapidly and have more time for the important projects that currently are neglected.

In sum, to describe the benefit your readers will gain from reading your communication, you must look from your readers' point of view at both the situation you are addressing and the communication itself. There is no task for which a reader-centered approach is more critical than for explaining the significance of your message to your audience.

GUIDELINE 2

STATE YOUR MAIN POINT

At the beginning of a communication it is often important to state the main point of your message. In earlier chapters, you encountered three major reasons for telling your main point at the start. First, by doing so you help your readers learn what they most want or need to know without requiring them to hunt extensively for it. You also increase the likelihood that they will actually notice your main point. On the job, people are often so busy that they set communications aside without completing them if they can't find the main point quickly. Finally, by stating your main point at the outset, you help your readers read efficiently by telling them the context within which you want them to view the details you will later provide. For example, by stating in her first paragraph that she was writing to request $175,000 to renovate the company's machine shop, a manager provided her readers with a context for understanding the detailed account that followed of problems encountered in the shop. Similarly, by telling his readers at the start of his report that he recommends redesigning an important circuit of a product his employer is now developing, an electrical engineer provided his readers with a context for reading his thorough discussion of his tests on the circuit.

Choose Your Main Point Thoughtfully

Choose the main point of your overall communication in the same way you choose the main point of any smaller part of your message. If you are responding to a request, your main point will probably be the answer to the major question your reader asked you to address. If you are writing on your own initiative, your main point might be the most important thing you want your reader to think or do after reading your communication.

Here are some sample statements of a writer's main point:

From the beginning of a memo written in a manufacturing company:
We should immediately suspend all purchases from Cleves
Manufacturing until they can guarantee us that the parts they supply
will meet our specifications.

From the beginning of a memo written within a food services company:
I request $1200 in travel funds to send one of our account executives
to the client's Atlanta headquarters.

From a research report:
The test results show that the walls of the submersible room will not
be strong enough to withstand the high pressures of a deep dive.

Provide Summaries for Longer Communications

In the three examples you've just read, the writers state just one main point, which they express in a single sentence. This works well in brief communications, but longer communications often contain many main points. Consequently, when writing something long, you should consider beginning with a summary that provides an overview of the principal points from every section of your communication.

Consider, for instance, the two-page feasibility report that Mike is writing concerning opening a new branch of his employer's home electronics stores in another town. In this report, Mike touches upon the following topics:

● Introduction to the possibility of a new branch
● Description of the significant characteristics of the people in the community (average age, income, spending habits)
● Analysis of competition in the community
● Possible sites
● Recommendation for full-scale investigation

For all of these topics, Mike includes at least one paragraph, and he uses several paragraphs for some topics. His summary is three sentences long and is included *within* the beginning of his memo.

In contrast, longer communications may have a separate summary that precedes the beginning of the body of the communication. These long summaries are often called *executive summaries* because they are used by decision-makers who may not read the rest of the communication, leaving that task to their advisers. These summaries are therefore almost always self-contained. This means, for instance, that in such a summary you must not only list your recommendations but also introduce the problem, tell how you conducted your investigation, describe the significant results, and answer questions about costs, benefits, implementation, and schedules.

Executive summaries conventionally do not contain anything that isn't also in the communication. To put it another way, the summary isn't the introduction to the communication but a separate component attached to the front of it. Following the summary, the body of the communication starts with a full-scale beginning that follows all the guidelines given in this chapter.

Figure 9–3 shows the executive summary from an 18-page report. Notice that the summary (like the report) begins with the main point. Also notice the large amount of detail that the author provides.

GUIDELINE
3

TELL YOUR READERS WHAT TO EXPECT

In addition to stating your main point (when appropriate), your beginning should tell readers what to expect in your communication. This advice echoes some given in Chapter 7: "Use forecasting statements." When given at the beginning of a communication, such statements should focus on the communication's organization and its scope.

Tell about Your Communication's Organization

The major reason for telling your readers about the *organization* of your overall communication is the same as for telling them about the organization of any of the communication's parts: to provide them with a general framework for understanding the connections among the various pieces of information you convey.

You can tell your readers about the organization of your communication in sentences:

> In this report, we state the objectives of the project, compare the three major technical alternatives, and present our recommendation. The final sections include a budget and a proposed project schedule.

Or, you can describe your organization in a list:

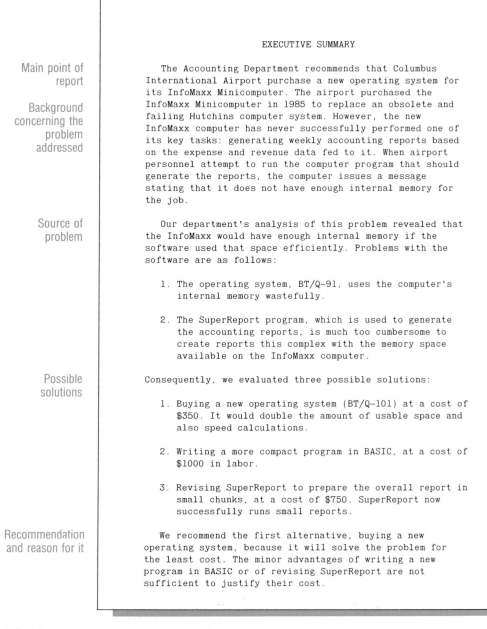

Main point of report

Background concerning the problem addressed

Source of problem

Possible solutions

Recommendation and reason for it

EXECUTIVE SUMMARY

The Accounting Department recommends that Columbus International Airport purchase a new operating system for its InfoMaxx Minicomputer. The airport purchased the InfoMaxx Minicomputer in 1985 to replace an obsolete and failing Hutchins computer system. However, the new InfoMaxx computer has never successfully performed one of its key tasks: generating weekly accounting reports based on the expense and revenue data fed to it. When airport personnel attempt to run the computer program that should generate the reports, the computer issues a message stating that it does not have enough internal memory for the job.

Our department's analysis of this problem revealed that the InfoMaxx would have enough internal memory if the software used that space efficiently. Problems with the software are as follows:

1. The operating system, BT/Q–91, uses the computer's internal memory wastefully.

2. The SuperReport program, which is used to generate the accounting reports, is much too cumbersome to create reports this complex with the memory space available on the InfoMaxx computer.

Consequently, we evaluated three possible solutions:

1. Buying a new operating system (BT/Q–101) at a cost of $350. It would double the amount of usable space and also speed calculations.

2. Writing a more compact program in BASIC, at a cost of $1000 in labor.

3. Revising SuperReport to prepare the overall report in small chunks, at a cost of $750. SuperReport now successfully runs small reports.

We recommend the first alternative, buying a new operating system, because it will solve the problem for the least cost. The minor advantages of writing a new program in BASIC or of revising SuperReport are not sufficient to justify their cost.

FIGURE 9–3   Summary of a Report Directed Primarily to Decision-Makers

This booklet covers the following topics:

- Principles of Sound Reproduction
- Types of Speakers
- Choosing the Right Speakers for You
- Installing the Speakers in Your Business

In some instances, you can indicate organization in your statement of your communication's relevance to your readers:

At our meeting on September 4, you expressed an interest in learning about the prospective market for each of the three major types of batteries that might be used in electrically powered cars in the next twenty years. In this memo, I describe each market and then compare them in terms of their likely profitability.

Tell about Your Communication's Scope

Readers often want to know from the beginning what a communication does and does not contain. Even if they are persuaded that you are addressing a subject relevant to them, your audience may still wonder whether you will discuss the specific aspects of the subject that they want to know about.

In many communications, you will have told your readers the scope of your communication when you tell them about its organization: when you list the topics the communication treats, you are indicating what it covers.

However, there will be times when you will need to include additional information about the scope of your communication. That happens when you want your readers to understand that you are not treating your subject comprehensively or that you are treating your subject from a particular point of view. For instance, you may be writing a troubleshooting guide that factory workers can use to solve certain problems with manufacturing robots they monitor. At the same time, many problems might require, for instance, the assistance of a computer programmer or an electrical engineer. In that case, you should tell your readers explicitly about the scope of your troubleshooting manual:

This manual treats problems you can correct by using tools and equipment normally available to you. It does not cover problems that require work by computer programmers or electrical engineers.

Of course, you must use your good judgment in deciding how much to say in your beginning about the organization and scope of your communications. For brief communications, readers don't need any information at all about these matters. But when reading longer ones, they often benefit from knowing at the start what lies ahead.

| GUIDELINE 4 | **ENCOURAGE OPENNESS TO YOUR MESSAGE** |

From reading other chapters in this book, you should be quite familiar with the notion that the members of your audience can adopt different strategies for responding to your message as they read it. For example, when reading recommendations from you, they can try to understand your arguments objectively, or they can seek aggressively to find flaws with your arguments. When they are reading a set of instructions you have prepared, they can strive to follow your directions in detail, or they can decide to attempt the procedure on their own, consulting your directions only if they become stumped.

The way you begin a communication strongly influences your audience's response to your message. You should begin in a way that encourages your audience to be receptive to your points.

Situations Vary

In your professional role, you generally will have no trouble encouraging a receptive response because you will usually communicate to fellow employees, customers, and others who earnestly desire the information and ideas you will provide. They are therefore likely to be open to your message.

However, a variety of circumstances can lead people to adopt a more negative initial attitude toward you or your message. For example, you may be writing a report that reveals shortcomings in a plan devised by your readers. Or you may be making a recommendation that, if followed, will have undesirable consequences for some members of your audience. Or you may be writing to a customer who is dissatisfied with the products or services provided by your company. In situations like these, you will need to take special care in the construction of the beginning of your communication if you are to obtain a fair hearing for your message.

How to Predict Your Readers' Initial Attitude

How do you decide whether you need to take extraordinary measures to obtain an open hearing for your message? The key is to use your knowledge of your readers to predict their initial reaction to your message. To do this, you might ask questions such as the following:

- Does my message contain bad news for my audience?
- Does my communication contain ideas or recommendations that will be unwelcome to my audience?
- Does my audience have any feelings of distrust, resentment, or competitiveness toward me, my department, or my company?
- Is my audience likely to be skeptical of my knowledge of my subject or of the situation?
- Is my audience likely to be suspicious of my motives?

If the answer to these or similar questions is yes, you should think about (or learn) the details of the feelings that are negatively shaping your audience's initial reactions to your communication. Then devise your beginning accordingly.

Some Strategies for Encouraging Openness

From situation to situation, the best strategy for preventing or counteracting an initial negative reaction will vary. However, here are three general strategies you can consider:

- **Present yourself as a partner, not a critic or competitor.** Present yourself as someone who is working with your readers to help solve a problem they want to solve or to achieve a goal they want to achieve. (See Guideline 1.)

- **Delay the presentation of your main point.** An initial negative reaction can cause the readers to aggressively devise counterarguments to each point that follows. Therefore, if you believe that your audience may react negatively to your main point, consider making an exception to Guideline 2, which tells you to include your main point in your beginning. If you delay the presentation of your main point, your readers may consider at least some of your points objectively before discovering your main point and reacting against it. For a full discussion of the strategy of delaying the main point, review the section in Chapter 5 on the direct and indirect patterns of organization.

- **Establish your credibility.** As you learned in Chapter 5, people are more likely to respond favorably to a message if they believe that the communicator is knowledgeable. Consequently, you can sometimes increase your audience's openness to your message if you begin by convincing them that you are expert in your subject and knowledgeable about the situation.

  Does this mean that you should establish your credentials in the beginning of every communication? No. In many situations, your readers will already consider you to be credible, as when you are writing to co-workers. Also, when you are presenting a position with which your readers already agree, your credentials aren't likely to matter much. Presenting your credentials needlessly will merely burden your readers with unnecessary information.

  With those qualifications in mind, however, you will still encounter situations where some mention of your credentials can increase your readers' openness to your message.

Sometimes Telling a Story Can Help You

Although the general strategies suggested above will often encourage openness in your readers, you should not employ the strategies mechanically. Always keep the particular attitudes, thoughts, and expectations of your audience in mind as you devise the beginning of a communication.

To do this, you might try organizing your knowledge of your readers by creating a story about them. Keep in mind that your story's purpose is to help you decide how to introduce your communication to your audience. It is not something that you would actually include in your communication. The central figure in your story should be your reader if you are communicating primarily to one person, or a typical member of your audience if you are addressing a group of people. Begin the story a few minutes before this person picks up your communication or enters the room where you will speak. Carry the story through the moment the person reads or hears your first words. This story can help you learn about your audience in a way that will assist you in deciding how best to introduce your communication.

A Sample Story

Here is a sample story, written by Jolene, a manager in an insurance company. Jolene wrote this story to help herself understand the readers of an instructional manual she is writing to teach new insurance agents how to use the company's computer system.

> It's Monday afternoon. After half a day of orientation meetings and tours, Bob, the new, green trainee, sits down at the computer terminal for the first time to try to learn this system. He was a French major who has never used computers. Now, in two hours, he is supposed to work his way through this manual and then enter some sample policy information. He feels rushed, confused, and quite nervous. He knows that the equipment is expensive and he does not want to damage it in any way.
>
> Despite his insecurity, Bob will not ask questions of the experienced agent in the next office because (being new to the company) he doesn't want to make a bad impression by asking dumb questions.
>
> Bob picks up the instruction manual for the SPRR program that I am writing: he hopes it will tell him quickly what he needs to know. He wants it to help him learn the system in the time allotted without breaking the machine and without asking embarrassing questions.

Through this story, Jolene focused on several important facts about her typical reader: the reader will be anxious, hurried, and highly uncertain. All these insights helped Jolene write the opening to her manual. Here are the first few sentences:

294 PART IV DRAFTING

This manual tells you how to enter policy information into our
SPRR system. It covers the steps for opening a file for a new policy,
entering the relevant information, revising the file, and printing a
paper copy for your permanent records.

Be sure to follow the instructions carefully, so that you can avoid
making time-consuming errors. At the same time, you should know
that the SPRR system is designed so that you cannot damage anyone
else's files or the system itself.

By means of her story, Jolene was able to identify her audience's probable
feelings and reduce their anxiety, thereby encouraging her readers to be more
open to her instructions.

GUIDELINE
5

PROVIDE NECESSARY BACKGROUND INFORMATION

Finally, as you draft the beginning of a communication you should ask yourself
whether your readers will need any general background information to understand the things you are going to tell them.

Here are examples of situations in which you may need to provide such
background information at the beginning of your communication:

- **Your readers must learn certain general principles to understand your
  specific points.** For instance, your discussion of the feasibility of locating
  a new plant in a particular city may depend upon a particular analytical
  technique that you will need to explain to your readers.

- **Your readers are unfamiliar with basic technical terms you will use.**
  For example, as a specialist in international trade, you may need to explain
  certain basic technical terms to the Board of Directors before you present
  them with your strategies for opening up foreign markets for the company.

- **Your readers are unfamiliar with basic features of the situation you
  are discussing.** For example, imagine that you are reporting to the Executive Directors of a large corporation about labor problems at one of the
  plants it recently acquired by taking over another company. To understand
  and weigh the choices that face them in this situation, the Directors will
  need an introduction to the plant and its labor history.

Some kinds of background information don't belong at the beginning of
your communication. Background explanations that pertain only to certain, specific places in your communication should be provided at the beginning of
those sections. In the beginning of your whole communication, include only
the general explanations that will help the audience overall.

HOW LONG SHOULD YOUR BEGINNING BE?

How long does the beginning of a particular communication need to be? In some situations, a good, reader-centered beginning takes many paragraphs—even many pages. However, in some situations it takes only a sentence or less. You need to tell your readers only the things that they didn't know on their own beforehand. The important thing is to be sure that when your readers finish reading or listening to your communication, they know each of the following things:

- Why they should read the communication (Guideline 1).
- What the main points of the communication are (Guideline 2).
- How your communication is organized and what its scope is (Guideline 3).
- All the background information necessary to understand and use the communication (Guideline 5).

If you are sure that your readers have all this information—and that you have encouraged them to receive your message openly (Guideline 4)—then you have written a complete beginning, regardless of how long or short it is.

Here is an opening prepared by a writer who followed all the guidelines in this chapter:

> In response to your memo dated November 17, I have called Goodyear, Goodrich, and Firestone for information about the ways they forecast their needs for synthetic rubber. The following paragraphs summarize each of those phone calls.

The following opening, from a two-paragraph memo, is even briefer. At first glance, it may seem to ignore all the guidelines given in this chapter.

> We are instituting a new policy for calculating the amount that employees are paid for overtime work.

This single sentence, however, exemplifies all the guidelines given in this chapter. From reading the topic of the memo (overtime pay), the particular people addressed by this writer would immediately understand the memo's relevance to them. This sentence also tells the main point of the memo (a new policy is being instituted). Furthermore, because the memo is only two paragraphs long, the scope of the memo is readily apparent without explicit comment about it. The brevity of this memo also suggests its organization, namely, a brief explanation of the new policy and nothing else. The writer correctly judged that his readers need no background information.

Figure 9–4 shows a relatively long beginning from a recommendation report written by a consulting firm hired to suggest ways to improve the food

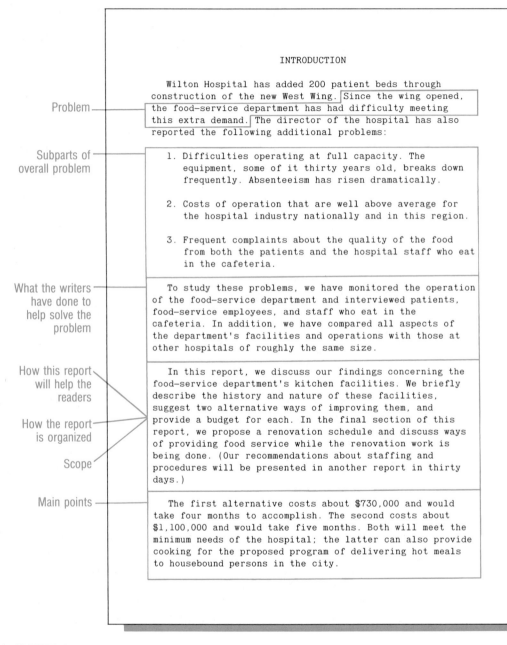

Problem

Subparts of overall problem

What the writers have done to help solve the problem

How this report will help the readers

How the report is organized

Scope

Main points

INTRODUCTION

Wilton Hospital has added 200 patient beds through construction of the new West Wing. Since the wing opened, the food-service department has had difficulty meeting this extra demand. The director of the hospital has also reported the following additional problems:

1. Difficulties operating at full capacity. The equipment, some of it thirty years old, breaks down frequently. Absenteeism has risen dramatically.

2. Costs of operation that are well above average for the hospital industry nationally and in this region.

3. Frequent complaints about the quality of the food from both the patients and the hospital staff who eat in the cafeteria.

To study these problems, we have monitored the operation of the food-service department and interviewed patients, food-service employees, and staff who eat in the cafeteria. In addition, we have compared all aspects of the department's facilities and operations with those at other hospitals of roughly the same size.

In this report, we discuss our findings concerning the food-service department's kitchen facilities. We briefly describe the history and nature of these facilities, suggest two alternative ways of improving them, and provide a budget for each. In the final section of this report, we propose a renovation schedule and discuss ways of providing food service while the renovation work is being done. (Our recommendations about staffing and procedures will be presented in another report in thirty days.)

The first alternative costs about $730,000 and would take four months to accomplish. The second costs about $1,100,000 and would take five months. Both will meet the minimum needs of the hospital; the latter can also provide cooking for the proposed program of delivering hot meals to housebound persons in the city.

FIGURE 9–4   Beginning for a Recommendation Report

service at a hospital. Like the brief beginnings quoted above, it is carefully adapted to its readers and situation. Figure 9–5 shows the long beginning of a 500-page service manual for the Detroit Diesel 53 engine, manufactured by General Motors.

CONCLUSION

If one part of a communication can be said to be more important than all the rest, then for most communications that part is the beginning. That's because the beginning can influence the ideas and attitudes your readers derive from the rest of your communication, and it can even determine whether or not they will read beyond your opening sentences.

This chapter has suggested that you start work on your beginnings by thinking about your readers' initial attitudes toward your message and by determining what you can say at the outset to help them readily understand and use what follows. This reader-centered approach will enable you to create beginnings that persuade your readers to pay careful attention, encourage them to treat your information and ideas open-mindedly, and help them read efficiently.

EXERCISES

1. Select a communication written to people in your field. This might be a letter, a memo, a manual, or a report (do not choose a textbook). Analyze the beginning of this communication in terms of the guidelines discussed in this chapter.

2. The instructions for many consumer products contain no beginnings at all. For instance, the instructions for some lawnmowers, cake mixes, and detergents simply provide a heading that says ''Instructions'' and then start right in. Find such a set of instructions and—in terms of the guidelines given in this chapter—evaluate the writer's decision to omit a beginning.

3. The following paragraphs are from the beginning of a report in which the manager of a purchasing department asks for better supplies from another department in the company that is providing unsatisfactory abrasives. Is this an effective beginning for what is essentially a complaint? Why or why not? Analyze this beginning in terms of the guidelines given in this chapter.

> I am sure you have heard that the new forging process is working well. Our customers have expressed pleasure with our castings. Thanks again, for all your help in making this new process possible.

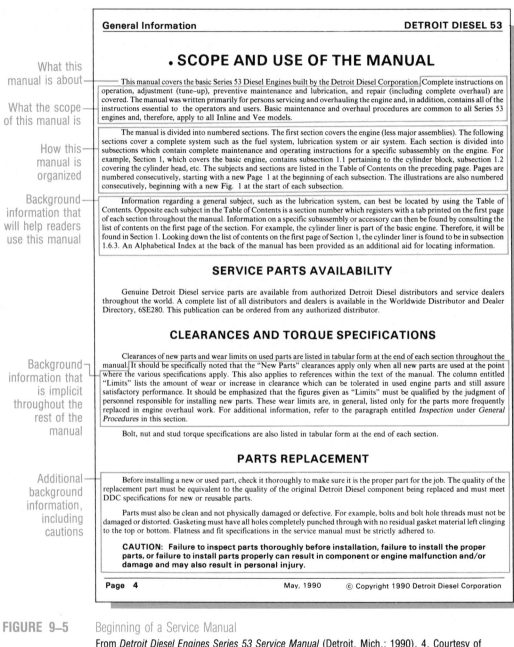

General Information   **DETROIT DIESEL 53**

• SCOPE AND USE OF THE MANUAL

What this manual is about

What the scope of this manual is

This manual covers the basic Series 53 Diesel Engines built by the Detroit Diesel Corporation. Complete instructions on operation, adjustment (tune–up), preventive maintenance and lubrication, and repair (including complete overhaul) are covered. The manual was written primarily for persons servicing and overhauling the engine and, in addition, contains all of the instructions essential to the operators and users. Basic maintenance and overhaul procedures are common to all Series 53 engines and, therefore, apply to all Inline and Vee models.

How this manual is organized

The manual is divided into numbered sections. The first section covers the engine (less major assemblies). The following sections cover a complete system such as the fuel system, lubrication system or air system. Each section is divided into subsections which contain complete maintenance and operating instructions for a specific subassembly on the engine. For example, Section 1, which covers the basic engine, contains subsection 1.1 pertaining to the cylinder block, subsection 1.2 covering the cylinder head, etc. The subjects and sections are listed in the Table of Contents on the preceding page. Pages are numbered consecutively, starting with a new Page 1 at the beginning of each subsection. The illustrations are also numbered consecutively, beginning with a new Fig. 1 at the start of each subsection.

Background information that will help readers use this manual

Information regarding a general subject, such as the lubrication system, can best be located by using the Table of Contents. Opposite each subject in the Table of Contents is a section number which registers with a tab printed on the first page of each section throughout the manual. Information on a specific subassembly or accessory can then be found by consulting the list of contents on the first page of the section. For example, the cylinder liner is part of the basic engine. Therefore, it will be found in Section 1. Looking down the list of contents on the first page of Section 1, the cylinder liner is found to be in subsection 1.6.3. An Alphabetical Index at the back of the manual has been provided as an additional aid for locating information.

SERVICE PARTS AVAILABILITY

Genuine Detroit Diesel service parts are available from authorized Detroit Diesel distributors and service dealers throughout the world. A complete list of all distributors and dealers is available in the Worldwide Distributor and Dealer Directory, 6SE280. This publication can be ordered from any authorized distributor.

CLEARANCES AND TORQUE SPECIFICATIONS

Background information that is implicit throughout the rest of the manual

Clearances of new parts and wear limits on used parts are listed in tabular form at the end of each section throughout the manual. It should be specifically noted that the "New Parts" clearances apply only when all new parts are used at the point where the various specifications apply. This also applies to references within the text of the manual. The column entitled "Limits" lists the amount of wear or increase in clearance which can be tolerated in used engine parts and still assure satisfactory performance. It should be emphasized that the figures given as "Limits" must be qualified by the judgment of personnel responsible for installing new parts. These wear limits are, in general, listed only for the parts more frequently replaced in engine overhaul work. For additional information, refer to the paragraph entitled *Inspection* under *General Procedures* in this section.

Bolt, nut and stud torque specifications are also listed in tabular form at the end of each section.

PARTS REPLACEMENT

Additional background information, including cautions

Before installing a new or used part, check it thoroughly to make sure it is the proper part for the job. The quality of the replacement part must be equivalent to the quality of the original Detroit Diesel component being replaced and must meet DDC specifications for new or reusable parts.

Parts must also be clean and not physically damaged or defective. For example, bolts and bolt hole threads must not be damaged or distorted. Gasketing must have all holes completely punched through with no residual gasket material left clinging to the top or bottom. Flatness and fit specifications in the service manual must be strictly adhered to.

CAUTION: Failure to inspect parts thoroughly before installation, failure to install the proper parts, or failure to install parts properly can result in component or engine malfunction and/or damage and may also result in personal injury.

Page 4   May, 1990   © Copyright 1990 Detroit Diesel Corporation

FIGURE 9–5   Beginning of a Service Manual

From *Detroit Diesel Engines Series 53 Service Manual* (Detroit, Mich.: 1990), 4. Courtesy of Detroit Diesel Corporation.

We are having one problem, however, with which I have to ask once more for your assistance. During the seven weeks since we began using the new process, the production line has been idle 28 percent of the time. Also, many castings have had to be remade. Some of the evidence suggests that these problems are caused by the steel abrasive supplies we get from your department. If we can figure out how to improve the abrasive, we may be able to run the line at 100 percent of capacity.

I would be most grateful for help from you and your people in improving the abrasive. To help you devise ways of improving the abrasive, I have compiled this report, which describes the difficulties we have encountered and some of our thinking about possible remedies.

10

ENDING A COMMUNICATION

DEFINING
OBJECTIVES

PLANNING

DRAFTING

EVALUATING

REVISING

Imagine you are talking with a friend. Suddenly, she turns around and leaves without any warning at all. Or, imagine that at the time set for the end of your class, your instructor simply stops talking and walks out of the room, without making any concluding remarks. Situations like these would probably leave you a little unsettled, a little uncertain. Why didn't your friend say goodbye? Why didn't your instructor bring the class to a close? What do they want you to think? To do?

As these examples suggest, the end of a conversation is very special. We expect other people to end their conversations with us in certain ways, and they have the same expectations of us. Similar expectations apply to much of the writing you will do on the job. To maintain an effective relationship with your readers, you need to satisfy those expectations.

The endings of on-the-job communications are important for several other reasons as well. First, an ending is a place of emphasis. People are more likely to remember a point made at the end of a communication than one made in the middle. Second, an ending creates your readers' final impressions while reading your communication. This final impression can be crucial in shaping their attitudes toward both you and your subject matter. Third, an ending is a transition. While reading or listening to your message, your readers are absorbed in it. The ending leads them out of the communication and into the larger stream of their activities. Consequently, it is a point where they naturally ask, "What should I do next?" The way you answer that question can influence their actions related to your subject matter.

For all these reasons, the endings of your communications deserve your special care and attention.

INTRODUCTION TO THE GUIDELINES

This chapter's guidelines describe a variety of ways to end the communications you prepare at work. Despite the important role that endings usually play, your best strategy sometimes will be to say what you have to say and then simply stop, *without* constructing a separate ending (Guideline 1). Other times, when you will need to create a special ending, you can use one or more of the strategies described in Guidelines 2 through 9. How will you know which strategies to use in a particular communication? First, think about your readers in the act of reading. Determine what you want them to think, feel, and do as they finish your communication. Then decide which strategy or strategies for endings are most likely to bring about that result. Second, look at what other people in your organization and your field have done when communicating in situations similar to yours. This will help you determine which kinds of ending your readers might be expecting and which kinds other writers have used successfully in similar circumstances.

GUIDELINE
1

AFTER YOU'VE MADE YOUR LAST POINT, STOP

As mentioned above, sometimes you should end your communications without doing anything special at all. That happens when you use a pattern of organization that brings you to a natural stopping place. Here are some examples:

- Proposals. You will usually end your proposals with a detailed description of what you will do and how you will do it. Because that's where your audience expects proposals to end, they will enjoy a sense of completion if you simply stop after presenting your last recommendation. Furthermore, by ending after your recommendations, you will have given them the emphasis they require.

- Formal Reports. When you prepare formal reports (those with covers, title pages, and bindings), you will often use a conventional pattern that ends either with your conclusions or your recommendations—both appropriate subjects for emphasis.

- Instructions. You will usually end instructions by describing the last step. Your audience will certainly enjoy a sense of completion after performing that step.

- Reference Manuals. Reference manuals usually have no endings because there is no single place where readers finish their reading. Readers seek particular sections and close the manual as soon as they have gotten the information they need. There would be no point in providing a special ending.

These are just some examples of communications that writers sometimes—but not always—end directly after making their last point. However, if your analysis of your purpose, readers, and situation convinces you that you should add something after your last point, read the remaining guidelines in this chapter and select from among the strategies they suggest those that will best help you meet your objectives.

GUIDELINE
2

REPEAT YOUR MAIN POINT

Because the end of a communication is a point of emphasis, you can use it to focus your readers' attention on one or more main points that you want to be foremost in their minds as they set your communication down.

Consider, for instance, the final paragraph of an article on "Preventing Wound Infections" that is directed toward family physicians.[1] The point made in this paragraph was stated in the abstract at the beginning of the article and then again in the fourth paragraph, where it was supported by a table. It was also referred to several other times in the article. Nevertheless, the writers considered it to be so important that in the final paragraph they stated it again:

Perhaps the most important concept to be gleaned from a review of the principles of wound management is that good surgical technique strives to maintain the balance between the host and the bacteria in favor of the host. The importance of understanding that infection is an absolute quantitative number of bacteria within the tissues cannot be overemphasized. Limiting, rather than eliminating, bacteria allows for normal wound healing.

You can use the same strategy in communications intended to help your readers make a decision or to persuade them to take a certain action. Here, for instance, is the final paragraph of a memo urging new safety measures:

I cannot stress too much the need for immediate action. The exposed wires present a significant hazard to any employee who opens the control box to make routine adjustments.

GUIDELINE 3

SUMMARIZE YOUR KEY POINTS

The strategy suggested by this guideline is closely related to the one you have just read about. The difference is that, in repeating a main point, you emphasize only that information which you consider to be of paramount importance. In summarizing, you are concerned that your audience has understood the general thrust of your communication.

Here, for example, is the ending of a 115-page book entitled *Understanding Radioactive Waste,* which is intended to help the general public understand the impact of the nuclear-power industry's plans to open new plants:[2]

It may be useful to the reader for us to now select some highlights, key ideas, and important conclusions for this discussion of nuclear wastes. The following list is not complete—the reader is encouraged to add items.

1. Radioactivity is both natural and manmade. The decay process gives radiations such as alpha particles, beta particles, and gamma rays. Natural background radiation comes mainly from cosmic rays and minerals in the ground.
2. Radiation can be harmful to the body and to genes, but the low-level radiation effect cannot be proved. Many methods of protection are available.
3. The fission process gives useful energy in the form of electricity from nuclear plants, but it also produces wastes in the form of highly radioactive fission products.

This list continues for thirteen more items, but this sample should give you an idea of how this author used the strategy of ending with a summary of key points.

Be sure to note that summaries at the end of a communication differ significantly from those at the beginning. A summary at the beginning addresses someone who has not yet read the communication. It must therefore devote space to information that is not of much concern at the end. For example, the beginning summary of a report on a quality-control study will establish the background of the study. In contrast, an ending summary usually focuses more sharply on conclusions and recommendations.

<table>
<tr><td>GUIDELINE
4</td><td>REFER TO A GOAL STATED EARLIER IN YOUR COMMUNICATION</td></tr>
</table>

Many communications begin by stating a goal and then describing or proposing ways to achieve it. If you end a communication by referring to that goal, you remind your audience of the goal and emphasize the focus of your communication. The following two examples show both the beginning and the ending of reports in which the ending refers to the beginning.

This first example comes from a seventeen-page proposal prepared by operations analysts in a company that builds customized, computer-controlled equipment used in print shops and printing plants:

Beginning

To maintain our competitive edge, we must develop a way of
supplying replacement parts more rapidly to our service technicians
without increasing our shipping costs or tying up more money in
inventory.

Ending

The proposed reform of our distribution network will help us meet
the needs of our service technicians for rapidly delivered spare parts.
Furthermore, it does so without raising either our shipping expenses
or our investment in inventory.

The second example is from *Biotechnology,* a journal concerned with the synthesis of new organisms that are commercially useful.[3] (Note that the highly technical language is appropriate because the writers are addressing people knowledgeable in this field.)

Beginning

Given the necessity of producing alcohol as an alternative fuel to
gasoline, especially in countries like Brazil, where petroleum is
scarce, it is important to have a yeast strain able to produce ethanol
directly from starchy materials.

Ending

We are convinced that the stable pESA transformants can be of technological value in assisting ethanol fermentation directly from starchy materials, and we have described the first step towards this end. Continuing this work, genetic crosses with different *Saccharomyces distaticus* strains are presently being carried out to introduce maltase and glucomylase genes into the stable transformants that secrete functional *a*-amylase.

GUIDELINE
5

FOCUS ON A KEY FEELING

Sometimes, you may want to focus your audience's attention not on a fact, but on a feeling. For instance, if you are writing instructions for a product manufactured by your employer, you may want your ending to encourage your audience's goodwill toward the product. Consider this ending of an owner's manual for a clothes dryer.[4] Notice how the last sentence provides no additional information but does seek to shape the readers' attitude toward the company.

> The GE Answer Center™ consumer information service is open 24 hours a day seven days a week.
>
> Our staff of experts stands ready to assist you anytime.

The following passage is the ending from a much different kind of communication, a booklet published by the National Cancer Institute that is addressed to people who have apparently been successfully treated for cancer but do not know how long the disease will remain in remission.[5] It, too, seeks to shape the readers' feelings.

> Cancer is not something anyone forgets. Anxieties remain as active treatment ceases and the waiting stage begins. A cold or cramp may be cause for panic. As 6-month or annual check-ups approach, you swing between hope and anxiety. As you wait for the mystical 5-year or 10-year point, you might feel more anxious rather than more secure.
>
> These are feelings that we all share. No one expects you to forget you have had cancer or that it might recur. Each must seek individual ways of coping with the underlying insecurity of not knowing the true state of his or her health. The best prescription seems to lie in a combination of one part challenging responsibilities that require a full range of skills, a dose of activities that seek to fill the needs of others, and a generous dash of frivolity and laughter.
>
> You still might have moments when you feel as if you live perched on the edge of a cliff. They will sneak up unbidden. But

they will be fewer and farther between if you have filled your mind with other thoughts than cancer.

Cancer might rob you of that blissful ignorance that once led you to believe that tomorrow stretched on forever. In exchange, you are granted the vision to see each today as precious, a gift to be used wisely and richly. No one can take that away.

GUIDELINE
6

TELL YOUR READERS HOW TO GET ASSISTANCE OR MORE INFORMATION

At work, a common strategy for ending a communication is to tell your audience how to get assistance or additional information. These two examples are from a letter and a memo:

> If you have questions about this matter, call me at 523–5221.

> If you want any additional information about the proposed project, let me know. I'll answer your questions as best I can.

By ending in this way, you not only provide your audience with useful information, but you also encourage them to see you as a helpful, concerned individual.

GUIDELINE
7

TELL YOUR READERS WHAT TO DO NEXT

Another strategy for effective endings is to tell your audience what you think should be done next. If more than one course of action is available, tell your audience how to follow up on each of them:

> To buy this equipment at the reduced price, we must mail the purchase orders by Friday the 11th. If you have any qualms about this purchase, let's discuss them. If not, please forward the attached materials, together with your approval, to the Controller's Office as soon as possible.

GUIDELINE
8

IDENTIFY ANY FURTHER STUDY THAT IS NEEDED

Much of the work that is done on the job is completed in stages. For example, one study might answer preliminary questions and, if the answers look promising, an additional study might then be undertaken. Consequently, one common way of ending is to tell the audience what needs to be found out next:

> This experiment indicates that we can use compound deposition to create microcircuits in the laboratory. We are now ready to explore the feasibility of using this technique to produce microcircuits in commercial quantities.

Such endings are often combined with summaries, as in the following example:

> In summary, over the past several months our Monroe plant has ordered several hundred electric motors from a supplier whose products are inferior to those we require in the heating and air conditioning systems we build. Not only must this practice stop immediately but also we should investigate the situation to determine why this flagrant violation of our quality-control policies has occurred.

GUIDELINE 9	**FOLLOW APPLICABLE SOCIAL CONVENTIONS**

All the strategies for ending that you have read about so far focus on the subject matter of your communications. It is also important for you to observe the social conventions that apply in your communication situation.

Some of the conventions that govern endings involve customary ways of closing particular kinds of communication. For example, letters usually end with some special gesture such as an expression of thanks, a statement that it has been enjoyable working with the reader, or an offer to be of further help if needed. In contrast, formal reports and proposals rarely end with such gestures.

Other conventions about endings are peculiar to the organizations in which they are found. For example, in some organizations, writers rarely end their memos with the kind of social gesture that is almost universally provided at the ends of letters. In other organizations, memos often end with such a gesture, and people who do not include such an ending risk seeming abrupt and cold. In still other organizations, ways of ending memos vary considerably from person to person.

Important conventions also apply to the personal relationships between you and your audience. Have they done you a favor? Thank them. Are you going to see them soon? Let them know that you look forward to the meeting.

CONCLUSION

The endings of communications you prepare at work can help you achieve four aims:

- Provide your readers with a sense of completion.
- Emphasize key material.
- Shape your readers' attitudes toward you and your subject matter.
- Direct your readers' attention to future action.

This chapter has described nine ending strategies. By considering your readers, your communication objectives and the customs that apply in your situation, you will be able to choose the strategy or combination of strategies most likely to succeed in the particular communication you are writing.

EXERCISES

1. The following paragraphs are the ending of a report to the U.S. Department of Energy concerning the economic and technical feasibility of generating electric power with a special type of windmill.[6] The windmills are called diffuser augmented wind turbines (DAWT). In this ending, the writer has used several of the strategies described in this chapter. Identify each of them.

Section 6.0

Concluding Remarks

We have provided a preliminary cost assessment for the DAWT approach to wind energy conversion in unit systems to 150 kw power rating. The results demonstrate economic viability of the DAWT with no further design and manufacturing know-how than already exists. Further economic benefits of this form of solar energy are likely through:

- Future refinements in product design and production techniques
- Economies of larger quantity production lots
- Special tax incentives

Continued cost escalation on nonrenewable energy sources and public concern for safeguarding the biosphere environment will surely make wind energy conversion by DAWT-like systems even more attractive to our society. Promotional actions by national policy makers and planners as well as industrialists and entrepreneurs can aid the emergence of the DAWT from its research phase to a practical and commercial product.

2. Describe the strategies for ending used in the following figures:

FIGURE	PAGE
1–8	24
2–7	61
2–8	62

5–4	151
5–7	165
5–8	167
5–13	177
5–14	178
5–15	182
7–9	225
20–5	539
22–4	593
23–5	616
24–3	642

3. Find examples of four of the nine types of endings described in this chapter by looking in magazines, textbooks, instruction manuals, and other publications. If you find an ending that uses more than one of these strategies, count it as more than one example. For each example, explain why the writer chose the ending strategy he or she used. Did the writer make a good choice? Explain why or why not.

WRITING SENTENCES

DEFINING
OBJECTIVES
PLANNING
DRAFTING
EVALUATING
REVISING

As a native speaker of English, you have incredible power over the language. And the language itself, tremendously flexible, awaits your command. To see just how much power you have, and just how flexible the language is, consider the following simple sentence.

Dogs eat red meat.

The information communicated in that sentence could also be conveyed in sentences with much different structures:

Dogs eat meat that is red.

Red meat is eaten by dogs.

The meat that dogs eat is red.

Red meat, that's what dogs eat.

Dogs are eaters of red meat.

Most of the sentences you write communicate much more information than does the short sentence, "Dogs eat red meat." Each of them could be written in a correspondingly larger variety of ways. Which choice is best? When writing sentences, you need to keep two distinct goals in mind. The first is to write each individual sentence so it is lively and easy to read, and the second is to write each sentence so it works together harmoniously with those that surround it. This chapter will help you pursue these two goals, and it will advise you about what to do in situations where these two goals conflict.

AVOIDING BUREAUCRATESE

This chapter's first four guidelines focus on developing sentences that are individually clear and direct. One enemy of clear, direct writing is the abstract, pompous, overly formal style that is often called *bureaucratese*. Bureaucratese is characterized by wordiness that buries significant ideas and information, by weak verbs that disguise action, and by abstract vocabulary that detaches meaning from the practical world of people, activities, and things. Here's an example:

According to good quality control practices in manufacturing any product, it is important that every component part that is a constituent of the aforementioned product be examined and checked individually when received from its supplier or other source but before the final, finished product is assembled. (46 words)

The writer simply means this:

> Good quality control requires that every component be checked individually before the final product is assembled. (16 words)

Another example:

> Over the most recent monthly period, there has been a large increase in the number of complaints that customers have made about service that has been slow. (27 words)

In plain English, the writer is saying:

> Last month, a lot more customers complained about slow service. (10 words)

Why do people write in bureaucratese? Many do it to impress their readers. They think it makes them sound knowledgeable or important. Others believe that the bureaucratic style is the way they are supposed to express themselves because they work in a company where that style is common. Even where common, however, bureaucratese is ineffective because it makes the reader dig and dig to determine what the writer is trying to say.

Achieving a clear, direct style is more than a considerate act performed by one person (the writer) on behalf of others (the readers). By increasing the efficiency with which your readers can comprehend your sentences, you also increase your chances of affecting them in the way you desire. Sentences that are needlessly difficult to understand are fatiguing and hard to concentrate on. Your readers may feel exasperated by having to work so hard to figure out what you are saying. If they do, they may begin skimming rather than reading closely, become inattentive, or simply put your communication down altogether, refusing to read it any more.

One of the most notable things about the decidedly *in*direct style of bureaucratese is that it is rarely used in one-on-one conversations. When we converse, we strive to say things as clearly as we can. Consequently, one way to write in a straightforward style is to build upon a conversational style. This advice doesn't mean that you should necessarily write the way you talk. Many occasions at work require a more formal style than people use in conversation. But figuring out how you would express your thoughts in conversation can lead you to a clear, direct expression of your ideas, one that you can then adapt to your written communication.

Guidelines 1 through 4 present specific strategies you can use to write in a style that is as direct and clear as conversation but still appropriate to on-the-job writing.

ELIMINATE UNNECESSARY WORDS

One excellent technique for achieving a straightforward style is to eliminate unnecessary words. Every word requires work from the reader. By eliminating unnecessary ones, you reduce the burden your prose imposes on your audience. Consider the following pair of sentences, in which the second takes thirty-eight percent fewer words than the first to convey essentially the same information. (Note that in this chapter the sample sentences are numbered for easy reference.)

> The physical size of the workroom is too small to accommodate this equipment. (1)

> The workroom is too small for this equipment. (2)

One way to eliminate unnecessary words is to avoid being *overly* precise. Certainly, you will want to use all the words necessary to make your meaning clear. But don't use any more. For example, instead of simply saying that the workroom is too small, the writer of Sentence 1 decided to specify the precise way in which the room is too small—in terms of its physical size. But the writer did not need to include the words *physical size:* when a person says that a room is too small for a piece of equipment, what feature of the room could he or she have in mind except its size?

A second way to eliminate unnecessary words is to look for places where you can substitute one word for several. Consider the following sentence:

> Due to the fact that the price of oil rose, Gulf Consolidated Services received many new orders for its fittings for drilling rigs. (3)

When hearing someone speak that sentence, we might not be troubled by the wordiness of the first five words. When reading, however, we would prefer to have them replaced by the single word *because*.

> Because the price of oil rose, Gulf Consolidated Services received many new orders for its fittings for drilling rigs. (4)

Here's another example:

> They do not pay any attention to our complaints. (5)

With a little more effort, the writer could have found the one word that contains the meaning of the six that he used:

> They ignore our complaints. (6)

Eliminating unnecessary words requires extra work on your part, but it saves work for your reader and also increases the clarity and forcefulness of your writing.

GUIDELINE
2

KEEP RELATED WORDS TOGETHER

A second strategy for achieving a clear, direct writing style is to keep related words together in your sentences. The two good reasons for doing so are both based solidly upon what is known about how people read.

Word Order Indicates Meaning

The first reason to keep related words together is that when we read English, we depend on word order to indicate meaning. For example, even though both of the following sentences have exactly the same words, they do not mean the same thing:

> The woman in the seat hit the man. (7)

> The woman hit the man in the seat. (8)

The placement of *in the seat* in Sentence 7 tells where the woman was, while the placement of the same words in Sentence 8 tells where the man was—or else where the woman hit him.

In some sentences, the words are arranged so that readers cannot know which of two possible meanings the sentence is supposed to have. Here is an example:

> We sent a note about the proposal to Chester. (9)

Unless we have some independent information about the situation, we cannot tell from this sentence whether the note went to Chester or whether the note went to someone else and it concerned a proposal made to Chester.

In most cases, however, a reader can figure out the meaning of the sentence even when a modifier (like *in the seat*) is separated from the word it is intended to modify. For example, most readers quickly figure out the meaning of the following sentence:

> A large number of undeposited checks were found in the file (10)
> cabinets, which were worth over $41,000.

According to the way our language works, that sentence tells us that the file cabinets were worth over $41,000. Yet, as we read we quickly reject this meaning. After all, it scarcely seems probable that the file cabinets were worth that much money, but it is very plausible that a large number of checks might be. Thus, the problem you would create by failing to keep related words together isn't so much that your readers will misunderstand the sentence as that they will have to pause in their reading to figure out your meaning—an effort you should have saved them.

Limitations of Short-Term Memory

The second reason you should keep related words together concerns the limitations of short-term memory. The short-term memory is the part of your mind that determines the meaning of each sentence as you read it. Short-term memory then passes these meanings along to your long-term memory, where the meanings you desire from the individual sentences are integrated with one another and with other things you know.

One of the most important features of short-term memory is that it has a very limited capacity, being able to hold only about five to nine bits of information at a time. If you write a sentence whose related words are too far apart, your readers will forget the earlier words by the time they reach the later words. Consider, for example, the following sentence:

> A new factory to produce chemicals for the OPAS system, which   (11)
> enables large manufacturers of business forms to make carbonless
> copy paper as part of their own manufacturing process, began to
> operate late last year.

If you are like most readers, you had to read that sentence twice to understand it. By the time you got to the verb of the sentence *(began),* you could not remember what the subject was. You had to move your eyes to the beginning of the sentence to remind yourself that the subject is *factory*. The large number of words following the subject pushes *factory* from your short-term memory before you reach *began*.

You should be careful not to misinterpret this explanation of the difficulty in Sentence 11. You *can* interject words and phrases between related words of your sentences. In fact, placing a phrase between related words is one way to emphasize the phrase. Consider the following sentence, in which the writer emphasizes a point by placing it between the first and second parts of the verb phrase (*managed* and *to make*):

> We managed, despite the recession, to make record profits   (12)
> that year.

As this sentence shows, brief interjections between related parts of a sentence can be very effective. In all sentences—whether they have interjections or not—keep related words close enough together that your readers can see immediately their relationship.

GUIDELINE 3	**FOCUS ON ACTION AND ACTORS**

Another way to achieve a clear, direct writing style is to focus your sentences on action and actors. Doing so will energize your writing. Two strategies for focusing on actors and action are to express action in your verbs and to use the active voice.

Express Action in Verbs

Most sentences are about action. Sales rise, equipment fails, vendors supply, managers approve. Clients praise or complain, and technicians advise. Yet, many people bury the action in nouns, adjectives, and other parts of speech, thereby suffocating their prose. Consider the following sentence:

> Our department accomplished the conversion to the new   (13)
> machinery in two months.

This sentence is indirect and vaguely irritating because the main action, *converting,* is expressed in a noun *(conversion)* and the verb is surrendered to the word *accomplished,* which is unnecessary to the meaning of the sentence. To improve the sentence, the writer need only move the main action from the noun to the verb:

> Our department converted to the new machinery in two months.   (14)

Not only is the new version briefer, it is more emphatic and livelier.

How to Spot Sentences That Don't Have Their Main Action in the Verb

To put action in your verbs, you need to identify sentences that have their main action somewhere else. Two kinds of sentences, both easily spotted, often displace the action from the verb.

The most common are sentences in which the verb is some form of the verb *to be:*

> This procedure is a protection against reinfection.   (15)

The major action in this sentence is *protecting,* but the main verb is *is.* The action can be put in the verb by rewriting the sentence in this way:

> This procedure protects against reinfection.   (16)

In the second kind of sentences that often displace the action from the verb, an important word ends in one of the following suffixes: *-tion, -ment, -ing, -ion, -ance.*

Consequently, I would like to make a <u>recommendation</u> that the   (17)
department hire two additional programmers.

The main action in Sentence 17 is *recommending,* but that action is expressed
in a noun, *recommendation,* rather than in the verb. To put the main action in
the verb, the writer could write:

Consequently, I <u>recommend</u> that the department hire two   (18)
additional programmers.

Although the action of a sentence is often displaced from the main verb by
use of the verb *to be* or by the use of words with the suffixes listed above, this
displacement can occur in other ways as well. Therefore, you must be alert to
any sentence in which the action is displaced from the verb. Consider, for
example, the following sentence:

In the third quarter, the soaring costs of raw materials caused a   (19)
<u>rise</u> in the prices of finished goods.

In Sentence 19, the action, *rising,* is placed in a noun, *rise.* The sentence can
be improved by the following revision, which places the main action in
the verb:

In the third quarter, the soaring cost of raw materials <u>raised</u> the   (20)
prices of finished goods.

How to Place the Action in the Verb   Once you have identified a sentence
that does not have its main action in the verb, you can follow this three-step
procedure for fixing it:

1.   Identify the main action of the sentence. Ask yourself, "What is happen-
     ing in this sentence that is really crucial?" In Sentence 15, that action is
     protecting, even though the verb is *is.*
2.   Identify the actor (the person or thing that performs that main action). In
     Sentence 15, the actor is the thing that does the protecting: *This procedure.*
3.   Start to say a sentence that begins with the actor and then immediately tells
     the main action: "This process protects . . ." In most cases, the rest of
     the words will follow automatically.

Use the Active Voice

A second way to focus your sentences on action and actors is to use the active
voice rather than the passive voice. To write in the active voice, place the
actor—the person or thing performing the action—in the subject position. Your

verb will then describe the actor's *actions*. In the active voice, the subject and the actor are the same. The subject acts.

Active Voice

Subject   Verb
↓   ↓
The consultant recommended these changes.   (21)
↑   ↑
Actor   Action

In the passive voice, the subject of the sentence and the actor are different. The subject is *acted upon* by the actor.

Passive Voice

Subject   Verb
↓   ↓
The changes were recommended by the consultant.   (22)
↑   ↑
Action   Actor

Figure 11–1 shows some additional examples of active and passive sentences.

By making the actor and the action the same, active verbs help readers see immediately what action a sentence describes and who performed it. This helps them read more quickly. The active voice also promotes efficient reading by

ACTIVE		
Actor Is Sentence Subject	**Action (Verb)**	**Object of Action**
We	purchased	the Korean ore.
Richard	could not find	the test results.
The project team	needs	our support.
PASSIVE		
Object of Action Is Sentence Subject	**Action (Verb)**	**Actor**
The Korean ore	was purchased	by us.
The test results	could not be found	by Richard.
Our support	is needed	by the project team.

FIGURE 11–1   Active and Passive Sentences

reducing the number of words used. For example, the first version of the statement about the consultant (Sentence 21) communicates the message in five
words, while the passive version (Sentence 22) communicates it in seven—an
increase of 40 percent.

In addition to promoting reading efficiency, the active voice has another
important advantage over the passive: it avoids the vagueness and ambiguity
that often characterizes passive voice. In the active voice, you must always tell
who the actor is; in the passive voice you don't need to. "The ball was hit" is
a grammatically correct sentence, even though it doesn't tell who or what hit
the ball. The writer of that passive sentence may know who did the hitting, but
the readers certainly don't. In an active version of that sentence, the writer
identifies the actor: "Linda hit the ball."

Avoiding the vagueness of the passive voice will be very important to you
in many situations at work.

Passive Voice   The operating temperatures must be checked daily to ensure that   (23)
                         the motor is not damaged.

Will the supervisor on the third shift know that he is the person responsible for
checking the temperature? Sentence 21 certainly allows him to think that someone else is, perhaps the supervisor of one of the other shifts.

When to Use the Passive Voice   Although you should generally use the active voice, the passive voice is preferable in some situations. Consequently, a
precise statement of the advice you should follow about the active voice is as
follows: use the active voice *unless there is good reason to use the passive*.

One place to use the passive voice occurs when you don't want to identify
the actor:

                         The lights on the third floor have been left on all night for the   (24)
Passive Voice   past week, despite the efforts of most employees to help us reduce
                         our energy bills.

Sentence 24 is from a memorandum in which the writer wants to urge all
employees to work harder at saving energy, but at the same time does not want
to cause embarrassment and resentment by pointing the finger at guilty parties.
Consequently, he has correctly used the passive voice to avoid identifying the
actors who have been leaving the lights on. Also, consider this sentence:

Passive Voice   I have been told that you are using the company telephone for an   (25)
                         excessive number of personal calls.

Perhaps the people who told the writer about that breach of the corporate telephone policy did so in confidence. If the writer judged that it is ethically acceptable to communicate this news to the reader without naming the people

who made the report, then she has used the passive voice effectively. (Be cautious, though, to avoid using the passive voice to hide an actor's identity when it is unethical to do so.)

A second good reason for using the passive voice is to focus your audience's attention on the object, not the actor—something you can do by placing the object in the subject position of the sentence. You will learn more about this use of the passive voice when you read Guideline 5.

How to Convert Passive Sentences to Active Ones   When you want to convert a passive sentence to an active one, you can use the same procedure that is described above for placing the action in the verbs. First, identify the main action and the actor who performs it. Next, start to say a sentence with the actor as the subject. Follow with a verb that describes the actor's action. In most cases, the rest of the sentence will flow naturally from there.

| GUIDELINE 4 | EMPHASIZE WHAT'S MOST IMPORTANT |

Another way to write clear and forceful sentences is to emphasize the most important information in them. The following paragraphs describe four ways you can do that.

Place the Key Information at the End of the Sentence

Generally, you can emphasize the most important information in a sentence by placing it at the end. In this respect, sentences differ significantly from paragraphs and larger units of prose, where the place of primary emphasis is the beginning (hence, the advice given in Chapter 8 to "Put the most important information first" in paragraphs, sections, and chapters).

As linguist Joseph Williams points out, you can verify that the end of a sentence is a place of emphasis by listening to yourself speak.[1] For example, read the following sentences aloud:

> Her powers of concentration are extraordinary.   (26)

> Last month, he topped his sales quota even though he was sick   (27)
> for an entire week.

As you speak those sentences, notice how you naturally stress the final word (*extraordinary*) in Sentence 26 and the final two words (*entire week*) in Sentence 27.

To obtain the emphasis that the final position in a sentence can give, you should construct or edit your sentences to place key information at the end. Sometimes that means moving other, less-important information to the beginning of the sentence:

Key Information

Before

The department's performance has been <u>superb</u> in all areas.   (28)

Revised

In all areas, the department's performance has been <u>superb</u>.   (29)

At other times, you can accomplish the same thing by moving the key information from the beginning of the sentence to the end:

Key Information

Before

The <u>bright exterior design is</u> one of the product's most appealing features to younger consumers.   (30)

Revised

One of the product's most appealing features to younger consumers is its <u>bright exterior design</u>.   (31)

Place Key Information in the Main Clause

For sentences with more than one clause, you can also emphasize the most important information by placing it in the main clause. Consider the following pair of sentences:

Although our productivity was down, our profits were up.   (32)

Although our profits were up, our productivity was down.   (33)

In Sentence 32, the emphasis is upon profits because *profits* is the subject of the main clause. Sentence 33 emphasizes productivity because that is the subject of the main clause. (Notice that in each of these sentences, the emphasized information is not only in the main clause but it is also at the end of the sentence.)

Highlight Key Information Typographically

Another way to highlight the most important information in a sentence is to use underlining and boldface printing:

While the CEO's endorsement letter certainly helped us obtain employee pledges to this year's United Way Campaign, we could have done even better if he had expressed his endorsement <u>enthusiastically</u>.   (34)

In the fifty-seven years we've been in business, we've **never** failed to pay a dividend to our shareholders.   (35)

Explicitly Tell What Information Is Key

A fourth way to emphasize key information is to announce its importance to your audience:

> Economists pointed to three important causes of the stock   (36)
> market's decline: uncertainty about the outcome of next month's
> election, a rise in inventories of durable goods, and—most
> important—signs of rising inflation.

At the beginning of this chapter, you read that when you are writing sentences you have the twin goals of making each sentence as clear and direct as possible and of coordinating each sentence with those around it.

The four guidelines you have just read concern ways of composing individual sentences that are as clear and direct as possible. Guidelines 5 and 6 shift your attention to the other goal of making each sentence fit into the context of the surrounding sentences.

GUIDELINE 5   SMOOTH THE FLOW OF THOUGHT FROM SENTENCE TO SENTENCE

When writing sentences, help your readers see how each one follows from the sentence before it. As you will recall, we read by building the small bits of information we obtain from individual sentences into the larger meaning of the overall communication. Consequently, as we read each sentence, we must figure out how the *new* information it presents to us relates to the *old* information we obtained from the preceding sentence. If we can discern that relationship immediately, we say that the writing flows smoothly. If we must pause to figure out that relationship, we rightly complain that the writing is choppy and difficult to follow.

As you think about the flow of thought in your writing, remember that sentence-to-sentence relationships that seem obvious to you may not be evident to your readers. As the writer, you gained your understanding of these relationships by deciding what you want to say. Your readers, in contrast, must rely solely upon the words you've placed on the page for clues about the relationship of one sentence to the next.

The following sections describe four things you can do to provide your readers with the help they need as they move from sentence to sentence: avoid needless shifts in topic, use transitional words, use echo words, and place your transitional and echo words at the beginnings of your sentences.

Avoid Needless Shifts in Topic

The simplest relationship between the information in two adjacent sentences is this: the first sentence says something about a particular topic, and the second says something more about the same topic.

The links of the drive chain must fit together firmly. They are   (37, 38)
too loose if you can easily wiggle two links from side to side
more than ten degrees.

Often a communicator speaks about the same thing in two adjacent sentences but the readers cannot see that both sentences are about the same thing, at least not immediately. You can help your readers rapidly detect the shared topic of adjacent sentences by carefully controlling the words you put in the subject position. That's because people usually perceive the topic of a sentence to be whatever is the grammatical subject of the sentence. If two of your sentences have the same topic, signal that similarity by giving them the same subject. For example, imagine that you had just written this sentence:

TOPIC (Subject)	COMMENT	
Our company's new inventory system	reduces our inventory costs considerably.	(39)

And suppose also that in your next sentence you wanted to communicate information that could be expressed either in this way:

TOPIC (Subject)	COMMENT	
Thousands of dollars	have been saved by the system this year alone.	(40)

or in this way:

TOPIC (Subject)	COMMENT	
The system	has saved thousands of dollars this year alone.	(41)

Would Sentence 40 or Sentence 41 be the best one to follow Sentence 39? Both contain essentially the same information. But Sentence 41 has the same topic—*system*—as Sentence 39, and Sentence 40 does not. Therefore, your readers would relate the old information in Sentence 39 more easily to the ''new'' information in Sentence 41 than to the same new information in Sentence 40:

Our company's new inventory system reduces inventory costs   (39, 41)
considerably. The system has saved thousands of dollars this year
alone.

The preceding example involves a pair of sentences in which you could aid your readers by using exactly the same topic *(systems)* in the subject position of the adjacent sentences. You will also encounter many situations in which you can avoid needless shifts in topic by keeping the same *general* topic in the subject position. Consider, for example, the following paragraph:

> The materials used to construct and furnish this experimental   (42–45)
> office are designed to store energy from the sunlight that pours
> through the office's large windows. The special floor covering
> stores energy more efficiently than wood. The heavy fabrics used
> to upholster the chairs and sofas also capture the sun's energy.
> Similarly, the darkly colored paneling holds the sun's energy
> rather than reflecting it as lightly colored walls would.

In this paragraph, the subject of the first sentence is *materials*. Although the same word is not the subject of the sentences that follow, the subjects of all those sentences are kinds of materials, namely the *special floor covering, heavy fabrics,* and *darkly colored paneling.* Thus, although the specific word placed in the subject position of the various sentences changes, the general topic does not.

You Can Use the Passive Voice to Maintain a Consistent Focus   One important implication of the preceding discussion is that you sometimes will be able to follow Guideline 5 only by using the passive voice. In the discussion of Guideline 3, you learned that it is generally desirable to use the active voice, not the passive. But you also learned that there are times when the passive is appropriate, even preferable to the active. One of those times is when the passive voice enables you to avoid a needless shift in the topic of two adjacent sentences. Consider the following paragraph:

> Tom works in the Paint Department. On Tuesday, he finished   (46–49)
> lunch late, so he took a shortcut back to his work station. Fifteen
> yards above the factory floor, a can of paint slipped off a scaffold
> and hit him on the left foot. Consequently, at the busiest part of
> the year, he missed 17 days of work.

The topic of most of the sentences in this accident report is "Tom" or "he." However, the third sentence shifts the topic from Tom to the can of paint. Furthermore, because the third sentence shifts, the fourth must also shift to bring the focus back to Tom. The writer could avoid these two shifts by rewriting the third sentence so that it is about Tom, not about the can of paint. That means making Tom the grammatical subject of the sentence, and, as a result, making the verb passive:

> He was hit on the left foot by a can of paint that slipped off a   (50)
> scaffold fifteen yards above the factory floor.

You Can Use Pronouns to Avoid Monotony You may wonder whether your prose will become monotonous if you keep the same topic in the subject position of adjacent sentences. Your writing *can* get monotonous if you follow this advice without imagination and sensitivity. One way you can avoid this monotony is to use pronouns. Words such as *he, she,* and *it* enable you to keep the same topic in the subject position of your sentences without keeping the same word there:

> A new paper-making machine is on order for the Chillicothe plant. It will produce carbonless copy paper, as well as copy machine and duplicator papers. (51–52)

Use Transitional Words

The preceding discussion explains how you can help your readers follow your flow of thought where two adjacent sentences are about the same topic. This situation arises often. However, in almost any communication most of the sentences shift the topic. Even so, every one of the sentences is related in some way to the sentence that precedes it—unless the communication is simply incoherent. Consequently, one of your major tasks as a writer is to indicate to your readers how the topic of each sentence relates to the different topic of the preceding sentence.

One way to do that is to include *transitional words* and phrases that explicitly state the relationship. Note some of the most commonly used words:

Links in time	after, before, during, until, while
Links in space	above, below, inside
Links of cause and effect	as a result, because, since
Links of similarity	as, furthermore, likewise, similarly
Links of contrast	although, however, nevertheless, on the other hand

Use Echo Words

Another way to guide your readers from one sentence to the next is to use echo words. Most sentences contain some word or phrase that recalls to the readers' minds some information they've already encountered. Such words and phrases might be called *echo words,* because they echo something the readers already know.

Consider the following example:

> Inflation can be cured. The cure appears to require that consumers change their basic attitudes toward consumption. (53–54)

In this example, the noun *cure* at the beginning of the second sentence echoes the verb *cured* in the first. It tells the readers that what follows will discuss the curing that they have just read about.

There are many kinds of echo words. In the example about curing inflation, the echo word is another version of the word being echoed (*cured; cure*). In other cases, the echo word can be a pronoun:

> We had to return the copier. Its frequent breakdowns were   (55–56)
> disrupting our work.

And sometimes an echo word is another word from the same "word family" as the word being echoed:

> I went to my locker to get my lab equipment. My oscilloscope   (57–58)
> was missing.

In this example, *oscilloscope* in the second sentence echoes *lab equipment* in the first because people know that an oscilloscope is a piece of lab equipment.

Finally, an echo word can be a word or set of words that recall some idea or theme expressed but not explicitly stated in the preceding sentence.

> The company also purchased and retired 17,399 shares of its   (59–60)
> $2.90 convertible, preferred stock at $5.70 a share. These
> transactions reduce the number of outstanding convertible shares
> to 635,250.

In this example, the words *these transactions* tell readers that what will follow in the sentence concerns the purchasing and retiring that were discussed in the preceding sentence.

You may have noticed that in each of the examples given so far the echo word is in the main clause of the sentence. Echo words can also be used effectively in subordinate clauses:

> The primary objective of the proposed campaign is to attract   (61–62)
> new members. For this campaign to be effective, the club's
> executive committee must carefully assess the willingness of the
> club's current members to undertake the projects that are
> suggested.

Furthermore, a sentence may contain more than one echo word.

> Because the park does not own enough horses, the stable   (63–64)
> manager must meet the demand for riding tours by making the
> horses work as much as seven hours a day, rather than a more
> reasonable four or five. The horses, laboring under this heavy

workload, grow tired and uncooperative by the end of the day, so that beginning riders have trouble getting their horses to perform.

Place Transitional Words and Echo Words at the Beginning of the Sentence

Transitional words and echo words will help your readers most if you use them at the beginning of your sentences. That's because, even while they are reading the beginning of a sentence, your readers already want to know what relationship that sentence has to the preceding one. To understand the difficulties you can create for readers by postponing the appearance of your echo words, imagine that you have just read this sentence in a report :

> The metal is then coated by the Barnhardt process. (65)

After completing that sentence, you begin the next. The first word you read is:

> Sales

What has the new information contained in the word *sales* to do with the old information in the preceding sentence? You can't tell. The next words are:

> have increased threefold

Do you know the relationship between the two sentences yet? You may have a guess, but you can't be sure. The next words are:

> in the year since

which are followed by:

> our engineers

Your suspicions about the relationship between the two sentences may now be very strong, but you still have only a guess. You know for sure only after you read the last three words of the sentence:

> modified this process.

Think how much easier your reading would have been had the second sentence been written so that the echo words appeared at the beginning, not the end, of the sentence:

> The metal is then coated by the Barnhardt <u>process</u>. Since <u>this</u> (65–66) <u>process</u> was modified by our engineers, sales have increased threefold.

GUIDELINE
6

VARY YOUR SENTENCE LENGTH AND STRUCTURE

If all the sentences in a sentence group have the same structure, two problems arise: monotony sets in and (because every sentence is basically alike) you lose the ability to emphasize major points and de-emphasize minor ones.

You can avoid such monotony and loss of emphasis in two ways. First, you can vary your sentence length. Longer sentences can be used to show the relationships among ideas. Shorter ones provide emphasis—if the sentences around them are longer:

<table>
<tr>
<td>Short
sentences
used for
emphasis</td>
<td>In April, many amateur investors jumped back into the stock market because they believed that another rally was about to begin. They noted that exports were increasing rapidly, which they felt would strengthen the dollar in overseas monetary markets and bring foreign investors back to Wall Street. Also, they observed that unemployment had dropped sharply, which they also predicted would be taken as an encouraging sign for the economy. They were wrong on both counts. Wall Street interpreted rising exports to mean that goods would cost more at home, and it predicted that falling unemployment would mean a shortage of workers, hence higher prices for labor. Where amateur investors saw growth, Wall Street saw inflation.</td>
</tr>
</table>

Second, you can vary your sentence structure. The grammatical subject of a sentence does not have to be the first word in the sentence. In fact, if it did, the English language would lose much of its power to emphasize more important information and to de-emphasize less important information. Here is a passage in which the topic remains the same from sentence to sentence, but the sentence structure varies:

<table>
<tr>
<td>Same subject
for all
sentences</td>
<td>Through hard work and chance, the present secretary of our organization has assumed a dangerously wide variety of responsibilities. He takes and distributes the minutes of our meetings, as secretaries customarily do. He (not the president) writes the agenda and he (not the treasurer) manages our bank account. Furthermore, just before last year's elections, he became the nominating committee, selecting the pool of people from which our new officers would be elected. In short, the secretary has become the most powerful person in the organization, one who has transformed our democratically run organization into a monarchy.</td>
</tr>
</table>

A SPECIAL NOTE ON TOPIC AND TRANSITIONAL SENTENCES

Most on-the-job communications have one class of sentences that deserve particular attention. These are the sentences that begin a paragraph, section, or larger unit. Such sentences perform two special functions. When they serve as topic sentences, they must establish not only their own focus but also the focus of the sentences that follow. When they provide transitions, they not only link themselves with the preceding sentence but they also link all the sentences that follow with the larger units of prose that have gone before.

Because of their special functions, these topic and transitional sentences often are *not* concerned with action. Consequently, they often use the verb *to be,* despite the advice to avoid that verb in most sentences:

> There are three reasons for the recent breakdown in communications between corporate headquarters and the out-of-state plants.

Nevertheless, these sentences can still be strengthened by following much of the other advice given in this chapter including the advice to eliminate unnecessary words and to use transitional and echo words at the beginning of a sentence. Another piece of advice that applies to topic and transitional sentences is to place information to be emphasized at the end. Compare the following sentences:

Weak

> Curtailing the use of outside vendors and relying more heavily upon our in-house printing facilities is our third recommendation.

Stronger

> Our third recommendation is to curtail the use of outside vendors and rely more heavily upon our in-house printing facilities.

In the stronger version, the writer states the recommendation at the end of the sentence, thereby giving it the emphasis it deserves. That arrangement also has the advantage of placing an echo word at the beginning of the sentence: the word *third* echoes a statement earlier in the report that announced that three recommendations would be given.

CONCLUSION

As a speaker of the English language, you have tremendous power over its vast resources for expressing ideas in sentences. Some writers misuse that power to write in a wordy, inflated, pompous style called bureaucratese. In contrast, you should seek to write sentences that are individually clear while still working together harmoniously.

Much of this chapter's advice focused upon achieving the first objective of writing clear, direct sentences. In particular, you have read suggestions for doing the following things:

- Eliminating unnecessary words
- Keeping related words together
- Focusing your sentences on action and actors
- Emphasizing what's most important

To aid you in creating sentences that fit together harmoniously, this chapter also provided additional suggestions for doing the following:

- Smoothing the flow of thought from sentence to sentence
- Varying your sentence length and structure

EXERCISES

1. Without altering the meaning of the following sentences, reduce the number of words in them.
   a. After having completed work on the data-entry problem, we turned our thinking toward our next task, which was the processing problem.
   b. Those who plan federal and state programs for the elderly should take into account the changing demographic characteristics in terms of size and average income of the composition of the elderly population.
   c. Would you please figure out what we should do and advise us?
   d. The result of this study will be to make total white-water recycling an economical strategy for meeting federal regulations.

2. Rewrite the following sentences in a way that will keep the related words together.
   a. This stamping machine, if you fail to clean it twice per shift and add oil of the proper weight, will cease to operate efficiently.
   b. The plant manager said that he hopes all employees would seek ways to cut waste at the supervisory meeting yesterday.
   c. About 80 percent of our pulp, to be made into linerboard and corrugated cardboard (much of it used for beverage containers), supplies our plants for manufacturing packages.
   d. Once they wilt, most garden sprays are unable to save vegetable plants from complete collapse.

3. Rewrite the following sentences to put the action in the verb. Follow the three-step procedure explained in the discussion of Guideline 3.
   a. The experience itself will be an inspirational factor leading the participants to a greater dedication to productivity.
   b. The system realizes important savings in time for the clerical staff.
   c. The implementation of the work plan will be the responsibility of a team of three engineers experienced in these procedures.
   d. Both pulp and lumber were in strong demand, even though rising interest rates caused the drying up of funds for housing.

4. Rewrite the following sentences in the active voice. Follow the three-step procedure explained in the discussion of Guideline 3.

   a. Periodically, the shipping log should be reconciled with the daily billings by the Accounting Department.

   b. Fast, accurate data from each operating area in the foundry should be given to us by the new computerized system.

   c. Since his own accident, safety regulations have been enforced much more conscientiously by the shop foreman.

   d. No one has been designated by the manager to make emergency decisions when she is gone.

5. For each of the sentences given below, do the following:

   a. Read the sentence.

   b. In the space provided, write in the number of the revisions that would most improve it. The possible revisions are as follows:

   (1) Eliminate unnecessary words.

   (2) Keep related words together.

   (3) Use the active voice.

   (4) Put action in verbs.

   c. Revise the sentence.

   _____ **1.** Marex Advanced Systems is engaged in the development of ink-jet printing equipment and other high-technology products.

   _____ **2.** Productivity will be increased greatly by the installation of robots on the assembly line.

   _____ **3.** She wanted to decide the question as to whether the Baltimore plant should try the process.

   _____ **4.** Our purpose is to achieve the identification of inefficiencies so that management can bring about their elimination.

   _____ **5.** During the past six months, 180 payroll-advance checks have been written by Jane.

   _____ **6.** Provisions should be included in the plans for the new office building for the handicapped.

   _____ **7.** Our Canadian affiliate will accomplish a doubling of software sales in the next eighteen months.

   _____ **8.** This is a machine that applies the welds twice as rapidly.

   _____ **9.** Because the employees drive so recklessly, the parking lot, when the dayshift leaves, especially on hot afternoons, becomes a dangerous place.

   _____ **10.** The driver's inability to find our office resulted in a delay of the delivery of the contracts, which had been sent by overnight express.

   _____ **11.** Our office used three weeks for the preparation of the proposal.

   _____ **12.** The circular chart recorder will aid the Process Control Department in controlling the process by providing a means by which its personnel can check to see that the speed has been properly set.

_____ **13.** The new billing procedure will be welcomed by our department because the accountants' workload will be reduced by it.

_____ **14.** In the event that you cannot attend the meeting, please call my secretary.

_____ **15.** I recommend that the electric stencil machine be moved to the scheduling office where decisions on production runs are made, for the following reasons.

_____ **16.** Recently that client gave him an invitation to make a presentation of his ideas for reorganizing their production line.

_____ **17.** Please draw your attention to the fact that your quarterly report is three months overdue.

_____ **18.** Vacation schedules are coordinated by the Assistant to the Plant Manager.

6. In three of the following pairs of sentences, the topic shifts from the first to the second sentence. Rewrite one or the other sentence so that the topics are the same.

a. "Grab" samplers collect material from the floor of the ocean. Rock, sediment, and benthic animals can be gathered by these samplers at rates as high as 8,000 tons per hour.

b. To fluoridate the drinking water, a dilute form of hydrofluorisilic acid is added directly to the municipal water supply at the main pump. An automatic control continuously meters exactly the right amount of the acid into the water.

c. Fourteen variables were used in these calculations. The first seven concern the volume of business generated by each sales division each week.

d. The city's low-income citizens suffer most from the high prices and limited selection of food products offered by commercial grocers. Furthermore, information concerning nutrition is difficult for many low-income citizens to find.

7. The sentences listed below form a single passage from a report. The report was prepared by a consulting firm for a company that wanted to learn about the potential for profit in manufacturing catalytic converters, which are pollution-control devices used to reduce harmful emissions from automobile engines. One type of converter is called a _pelleted_ converter because it contains many pellets of a solid material. The other type of converter is called a _monolithic_ converter because it contains a single large cylinder of the material.

Without changing the order of the sentences, rewrite them in a way that will guide readers smoothly from sentence to sentence. As you revise the sentences, make notes in the margin to tell which of the guidelines given in this chapter you have used. You may rework the sentences freely, but try not to leave out any ideas. Some sentences may not have to be rewritten. To start your work on each sentence, decide what you want its topic to be.

TO: Wally Nugent
            Don McNeal

FROM: Kent Bradshaw

SUBJECT: Response to Your Request for Names of Courses in
            Management Fundamentals

I apologize for the delay in following up our conversation;
however, it turns out that fewer courses are available in
management fundamentals than I expected, so it took some
digging. The digging uncovered only two courses that I feel
comfortable recommending.

>   First choice is the BASIC MANAGEMENT FOR THE NEWLY
>   APPOINTED MANAGER seminar offered by the University of
>   Michigan. This is a three-day program designed as an
>   introduction and a brush-up. I have attached a
>   description of it for your review.
>
>   The second choice is a public session of THE
>   EXCEPTIONAL MANAGER seminar offered by the Forum
>   Corporation. The next offering is in Chicago on
>   September 19 through 21. One of the outcomes of this
>   seminar is a personal action plan, which may not be
>   useful to you if your primary purpose in taking a
>   course is to show your superiors that you have had
>   formal training in management and theory;
>   nevertheless, this course is a good one, and you
>   should consider it.

In addition, a few supervisory training programs are offered
at local universities; however, I believe these programs
focus on too low a level of management to meet your needs.
They tend to be directed to production supervisors.

None of the courses I found has the emphasis on the financial
aspects of management you want. If you wish to look at some
basic finance courses, please let me know. I will be glad to
help you enroll in either of the courses listed above, or to
help you get further information if you need it.

FIGURE 11–2   Memo for Exercise 8

a. Catalytic converters are required equipment on almost all new cars sold in the United States.

b. Removing undesirable emissions, particularly sulfates, from the cars' exhaust gases is the purpose of those converters.

c. More efficient operation is demonstrated by some catalytic converters than by others.

d. The Environmental Protection Agency (EPA) has recently compared two converter designs.

e. The catalyst is in pellets (like mothballs) in the first design, called the pelleted converter.

f. The second design, called a monolithic converter, has its catalyst in a single cylinder, like an ancient Egyptian monolith.

g. Cars equipped with each type of converter were driven by the EPA over a standardized route in order to compare the two designs.

h. Sulfates were stored by both converters when used at the low speeds of city driving, whereas both emitted sulfates when used at the higher speeds of highway driving.

i. The pelleted converter exhibited a greater sulfate storage capacity.

j. At low speeds, only 23 percent of the sulfur in the gasoline was released by the pelleted converter.

k. The monolithic converter released 92 percent at low speeds.

l. The pelleted converter released a much greater proportion of its stored sulfur when used at higher speeds.

m. A characteristic desired by the EPA is possessed by the pelleted converter: it is able to store sulfates in the more densely populated urban areas, where people drive slowly, and then release those sulfates in the less densely populated suburban and rural areas, where people drive faster.

8. As you read the memo in Figure 11–2, circle the various devices used by the writer, Mr. Bradshaw, to help his readers move from one sentence to the next. In the margin, explain which of the guidelines in this chapter have been followed by the writer in each case.

12

CHOOSING WORDS

DEFINING
OBJECTIVES
PLANNING
DRAFTING
EVALUATING
REVISING

David, a recent college graduate, is looking over a draft of the first report he has written on the job. As he reads, his word choices worry him. "Should I use more sophisticated terms or more ordinary ones?" he asks. "Is my diction too formal or too informal? Am I using too many technical terms or too few?"

David's worries are perfectly natural. Choosing words can be a very challenging task even for experienced employees. At the same time, the words you choose can have a significant impact upon the success of your writing.

YOUR AIMS WHEN CHOOSING WORDS

The key to choosing words, of course, is to think about your readers and purpose. First, you need to select words that will enable your readers to grasp your meaning quickly and efficiently. Also, you need to choose words that express your meaning as precisely as possible. Unfortunately, these two goals sometimes conflict, as when the word that most precisely expresses your meaning is a technical or specialized term your readers won't understand. Your third goal is to shape your reader's attitudes—not only about your subject matter but also about you. As David is well aware, the words that writers select can make the writers appear to be masters of their subject or bumbling incompetents, like people dedicated to assisting their readers or inconsiderate, self-centered individuals.

The strategies for achieving these on-the-job goals of understandability, precision, and persuasive impact may differ somewhat from the strategies for word selection you are studying in other college courses. For instance, in literature and composition classes your writing may be intended to increase your readers' sensitivity to various aspects of human experience. Accordingly, you may be encouraged to choose colorful, out-of-the ordinary words that will create new impressions on your readers' imaginations.

At work, however, your purpose for writing will usually be much more practical: to help your readers make a business decision, perform a procedure, and so on. Consequently (as you will read below), you should usually pick plain words, not fancy ones, and you should select words your readers already know, not ones that will send them scurrying to the dictionary. This doesn't mean that the advice given in other courses is incorrect, just that it usually doesn't apply to on-the-job writing.

This chapter presents six guidelines that will help you on the job as you search for words that will shape your readers' attitudes favorably while communicating your message understandably and precisely.

GUIDELINE 1 USE CONCRETE, SPECIFIC WORDS

One of the most important strategies for choosing words is to pick concrete, specific ones. Almost anything can be described in relatively abstract and general words, or relatively concrete and specific ones. You may say that you are writing on a piece of *electronic equipment,* or that you are writing on an *electronic typewriter* or a *word processor.* You may say that your employer produces *consumer goods* or that it makes *men's clothes.*

When ranked according to their degree of abstraction, groups of related words form hierarchies. Figure 12–1 shows such a hierarchy where the most specific terms identify concrete items we can perceive with our senses; Figure 12–2 shows a hierarchy where all the terms are abstract, but some are more specific than others.

When writing at work, you can usually strengthen your writing by using concrete and specific words rather than abstract and general ones. There are two important reasons for preferring concrete, specific words.

First, concrete, specific words make it easier for your readers to understand precisely what you mean. If you say that your company produces television shows for a *younger demographic segment,* they won't know whether you mean *teenagers* or *toddlers.* If you say that you study *natural phenomena,* your readers won't know whether you mean *volcanic eruptions* or *the migration of monarch butterflies.*

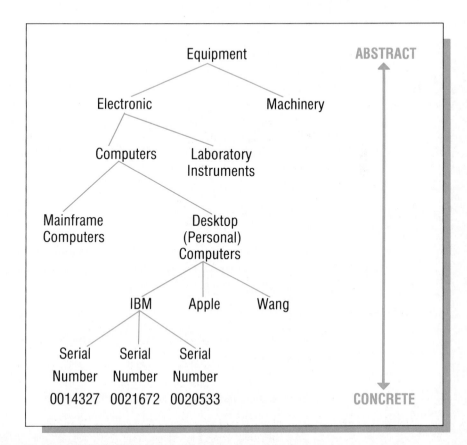

FIGURE 12–1   Hierarchy of Related Words that Move from Abstract to Concrete

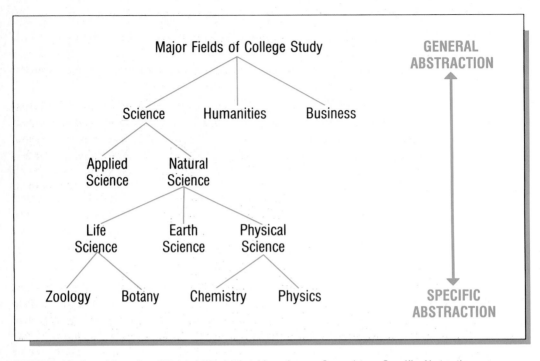

FIGURE 12–2   Hierarchy of Related Words that Move from a General to a Specific Abstraction

Such vagueness can hinder readers from getting the information they need in order to make decisions and take action. Consider the following sentence from a memo addressed to an upper-level manager who wanted to know why production costs were up.

Original

The cost of one material has risen recently.

This sentence doesn't give the manager the information she needs to take any remedial action. In contrast, the following sentence, using specific words, does help by suggesting more precisely what the situation is.

Revised

The cost of the bonding agent has tripled in the past six months.

As another example, look at the original and revised versions of a sentence from a letter to state officials trying to decide whether to approve a series of construction projects proposed by an Alaskan mining company.

Original

Some of the planned construction may affect animals.

Revised

Building the planned fence along the west side of the Hilton River access road may interrupt the natural migration of caribou, causing many to starve.

Of course, abstract and general terms do have important uses. For example, in scientific, technical, and other specialized fields, writers often are concerned with making general points, describing the general features of situations, and providing general guidance for action. Your objective when choosing words is not to avoid abstract and general words altogether, but to avoid using words that are more abstract or more general than your purpose requires.

GUIDELINE 2

USE SPECIALIZED TERMS WHEN—AND ONLY WHEN—YOUR READERS WILL UNDERSTAND THEM

A second way to make your writing clear and effective is to use the specialized terms of your own specialty wisely. In some situations, specialized terms help you communicate effectively because they convey precise meanings very economically. Many have no exact equivalent in everyday speech and would take many sentences or even paragraphs to explain to someone who doesn't know them. These specialized terms can also help to establish your credibility with your fellow specialists: when you use the special vocabulary of your field accurately, you show your audience that you are adept in that field.

On the other hand, if you use technical terms when communicating to people who are unfamiliar with them, you will make your messages very difficult to understand and use. Consider the following sentence:

The major benefits of this method are smaller in-gate connections, reduced breakage, and minimum knock-out—all leading to great savings.

Although this sentence would be perfectly clear to any manager who works in a foundry that manufactures parts for automobile engines, it would be unintelligible to most other people because of the use of the specialized terms *in-gate connections* and *knock-out*.

How to Identify Words Your Readers Won't Know

When seeking to identify words you need to avoid, be sure to consider all specialized terms you use from your *readers'* point of view. Many recent college graduates are so used to talking with their instructors and with other students in their majors that they think the specialized terms used in their fields are more widely understood than they actually are. On the job, remind yourself that you will often address people who are not in your field.

The task of identifying words readers won't know can be complicated by the fact that in many fields—perhaps including yours—some of the specialized terms are widely known but others are not. For instance, most people are familiar enough with chemistry to know what an *acid* is, and many have some sense of what a *base* is. But many fewer know what a *polymer* is. Similarly, with respect to computers, many know what *disk drives* and *programs* are, but few know what an *erasable, programmable, read-only memory* is. Thus when writing to people who are not familiar with your specialty, you must distinguish between those technical terms your audience knows and those they don't. To do this, you must have a fairly specific knowledge of your readers, especially of their level of familiarity with your special field.

How to Explain Unfamiliar Terms If You Must Use Them

So far, the discussion of this guideline has advised you to avoid using specialized terms when writing to readers who do not understand them. However, there will be times when you *must* use specialized terms unfamiliar to your readers. This might happen, for example, when you have a large audience composed of some people in your field and some people outside your field, or when you are explaining an entirely new subject not familiar to any of your readers.

When you determine that there is a good reason to use unfamiliar words, explain them carefully. The amount of explanation unfamiliar words need can vary greatly. Sometimes you simply define them in ordinary terms.

> On a boat, a *line* is a rope or cord.

> The *exit gate* consists of two arms that hold a jug while it is being painted, then let it proceed down the production line.

At other times, you will need to use somewhat more elaborate procedures. Two that are especially useful in writing on the job are to make a classical definition and to use an analogy. The essential strategy of both is to help your audience understand the unfamiliar by describing it in terms of things they know well.

Classical Definition   To write a classical definition, you need to identify three things:

- The thing being defined
- Some familiar group of things to which the thing being defined belongs
- The key distinction between the thing being defined and the other things in the group

The following list provides examples of classical definition:

Word	Group	Distinguishing Characteristic
burrow	hole in the ground	dug by an animal for shelter and habitation
crystal	solid	in which the atoms or molecules are arranged in a regularly repeated pattern
modem	electronic device	that permits computers to talk with one another over the telephone
vine	plant	that has a stem too flexible or weak to support itself

Of course, a definition can work only if the readers are already familiar both with the group to which the defined item belongs and with the terms that specify the key distinction of the defined item. For example, a reader would need to know what *slurry* means to understand this definition: "Pulp is a slurry made of fiber, usually from wood."

This does not mean, however, that you should always use ordinary rather than technical terms for your definitions. When defining something for an audience familiar with your specialty, you should employ the specialized terms they already know to define those they don't. Thus, it would be appropriate for you to say that "wood sorrel is a trifoliate legume" when you are addressing a botanist or any other person who would understand the terms *trifoliate* ("having three leaves") and *legume* ("one of the dicotyledonous plants having a simple superior ovary and dehiscing into two valves with the seed attached to the ventral suture"). But it would not be appropriate for you to give such a definition to people unfamiliar with botanical terms.

Analogy   Another way to help your readers understand an unfamiliar word is to use analogy. In an analogy, you explain an unfamiliar thing by comparing it with something your readers know well.

An *atom* is like a miniature solar system in which the nucleus is the sun and the electrons are the planets that revolve around it.

A *computer memory* is like a filing cabinet in which information is stored in various drawers.

To explain complex ideas, you can extend your analogies for a paragraph or more. In lengthier analogies, it is sometimes important to tell your readers how the thing you are defining differs from the thing they are familiar with—

so that they do not draw incorrect inferences from the analogy. Nevertheless, in an analogy, the emphasis is upon similarities.

Similarities

   The suspension system on the monorail is very similar to that used on a passenger automobile. At the front and rear of each car, an automotive-type axle is attached to the body by a swivel-bar assembly that permits vertical and lateral movement of the car with respect to the axle. Shock absorbers like those used on an automobile damp vertical and lateral movement. The pneumatic tires are identical to those used on some buses.

Important difference

   This suspension system differs from that of an automobile in one important respect. On each axle, two guide wheels are mounted horizontally between the tires. These guide wheels press against an I-beam that is laid along the middle of the track. By following the curves of the I-beam, the guide wheels steer the cars along their route.

Classical definitions and analogies are not the *only* methods of explaining the unfamiliar, but they are often very useful in on-the-job writing.

| GUIDELINE 3 | **USE WORDS ACCURATELY** |

   Whether you use specialized terms or everyday ones, whether you use abstract and general terms or concrete and specific ones, you must be careful to use all your words accurately. This point may seem obvious, but inaccurate word choice is all too common in on-the-job writing. For example, people often confuse *imply* (meaning to *suggest* or *hint,* as in ''He implied that the operator had been careless'') with *infer* (meaning *to draw a conclusion based upon evidence,* as in ''We infer from your report that you do not expect to meet the deadline''). It's critical that you avoid such errors. They distract your readers from your message by drawing their attention to your problems with word choice, and they may lead your readers to believe that you are not skillful or precise in other areas—such as laboratory techniques or analytical skills.

   How can you ensure that you use words accurately? There's no easy way. Consult a dictionary whenever you are uncertain. Be especially careful when using words that are not yet part of your usual vocabulary. Also, pay careful attention to the way words are used by other people. Figure 12–3 lists some pairs of words that are often confused with one another.

Words Often Confused

adapt, adopt	**Adapt** means "to adjust or make suitable for a certain use." *When you travel overseas, you will need to bring a special attachment to **adapt** your personal computer to the voltages used in European homes and offices.* **Adopt** means "to select or take up as one's own." *The corporation has **adopted** a policy concerning the protection of company secrets.*
affect, effect	**Affect** is a verb meaning "to influence." *Temperature will **affect** the seals.* **Effect** is a noun meaning "result" or a verb meaning "to cause" or "to bring about." *They studied the **effect** of the acid on tin.* *They **effected** many improvements.* (This could be simplified to *They made many improvements.*)
already, all ready	**Already** means "before the time specified." *The engineers had **already** completed the design.* **All ready** means "completely set or prepared." *The components are **all ready** for final assembly.*
altogether, all together	**Altogether** means "entirely" or "thoroughly." *The control panel is **altogether** too small.* **All together** means "in a single group." *The elk were **all together** at the bend in the river.*
among, between	Usually, **among** is used with three or more things. *KL-Grip is the strongest **among** the several glues available for this use.*

FIGURE 12–3   Words Often Confused with One Another

Usually, **between** is used with two things.

The scientists had difficulty choosing between these two research methods.

complement, compliment

Complement means "to fill out," "complete," or "to mutually supply each other's lacks."

The two systems complement one another effectively.

Compliment means "to praise."

We complimented him on his plan.

continual, continuous

Continual means "happening again and again."

This machine requires continual adjustments.

Continuous means "occurring without interruption."

The continuous action of the river for two million years has worn a deep canyon in the bedrock.

eminent, imminent

Eminent means "distinguished."

Dr. Korf is an eminent researcher.

Imminent means "about to take place."

Feeling that a breakdown was imminent, he recommended that extra safety precautions be taken.

explicit, implicit

Explicit means "stated directly and unambiguously."

She gave us explicit orders to shut down the line at 9 p.m.

Implicit means "implied" or "expressed indirectly."

Though the commission didn't name the company that was suspected of illegally discharging the waste, the company's identity was implicit in the commission's report.

foreword, forward

Foreword is a noun meaning "a brief introductory section at the beginning of a report, manual, or similar communication."

FIGURE 12–3   *(continued)*

*In the **foreword** of its report, the design team identified the three sources of funding for its work.*

Forward is an adjective or adverb meaning "at or near the front."

*To engage the gears, push the handle to the **forward** position. (adjective)*
*The conveyor belt carries the chassis **forward** to the painting booth. (adverb)*

imply, infer

Imply means "to suggest" or "to hint."
*He **implied** that the operator had been careless.*

Infer means "to draw a conclusion based on certain evidence."
*We **infer** from your report that you do not expect to meet the deadline.*

its, it's

Its is the possessive form of it.
*That is **its** control panel.*

It's is a contraction of it is.
It's a significant difference.

principal, principle

Principal means "primary" or "main."
*Metal fatigue was the **principal** cause of the break.*

Principle means "basic law, rule, or assumption."
*In dealing with clients, we observe three **principles**.*

stationary, stationery

Stationary means "fixed in one position."
*This pole is supposed to remain **stationary**.*

Stationery means "paper and envelopes for writing and sending letters."
*The company had a graphic designer create new **stationery** for its official correspondence.*

FIGURE 12–3   *(continued)*

<table>
<tr><td>GUIDELINE
4</td><td>CHOOSE WORDS WITH APPROPRIATE ASSOCIATIONS</td></tr>
</table>

Guidelines 1, 2, and 3 have provided you with advice related to the literal or dictionary meanings of words. At work, you must also consider the associations your words have for your readers. Two words that the dictionary describes as synonyms can create much different impressions. For example, according to the dictionary, *flatfoot* and *police detective* are synonyms, but they connote much different things: *flatfoot* suggests a plodding, perhaps not very bright cop, while *police detective* suggests a highly trained professional.

At work, two kinds of associations that you should be especially sensitive to are *connotation* and *register*.

Consider the Connotations of the Words You Use

Connotation is the extended or suggested meaning that a word has beyond its literal meaning. The distinction between *flatfoot* and *police detective* involves the connotations of the two words. For another example, consider *plump* and *fat*. To say that person is plump is to suggest that he is also pleasing and pleasant. To say that he is fat is to suggest that he is overweight and unattractive. Note that part of a word's connotation is often the implied attitude of the writer toward the person or thing being described. A flatfoot or a fat person is probably not regarded very highly by the writer, while a police detective or plump person probably is.

Verbs, too, have associated meanings. For instance, to *suggest* that someone has overlooked a key fact is not the same as to *insinuate* that she has. To *devote* your time to working on a·client's project is not the same as to *spend* your time on it. To *think about* a situation is not the same as to *evaluate* it.

The connotations of your words can shape your audience's perceptions of your subject matter. To demonstrate this effect, researchers Raymond W. Kulhavy and Neil H. Schwartz wrote two versions of a description of a company.[1] The two versions differed from one another in only seven words scattered throughout the two hundred and forty-six of the entire description. In one version, the seven words suggested flexibility, such as *asked* and *should*. In the second version, those seven words were replaced by seven others that suggested stiffness, such as *required* and *must*. For instance, consider the following sentence:

> Our sales team is constantly trying to locate new markets for our various product lines.

In the second version of this sentence, the researchers replaced the flexible word *trying* with the stiff word *driving*. None of the substitutions changed the facts of the overall passage.

The researchers then gave one version of the company description to half the people in a group, and the second version to the other half. After reading, both halves were asked several questions about the company. People who read the flexible version believed such things as that the company would actively commit itself to the welfare and concerns of its employees, voluntarily participate in affirmative action programs for women and minorities, receive relatively few labor grievances, and pay its employees well. People who read the flexible versions said that they would recommend the company to a friend as a place to work. People who read the strict version reported having opposite impressions of the company. That people's impressions of the company could be affected so dramatically by just seven nonsubstantive words demonstrates the great importance of paying attention to the connotations of the words you use.

Consider the Registers of the Words You Use

Linguists identify a second type of association that words have, called *register*. A word's register is the type of communication in which one expects the word to appear. At work, you need to use words whose register matches the type of communication you are preparing. In an advertisement, you might say that your restaurant gives *amazingly* good service, but you would not say the same thing about your engineering consulting firm in a letter to a prospective client. The word *amazingly* has the register of consumer advertising but not of letters to business clients.

If you accidentally choose words with the wrong register, you may give the impression that you don't fully grasp how business is conducted in your field and your credibility can be lost. As you choose words, be sensitive to the kinds of communications in which you usually see them used.

GUIDELINE 5

CHOOSE PLAIN WORDS OVER FANCY ONES

Another way to make your writing easy to understand and effective is to avoid using fancy words where plain ones will do just as well. At work, people often do just the opposite, perhaps because they think fancy words sound more official or make the writers seem more knowledgeable. The following list identifies some of the more commonly used fancy words; it includes only verbs but might have included nouns and adjectives as well.

Fancy Verbs	Common Verbs
ascertain	find out
commence	begin
compensate	pay
constitute	make up

Fancy Verbs	Common Verbs
endeavor	try
expend	spend
fabricate	build
facilitate	make easier
initiate	begin
prioritize	rank
proceed	go
terminate	end
transmit	send
utilize	use

There are two important reasons for preferring plain words over fancy ones. First, because the plain words are more familiar, readers can read more efficiently. The readers don't have to consult a dictionary to learn that *prevaricator* means *liar* or that *prognosticate* means *predict*. Further, research has shown that even if your readers know both the plain word and its fancy synonym, they will still comprehend the plain word more rapidly.[2]

Second, by using plain words, you reduce the risk of creating a bad impression. If you use words that make for slow, inefficient reading, you may annoy your readers, who are depending upon your assistance, and you may cause them to conclude that you are behaving pompously, showing off, or trying to hide a lack of ideas and information behind a fog of fancy terms. Consider, for instance, the effect of the following sentence, which one writer included in a job application letter:

> I am transmitting the enclosed resume to facilitate your efforts to determine the pertinence of my work experience to your opening.

Don't misunderstand this guideline, however. It doesn't suggest that you should use only simple language at work. When addressing people with vocabularies comparable to your own, use all the words at your command, provided that you use them accurately and appropriately. This guideline merely cautions you against using needlessly inflated words that bloat your prose and may open you to criticism from your readers.

GUIDELINE
6

AVOID UNNECESSARY VARIATION OF TERMS

Most people have been taught at one time or another that they will write or speak monotonously if they use the same word over and over again. As a result, they use a variety of terms for the same thing. What they call a *dog* in one sentence, they call a *hound* or *cur* in the next. There are, however, at least two good reasons for calling a dog a *dog* throughout the communications you prepare at work.

The first reason is ease of recognition. People recognize and understand a term on its second appearance more quickly than they do an entirely new word for the same thing. If you've been discussing *structures* in a report but abruptly begin referring to *buildings,* your readers may have to pause to decipher whether you intend for *structures* and *buildings* to mean the same thing, or for one to be a subcategory of the other, or for the terms to refer to related but separate things. Or, your readers may simply fail to recognize that there might be any connection at all between the two terms.

The second reason to avoid unnecessary variation of terms is to avoid an ambiguous reference, in which it is unclear which item a term refers to. Consider the following example:

> We have a new typewriter, for which we bought a special ribbon
> that lifts errors off the page. This item was very reasonably priced.

It is not clear in the second sentence whether the item referred to is the typewriter or the ribbon. In this case, variation in the terms used to designate something leads to an ambiguity that is not settled by the sentences themselves.

A common kind of ambiguous reference arises from the careless substitution of pronouns for nouns.

> This job has much slack time, which the operator spends cleaning
> up. It takes much getting used to.

In the second sentence, *it* could refer to the job, the slack time, the cleaning up, or even to the whole idea that the job entails slack time that is spent in cleaning up.

Nevertheless, communications can become monotonous if the same word is used repeatedly throughout a sentence or adjacent sentences. One way to avoid that monotony is to use carefully selected pronouns for variation. Compare the following sentences:

> If the operators encounter a problem, the operators should report
> the problem directly to the shift supervisor.

Pronouns
> If the operators encounter a problem, they should report it directly
> to the supervisor.

However, when your choice is between monotony and ambiguity, you should choose monotony. It is better to use the same word repeatedly to keep your meaning clear than to use a lively array of words that makes your meaning difficult to decipher.

CONCLUSION

The words you use can make a great deal of difference to the success of your writing. You should strive to use words that are clear and readily understandable by your readers, that convey your meaning precisely, and that foster the impressions and attitudes you desire your audience to adopt. This chapter has suggested several things you can do to choose words wisely. Underlying all these suggestions is the advice that you consider your words from your readers' point of view.

SPECIAL
TOPIC

AVOIDING SEXIST AND DISCRIMINATORY LANGUAGE

Sexist language is language that suggests people's qualities and abilities are determined by their sex. Similarly, discriminatory language depicts people in terms of stereotypes based on race, age, ethnic background, physical handicap, or similar factors. Many employers make a concerted effort to avoid both of these types of language.

The concern of employers with eliminating sexist and discriminatory language is part of the larger effort to end discriminatory practices of all sorts in business organizations. Employers have moral, practical, and legal reasons for wanting to do this.

- **Moral.** Many employers believe that all people should be treated equally. Although the stereotype of business executives portrays them as more interested in profits than people, most are very concerned about the individuals who work for them.
- **Practical.** By giving all their employees equal opportunities to contribute and advance, employers stand to benefit from the hard work, good ideas, and leadership ability of their entire work force, not just a portion of it. In addition, employers realize that sexist and discriminatory language offends many potential customers and clients.
- **Legal.** Federal laws such as the Civil Rights Act, the Equal Pay Act, and the Rehabilitation Act require employers to avoid discrimination. Numerous state and local laws impose similar legal requirements. One consequence of these laws is that employers must write job descriptions very carefully to avoid even hinting that certain positions are open only to women or to men, or that men and women will be treated differently in

these jobs. People who seek government contracts must certify—and be able to prove—that they do not discriminate.

To end discriminatory practices, employers promote equality not only in their job descriptions and advertisements but also in the routine communications their employees prepare daily on the job. Here are six strategies you can use for avoiding bias when you write and speak.[3]

Avoid Describing People in Terms of Stereotypes

Stereotypes suggest that all members of a group share the same characteristics. They tell us, for instance, that women are emotional and spontaneous while men are rational and reserved. Stereotypes fail to notice that all individuals are different. To avoid falling into the use of stereotypes, describe each person in terms of his or her specific characteristics. In your reports, sales presentations, policy statements, and other communications, avoid giving examples that rely upon or reinforce stereotypes.

Identify a Person's Sex, Race, or Similar Characteristics Only When Relevant

When we identify a person as a man or woman, or when we identify a person as a member of some racial, religious, ethnic, or other group, we seem to indicate that the fact is an important one. To do so is to think of the person as a member of the group rather than as an individual. One way to test the reasonableness of group identification is to see if the statement is one we would make if we were talking about someone in the majority group. If you wouldn't say, ''The suggestion for improving our accounting system was made by Jane, a person in perfect physical condition,'' then you shouldn't say, ''The suggestion for improving our accounting system was made by Margaret, a handicapped person.'' If you wouldn't say, ''The Phoenix office is managed by Brent, a hardworking white person,'' then don't say, ''The Phoenix office is managed by Terry, a hardworking Mexican-American.''

Avoid the Word *Man* When Referring to People of Both Sexes

In English, the word *man* has traditionally been used to refer to people in general, as in statements like this: ''Man does not live by bread alone.'' Thus people sometimes talk about *man-made materials,* even though many synthetic materials were created by women, and they talk about *mankind's scientific achievements* even though many scientists are women.

This usage is becoming unpopular, largely because it is viewed as a way of perpetuating sterotypes of male and female roles. Critics of this usage argue that the language people use plays such a powerful part in shaping their perceptions of reality that women cannot achieve equality in business and other settings until people start using language that gives women equality with men.[4]

Whether you agree with that line of argument or not, you should realize that using *man* to refer to both men and women will offend many people, both male and female, and that their negative reaction can seriously interfere with the effectiveness of your communications.

When seeking to avoid using *man* in this way, you should be especially alert to job titles and similar terms that include the word *man*.

Avoid	Use
Businessman	Business person, manager, executive
Fireman	Firefighter
Mailman	Mail carrier
Salesman	Salesperson, sales representative

You should also be aware of adjectives that incorporate the word *man*.

Avoid	Use
Man-hours	Working hours, effort hours
Man-made	Artificial, synthetic
Manpower	Work force
Man-sized job	Large job, challenging job
Workman's compensation	Worker's compensation

Avoid Salutations that Imply the Reader of a Letter Is a Man

Once common in business correspondence, the salutations "Dear Sir" and "Gentlemen" are now rarely used. Instead, people prefer salutations that don't imply that the reader is male. For example, you can use the title of the department or company you are addressing: "Dear Personnel Department" or "Dear Switzer Plastics Company." Or you can use a job title: "Dear College Recruiter" or "Dear Supervisor." Or you can use a descriptive word: "Dear Customer," "Dear Subscriber," or "Dear Recipient."

Avoid Using Sex-Linked Pronouns When Referring to People of Both Sexes

When you are writing a sentence that might apply to people of either sex avoid using pronouns such as *he* or *she* that seem to indicate you have in mind persons of only one sex.

One way to avoid using sex-linked pronouns is to write your sentences in the plural:

Original

This survey shows that the consumer is very worried that he won't get quick and courteous service during the warranty period.

Revised

This survey shows that consumers are very worried that they won't get quick and courteous service during the warranty period.

Original

Our supermarkets cater to the affluent shopper. She looks for premium products and appreciates an attractive shopping environment.

Revised

Our supermarkets cater to affluent shoppers. They look for premium products and appreciate an attractive shopping environment.

Another strategy is to use *he or she* and *his or her:*

Original

Before the owner of a new business files the first year's tax returns, he might be wise to seek advice from a certified public accountant.

Revised

Before the owner of a new business files the first year's tax returns, he or she might be wise to seek advice from a certified public accountant.

Refer to Individual Men and Women in a Parallel Manner

Another way to eliminate discriminatory language is to refer to all individuals in the same way, regardless of their gender. For example, if you use full names for people of one sex, use full names for people of the other:

Original

Christopher Sundquist and Ms. Tokagawa represented us at the trade fair.

Revised

Christopher Sundquist and Anna Tokagawa represented us at the trade fair.

Similarly, if you use only first or last names for people of one sex, do the same for people of the other sex:

Original

The two leaders of the project are Sheila and Bledsoe.

Revised

The two leaders of the project are Neilson and Bledsoe.
or
The two leaders of the project are Sheila and Chip.

Also, if you use courtesy titles *(Mr., Ms.)* for one sex, do the same for the other.

What about Miss, Mrs., and Ms.?

People in business, as well as students, are sometimes confused about whether to address or refer to a woman with the traditional terms *Miss* or *Mrs.* or the newer term *Ms.* In business, *Ms.* is becoming increasingly popular. People argue that using the older terms in business settings represents a special kind of sexism because it makes it appear that a woman's marital status is relevant to her performance of her job or to the way her co-workers and other business contacts should view her. In contrast, all men, whether married or single, are addressed by the single term *Mr.*

The term *Ms.* is now so widely used that employing the other terms is considered to be insensitive—or at least old fashioned—in many companies and many regions of the country. At the same time, some women prefer to be called either *Mrs.* or *Miss.* Courtesy dictates that you follow an individual's preferences about the name by which he or she is called.

As a rule of thumb, it is best to use the term *Ms.* when communicating with or about a woman you don't know. When communicating with or about a woman you do know, use the term she prefers.

EXERCISES

1. In the memo shown in Figure 12–4, identify places where the writer has ignored the guidelines given in this chapter. You may find it helpful to use a dictionary. Then write an improved version of the memo by following the guidelines in this chapter and in Chapter 11 (''Composing Sentences'').

2. Create a one-sentence, classical definition for a word used in your field that is not familiar to people in other fields. The word might be one that people in other fields have heard of but cannot define precisely in the way

MEMO

July 8, 19—

TO: Gavin MacIntyre, Vice President, Midwest Region

FROM: Nat Willard, Branch Manager, Milwaukee Area Offices

The ensuing memo is in reference to provisions for the cleaning of the six offices and two workrooms in the High Street building in Milwaukee. This morning, I absolved Thomas's Janitor Company of its responsibility for cleansing the subject premises when I discovered that two of Thomas's employees had surreptitiously been making unauthorized long-distance calls on our telephones.

Because of your concern with the costs of running the Milwaukee area offices, I want your imprimatur before proceeding further in making a determination about procuring cleaning services for this building. One possibility is to assign the janitor from the Greenwood Boulevard building to clean the High Street building also. However, this alternative is judged impractical because it cannot be implemented without circumventing the reality of time constraints. While the Greenwood janitor could perform routine cleaning operations at the High Street establishment in one hour, it would take him another ninety minutes to drive to and fro between the two sites. That is more time than he could spare and still be able to fulfill his responsibilities at the High Street building.

Another alternative would be to hire a full-time or part-time employee precisely for the High Street building. However, that building can be cleaned so expeditiously, it would be irrational to do so.

The third alternative is to search for another janitorial service. I have now released two of these enterprises from our employ in Milwaukee. However, our experiences with such services should be viewed as bad luck and not effect our

FIGURE 12–4   Memo to Be Used in Exercise 1

Gavin MacIntyre   —2—   July 8, 19—

decision, except to make us more aware that making the optimal selection among companies will require great care. Furthermore, there seems to be no reasonable alternative to hiring another janitorial service.

Accordingly, I recommend that we hire another janitorial service. If you agree, I can commence searching for this service as soon as I receive a missive from you. In the meantime, I have asked the employees who work in the High Street building to do some tidying up themselves and to be patient.

FIGURE 12–4 *(continued)*

specialists in your field do. Underline the word you are defining. Then circle and label the part of your definition that describes the familiar group of things that the defined word belongs to. Finally, circle and label the part of your definition that identifies the key distinction between the defined word and the other things in the group. (Note that not every word is best defined by means of a classical definition, so it may take you a few minutes to think of an appropriate word for this exercise.)

3. Create an analogy to explain a word used in your field that is unfamiliar to most readers. (Note that not every word is best defined by means of an analogy, so it may take you a few minutes to think of an appropriate word for this exercise.)

13

USING VISUAL AIDS

DEFINING
OBJECTIVES
PLANNING
DRAFTING
EVALUATING
REVISING

During her first two weeks on the job, Sandra has been assigned to learn as much as she can about her department's work. She's having fun. She is visiting each of her co-workers, one by one, so that they can tell her in detail about their specialties. And she is reading lots of things those people have written, including proposals, reports, and even a few published articles.

When she began this reading, she was struck by the abundance of tables, graphs, drawings, and other visual aids that she found. Her co-workers use these devices much more often than she and her classmates did in college. Furthermore, as a reader, she likes the visual aids a great deal. They make writing more informative, easier to use, and more persuasive than it would otherwise be.

Now, she wants to know what she must do to use visual aids effectively in her own writing. The four guidelines in this chapter tell her. Like Sandra, you can apply these *general* guidelines to all types of visual aids, including tables, graphs, drawings, flow charts, budget statements, or any of the many others used on the job. In the next chapter, you will find *specific* advice about employing and constructing twelve kinds of visual aids that you are likely to find useful at work.

MORE THAN JUST AIDS

Before reading the guidelines in this chapter, you need to know one important fact about the name *visual aids*. It's misleading. It suggests that in the writing you do at work, your *words* are primary and the visual *aids* are merely assistants.

Nothing could be further from the truth. When used creatively and effectively, visual aids are an integral part of communications. They can carry some parts of a message more effectively than prose can. In fact, in some situations, visual aids can carry the *entire* message. No words are needed at all.

For instance, if you've ever flown, you may recall reaching into the pocket on the back of the seat ahead of you to pull out a sheet of instructions for leaving the plane in an emergency. Many airlines use instruction sheets that are wordless. Figure 13–1 shows an example. Figure 13–2 shows the instructions that one company provides for unpacking and setting up an electric typewriter. These, too, communicate without words.

Although you may never create a communication that relies solely on visual aids, you should remember that visual aids are powerful communication tools, not mere decorations or supplements.

COMPUTERS AND VISUAL AIDS

As you read this chapter, you should also keep in mind that its advice will help you not only with visual aids you prepare by traditional methods but also those you might create on a computer. That's because many computer programs for

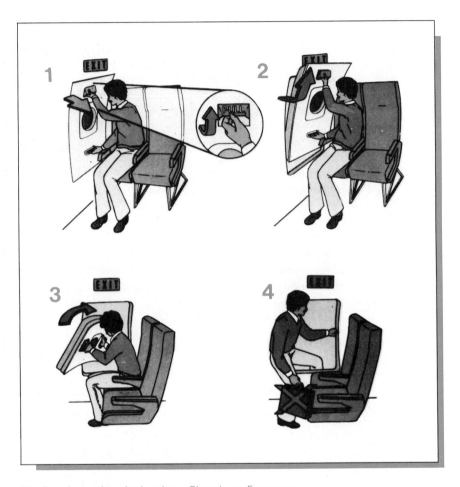

FIGURE 13–1   Wordless Instructions for Leaving a Plane in an Emergency
Courtesy of USAir.

making visual aids leave essential design decisions to you. For instance, if you are using a program to make a line graph, you will still have to decide which variable to place on the horizontal axis and which on the vertical axis. You will still have to decide what intervals to use for your variables, and what your labels should say. This chapter and the next provide you with advice about such basic design decisions.

Furthermore, if you use a computer program that *doesn't* leave basic design decisions to you, you will have to determine whether the designs created by the program are good ones. This chapter's advice will help you distinguish a good design from a bad one. If you determine that the program's standard designs aren't good, you should obtain a different program or make the visual aids some other way.

FIGURE 13-2 Wordless Instructions for Setting Up a Typewriter
Courtesy of IBM.

Computer programs for visual aids may present one other difficulty. Many allow you to create fancy visual aids. In some situations, those extra flourishes can help you achieve your communication objectives. However, some people become so enthralled with a program's ability to create special effects that they forget the purpose and readers of their communications. As a result, they make overly elaborate visual aids that diminish the effectiveness of their message. By following the advice in this chapter, you can avoid such difficulties.

GUIDELINE 1   LOOK FOR PLACES WHERE VISUAL AIDS WILL HELP YOU ACHIEVE YOUR COMMUNICATION OBJECTIVES

The first step in using visual aids effectively is to *search actively* for places where they can help you achieve your communication objectives. The following paragraphs describe some of the many things visual aids can help you do.

- **Show what something looks like.** In every field, people describe physical objects: machines, experimental apparatuses, geological formations, internal organs of animals and people, to name just a few examples. Visual aids can be very helpful in such situations. For instance, an engineer designed a new hinge that can be used in spacecraft, portable bridges, and many other applications. Next he wanted to explain the design and its virtues to other engineers who might want to use it. By means of the diagrams shown in Figure 13–3, he described the hinge much more clearly than he could with words.

- **Show how to do something.** In many circumstances, visual aids provide the best form of instructions. For example, a writer needed to explain how to remove a piece of paper jammed in a photocopy machine. Using the drawing shown in Figure 13–4, she did that clearly and economically.

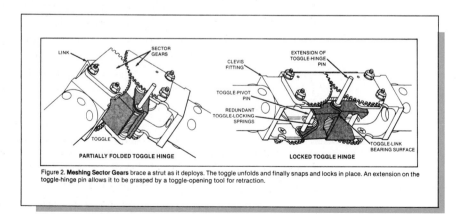

Figure 2. **Meshing Sector Gears** brace a strut as it deploys. The toggle unfolds and finally snaps and locks in place. An extension on the toggle-hinge pin allows it to be grasped by a toggle-opening tool for retraction.

FIGURE 13–3   Drawing that Shows How Something Looks
From "Toggle Hinge for Deployable Struts," *NASA Tech Briefs* 9 (Fall 1985): 132.

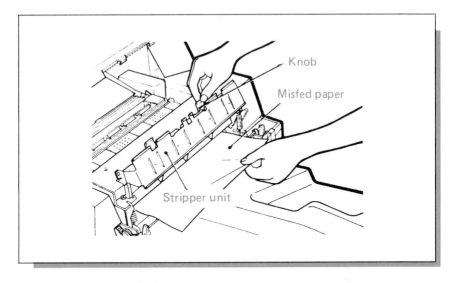

FIGURE 13–4 Drawing that Shows How to Do Something
Courtesy of Toshiba Corporation.

Similarly, you can use drawings, photographs, and other visual aids to show your readers how to do something.

● **Show how something is organized.** At work, you may need to describe organizational relationships that are difficult to explain in prose, such as the relationships among the various departments and divisions of your employer's company or among the parts of a computer system. Visual aids can help. For example, a group of computer researchers integrated a series of computer programs into a single system for helping engineers design manufacturing processes. By using the diagram shown in Figure 13–5, they were able to provide their readers with an excellent understanding of the complex relationships among these programs.

● **Clarify the relationships among numerical data.** On the job, you may need to describe the relationships among various pieces of data. The data may be from laboratory research, survey research, or other forms of information gathering. Visual aids can help you make those relationships immediately clear to your readers.

For example, in an effort to understand the postoperative complications experienced by some patients after surgery, a team of researchers monitored two components of a surgery patient's blood for 23 days. Although they could have presented their data in prose, they used instead the very effective graph shown in Figure 13–6.

● **Support your arguments.** You can also use visual aids to present information in support of your persuasive points. For example, the manufacturer of a plastic insulating material wanted to persuade greenhouse owners that they could greatly reduce their winter heating bills by covering their

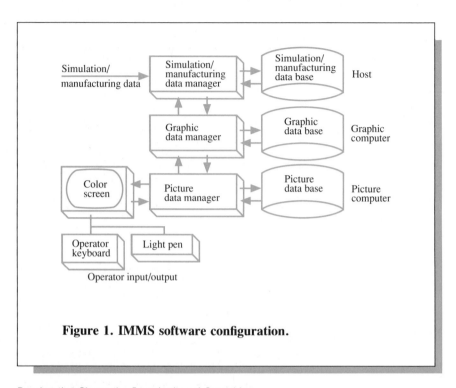

Figure 1. IMMS software configuration.

FIGURE 13–5 Drawing that Shows the Organization of Something

From H. Engelke et al., "Integrated Manufacturing of Modeling System," *IBM Journal of Research and Development* 29 (July 1985): 350. Copyright 1985 by International Business Machines Corporation.

greenhouses with plastic sheets. To emphasize the effectiveness of the plastic, the manufacturer used the bar graph shown in Figure 13–7 (p. 368).

● **Make detailed information easy to find.** For many tasks, visual aids are much easier than prose for readers to use. For instance, a manufacturer of photographic film wanted to tell its customers how to develop their own negatives. One crucial piece of information the customers need is the length of time they should leave the film in the developer solution. Unfortunately, there is no single time for all situations because developing time depends upon the type of developer used, the temperature of the developer, and the size of the developer tank.

Therefore, the writers decided to present the information in the table shown at the top of Figure 13–8 (p. 368). It enables readers to quickly find the correct developing time for the particular developer, temperature, and tank they are using.

To take full advantage of the powerful assistance visual aids can provide, begin searching for places to use them when making your very first plans for

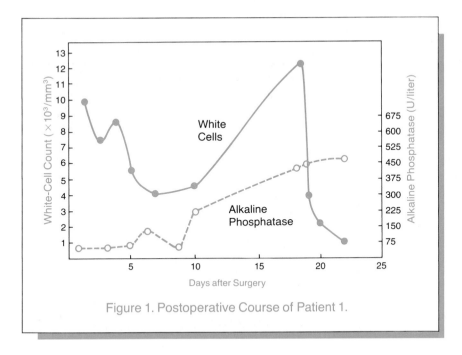

Figure 1. Postoperative Course of Patient 1.

FIGURE 13–6   Graph that Shows the Relationships among Separate Pieces of Data

From Michael Thaler, Arie Shamiss, Shlomit Orgad, Monica Huszar, Naomi Nussinovitch, Simcha Meisel, Ephraim Gazit, Jacob Lavee, and Aram Smolinsky, ''The Role of Blood from HLA-Homozygous Donors in Fatal Transfusion-Associated Graft-Versus-Host Disease After Open-Heart Surgery,'' *New England Journal of Medicine* 321 (1989): 25.

writing. For instance, if you are making an initial outline for your communication, decide then what visual aids you will use. Early planning will help you coordinate your prose and visual aids from the beginning, thereby reducing the amount of revision. Also, if you need to ask an artist or photographer to prepare some of your visual aids, early planning will ensure that this person has plenty of time to do the work before your deadline.

One caution is in order. Although visual aids can help you immensely, you should guard against putting them in thoughtlessly. It is possible to overuse visual aids. Be sure that each one has a specific purpose and is carefully designed to achieve that purpose.

GUIDELINE
2

CHOOSE VISUAL AIDS APPROPRIATE TO YOUR OBJECTIVES

Once you have decided where to use visual aids, you must decide which type to use. There are many types of visual aids, and most information can be presented in more than one type. For example, numerical data can often be presented in a table, bar graph, line graph, or pie chart. The components of an

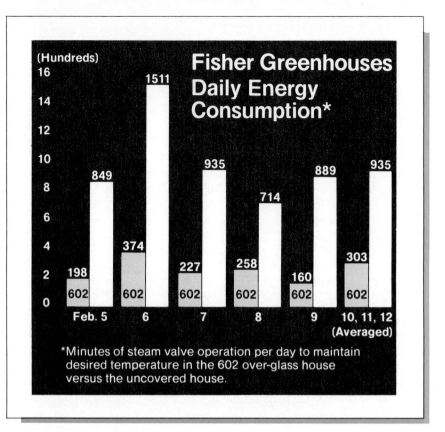

FIGURE 13–7 Bar Graph that Supports an Advertising Claim for Plastic Insulation
Courtesy of Monsanto Company.

KODAK Developer	Developing Time (in Minutes)									
	SMALL TANK (Agitation at 30-Second Intervals)					LARGE TANK (Agitation at 1-Minute Intervals)				
	65°F (18°C)	68°F (20°C)	70°F (21°C)	72°F (22°C)	75°F (24°C)	65°F (18°C)	68°F (20°C)	70°F (21°C)	72°F (22°C)	75°F (24°C)
HC-110 (Dil B)	8½	7½	6½	6	5	9½	8½	8	7½	
D-76	9	8	7½	6½	5½	10	9	8	7	
D-76 (1:1)	11	10	9½	9	8	13	12	11	10	
MICRODOL-X	11	10	9½	9	8	13	12	11	10	
MICRODOL-X (1:3)*	—	—	15	14	13	—	—	17	16	
DK-50 (1:1)	7	6	5½	5	4½†	7½	6½	6	5½	
HC-110 (Dil A)	4½†	3¾†	3¼†	3†	2½†	4¾†	4¼†	4†	3¾†	

*Gives greater sharpness than other developers shown in table.

†Avoid development times of less than 5 minutes if possible, because poor uniformity may result.

Note: Do not use developers containing silver halide solvents.

FIGURE 13–8 Table Used to Help Readers Find a Particular Piece of Information
Courtesy of Eastman Kodak Company.

electronic instrument can be represented in a photograph, sketch, block diagram, or electronic schematic.

How can you determine which type of visual aid will most effectively communicate a certain point to your particular readers? First, consider the objectives of your visual aid, thinking about those objectives in exactly the same way you think about the objectives of your overall communication: identify the task you want the visual aid to enable your readers to perform while reading your visual aid, then decide how you want the visual aid to affect your readers' attitudes.

Consider Your Readers' Tasks

Different visual aids are suited to different reading tasks. For instance, Ben has surveyed people who graduated over the past three years from three departments in his college. Now Ben wants to tell those same people what he learned from them about their average starting salaries. Ben could do this with a table, bar graph, or line graph. Which would be best? The answer depends upon the way Ben wants his readers to be able to use his results.

If Ben's purpose is simply to enable the alumni to learn the average starting salary of people who graduated in their year from their department, he could use a table (see Figure 13–9). If Ben wants them to be able to see at a glance how the average starting salary in their department compared with those in that same year from other departments, he could use a bar graph. And if Ben wants them to be able to see how the average starting salary in their department changed over the years and to compare that change with the changes experienced by the other departments, he could use a line graph.

For a quick summary that will help you choose among tables, bar graphs, pictographs, line graphs, and pie charts, see Figure 13–10. In Chapter 14 you will find additional information about the types of reading tasks that are served well by these types of visual aids and also by seven others: photographs, drawings, diagrams, flow charts, organizational charts, schedule charts, and budget statements.

Consider Your Readers' Attitudes

In addition to thinking about your readers' tasks, you should also think about the way you want to affect your readers' attitudes: pick the type of visual aid that most quickly and dramatically communicates the evidence that supports your persuasive point. Suppose, for instance, that you want to show that because of a design change you recommended, your company has had to make many fewer service calls on one of its products. Your evidence is a tally of the

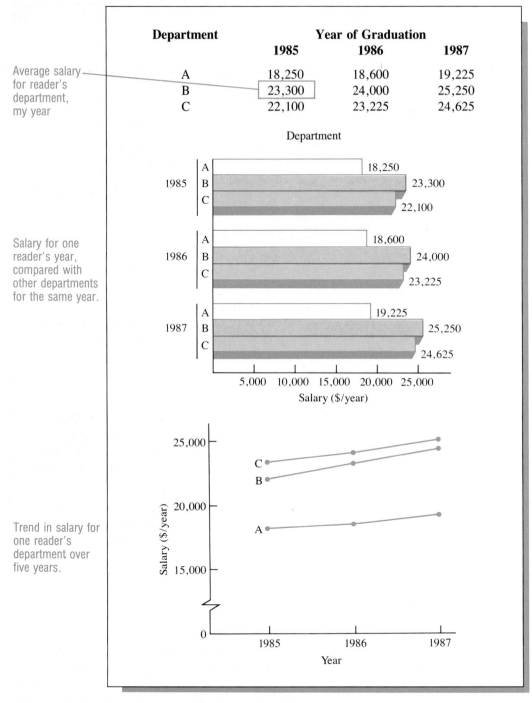

Average salary
for reader's
department,
my year

Salary for one
reader's year,
compared with
other departments
for the same year.

Trend in salary for
one reader's
department over
five years.

Department	Year of Graduation		
	1985	1986	1987
A	18,250	18,600	19,225
B	23,300	24,000	25,250
C	22,100	23,225	24,625

FIGURE 13–9   Three Ways of Showing Average Starting Salaries for Three Departments over Three
Years

Type	Typical Application	Notes
Table	Helps readers find particular facts in large sets of information	Not good for emphasizing trends or making comparisons unless only a few points are involved.
Bar Graph	Helps readers see at a glance how great the differences are between quantities	Can also show trends if readers can easily draw lines mentally between the types of the relevant bars
Pictograph	A type of bar graph that makes data more vivid by calling to mind the people and things discussed	In some situations, may be inappropriate to readers
Line Graph	Highlights trends and interactions between variables	Can display complex relationships more clearly than other types of visual aids listed here
Pie Chart	Shows how a whole is divided up into parts	Often difficult for readers to compare the size of wedges accurately

FIGURE 13–10   Advice for Choosing among Various Common Ways for Displaying Numerical Data

number of calls made in each of the six months before the design change and each of the six months after the change. As Figure 13–11 shows, if you present that data in a table, your readers will have to do a lot of subtracting to appreciate the extent of the decrease. If you present the data in a line graph, however, your readers will be able to see the extent of the decrease at a glance.

Of course, readers will not find your visual aids either useful or persuasive unless you use ones they know how to read. Although some types are familiar to us all, others are much more specialized. See Figure 13–12 (pp. 374–75)

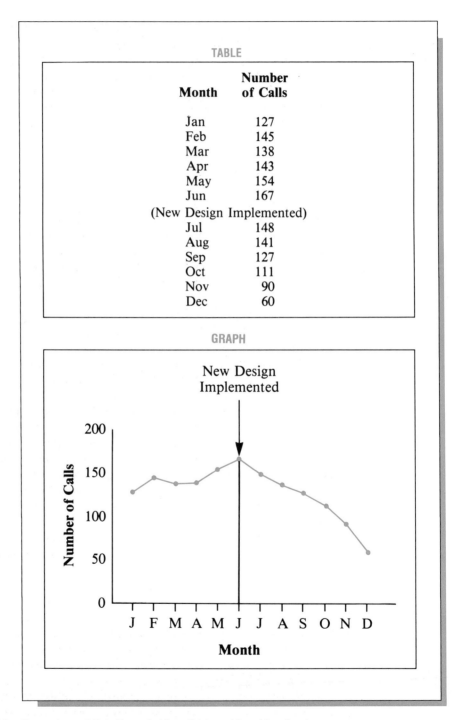

TABLE

Month	Number of Calls
Jan	127
Feb	145
Mar	138
Apr	143
May	154
Jun	167
(New Design Implemented)	
Jul	148
Aug	141
Sep	127
Oct	111
Nov	90
Dec	60

GRAPH

FIGURE 13–11   Comparison of Data Presented in a Table and in a Line Graph

for some specialized examples. You may be learning to create these or other special types of visual aids in courses in your major department. Although they are very informative to people who understand the symbols and conventions used, they will only baffle others. Be especially careful to avoid making the mistake of assuming that everyone who works for your employer knows how to read the specialized figures commonly used in your field or department. This may not be the case.

GUIDELINE 3

MAKE YOUR VISUAL AIDS EASY TO UNDERSTAND AND USE

Having chosen the *type* of visual aid you will use, you must design the aid itself. When doing so, remember that, like your prose, your visual aids should be easy for your readers to understand and use. The following paragraphs describe four strategies for achieving this goal: design your visual aids to support your readers' tasks, make them simple, label the important content clearly, and provide informative titles.

Design Your Visual Aids to Support Your Readers' Tasks

The most important step in designing visual aids that are easy to understand and use is to use this familiar strategy: imagine your readers in the act of using your visual aid and then design it accordingly.

For example, in drawings or photographs for step-by-step instructions, show objects from the same angle that your readers will see them when trying to perform the actions you describe. Likewise, in a table, arrange the columns and rows in an order useful to your readers when they are trying to find a particular piece of information. Maybe that means you should arrange the columns and rows in alphabetical order, according to a logical pattern, or according to some other system. Use whatever arrangement your readers will find most efficient.

Similarly, when designing any other type of visual aid, remember that you are trying not only to display information but to help your audience to understand and use the information.

Make Your Visual Aids Simple

A second strategy for making your visual aids easy to understand and use is to simplify them. Partly that means avoiding the temptation of cramming too much information into them. Sometimes, two or three visual aids can communicate the same information more effectively than one.

Simplifying visual aids also means removing unnecessary details. Like unnecessary words in prose, superfluous details in visual aids create extra, unproductive work for readers and obscure the really important information. Figure

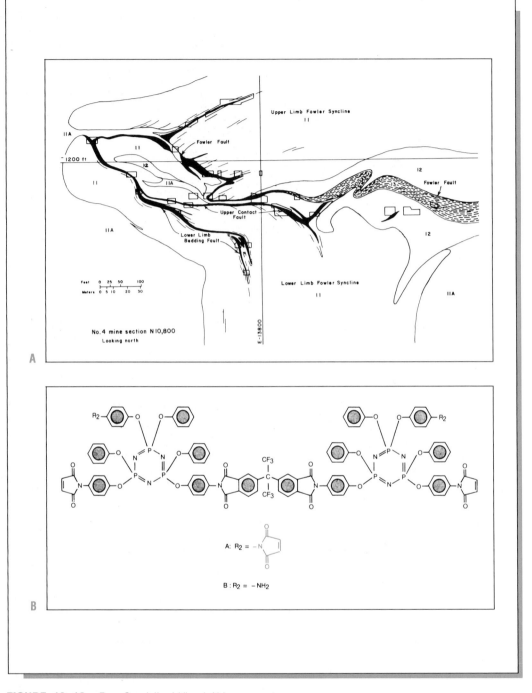

FIGURE 13–12 Four Specialized Visual Aids

Diagram A, which displays plans for a mine, is shown by the courtesy of St. Joe Mineral Company. *Diagram B,* which shows chemical structure, is from "Imide Cyclotriphosphanzene/ Hexafluorisopropylidene Polymers," *NASA Tech Briefs,* (Fall 1985): 88. *Diagram C,* a computer-generated model, is from J. J. Sojka, et al., "Diurnal Variation of the Dayside, Ionospheric, Mid-

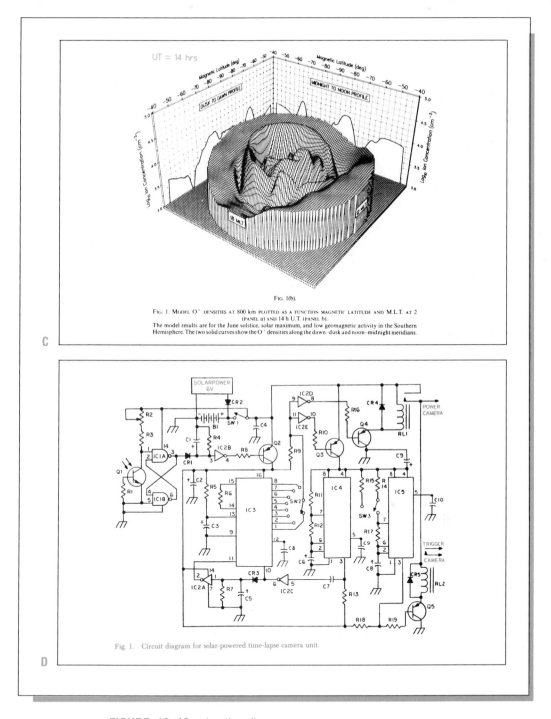

C

FIG. 1(b).

FIG. 1. MODEL O⁻ DENSITIES AT 800 km PLOTTED AS A FUNCTION MAGNETIC LATITUDE AND M.L.T. AT 2 (PANEL a) AND 14 h U.T. (PANEL b).

The model results are for the June solstice, solar maximum, and low geomagnetic activity in the Southern Hemisphere. The two solid curves show the O⁻ densities along the dawn–dusk and noon–midnight meridians.

D

Fig. 1. Circuit diagram for solar-powered time-lapse camera unit.

FIGURE 13–12 *(continued)*

Latitude Trough in the Southern Hemisphere at 800 km: Model and Measurement Comparison,'' *Planetary and Space Science* (December, 1985): 1378. Reprinted with permission of Pergamon Press, Inc. *Diagram D,* ''A Solar-Powered Time-Lapse Camera to Record Wildlife Activity,'' by Frank Montalbano III, et al. *Wildlife Society Bulletin* 13: 178–82.

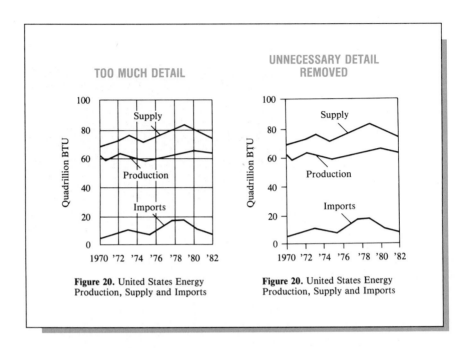

FIGURE 13–13 A Line Graph with and without Grid Lines

Based on a graph without grid lines in U.S. Bureau of Census, *Statistical Abstract of the United States,* 106th ed. (Washington, D.C.: U.S. Government Printing Office, 1986), 552.

13–13 illustrates the way that removing extraneous detail can improve a visual aid. In the graph on the left, the long grid lines obscure the plotted lines, which are the truly important information in this visual aid. In the version on the right, the shape of the curves emerge much more clearly because the long grid lines are replaced by short tick marks along the horizontal and vertical axes.

As another example, Figure 13–14 shows two photographs and a line drawing of the same experimental apparatus used in new-product research. The main parts of the apparatus are obscured in the less effective photograph because it includes much irrelevant material in the background. This material is removed from the more effective photograph by a screen placed behind the apparatus. The drawing (made by tracing the more effective photograph) removes even more detail, showing only the essential material.

As these two examples suggest, you will find great variation from situation to situation in the kinds of details you need to eliminate. The general procedure, however, remains the same: find and eliminate any details not needed to understand and use your visual aid.

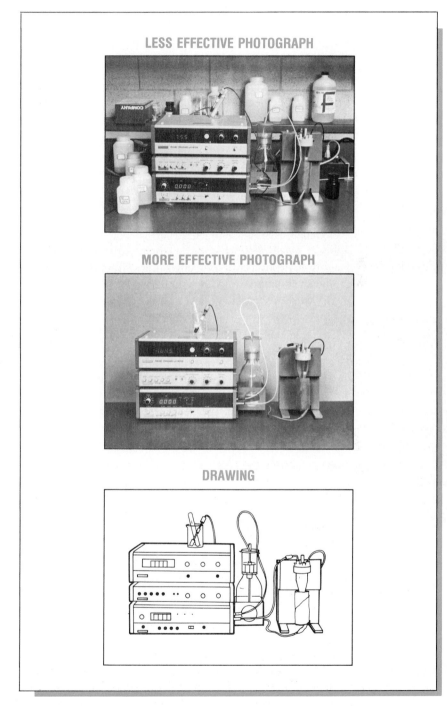

FIGURE 13–14   Two Photographs and a Drawing of the Same Apparatus
Photograph courtesy of Dr. Alan Mills.

Label the Important Content Clearly

While it is important for you to eliminate unnecessary details from your visual aids, it is also critical to include labels for the important content. Labels help your readers find the things they seek. Labels also help people know what they are seeing when they read a figure. For instance, consumers looking at your drawing of a new product will want you to identify the various features of the product.

To create labels, first determine what parts need labeling. In some cases, this is easy to do. In a table, for instance, every row and column usually needs a heading. In other cases, you should label every part that your readers will need to find in a particular visual aid. On the other hand, avoid labeling parts your readers won't be looking for. Unnecessary labels clutter a visual aid and make it difficult to understand and use.

Some visual aids don't need labels. For instance, the title for your visual aid may make clear what its important parts are (for example, ''Figure 3–2. Dents Caused by Hail''). At other times, the things you say when referring to the figure in your prose will serve the same purpose.

After you've decided that a certain part needs a label, choose the appropriate word or words and place them where they are easy to see. If readers might be unsure what part is identified by the label, draw a line from the label to the part. Figure 13–15 shows a drawing and a line graph that are both clearly labeled. Notice that the writer of the line graph was able to place labels close enough to most of the plotted lines so that readers would know for certain which line each label referred to; for the two labels that might have been ambiguous (''Hydro'' and ''Nuclear''), the writer used arrows.

Provide Informative Titles

Titles serve much the same purposes as labels. They help your audience find the visual aids they are looking for and know what the visual aids contain when they find them. Titles typically include both a number (for example, ''Figure 3'' or ''Table 6–11'') and a description (''Interest Rates for Consumer Loans in Six Metropolitan Areas'' or ''Effects of Temperature on the Strength of M312'').

To help your audience understand and use your visual aids, you should make the descriptive part of your titles as brief—yet informative—as possible. Select the key words that describe in specific terms what the visuals show:

Site Plan for the Creighton Landfill

Projected Increases in Peak Demand for Electricity
over the Next Ten Years

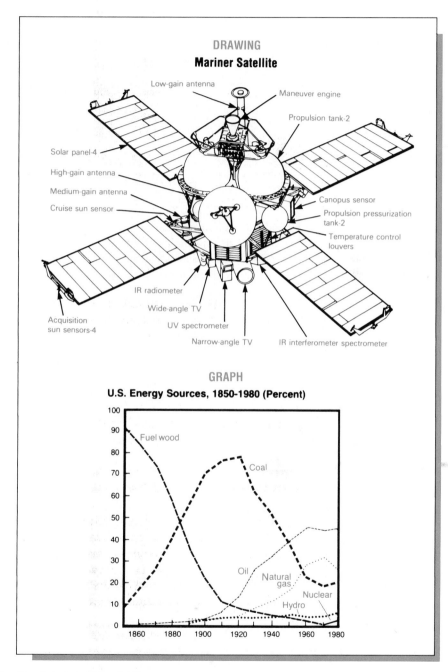

FIGURE 13–15 Clearly Labeled Visual Aids

Graph courtesy of Oak Ridge Associated Universities. Drawing from Edward Clinton Ezell and Linda Neuman Ezell, *On Mars: Exploration of the Red Planet* (Washington, D.C.: National Aeronautics and Space Administration, 1984), 146.

Leave out redundant or vague words. Don't use a title like this one:

Table 3
Table of Information about Three Amplifiers

This title would be more useful if it were changed:

Table 3
Comparison of Three Amplifiers

Visual aids are numbered consecutively, either in one long sequence through the entire communication or with a new sequence in each chapter (as in this book). Tables are usually assigned a separate sequence from that for all the other visual aids, which are called *figures*. Consequently, there will be a Table 1 and a Figure 1, a Table 2 and a Figure 2, and so on.

According to custom, the numbers assigned to figures are usually arabic (1, 2, 3, 4, and so on), while the numbers assigned to tables are either arabic or roman (1, 2, 3, 4, or I, II, III, IV). In communications where the numbering sequences begin anew in each chapter or section, the numbers contain two parts: the number of the chapter and the number of the figure or table within the sequence of that chapter. Thus, Figure 3–7 is the seventh figure in Chapter 3 and Figure 7–3 is the third figure in Chapter 7.

Conventions about where to place titles vary. In typewritten communications, titles are almost always placed *above* tables and *below* figures. Generally, they are centered between the margins. In communications prepared on a computer or typeset by a printer, titles are always placed so that they are prominent and attractive, but their location varies considerably from communication to communication. Sometimes they appear above the visual aids, sometimes below them, and sometimes within them. Within a single typeset communication, of course, all tables share a consistent placement of titles, and a consistent placement is also used for figures, though the placements for table titles and figure titles are not necessarily the same.

Long reports, proposals, and instruction manuals sometimes have special tables of contents that list the number, title, and location of each visual aid.

You may find that your employer or client has rules about how titles are to be written and where they are to be placed. Even without such explicit regulations, your audience may have strong expectations concerning titles. To learn what these are, you can ask other people and scan communications similar to the one you are writing.

Finally, you should note that sometimes you *don't* need to provide a title for a visual aid. That happens, for instance, when you are including a very short table in your text in a way that makes perfectly clear what it contains, as happens in the following example:

> You will be pleased to see how well our top four salespersons did in June:

Chamberlain	$127,603
Tonaka	95,278
Gonzales	93,011
Albers	88,590

Similarly, the visual aids in brochures are often untitled; untitled figures are much rarer in reports and proposals.

One important implication of Guidelines 1 (''Look for Places Where Visual Aids Will Help You Achieve Your Communication Objectives''), 2 (''Choose Visual Aids Appropriate to Your Objectives''), and 3 (''Make Your Visual Aids Easy to Understand and Use'') is that you should resist the temptation to let your decisions about visual aids be determined by what you have handy or what you have done in some other document. If you were to do either of those things, you would be taking a writer-centered approach to visual aids, one in which you choose the type that is easiest for *you* to prepare, not the one that is most likely to be useful or persuasive to your *readers*.

GUIDELINE 4   FULLY INTEGRATE YOUR VISUAL AIDS WITH YOUR PROSE

The first three guidelines in this chapter tell you how to plan and design visual aids. They focus on your visual aids in isolation from the rest of your communication. In contrast, this guideline asks you to think about your visual aids from the point of view of your readers in the act of reading your overall communication. It suggests that you integrate your visual aids with your prose so that they work together harmoniously to create a single, unified message. Three ways to do that are to introduce your visual aids in your prose, state the conclusions you want your readers to draw, and make your visual aids easy to find.

Introduce Your Visual Aids in Your Prose

When people read sequentially through a written communication, they read one sentence then the next, one paragraph then the next, and so on. When you want the next element they read to be a visual aid rather than a sentence or a paragraph, carefully direct the readers' attention from your prose to the visual aid to indicate what they will find in the visual aid.

You might do these things in two separate sentences:

> In a market test, we found that the Radex was much more appealing than Talon, especially among rural consumers. See Figure 3.

Or you might do both in a single sentence:

> As Figure 3 shows, Radex was much more appealing than Talon, especially among rural consumers.

or

> Our market test showed that Radex was much more appealing than Talon, especially among rural consumers (Figure 3).

Sometimes, your introduction to a visual aid will have to include information your readers or listeners need in order to understand or use the visual aid. For example, the writers of an instruction manual explained in this way how to use one of their tables:

> In order to determine the setpoint for the grinder relay, use Table 1. First, find the grade of steel you will be grinding. Then, read down column two, three, or four, depending upon the grinding surface you are using.

Whatever kind of introduction you make to a visual aid, place it at the exact point where you would like your readers to focus their attention on it.

State the Conclusions You Want Your Readers to Draw

An additional way to integrate your visual aids with your prose is to state explicitly any conclusion you want your readers to draw from your visual aid. A visual aid presents your readers with a set of facts. Unless you tell what conclusion you want them to draw from those facts, they may draw a different one than you want them to.

Consider, for example, the way a writer discussed a graph that showed how many orders he thought his company would receive for its rubber hoses over the next six months. The graph showed that there would be a sharp decline in orders from automobile plants, and the writer feared that his readers might focus on that fact. However, the main point the writer wanted to make was quite different, so he wrote the following statement:

> As Figure 7 indicates, our outlook for the next six months is very good. Although we predict fewer orders from automobile plants, we expect the slack to be taken up by increased demand among auto parts outlets.

You might find it helpful to think of the sentences in which you explain a visual aid's significance as a special kind of topic sentence. Just as in the topic sentence at the head of a paragraph, you can tell your audience the point to be derived from the various facts that follow.

Make Your Visual Aids Easy to Find

What happens when your readers come to a statement in which you ask them to look to a certain visual aid? Instead of going on to read the next sentence, as they would usually do, the readers lift their eyes from your prose and search for the visual aid. You want to make that search as short and simple as possible.

To do that, you must put your visual aids where your readers can locate them quickly. The ideal location is on the same page as the prose that accompanies the visual aid. Not only will the visual aid be easy to find, but your readers will be able to look back and forth between the prose and visual aid, if necessary.

However, you may not always have room for your visual aid on the same page as the prose that introduces it. If your communication has text on facing pages, try to put the visual aid on the page facing its accompanying prose. If your communication does not have facing pages of text, try to put your visual aid on the next page following its introduction. If you place the figure farther away than that (for instance in an appendix), you can help your readers by providing the number of the page on which the figure may be found:

> A detailed sketch of this region of the new building's floor plan is shown in Figure 17 in Appendix C (page 53).

CONCLUSION

Visual aids can greatly increase the clarity and impact of your written communications. To use visual aids well, you need to follow the same reader-centered strategy that you use when writing your prose: think about the tasks your readers will perform while reading and think about the ways you want your communication to shape your readers' attitudes. Doing so will enable you to decide where to use visual aids, determine the most effective types of visual aids to use, make them easy to understand and use, and integrate them successfully with your prose.

EXERCISES

1. Figure 13–16 shows a table containing information about enrollments in institutions of higher education in the United States. A writer might use that information in a variety of ways, including the four listed below. For each of these four uses, decide whether the writer should leave the information in a table or present the relevant facts in a bar graph, pictograph, line graph, or pie chart.

NO. **250.** ENROLLMENT IN INSTITUTIONS OF HIGHER EDUCATION, BY SEX, AGE, AND ATTENDANCE STATUS, 1970 TO 1987, AND PROJECTIONS, 1992 AND 1997

[As of fall]

SEX AND AGE	NUMBER (1,000)						PERCENT PART-TIME					
	1970	1980	1985	1987	1992, proj.	1997, proj.	1970	1980	1985	1987	1992, proj.	1997, proj.
Total	8,581	12,097	12,247	12,544	12,408	12,173	32.2	41.3	42.2	42.5	44.3	44.5
Male	5,044	5,874	5,818	5,881	5,845	5,688	30.5	37.2	38.0	38.6	39.6	39.8
14 to 17 years old	129	99	121	91	81	90	3.9	15.2	15.7	9.9	9.9	10.0
18 and 19 years old	1,349	1,375	1,230	1,309	1,309	1,367	6.2	10.6	9.9	10.3	10.1	10.5
20 and 21 years old	1,095	1,259	1,216	1,089	1,088	1,008	9.6	12.2	15.5	14.9	14.8	14.7
22 to 24 years old	964	1,064	1,048	1,080	977	878	32.6	35.4	30.3	32.6	32.8	32.7
25 to 29 years old	783	993	991	1,016	935	867	58.2	61.9	60.1	59.0	59.0	59.1
30 to 34 years old	308	576	574	613	653	599	76.6	77.6	73.9	72.4	72.4	72.5
35 years old and over	415	507	639	684	801	878	81.9	84.8	84.8	83.2	83.3	83.1
Female	3,537	6,223	6,429	6,663	6,563	6,485	34.6	45.2	46.1	45.9	48.4	48.7
14 to 17 years old	129	148	113	119	104	115	9.3	11.5	10.6	10.9	10.6	11.3
18 and 19 years old	1,250	1,526	1,370	1,455	1,299	1,379	8.0	11.4	11.4	12.3	12.9	12.8
20 and 21 years old	785	1,165	1,166	1,135	1,136	1,043	16.3	17.9	18.7	18.9	19.0	19.0
22 to 24 years old	493	925	885	968	864	774	47.4	43.8	40.1	40.9	40.9	40.8
25 to 29 years old	292	878	962	931	854	793	72.6	73.6	68.8	67.0	67.9	67.8
30 to 34 years old	179	667	687	716	755	693	84.8	79.6	76.7	75.6	76.2	76.2
35 years old and over	409	913	1,246	1,339	1,553	1,686	85.6	87.5	80.0	81.8	82.3	82.3

Source: U.S. Dept. of Education, National Center for Education Statistics, *Projections of Education Statistics*, annual.

FIGURE 13–16   Table to Be Used in Exercise 1

From U.S. Bureau of the Census, *Statistical Abstract of the United States,* 109th ed. (Washington, D.C.: U.S. Government Printing Office, 1989), 150.

a. To show how steadily and dramatically the number of women over age 35 enrolled in college has risen since 1970 and is expected to continue to rise through 1997.

b. To provide a reference source in which researchers can find the numbers of men (or women) in a particular age group who were enrolled in college in a particular year.

c. To enable readers to make a quick, rough comparison of the 1987 enrollments of men in each age group with the 1987 enrollments of women in each age group.

d. To allow educational planners to see what proportion of the males (or females) who will enroll in 1997 are expected to be from each of the age groups. (To present this information, the writer would have to perform some mathematical calculations based upon the data supplied in Figure 13–16.)

CHAPTER 14

CREATING TWELVE TYPES OF VISUAL AIDS

Drawings

    How to Create a Drawing

Diagrams

    How to Create a Diagram

Flow Charts

    How to Create a Flow Chart

Organizational Charts

    How to Create an Organizational Chart

Schedule Charts

    How to Create a Schedule Chart

Budget Statements

    How to Create a Budget Statement

EXERCISES

DEFINING
OBJECTIVES
PLANNING
DRAFTING
EVALUATING
REVISING

In the preceding chapter, you read four guidelines for using visual aids effectively. Those guidelines can help you succeed with any of thousands of types of visual aids used on the job, whether they are the common ones such as line graphs and photographs that we can all understand, or whether they are specialized ones that can be created and interpreted only by people with special training.

In this chapter, you will find additional, detailed advice about choosing and constructing twelve common types of visual aids. These twelve were chosen because they are widely useful in the kinds of communications that most college graduates write on the job. To make the chapter as helpful as possible for you, it is organized into twelve separate discussions, one for each type of visual aid. This organization will let you use the chapter as a reference source in which you can quickly find complete information about the type of visual aid you want to use or study.

The twelve types of visual aids are tables, bar graphs, pictographs, line graphs, pie charts, photographs, drawings, diagrams, flow charts, organizational charts, schedule charts, and budget statements.

TABLES

The table is one of the most versatile and widely used visual aids. When you sit down to breakfast, there is one on the side of your cereal box, telling you the nutritional value of the contents. When you read a technical or scientific report, you will probably find a table presenting the researcher's results. When you buy a stereo or piece of laboratory equipment, you will probably discover a table in the owner's manual that tells the specifications and capabilities of your purchase.

Tables are used so often because they can help writers achieve several common objectives. For example, they are an excellent tool for presenting groups of detailed facts in a concise and readable form. To see how well tables do this, look at Figure 14–1, which shows a table and then the prose that gives the same information as the *first three columns* of the table. To present the information contained in all eleven columns of the table, the prose needs to be nearly three times as long. Even as it is, the prose is virtually unreadable because it is so dense with facts.

Tables also help readers find particular facts quickly. To demonstrate this point, use the prose and then the table in Figure 14–1 to find out the total amount of money spent in the U.S. on health care in 1980. The superiority of the table would be even more evident if you tried to find and compare two or more facts, in an effort, for instance, to determine how the percentage of the gross national product spent on health care services changed from 1970 to 1980.

Tables can be just as effective at presenting information in words as in numbers. For example, tables are often used in reports and advertising to

TABLE

No.135. National Health Expenditures: 1970 to 1986

[Includes Puerto Rico and outlying areas]

YEAR	TOTAL [1] Total (bil. dol.)	Per capita (dol.)	Per cent of GNP [2]	HEALTH SERVICES AND SUPPLIES — Private Total [3] (bil. dol.)	Direct patient payments Total (bil. dol.)	Per cent of total private	Insurance premiums [4] (bil. dol.)	Public Total [5] (bil. dol.)	Per cent of total health exp.	Medical payments Medicare (bil. dol.)	Public assistance (bil. dol.)
1970	75.0	349	7.4	44.7	26.5	59.3	16.9	24.9	33.2	7.5	6.3
1971	83.5	384	7.6	48.9	28.1	57.4	19.3	28.4	34.0	8.3	8.1
1972	94.0	428	7.7	55.3	30.6	55.4	22.2	32.1	34.1	9.1	9.1
1973	103.4	467	7.6	60.7	33.3	54.8	24.7	35.8	34.6	10.1	10.3
1974	116.1	521	7.9	65.8	36.2	55.0	27.9	42.9	36.9	13.1	12.1
1975	132.7	590	8.3	73.0	38.1	52.2	33.2	51.3	38.6	16.3	15.1
1976	150.8	665	8.5	84.5	42.0	49.8	40.4	57.3	38.0	19.3	16.9
1977	169.9	743	8.5	96.7	46.4	48.1	48.0	64.0	37.7	22.5	18.9
1978	189.7	822	8.4	106.6	50.7	47.6	53.6	73.3	38.7	25.9	21.1
1979	214.7	921	8.6	120.4	55.8	46.4	62.0	83.8	39.0	30.3	24.3
1980	248.1	1,054	9.1	138.7	63.0	45.5	72.6	97.5	39.3	36.8	28.1
1981	287.0	1,207	9.4	160.6	72.6	45.2	84.4	113.2	39.4	44.8	32.3
1982	323.6	1,348	10.2	182.2	79.6	43.7	98.7	127.1	39.3	52.4	34.9
1983	357.2	1,473	10.5	202.8	88.7	43.7	109.7	138.9	38.9	58.9	37.7
1984	391.1	1,597	10.4	224.7	98.4	43.8	121.5	150.8	38.6	64.4	38.3
1985	422.6	1,710	10.6	240.7	105.3	43.7	130.1	166.5	39.4	72.3	42.2
1986	458.2	1,837	10.9	262.5	116.1	44.2	140.7	179.5	39.2	77.7	45.8

[1] Includes medical research and medical facilities construction. [2] GNP = Gross national product; see table 688. [3] Includes other sources of funds not shown separately. [4] See footnote 3, table 136. [5] Includes other programs, not shown separately.

Source: U.S. Health Care Financing Administration, *Health Care Financing Review*, Summer 1987.

PROSE

(Corresponds to First Three Columns Only)

National Health Care Expenditures: 1970 to 1986

In the Summer 1987 issue of Health Care Financing Review, the U.S. Health Care Financing Administration published figures that show how much the nation spent on health care during selected years from 1970 through 1986. To calculate the total amount expended during these years, the Health Care Financing Administration included the following: direct private payments and insurance premiums for health care; public medical payments by medicare and public assistance programs; and expenditures for medical research and medical facilities construction.

In 1970, it spent $75.0 billion, which was $349 per capita and 7.4% of the gross national product (GNP). (To learn what the gross national product was for 1970 and for each of the years mentioned below, see Table 13.6.) In 1971, it spent 83.5 billion, which was $384 per capita and 7.6% of the GNP. In 1972, it spent $94.0 billion, which was $428 per capita and 7.7% of the GNP. In 1973, it spent $103.4 billion, which was $467 per capita and 7.6% of the GNP. In 1974, it spent $116.1 billion, which was $521 per capita and 7.9% of the GNP. In 1975, it spent $132.7 billion, which was $590 per capita and 8.3% of the GNP.

In 1976, it spent $150.8 billion, which was $665 per capita and 8.5% of the GNP. In 1977, it spent $169.9 billion, which was $743 per capita and 8.5% of the GNP. In 1978, it spent $189.7 billion, which was $822 per capita and 8.4% of the GNP. In 1979, it spent $214.7 billion, which was $921 per capita and 8.6% of the GNP. In 1980, it spent $248.1 billion, which was $1,054 per capita and 9.1% of the GNP.

In 1981, it spent $287.0 billion, which was $1,207 per capita and 9.4% of the GNP. In 1982, it spent $323.6 billion, which was $1,348 per capita and 10.2% of the GNP. In 1983, it spent $357.2 billion, which was $1,473 per capita and 10.5% of the GNP. In 1984, it spent $391.1 billion, which was $1,597 per capita and 10.4% of the GNP.

In 1985, it spent $422.6 billion, which was $1,710 per capita and 10.6% of the GNP. In 1986, it spent $458.2 billion, which was $1,837 per capita and 10.9% of the GNP.

FIGURE 14–1 Comparison of a Table and Prose for Presenting Detailed Information

From U.S. Bureau of the Census, *Statistical Abstract of the United States,* 109th ed. (Washington, D.C.: U.S. Government Printing Office, 1989), 90.

PRESS BRAKE TYPES COMPARED

Characteristic	Mechanical	Hydraulic		Hydromechanical	
		Down-acting	Up-acting	Conventional	Toggle
Operating speed	Good	Fair[a]-good	Good	Good	Fair
Ram levelness	Excellent	Good[a]-Excellent	Excellent	Excellent	Fair
Stroke accuracy	Excellent	Poor[a]-Excellent	Excellent	Excellent	Fair
Control, stroke reversability	Poor	Good	Excellent	Good	Good
Time for tool installation and setup	Fair	Good	Excellent	Good	Good
Setup skill required	High[b]	Medium	Low[c]	Medium	Medium
Overload resistance	Poor	Good	Good	Good	Good
Initial cost	High	Medium	Medium	High	Medium
Maintenance cost	High[d]	Medium	Low[e]	Medium-High	Medium
Share of market, percent	10	45	8	1	35

a. Conventional style; computerized controls upgrade capabilities.
b. Clutch slippage adjustment requires high skill level; remaining setup needs medium skill.
c. Operators quickly learn "feel" of the press.
d. Most wear, lubrication, and adjustment points.
e. Short hydraulic circuits ease fault or line tracing and maintenance.

FIGURE 14–2   A Table Using Words to Compare Types of Press Brakes
From Larry Conley, "Progress in Press Brake Design," *Welding Design & Fabrication* 58 (September 1985):54. Reprinted with permission from *Welding Design & Fabrication*, a Penton Publication, copyright 1985.

compare the features of competing products or processes (Figure 14–2). And they are often used to display information in the troubleshooting sections of instruction manuals (Figure 14–3).

How to Construct a Table

To create a table, you systematically arrange information in rows and columns. Figure 14–4 shows the basic structure of a table.

You should adjust this basic structure as needed to create an attractive visual aid that your readers will find easy to use. For example, if your table is crowded enough that your readers will need help reading down the columns, you can separate the columns (or groups of columns) with vertical lines, called *rules* (see the table in Figure 14–1). To help your audience read across the rows, you can place a horizontal rule or leave a blank line after every five rows or so. Of course, if you are creating a table where the rows are grouped

Trouble Shooting Chart

Problem	Cause	Remedy
1 Engine fails to start	A Blade control handle disengaged	A Engage blade control handle.
	B Check fuel tank for gas	B Fill tank if empty.
	C Spark plug lead wire disconnected.	C Connect lead wire.
	D Throttle control lever not in the starting position	D Move throttle lever to start position.
	E Faulty spark plug	E Spark should jump gap between control electrode and side electrode. If spark does not jump, replace the spark plug.
	F Carburetor improperly adjusted, engine flooded	F Remove spark plug, dry the plug, crank engine with plug removed, and throttle in off position. Replace spark plug and lead wire and resume starting procedures.
	G Old stale gasoline	G Drain and refill with fresh gasoline.
	H Engine brake engaged	H Follow starting procedure.
2 Hard starting or loss of power	A Spark plug wire loose	A Connect and tighten spark plug wire.
	B Carburetor improperly adjusted	B Adjust carburetor. See separate engine manual.
	C Dirty air cleaner	C Clean air cleaner as described in separate engine manual.
3 Operation erratic	A Dirt in gas tank	A Remove the dirt and fill tank with fresh gas.
	B Dirty air cleaner	B Clean air cleaner as described in separate engine manual.
	C Water in fuel supply	C Drain contaminated fuel and fill tank with fresh gas.
	D Vent in gas cap plugged	D Clear vent or replace gas cap.
	E Carburetor improperly adjusted	E Adjust carburetor. See separate engine manual.
4 Occasional skip (hesitates) at high speed	A Carburetor idle speed too slow	A Adjust carburetor. See separate engine manual.
	B Spark plug gap too close	B Adjust to .030″.
	C Carburetor idle mixture adjustment improperly set	C Adjust carburetor. See separate engine manual.

FIGURE 14–3   A Table Used to Provide Troubleshooting Information
Courtesy of MTD Products Incorporated.

logically, you can insert the blank lines between the logical groupings rather than at arbitrarily fixed intervals. If your table has too many columns to fit legibly and attractively on the page, turn the table sideways so that its bottom faces to the right. Finally, note that fancier variations of the basic design may be used where attractiveness is especially important, as in advertising brochures and annual reports to stockholders.

When deciding how to display your information within the framework of your table, you have several basic decisions to make:

● **How to order the rows and columns.** Consider how your readers will use the table and what you want to emphasize. For example, when making a table that shows the output of paper by the nation's largest paper companies, you would use an alphabetical order if your readers will seek

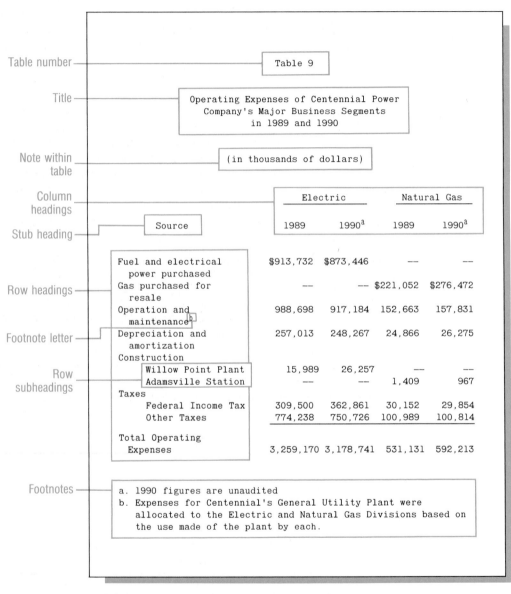

FIGURE 14–4 Structure of a Typical Table (Note that the area containing the row headings is called the stub.)

information about the output of specific companies. However, if your readers' primary interest is finding out which companies have the highest total output, you should order the companies from the highest to the lowest in terms of amount of paper produced.

● **Which labels to use for the columns and which for the rows.** Usefulness, persuasive purpose, and attractiveness can be the primary considerations when deciding which labels to use for the columns and which for
rows. Take advantage of your readers' normal left-to-right eye movement
to emphasize the information that is most useful to them or the information
that you most want to emphasize. For example, a company that manufactures welding supplies wanted to use the table shown in Figure 14–5 to
encourage potential customers to make a point-by-point comparison of its
welding wire with ordinary welding wires. Therefore, the company's writers labeled the rows with the features that are the points of comparison,
and they used the names of the products to label the columns. Because of
this arrangement, the readers' normal left-to-right eye movement causes
them to read the name of a feature, the pertinent data for the competitor's
products, then the pertinent data for the company's product. With this pattern of reading, the desired point-by-point comparison results.

In contrast, when Mitsubishi created the table shown in Figure 14–6,
the company wanted to emphasize that its car offered many features that
its competitors' cars did not. Therefore, Mitsubishi's writers used the
names of the cars to label the rows and the names of the features to label
the columns. With this arrangement, people's normal left-to-right eye
movement causes them to read across the perfect string of ''Yes's'' for the
Mitsubishi car.

In many situations, however, the direction of the readers' eye movements is not the major consideration when deciding which labels to use for
the columns and which for the rows. For instance, if one set of labels will

	Flux Cored Welding vs. Hi-Dep II (1/4"-3/8" Fillet and Downhand Butt Welds)	
	Cored Wire Gas-Shielded or Gasless	**Solid Wire Hi-Dep II**
Wire Deposition Rate	18 lbs/hr	20 lbs/hr
Deposition Efficiency	.85	.98
Weld Metal Deposition Rate	15.3 lbs/hr	19.6 lbs/hr
Weldor Duty Cycle	35%	40%*
Weld Deposited-8 hr day	42.8 lbs	62.7 lbs

*No slag chipping required

FIGURE 14–5   Table Arranged to Encourage a Point-by-Point Comparison
Courtesy of L-Tec Welding & Cutting Systems.

STANDARD MODELS	ENGINE	ELECTRONIC SUSPENSION WITH RIDE HEIGHT & SPRING CONTROL, OPT.	ABS, ELECTRONIC ANTI-LOCK BRAKES, OPT.	4-WHEEL DISC BRAKES	SPEED-VARIABLE POWER STEER-ING, W/DRIVER SELECT MODE	FULLY AUTOMATIC CLIMATE CONTROL	THEFT DETERRENT SYSTEM	SPLIT FOLD-DOWN REAR SEAT
MITSUBISHI SIGMA	3.0-liter EFI V-6	Yes	Yes	Yes	Yes	Yes	Yes	Yes
NISSAN MAXIMA GXE	3.0-liter EFI V-6	No	No	Yes	No	No	Yes	Yes
TOYOTA CRESSIDA	2.8-liter EFI Inline-6	No	No	Yes	No	Yes	Yes	No
ACURA LEGEND	2.7-liter EFI V-6	No	No	Yes	No	No	No	No

*MFRS. SUGG. RETAIL PRICE AND FEATURE COMPARISON OF BASE MODELS WITH AUTO. TRANS. ACTUAL PRICE SET BY DLRS. TAX, LICENSE, FREIGHT, DLR. OPTIONS AND CHARGES EXTRA

FIGURE 14–6 Table Arranged to Emphasize All the Features Offered by One Company's Product Courtesy of Mitsubishi.

each take several words, and another is brief, then you can make a more attractive and convenient table by using the long labels for the rows because it is easier to read long labels when they are placed horizontally than when they are placed vertically (something that often must be done when long labels are used for rows).

● **How to align entries in the columns.** To align numerical entries, use the units column or the decimal point:

    23,418   2.79
     5,231   618.0
        17   13.517

Align prose on the left-hand margin:

    Acceptable
    Marginally Acceptable
    Unacceptable

Alternatively, you can center the prose within the column:

        Acceptable
    Marginally Acceptable
        Unacceptable

● **Where to place special notes.** In some tables, you will need to include notes that do such things as explain the labels or cite the sources of your information. If your notes are short enough, you can place them next to the appropriate title or headings in your table. Otherwise, place them in footnotes at the bottom of your table. Use lowercase letters to label the footnotes if your readers might otherwise confuse a superscript footnote number with a mathematical exponent. For example, "14^2" could be read as "fourteen squared" rather than the number *14* followed by a superscript footnote. Using a lowercase letter avoids that confusion: 14^b.

If your table displays numerical data, be sure to indicate the units you are using: dollars, kilograms, percentage, and so on. When all the entries in a long column or row use the same units, it's usually best to identify the units in the heading and thereby avoid cluttering the table with repeated information.

How to Use Informal Tables

Sometimes you can create tables that are much simpler than those just de-scribed. These "informal tables" are useful where the sentence or sentences that precede the table provide the readers with all the information they need to interpret it or if the interpretation is otherwise obvious.

Here is an example showing how an informal table is built into a para-graph:

Even more important, the sales figures demonstrate how our investment in technical research has helped several of our divisions become more competitive. The following figures show the increase in the market share enjoyed since the same quarter last year by three major divisions that have extensive research programs:

Strauland Microchips	7%
Minsk Machine Tools	5%
PTI Technical Services	4%

Note that such tables work only when you present a single column of facts and when your readers will understand your table without a title, column head-ings, or notes. If you think that such an informal table might confuse your readers, even momentarily, then use a formal table instead.

BAR GRAPHS

Like a table, a bar graph can represent numerical quantities. However, a bar graph does it with rectangles called *bars*. The greater the quantity, the longer the bar. Some uses of bar graphs are as follows:

● **To compare quantities at a glance.** Because the bars are drawn to scale, bar graphs can help readers tell immediately not only which quantities are

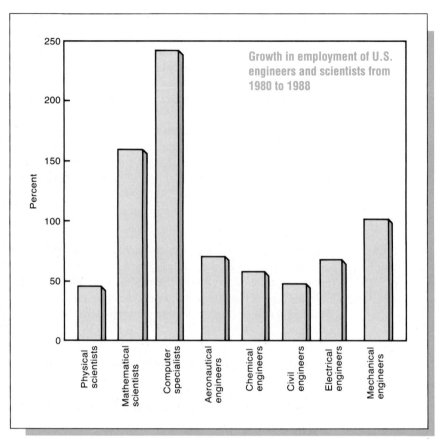

FIGURE 14–7   Bar Graph Showing Comparison
Based on *U.S. Scientists and Engineers: 1988* (Washington, D.C.: National Science Foundation, 1988), 6.

the larger or smaller, but also how great the differences are (see Figure 14–7).

● **To show trends.** If a series of bars is used to represent a quantity (such as health expenditure in the U.S) at particular times, readers will be able to determine how the quantities change (or don't change) over time (see Figure 14–8).

● **To indicate the composition of a whole.** For instance, a bar that represents the entire revenue of a company might be subdivided to indicate the portion of its revenue that derives from various sources. A series of such subdivided bars can show the changing (or unchanging) composition of the company's revenue over time (see Figure 14–9).

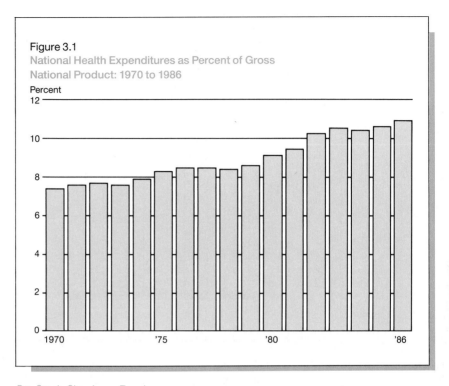

Figure 3.1
National Health Expenditures as Percent of Gross
National Product: 1970 to 1986

FIGURE 14–8   Bar Graph Showing a Trend

From U.S. Bureau of the Census, *Statistical Abstract of the United States,* 109th ed. (Washington, D.C.: U.S. Government Printing Office, 1989), 91.

How to Construct a Bar Graph

Begin by drawing your axes so that your graph will be roughly square. Along one axis place tick marks at regular intervals to indicate quantities ($5 million, $10 million, etc.; 50 psi, 100 psi, etc.). Plan the tick marks so that the longest bar will extend nearly to the end of its parallel axis.

You may extend the bars vertically or horizontally—vertical bars are often used for height and depth, whereas horizontal bars are often used for distance, length, and time. Order the bars to suit your purpose. For example, if you want your readers to discern rank orders, arrange the bars in order of length. If you want your readers to make ready comparisons among subgroups of quantities, group the appropriate bars together.

Generally, your readers will find it easier to use labels placed next to the bars than to use labels provided in a separate key. One exception is when you use the same groups of bars repeatedly, as in Figure 14–10 (p. 399). Then, use distinctive hatching, shading, or coloring to distinguish each repeating category within the group of bars and provide a key to those categories.

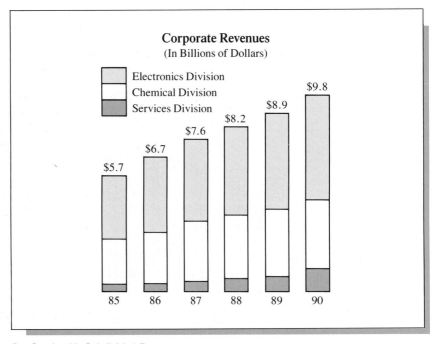

FIGURE 14–9   Bar Graph with Subdivided Bars

Generally, numbers indicating the exact quantity represented by each bar (as in Figure 14–7) are *not* necessary and you can simplify your bar graph by omitting them. Provide exact quantities only if your readers need them.

How to Avoid Misleading Your Readers

When making bar graphs, you need to avoid accidentally misleading your readers. One common mistake is to omit the zero point on the axis that shows quantity. Compare the two bar graphs shown in Figure 14–11. In the bar graph on top, the difference between the two bars seems very significant because it amounts to two-thirds the length of the longest bar. In contrast, the difference in the graph on the bottom looks much less significant. Both graphs, however, represent the same data. The difference between the bars on the top is exaggerated because the quantity scale begins not at zero, as it should, but at 85 percent.

If you simply cannot use the entire quantity scale, indicate that fact to your readers, perhaps by using hash marks to signal a break in the quantity axis and in the bars themselves (see Figure 14–12, p. 400).

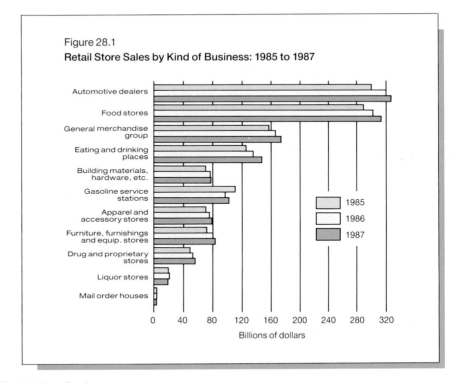

FIGURE 14–10   Multibar Graph
From U.S. Bureau of the Census, *Statistical Abstract of the United States,* 109th ed. (Washington, D.C.: U.S. Government Printing Office, 1989), 750.

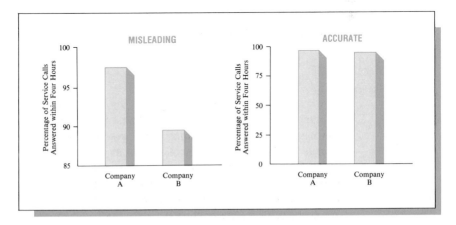

FIGURE 14–11   Comparison of Bar Graphs with and without a Zero Point
Based on data provided in Edwin E. Mier and John Bush, "Rating the Long-Distance Carriers," *Data Communications* 13 (August 1984): 109. Copyright 1984 McGraw-Hill, Inc. All rights reserved.

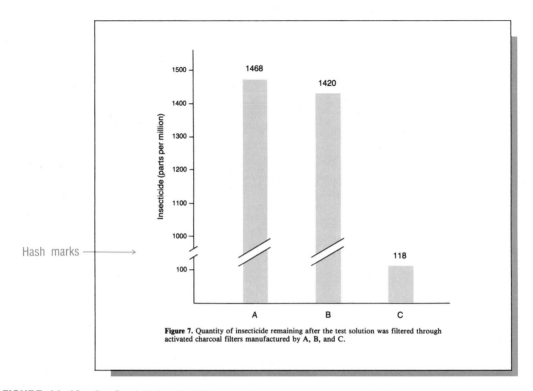

Hash marks
Insecticide (parts per million)
1500
1400
1300
1200
1100
1000
100
1468
1420
118
A
B
C
Figure 7. Quantity of insecticide remaining after the test solution was filtered through activated charcoal filters manufactured by A, B, and C.

FIGURE 14–12 Bar Graph Using Hash Marks on the Vertical Axis and on the Bars

Another way to avoid misleading your readers is to make all bars the same width. If you vary the width, readers may compare either the widths or the areas of the bars, rather than their lengths, and thereby draw incorrect conclusions about the quantities they are comparing.

PICTOGRAPHS

Pictographs are a special kind of bar graph in which the bars are replaced by drawings that represent the thing being described. Figure 14–13 provides an example in which the number of barrels of oil used per capita by the United States is represented by pictures of oil barrels. The chief advantage of the pictograph is that it uses drawings to symbolize concretely the quantities you are talking about in your graph.

You will find pictographs especially useful where you want to do one or both of the following:

- **Emphasize the practical consequences of the data represented.** A pictograph that uses silhouettes of people to represent the workers who will

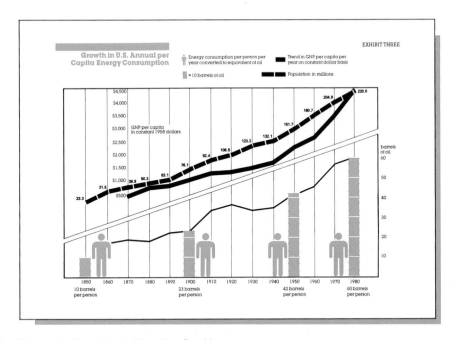

FIGURE 14–13   Pictograph (Combined with a Line Graph)

From Standard Oil Company, *Energy Adventures* (Cleveland, Ohio: Standard Oil Company, 1983), Exhibit 3. Courtesy of Oak Ridge Associated Universities. Courtesy B. P. America.

be employed in a new plant helps to emphasize that the plant will bring a practical benefit to individual men and women in a community.

● Make your data visually interesting and memorable. Visual interest may be especially important when your communication is addressed to the general public. In some situations at work, however, your readers expect a more abstract representation of information and would consider pictographs to be inappropriate.

How to Create a Pictograph

The procedure for creating a pictograph is nearly identical to that for creating a bar graph. The difference is that you draw pictures instead of rectangles to represent quantities.

How to Avoid Misleading Your Readers

Like bar graphs, pictographs can mislead the reader if they are not drawn properly. Consider, for example, the two pictographs shown in Figure 14–14, which both show the average percentage of an apple harvest that a grower

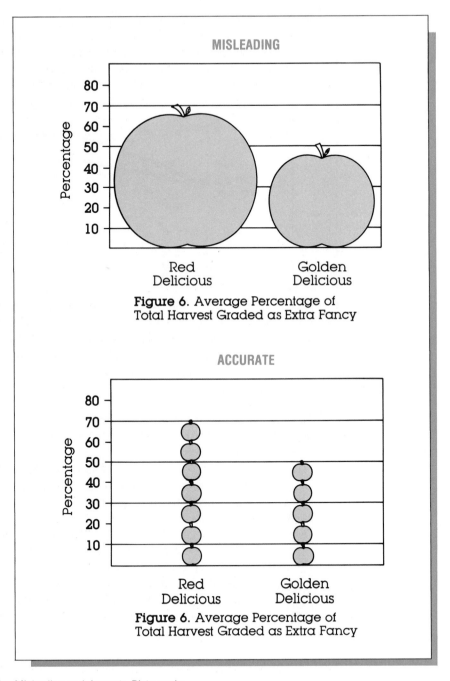

FIGURE 14–14 Misleading and Accurate Pictographs

Based on data from Paul J. Tvergyak, "Branching Out," *American Fruit Grower* 106 (February 1986): 56.

should expect to be graded Extra Fancy. The pictograph on the top makes the percentage for red delicious apples seem at least twice as great as the percentage for golden delicious, even though the actual difference is only 20 percent. That's because the picture for the red delicious apples is larger in height *and* width, so that its area is much greater. In the pictograph on the bottom, the column for the red delicious apples is greater in height alone. When using pictographs, you can avoid misleading your readers if you keep all of your pictures the same size and use more of them to represent greater quantities.

LINE GRAPHS

A line graph shows how one quantity changes as a function of changes in another quantity. You can use line graphs in many ways, including the following:

● **To show trends and cycles.** When you want to show a pattern of change over time, line graphs can be very helpful—especially when compared with a table. For an example, see Figure 14–15, which presents both a table and a line graph that show how the amount of intercity freight carried by railroads and motor vehicles changed over a fourteen-year period.

● **To compare trends.** Line graphs are also very useful for showing readers how two or more trends compare with one another. For instance, the writers of the graph shown in Figure 14–16 used one to create a favorable impression of 304 stainless steel by showing that its price has risen much more slowly than either the price of finished steel mill products or the consumer price index.

● **To show how two or more variables interact.** Figure 14–17 (p. 406) shows two line graphs that display interactions between variables. The graph at the top shows how the price of natural gas interacts with the amount consumed: price and usage move in opposite directions, so that when one goes up, the other goes down. The graph at the bottom shows the results obtained in one experiment by three scientists who were studying the causes and prevention of cancer. The scientists wanted to learn what could happen if they treated a certain kind of cancer cell with Interleukin-3, a peptide that promotes growth. In their experiment, they measured three variables at each of five concentrations of Interleukin-3.

How to Create a Line Graph

In line graphs, you generally show how variation in one thing (called the *dependent variable*) is affected by variation in another thing (the *independent variable*). For example, the line graph shown at the bottom of Figure 14–17

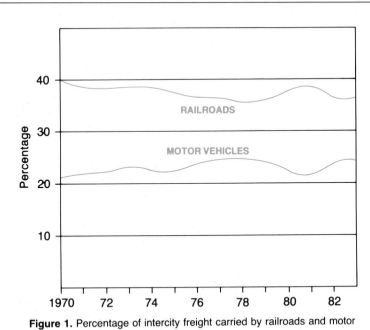

Figure 1. Percentage of intercity freight carried by railroads and motor
vehicles, 1970–1982.

Table 1
Percentage of Intercity Freight Carried by
Railroads and Motor Vehicles, 1970–1983

	Railroads	Motor Vehicles
1970	39.82	21.28
1971	38.54	21.53
1972	38.21	21.68
1973	38.44	22.63
1974	38.52	22.38
1975	36.74	21.97
1976	36.33	23.16
1977	36.15	24.06
1978	35.18	24.28
1979	36.03	23.64
1980	37.47	22.32
1981	37.90	21.94
1982	35.81	23.21
1983	35.93	23.62

FIGURE 14–15 Comparison of a Line Graph and Table

Based on data from U.S. Bureau of the Census, *Statistical Abstract of the United States,* 105th ed.
(Washington, D.C.: U.S. Government Printing Office, 1985), 589.

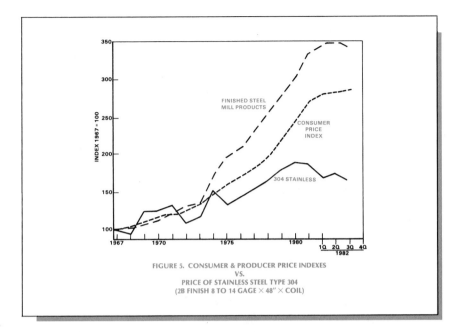

FIGURE 5. CONSUMER & PRODUCER PRICE INDEXES
VS.
PRICE OF STAINLESS STEEL TYPE 304
(2B FINISH 8 TO 14 GAGE × 48″ × COIL)

FIGURE 14–16   Line Graph Comparing Trends

From Ronald S. Paul, "Materials Research and Technological Progress," *Fourth Technology Briefing: Focus on Materials Technology for the '80s* (Columbus, Ohio: Battelle Press, 1982), 11. Courtesy of R. K. Pitler, Allegheny Ludlum Steel Corporation.

shows how three factors (dependent variables) change when the concentration of Interleukin-3 is changed (independent variable). Line graphs almost always show the dependent variable on the vertical axis and the independent variable on the horizontal one. Time is usually treated as an independent variable so it goes on the horizontal axis.

Mark off each axis at regular intervals, using labeled tick marks. Usually your graph will be less cluttered if you make these tick marks short (but still clearly visible). In some situations, however, your readers may find it easier to read your line graph if you extend the tick marks all the way across the graph to form a grid. If you do this, make the grid with a thinner line than you use to represent the quantities you are describing. The heavier lines that represent your data will then stand out against the lighter grid lines.

Generally, the vertical axis should indicate a zero point. Otherwise readers can be misled about the proportions of change shown in the graph (see Figure 14–18). If you don't begin the vertical axis at zero, alert your readers to that fact by using hash marks or similar symbols to indicate that the scale is not continuous from zero (see Figure 14–19). Where you need to indicate some negative quantities, such as losses, or degrees below zero, use both positive and negative values on whichever of the axes needs them.

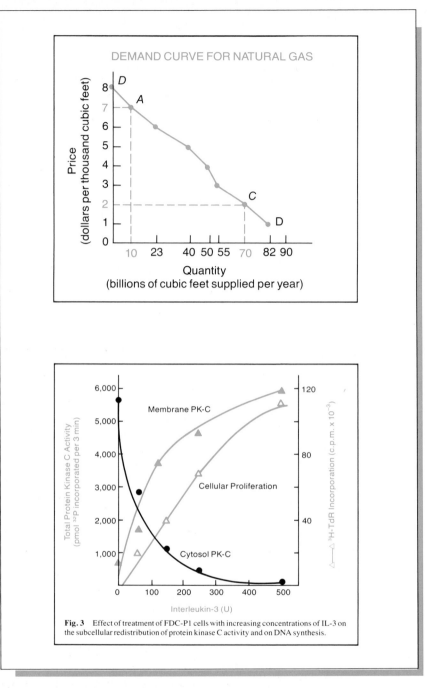

FIGURE 14–17 Line Graphs Showing the Interaction between Two Variables

Top graph from William J. Baumol and Alan S. Blinder, *Economics: Principles and Policy,* 2nd ed. (San Diego: Harcourt Brace Jovanovich, 1982), 54. Reprinted by permission of the Publisher. Bottom graph from William L. Farrar, Thomas P. Thomas, and Wayne B. Anderson, "Altered Cytosol/Membrane Enzyme Redistribution on Interleukin-3 Activation of Protein Kinase C," *Nature* 315 (1985): 237. Reprinted by permission. Copyright 1985 by Macmillan Journals Limited. Courtesy of Dr. William L. Farrar, National Cancer Institute.

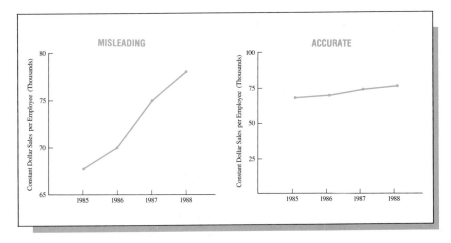

FIGURE 14–18   Comparison of Line Graphs with and without a Zero Point.

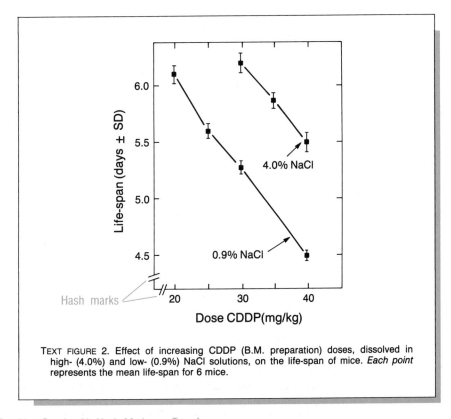

TEXT FIGURE 2. Effect of increasing CDDP (B.M. preparation) doses, dissolved in high- (4.0%) and low- (0.9%) NaCl solutions, on the life-span of mice. *Each point represents the mean life-span for 6 mice.*

FIGURE 14–19   Line Graph with Hash Marks on Two Axes
From Steiner Aamdal, Oystein Fodstad, Olav Kaalhus, and Alexander Pihil, "Reduced Antineoplastic Activity in Mice of Cisplatin Administered with High Salt Concentration in the Vehicle," *Journal of the National Cancer Research Institute* 73 (1984): 745.

Remember to label the plotted lines on your graph. If possible, place the labels immediately next to the lines. Readers find it easier to use labels there than to consult labels provided in a separate key. If the plotted lines cross or might otherwise become confused with one another, use a solid line for one, a dashed line for the second, and so on.

PIE CHARTS

Pie charts are unsurpassed in their ability to depict the composition of a whole—for example, to show how much each of several food sources contributes to the total amount of dietary fat consumed by the average American (see Figure 14–20).

How to Create a Pie Chart

To create a pie chart, you draw a circle and draw lines that slice it into wedges. Each wedge occupies a portion of the circle's circumference proportional to the amount of the total pie that the wedge represents. Arrange the wedges in a way that helps your audience determine the rank order of the wedges and compare the relative sizes of particular wedges. For example, start the largest wedge at the twelve o'clock position and measure clockwise. Then place the remaining wedges in descending order of size. Of course, in some situations, you may need to arrange your slices in some other way to make your points effectively or to achieve a visually pleasing design.

Label each wedge with its title and percentage. Depending upon the size of the wedge, you might place the labels inside or outside the circle.

PHOTOGRAPHS

With a photograph, you can show your readers exactly what they would see if they personally were to look at an object. Because photographs can provide a realistic view of an object's appearance, they help you achieve a variety of communication purposes:

- **To show the appearance of something the readers have never seen.** Perhaps it's the inside of a human heart, the surface of one of the moons of Saturn, the new building purchased in another city, or a new product your company has just begun manufacturing.
- **To show the condition of something.** Photographs can help portray the condition of an object when that condition is indicated by the object's

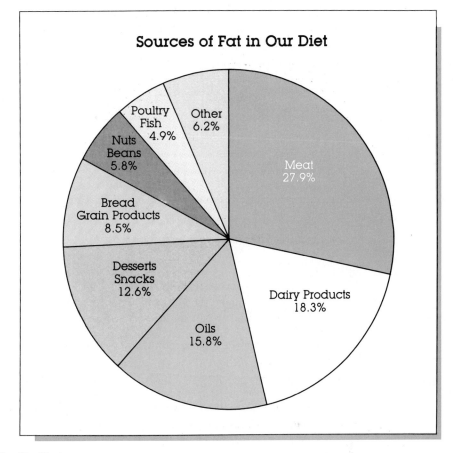

FIGURE 14–20   Pie Chart

Data from Bonnie Liebman, "What America Eats," *Nutrition Action Healthletter* 12 (December 1985): 4. Reprinted from *Nutrition Action Healthletter,* which is available from the Center for Science in the Public Interest, 1501 16th Street, N.W., Washington, D.C., 20037, for $20/yr; copyright 1986.

appearance. Maybe you want to show the result of a treatment of a skin ailment with a new drug your company is developing, the damage to a shipment of your products caused by improper handling, the progress being made on the construction of a new building being erected by your employer, or the difference between competing steel products (see Figure 14–21).

● **To help the readers recognize something.** For instance, in a lab manual for operators of a steel mill, you might include a series of photographs that would enable them to identify the imperfections they might encounter in sheet steel (see Figure 14–22).

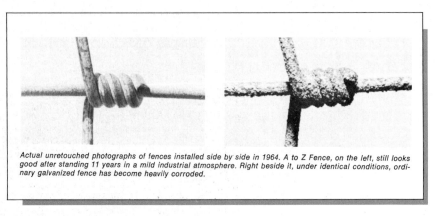

Actual unretouched photographs of fences installed side by side in 1964. A to Z Fence, on the left, still looks good after standing 11 years in a mild industrial atmosphere. Right beside it, under identical conditions, ordinary galvanized fence has become heavily corroded.

FIGURE 14–21   Photograph Used to Show the Condition of Something
Courtesy of Midwestern Steel Division, Armco, Inc.

● **To help your audience find something.** In an instructional manual, for instance, you might show a photograph of the outside of a power unit, labeling each major part the readers will need to know if they are to operate the machinery properly (see Figure 14–23).

How to Create a Photograph

The following suggestions will help you create effective photographs to achieve your communication objectives:

● **Choose an appropriate angle of view.** For instance, if you want to help your readers recognize or find something, use the same angle of view that the readers would take when looking for it.

● **Eliminate unnecessary or distracting detail.** The camera will record whatever is in its field of view, including even the most irrelevant objects. If objects near the one you want to show are irrelevant, remove them or at least screen them from view before taking the photograph. Often the photograph will include unnecessary material above, below, or beside your subject. You can crop (trim away) that extra material to focus your readers' attention more sharply on the object you want to show them.

● **Ensure that all the relevant parts show clearly.** Don't let important parts be hidden (or half-hidden) behind other parts. Be sure that the important parts are in focus.

● **Provide whatever labels your audience will need.** Be sure that all labels stand out sufficiently from the background of the photograph. One way to do that is to affix them to white strips of paper that you then paste on the photograph (see Figure 14–23).

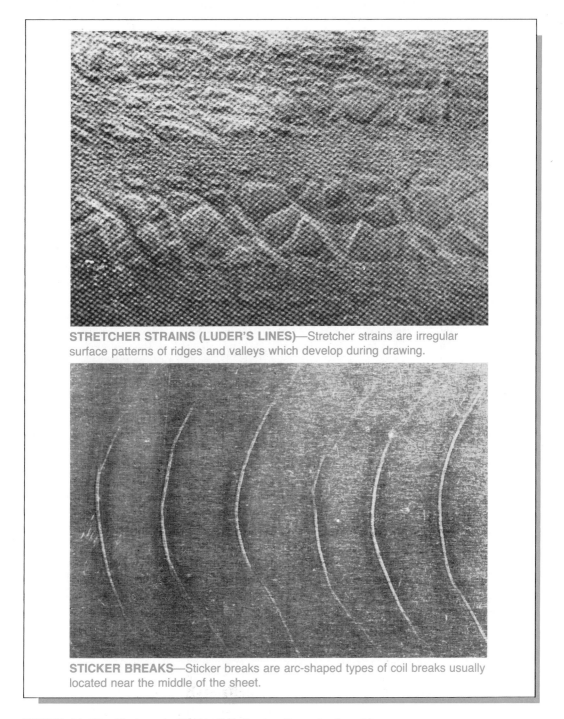

STRETCHER STRAINS (LUDER'S LINES)—Stretcher strains are irregular surface patterns of ridges and valleys which develop during drawing.

STICKER BREAKS—Sticker breaks are arc-shaped types of coil breaks usually located near the middle of the sheet.

FIGURE 14–22 Photographs Used to Help Readers Recognize Something

From Society of Automotive Engineers, ''Classification of Major Visible Imperfections in Sheet Steel,'' *SAE Handbook* (Warrendale, Penn.: Society of Automotive Engineers, 1980), Part 1, 4.24–4.25.

FIGURE 14–23   Photograph That Helps Readers Find Something
Courtesy of Manufacturing Engineering Department, Miami University (Ohio).

DRAWINGS

Like photographs, drawings can show how something looks. Drawings have decided advantages over photographs in certain situations. You can prepare them quickly, unless the skills of a professional artist are needed. Drawings can emphasize details that are important and omit ones that are distracting. Drawings can also show things that photographs cannot, such as the inner parts of something. Figure 14–24 shows a drawing portraying an external view, as well as examples of three types of drawings showing internal parts: cutaway, cross-section, and exploded view.

How to Create a Drawing

When choosing the angle of view from which your drawing will show the object, select the one that your readers will find most helpful. In many cases, this will be a "three-cornered" view, which shows, for example, two sides and the top of an object portrayed in perspective. Such a view helps people grasp at a glance what the object looks like and how its major parts fit together. If you are preparing your drawing for a set of instructions, show the object from the same point of view that your readers will have when working with the object.

When selecting the details you will include, remember that your purpose is usually not to produce a realistic portrait of the object but to draw attention to its significant parts or features. In situations where you want to make a relatively realistic drawing, you might trace the important elements from a photograph. To emphasize significant details, you can do such things as depict them slightly larger than they actually are, draw them with a wider line than you use for other parts, or point them out with labels.

DIAGRAMS

A diagram is much like a drawing except that it relies heavily on symbols to communicate. Whereas drawings convey more or less accurately the actual appearance of the things they portray, diagrams depict subjects more abstractly. For instance, Figure 14–25 shows the water cycle. Although it uses drawings of trees and clouds, for instance, the diagram portrays no actual scene. Figure 14–26 is a diagram that uses symbols that are even more abstract to describe a system used by some companies in the food industry to process fruits and fruit fillings.

How to Create a Diagram

Because diagrams can be used for so many different purposes and rely so heavily upon your creativity for their design, there are few specific guidelines for making diagrams. The general suggestions—to think about your readers and your purpose and to use your common sense—are already quite familiar to you. Nevertheless, the following suggestions will help:

- **Decide exactly what you want to show.** What are the objects or events involved? What are the important relationships among them? These preliminary decisions may take a substantial amount of thought.

- **Create an appropriate means to represent your subject.** You can represent objects and events with geometric shapes, or with sketches that suggest their appearance. You can show relationships among objects and events by the way you place the representative shapes relative to each other and by drawing arrows between them. Figures 14–25 and 14–26 illustrate these techniques. For ideas about how to represent your material, you may find it helpful to look at communications similar to yours that other people have written.

- **Provide the explanations people need in order to understand your diagram.** You may provide necessary explanations in the diagram itself, in a separate key, in the title, or in the accompanying text.

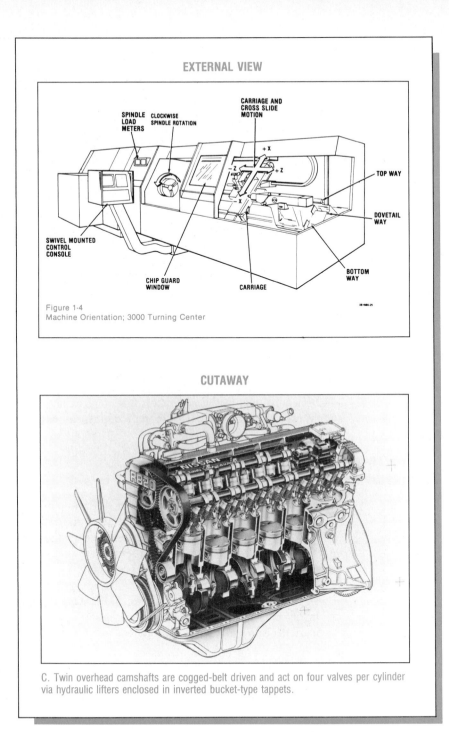

EXTERNAL VIEW

SPINDLE LOAD METERS

CLOCKWISE SPINDLE ROTATION

CARRIAGE AND CROSS SLIDE MOTION

+ X

− Z

+ Z

X

TOP WAY

DOVETAIL WAY

SWIVEL MOUNTED CONTROL CONSOLE

CHIP GUARD WINDOW

CARRIAGE

BOTTOM WAY

Figure 1-4
Machine Orientation; 3000 Turning Center

CUTAWAY

C. Twin overhead camshafts are cogged-belt driven and act on four valves per cylinder via hydraulic lifters enclosed in inverted bucket-type tappets.

FIGURE 14–24 Four Types of Drawings

External View from White Consolidated, *Operator's Manual, 3000 Turning Center, SWINC System CNC* (Cincinnati, Ohio: White Consolidated, 1982), 5. Permission granted by DeVlieg–Sundstand Machine Tool Group. Cutaway from Jack Yamaguchi, "Active Suspension Incorporates Rear Wheel Steering Effect," *Automotive Engineering* 93 (October 1985): 107. Reprinted with permission. Copyright *Automotive Engineering* magazine, October 1985, Society of Automotive Engineers, Inc.

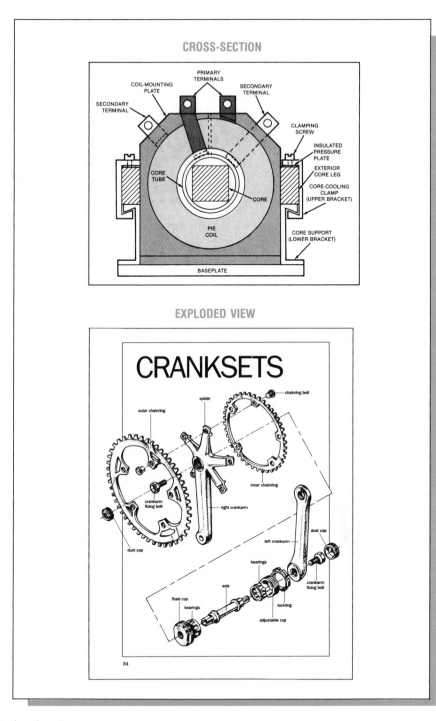

FIGURE 14–24 *(continued)*

Artwork courtesy of Nissan Motor Company, Ltd. Cross-Section from ''High Efficiency, Low-Weight Power Transformer,'' *NASA Tech Briefs* 9 (Fall 1985): 156. Exploded View reprinted from *Bicycling Magazine's Complete Guide to Bicycle Maintenance and Repair* (Emmaus, Penn.: Rodale Press, 1990), 84. Copyright 1990 by Rodale Press, Inc. Permission granted by Rodale Press, Inc., Emmaus, PA, 18098.

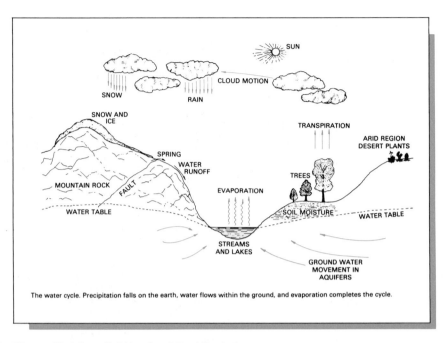

SUN

CLOUD MOTION

SNOW

RAIN

SNOW AND ICE

TRANSPIRATION

ARID REGION DESERT PLANTS

SPRING

WATER RUNOFF

TREES

MOUNTAIN ROCK

FAULT

EVAPORATION

SOIL MOISTURE

WATER TABLE

WATER TABLE

STREAMS AND LAKES

GROUND WATER MOVEMENT IN AQUIFERS

The water cycle. Precipitation falls on the earth, water flows within the ground, and evaporation completes the cycle.

FIGURE 14–25   Diagram That Does Not Use Specialized Symbols
From Raymond L. Murray, *Understanding Radioactive Waste* (Columbus, Ohio: Battelle Press, 1989), 100. Courtesy of Battelle Memorial Institute.

FLOW CHARTS

Flow charts are an excellent means of representing the succession of events in a process or procedure. The simplest flow charts use rectangles, circles, diamonds, or other geometric shapes to represent events, and arrows to show the progress from one event to another. Sketches suggesting the appearance of objects also can be used (see Figure 14–27, p. 418).

Some technical fields, such as systems analysis, have developed special techniques for their flow charts (see Figure 14–28). For example, they use agreed-upon sets of symbols to represent specific kinds of events and outcomes, and they follow agreed-upon rules for arranging these symbols on the page. If you are in a field that uses such specialized flow charts, include them in communications written to your fellow specialists. Remember, however, that you should not use them when writing to people outside your specialty, who will probably not know how to read them.

How to Create a Flow Chart

A few conventions govern the creation of flow charts. The labels that identify the activities are placed *inside* the boxes that represent those activities. Boxes are arranged so that activity flows from left to right, or from top to bottom, or

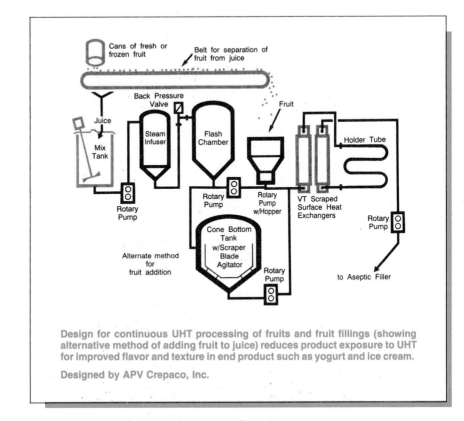

Design for continuous UHT processing of fruits and fruit fillings (showing alternative method of adding fruit to juice) reduces product exposure to UHT for improved flavor and texture in end product such as yogurt and ice cream.

Designed by APV Crepaco, Inc.

FIGURE 14–26   Diagram That Uses Abstract, Nonspecialized Symbols
From "Upgrades for Continuous UHT Processing of Viscous Products," *Food Engineering* 58 (August 1986): 119.

both. This corresponds with a person's normal eye movements when reading. If your flow chart requires more than one line of shapes and arrows, begin the second and subsequent lines, like the first line, at the left-hand margin or at the top of the chart (as in Figure 14–27).

As you create a flow chart, be patient. It may take you several drafts to get the boxes the right size, to place the labels neatly in them, and to arrange the shapes and arrows in an attractive and readily understandable way.

ORGANIZATIONAL CHARTS

An organizational chart uses rectangles and lines to represent the arrangement of people and departments in an organization. It reveals the organization's hierarchy, indicating how the smaller units (such as departments) are combined to create larger units (such as divisions). It also indicates who reports to whom

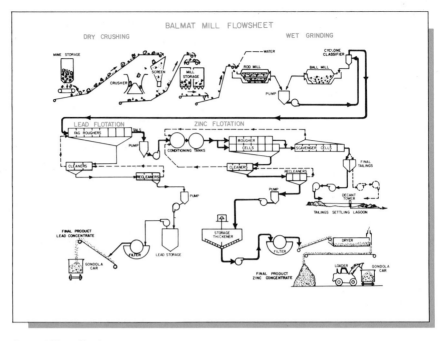

FIGURE 14–27 General Flow Chart
Courtesy of Zinc Corporation of America.

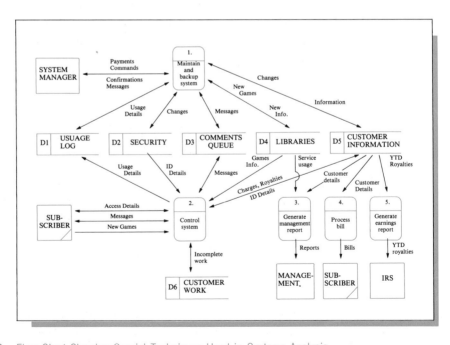

FIGURE 14–28 Flow Chart Showing Special Techniques Used in Systems Analysis
From William S. Davis, *Systems Analysis and Design: A Structured Approach* (Reading, Mass.: Addison-Wesley, 1983), 187. Copyright 1983. Addison-Wesley Publishing Company, Inc., Reading, Mass. Reprinted with permission.

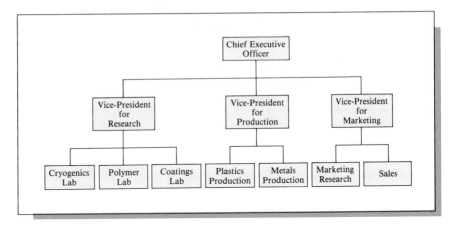

FIGURE 14–29   Organizational Chart

and who gives direction to whom (see Figure 14–29). Some uses of an organizational chart are as follows:

- To show the scope and arrangement of an organization
- To show the formal lines of authority and responsibility in an organization
- To provide a map of an organization, so that readers can readily locate the particular people they want to contact

How to Create an Organizational Chart

Because of the hierarchical nature of most business organizations, organizational charts usually are pyramidal. You do not need to show every part of the organization, only those parts and subparts relevant to your particular readers. Sometimes, you may need to represent more than one kind of relationship. You can do this by using different lines for the different relationships. For instance, you might use solid lines to indicate relationships of direct authority and responsibility and dashed lines to signify relations of close consultation or cooperation.

SCHEDULE CHARTS

A schedule chart identifies the major steps in a project and tells when they will be performed. As Figure 14–30 illustrates, a schedule chart enables readers to see what will be done, when each activity will start and end, and when more than one activity will be in progress simultaneously. (This particular type of schedule chart is called a *Gantt chart*, named for the inventor)

Schedule charts are often used in project proposals to show the proposer's plan of work. You can also use schedule charts in progress reports to show

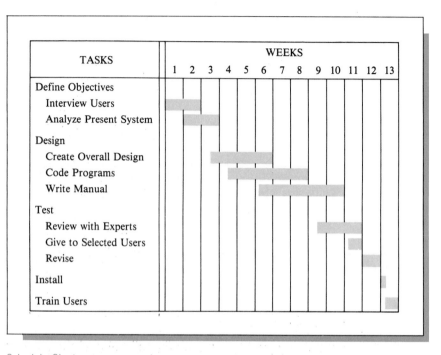

FIGURE 14–30   Schedule Chart

what you have accomplished and what you have left to do, and you can use them in final reports to describe the process you followed while working on a project that you have now completed.

How to Create a Schedule Chart

One of the principal considerations in creating a schedule chart is deciding how much detail to include, something you can determine only on a case-by-case basis in light of your purpose and your audience's needs and expectations. Tasks are always listed along the vertical axis, with indentation used to distinguish subtasks from major tasks. The intervals (weeks, months) are usually marked off with vertical lines that help readers see exactly when tasks begin and end.

BUDGET STATEMENTS

A budget statement is a table that shows how money will be gained or spent. It may be very simple or very elaborate.

On the job, you can use budgets in the following situations:

● **To explain the expenses involved with a project or purchase.** might do this when requesting funds (Figure 14–31) or when reporting on the financial feasibility of a particular course of action.

```
          The most efficient and accurate way to obtain the
      readings in this experiment is to use electronic
      instrumentation that will enter the results directly into
      a computer. To do that, we will need to purchase the
      following equipment:

          HPIP Converter                    $3,200
          Digital Plotter                    1,950
          Current Source                     3,000
          Digital Thermometer                1,100
          Miscellaneous Components             375
            to Create Interface Board       _____
                                             9,625
```

FIGURE 14–31 Simple Budget Statement

- **To summarize the savings to be realized by following a recommendation you are making.** Because profits are a major goal of many organizations, budget statements are often prepared for this purpose.
- **To report the costs that have been incurred by a project for which you have responsibility.** Figure 14–32 shows an example of a budget statement prepared for this purpose.
- **To explain the sources of revenue associated with some project or activity.** For instance, you may use a budget statement when reporting your department's sales of products and services.

How to Create a Budget Statement

The following steps will enable you to prepare simple, informative budget statements. (When accountants, auditors, and specialists in various other business fields prepare budgets, they sometimes use specialized conventions that are not discussed below.)

1. Divide your page into two vertical columns.
2. In the left-hand column, list the major categories of *expense*. If your audience will want the additional detail, list the principal expenses found under each major category. Be sure to arrange your overall list so that the major categories are most prominent.
3. In the right-hand column, write the amount of each expense. Align these amounts on the decimal points.
4. Clearly indicate the total.
5. If you want to show *income* also, repeat Steps 2 through 5 for it.

```
                         PROJECT COSTS

        Equipment
            4 KRN 3781 Robots ($37,000 apiece)           $148,000

            1 Microcomputer with hard disk                  6,500
            4 Power supplies ($5,100 apiece)               20,400

        Construction (includes labor and supplies)
            Rewiring                                         7,100
            Constructing pads for mounting robots            3,000

        Initial Programming (ten days at $125 per           1,250
          day)

        Travel for Installation Personnel
            Airfare (two round trips)                          400
            Car rental                                         200
            Living expenses (two people for five days at     1,500
              $150 per day)
                                                         _____

                                         TOTAL           $188,350
```

FIGURE 14–32   Detailed Budget Statement

EXERCISES

1.   In response to a survey, 13,586 employees of the General Electric Company who had earned college degrees each listed the college courses they found most valuable in their careers, and those they found least valuable.[1] You are to present the results (which are given below) first in a table and then in a bar graph that would be useful to college juniors and seniors to help them decide what courses to take next term.

Begin by imagining a typical student preparing to register. The student has a course catalog, a schedule showing when classes are available, and the table or bar graph you are preparing. Your job is to decide how to make a table (and then a bar graph) that will present the survey results as helpfully as possible.

In the paragraphs below, the survey results for engineering graduates are presented separately from those for nonengineering graduates. You will have to decide whether you will help your readers more by making one table or two (and one bar graph or two). Be prepared to show your table(s) and bar graph(s) to your classmates and explain why you designed each the way you did.

When asked which courses they found most valuable in their careers, 55.21% of the engineering graduates included physics in their list of valuable courses; 53.84% engineering; 21.60% economics; 58.40% English communication; and 72.21% mathematics. Of the nonengineering graduates, 43.67% listed business; 73.68% English communication; 55.59% economics; 25.55% psychology; 25.00% physics; 53.24% mathematics; and 33.08% accounting.

When asked which courses they found least valuable in their careers, 1.99% of the engineering graduates listed miscellaneous humanities; 59.84% foreign language; 12.11% miscellaneous science; 14.74% miscellaneous business; 46.81% history; 20.42% government; 25.10% chemistry; 45.16% engineering; and 23.32% economics. Of the nonengineering graduates, 22.09% indicated miscellaneous humanities; 55.30% miscellaneous business; 52.24% foreign language; 52.45% history; 48.15% miscellaneous science; 18.78% government; 18.04% chemistry; and 17.70% physics.

2. Figure 14–33 presents a table showing how much crude oil was produced by the world's leading producers in the years 1983 through 1987. Using

MINERALS IN THE WORLD ECONOMY

Table 37.—Leading world producers of crude oil[1]

(Million 42-gallon barrels)

Country	1983	1984	1985	1986[p]	1987[e]
U.S.S.R.	4,530	4,500	4,380	4,520	4,590
United States	3,171	3,250	3,274	3,168	[2]3,047
Saudi Arabia[3]	1,657	1,702	1,237	1,841	[2]1,536
China	774	830	874	954	978
Mexico	973	983	960	886	[2]927
United Kingdom	809	885	894	884	[2]858
Iran	891	794	803	686	854
Iraq	401	438	521	617	765
Venezuela	657	[r]660	614	654	[2]664
Canada	495	526	538	537	[2]561
United Arab Emirates (Abu Dhabi, Dubai, Sharjah)	420	391	439	500	[2]548
Indonesia	490	517	484	507	509
Kuwait[3]	385	424	374	519	[2]497
Nigeria	452	508	544	534	[2]487
Libya	[r]405	[r]406	392	389	368
Total	[r]16,509	[r]16,820	16,328	17,196	17,189
Other	[r]2,824	[r]3,020	3,253	3,204	3,286
Grand total	[r]19,333	[r]19,838	19,581	20,400	20,475

[e]Estimated   [p]Preliminary   [r]Revised
[1]Table includes data available through Oct. 31, 1988.
[2]Reported figure.
[3]Includes the country's share of production from the Kuwait-Saudi Arabia Divided Zone.

FIGURE 14–33 Table for Use with Exercise 2

From U.S. Department of the Interior, *Minerals Yearbook, 1987* (Washington, D.C.: U.S. Government Printing Office, 1987), vol. III, 41. Courtesy of the Bureau of Mines.

that information, make the following visual aids. Be sure to provide all appropriate labels and a title for each.

a. A bar graph showing how much crude oil was produced in 1987 by each of the five countries that produced the most in that year.

b. A pictograph showing how much crude oil was produced by Nigeria in each of the five years. Because the sample pictograph in Figure 14–13 uses barrels to represent quantities of oil, use some other symbol for your pictograph.

c. A line graph that shows trends in the crude oil production in the following countries: the United States, the United Kingdom, China, Mexico, and Iraq.

d. A pie chart that shows the proportions of the world's crude oil produced in 1985 by the following countries. For your convenience, the correct percentages are provided below:

Canada	3%
China	4%
Mexico	5%
Saudi Arabia	6%
United Kingdom	5%
United States	17%
U.S.S.R.	22%
All other countries	38%

3. In textbooks, journals, or other publications related to your major, find and photocopy three photographs. For each photograph, answer the following questions:

a. What does the photograph show, who is the intended audience, and how is the reader supposed to use or be affected by the photograph?

b. What angle of view has the photographer chosen and why?

c. What, if anything, has the writer done to eliminate unnecessary detail?

d. Do all the relevant parts of the subject show? If not, what is missing?

e. Has the writer supplied helpful labels? Would any additional labels help? Are any of the labels supplied by the writer unnecessary?

f. Could the writer have achieved his or her purpose more effectively by using a drawing or a diagram instead of a photograph? Why or why not?

4. Create a drawing that you could include in a set of instructions for operating one of the following pieces of equipment. Tell who your readers are and, if it isn't obvious, how they will use the product. Be sure to label the significant parts and include a figure title.

- Typewriter
- An instrument or piece of equipment used in your field
- Power lawnmower
- Clock radio
- The subject of a set of instructions you are preparing for your writing class

- Some other object that has at least a half-dozen parts that are important to show in a set of instructions

5. Draw a diagram that you might use in a report, proposal, instruction manual, or other communication that you would write on the job. The diagram might be an abstract representation of some object, design, process, or other subject. You can create a diagram that requires special symbols, but you do not have to. Tell when you might include the diagram in something you would write on the job. Also, tell what your readers would use the diagram for. Be sure to provide all appropriate labels and a figure title.

6. Create a flow chart on one of the following processes and procedures. Tell when you might use the flow chart and why. Be sure to provide appropriate labels and a figure title. Do not show more than sixteen steps. If your process has more, show the major steps, not all the substeps.

- Applying for admission to your college
- Changing the spark plugs in a car
- Making paper in a paper mill. Begin with the trees in the forest
- Preparing to make an oral presentation in class or on the job. Your presentation should include visual aids. Start at the point where you decide that you want to speak or are given the assignment to do so
- A process or procedure that you will describe in a communication you are preparing for your writing class
- Some process or procedure that is commonly used or important in your field

7. Create an organizational chart for some organization that has at least three levels. You may choose any type of organization, including a club you belong to or a company that employs you. You may also visit some office, store, or business, asking them to provide you with information about their organizational hierarchy.

8. Make a schedule chart that describes your schedule for working on an assignment you are preparing for your writing class. The chart should cover the period from the date you received the assignment to the date you turn the assignment in. Be sure to include all your major activities, such as planning your work, gathering your information, bringing a draft to class for review, and so on.

9. Create a budget statement from the following data concerning the monthly costs that an electronics company would incur if it opened an additional service center in a new city. Remember to group related expenses and to provide a total.

> Salary for central manager $2450. Rent $600. Business tax (prorated) $50. Electricity (year-round average) $180. Water $45. Receptionist's salary $700. Office supplies $200. Car lease $220. Salary for technician $1800. Salary for technician's assistant $1300. Supplies for technician $300. Travel for monthly trip to main office by center manager $300. Car driving expenses $250. Telephone $130. Depreciation on equipment $250.

15

DESIGNING PAGES

DEFINING
OBJECTIVES
PLANNING
DRAFTING
EVALUATING
REVISING

Y ou build your written messages out of *visual* elements. These visual elements are dark marks printed on a lighter background: words and sentences and paragraphs; drawings and graphs and tables. They are *seen* by readers before they are read and understood.

THE IMPORTANCE OF GOOD PAGE DESIGN

When you type or print your communications, you can arrange the visual elements in many different ways—using wide margins or narrow ones, aligning your text in one column or in two or more columns, placing your prose and illustrations here or there on the page. Your page design—the way you arrange your visual elements—can greatly affect the success of your communications. Good page design helps you achieve your purposes in the following ways:

- **Good page design helps readers read efficiently.** To read a page, you (and your readers) move your eyes from one place to another. Sometimes you move your eyes in a simple, rhythmic pattern, as when reading sequentially through a text—reading one phrase, then the next, and so on. However, you often move your eyes in more complicated patterns, as when scanning for certain facts, when comparing facts presented in different paragraphs, or when looking back and forth between a paragraph and a drawing that illustrates it. Poorly designed pages confuse and frustrate readers by failing to tell them where to focus their eyes next. Well-designed pages guide the readers' eyes around the page so that the readers can read easily and efficiently.

- **Good design emphasizes the most important contents.** By varying such things as the size and placement of the visual elements on the page, a good design establishes a visual hierarchy. With this hierarchy, you can guide your readers' attention first to one element, then another, and you can emphasize the most important material by drawing the readers' eyes there first.

- **Good design encourages readers to feel good about a communication.** Page design affects readers' feelings. You've surely seen pages—perhaps in a textbook or a set of instructions—that struck you as uninviting, even ugly. As a result, you may have been reluctant to read them. And undoubtedly you've seen other pages that you have found to be attractive. These pages you've probably approached more eagerly, more receptively. A good design creates a favorable impression on its readers, making them feel good toward both the communication and its writer.

This chapter describes an easy-to-follow method you can use at work to design pages that will affect your readers in the ways you want.

SINGLE-COLUMN AND MULTICOLUMN DESIGNS

Page designs are classified according to the number of columns they have. You are already quite accustomed to using a single-column design because it is the one used in college term papers. You type from one margin to the other, and you center most other material, such as tables and indented quotations, between the margins. In multicolumn formats, the area between the margins is divided into two or more columns. Multicolumn formats are used in many popular magazines, such as *Time, Sports Illustrated,* and *Mademoiselle,* as well as in many professional periodicals, such as *Textile World, International Journal of Biochemistry,* and the various *Proceedings of the Institute of Electrical and Electronics Engineers.* Figure 15–1 shows multicolumn page designs from a variety of publications.

For much of the writing you do at work, a single-column page design will work quite well. In fact, in most situations it's what your readers and employers will expect. However, in some situations you will be able to write much more effectively by using multicolumn designs. Why? First, people can read shorter lines of type more easily than they can read longer lines. When two or more columns are used on a page, the lines are shortened and reading becomes easier. Second, in multicolumn formats, writers have more flexibility to create varied and interesting designs and thereby to direct their readers' attention to the material they want to emphasize. Third, by using two or more columns, writers can often help their readers see readily the relationships among various pieces of information, as in instruction manuals where the text describing each step and the illustration showing that step can be placed directly opposite each other in adjacent columns.

For these reasons, multicolumn formats are especially useful in instructions, technical bulletins, product brochures, company newsletters, and advertisements. They are also useful in sections of reports and proposals where you can present your information in tabular form.

Because you are already familiar with one-column page design, this chapter concentrates on multicolumn designs.

SCISSORS AND TAPE, COMPUTER PROGRAMS, AND GRAPHIC ARTISTS

You can create these multicolumn page designs in two distinct ways. The traditional method is to prepare your prose on a typewriter and draw your illustrations by hand. Then, using scissors and tape, you place the various elements in the desired locations on your page.

Recently, many computer programs have become available that allow you to type your text into a computer and arrange it in a multicolumn framework that you design right on the computer screen. These programs have the advantage of making it much easier for you to "play" with the page design by trying out alternative arrangements of your material. Also, they produce printed cop-

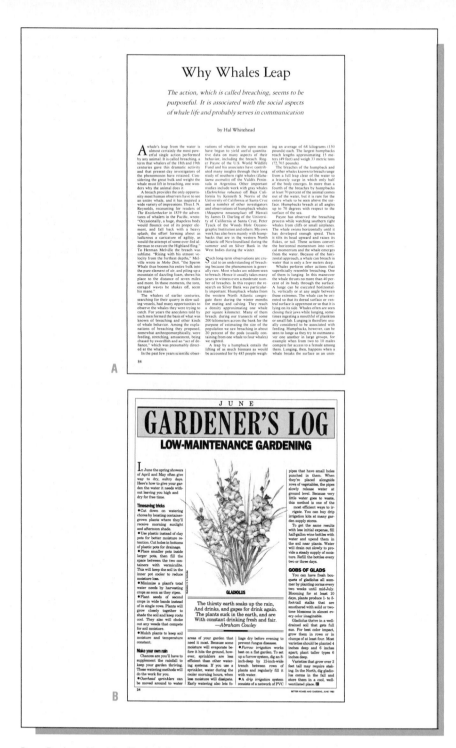

FIGURE 15–1 Page Designs Used by Typical Popular and Professional Periodicals

A, From Hal Whitehead, "Why Whales Leap," *Scientific American* 252 (March 1985): 84. Copyright © 1985 by Scientific American, Inc. All rights reserved. B, From "Low-Maintenance Gardening," *Better Homes and Gardens* 63 (June 1985): 24. Copyright Meredith Corporation, 1985. All rights reserved. C, From John W. Balde, "Status and Prospects of Surface Mounted

FIGURE 15–1 *(continued)*

Technology,'' *Solid State Technology* 29 (June 1986): 99. Reprinted with permission of *Solid State Technology,* published by Technical Publishing, a company of Dun & Bradstreet. D, From Stephen Ohr, ''VME Card Gears VAX for Data Acquisition,'' *Electronic Design* 34 (June 1986): 49.

ies that are neater than most people can readily achieve with scissors and tape. To incorporate your visual aids into your page using some of these programs, you leave blank spaces in your text where the visual aids can later be pasted in. Other programs allow you to incorporate the visuals directly into your computer file, so that the text and visual aids can be printed together. These visual aids might be ones that you have drawn using the computer or they might be ones you have copied electronically into the computer from a piece of paper; the device that does this copying is called a *scanner*. Figure 15–2 (pp. 434–35) shows multicolumn pages from a corporate newsletter and an instruction manual that were prepared using a word-processing program run on a desktop computer. Both were printed using laserprinters.

The guidelines for page design presented in this chapter are equally valid whether you are preparing your pages with a typewriter, scissors and tape, or whether you are preparing them with a computer. All the guidelines are fashioned to be usable by people with no training in art or graphic design.

It is possible that on the job you will receive page-design help from a professional graphic designer. This is especially likely if you write reports to stockholders, instructions to be packed with your company's products, or other communications through which your employer wants to make the best possible impression. Even when you have artistic help, however, you will benefit from knowing the guidelines for page design described below. Knowledge of them will help you work smoothly, productively, and knowledgeably with the graphic designer.

THE VISUAL ELEMENTS OF A PAGE

This chapter's first two guidelines describe a two-step strategy that graphic designers use to design pages. To follow these steps, you need to adopt the graphic designers' way of looking at a page. According to this view, every pages has four kinds of material:

- Text. Your sentences and paragraphs.
- Headings and Titles.
- Visual Aids. Tables, graphs, drawings, and so on.
- White Space. The blank areas of the page that contain no printed information.

When looking at a page, graphic designers see each of these elements in the abstract—as one of the building blocks for a page—not as particular words or illustrations. You need to do the same thing to follow the two-step strategy for designing pages that is explained in the first two guidelines.

CREATE A GRID TO SERVE AS THE VISUAL FRAMEWORK FOR YOUR PAGES

As you design your pages, your aim is to create simple, clear, consistent, and meaningful relationships among the prose, visual aids, and headings and titles. The first step in doing this is to decide which areas on your page will be occupied by your visual materials and which will be devoted to the white space that defines the borders of those materials. Graphic designers do this by drawing a *grid* of vertical and horizontal lines. The designers place their visual materials within the rectangles formed by these lines. The areas around the rectangles are left to white space.

A Grid You Know Well

Actually, the process of creating a grid is already familiar to you. You use a grid every time you type a term paper. When you establish your four margins (top, bottom, left, right), you create a *communication area* to be occupied by your prose and other visual materials, as shown in Figure 15–3 (p. 436). The margins are devoted to white space.

You can create more complex grids for your term papers by adding other grid lines. For example, if you wish to include indented lists or indented blocks of quotations, you establish additional grid lines for the left-hand and right-hand borders of that material. Similarly, if you indent the first line of each paragraph, you are in effect creating another grid line that defines the left-hand border of the first lines of all your paragraphs. Figure 15–4 (p. 437) shows the grid lines of a typical term paper.

How to Create a Multicolumn Grid

To create a multicolumn grid, you follow a process similar to the one you use to create the single-column grid for a term paper. First, establish your margins to define the communication area of your page. Next, divide this communication area by means of narrow, vertical *gutters* of white space. To be effective, gutters must be at least as wide as three letters of your text. Figure 15–5 (p. 438) shows pages with gutters for two-column and three-column grids.

USE THE GRID TO COORDINATE RELATED VISUAL ELEMENTS

Once you have created a basic grid, you can provide a clear, well-organized relationship among your prose, visual aids, and headings by aligning their edges against the appropriate grid lines. To establish a visual connection among parallel or related blocks of material, you simply align them with the same grid line—just as you align all the paragraphs in a term paper with the grid lines of

CBISness

Volume 8, Number 6
June 1990

CBIS supports wristwatch message service

CBIS will provide the "entire software infrastructure" for a new wristwatch message service that will be launched in July.

The new service, called RE-CEPTOR, uses specially made Seiko timepieces called Message-Watches to display numeric or word messages on a wrist-worn timepiece. RECEPTOR will be marketed to consumers as well as business people.

CBIS will be working with AT&E Corp., San Francisco, which developed the RECEPTOR concept. At an estimated cost of $12.50 for monthly service and $200-300 for a watch, the price is less than regular paging service, said Steve Lecoy, CBIS account executive.

Subscribers get much more with RECEPTOR than they receive with an ordinary pager, he said. Extras include:
• "Roaming," or the ability to receive messages while traveling from one service area to

To send a message:
A caller dials the Message-Watch telephone number. The call is carried over telephone lines to a message control center, which offers the caller a variety of short messages and the option of relaying a phone number. The message is sent to FM radio stations.

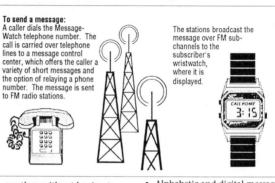

The stations broadcast the message over FM sub-channels to the subscriber's wristwatch, where it is displayed.

another, without having to tell another area's paging provider to activate service
• A watch whose accuracy is checked and adjusted 48 times daily through contact with the National Bureau of Standards' atomic clock in Colorado

• Alphabetic and digital message functions
• Battery life of a year, instead of a month for ordinary pagers
• A selection of MessageWatch styles and colors.

(Continued on page 9)

CBIS launches banking product

CBIS has introduced Docu-Banc, a document processing system that enables banks to easily vary the formats and messages on customer statements.

DocuBanc is the second major product in CBIS's strategy to provide comprehensive document and image management solutions to the financial industry. In April, 1989, CBIS introduced ImageBanc, which prints small images of canceled checks on a single sheet of paper and in so doing creates checkless statements.

(Continued on page 7)

CBIS will supply Kelly Services with imaging technology

CBIS will supply Kelly Services Inc. with software that will enable Kelly to print images of time cards and other documents on invoices it sends to its 180,000 customers.

The contract with Kelly calls for CBIS's Composed Applications Group to be the systems integrator and software supplier for Kelly Services' imaging technology. The Kelly project represents a major application of CBIS's image capture and processing software outside the financial industry. Our software is now used for image-based check sorting and checking account statement preparation at two of the nation's top 10 banks.

The Kelly system, called an

Image Statement Rendering system (ISR), will scan documents required by customers and store their images on optical disks. In addition to printing invoices rapidly, the system will support on-line inquiry, display, and slow-speed printing. The system will interface fully

(Continued on page 4)

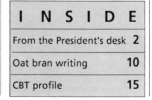

I N S I D E	
From the President's desk	**2**
Oat bran writing	**10**
CBT profile	**15**

FIGURE 15–2   Two Multicolumn Pages Created with Desktop Computers

Adding A New Manufacturer
To The Inventory

Locate the bottom of the list of all manufacturers

1. Select *Show All Records* in the Organize menu.

2. Drag the box in the up-down scroll bar down to the bottom arrow.
   The screen will quickly move to the bottom where you will see blank lines.

   Do not be concerned right now about entering the new manufacturer in alphabetical order. The computer will put the manufacturer in its correct place later.

Enter the information for the first item

1. Click the *Co* cell in the first blank line at the end of the inventory list.

2. Type the manufacturer code for the new item.
   You will see the code at the top of the screen where the cursor is flashing as shown in Figure 1-28 below.

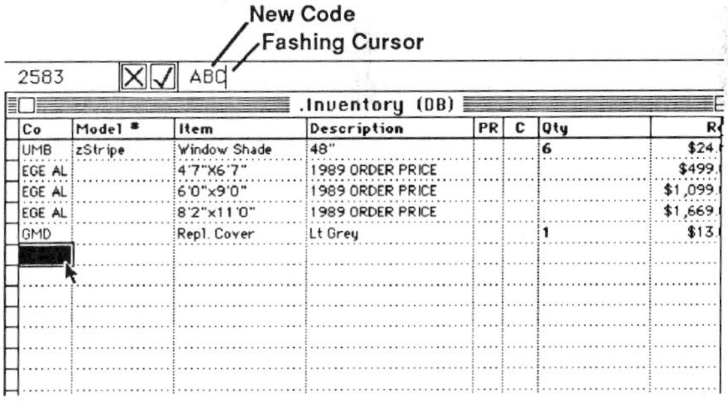

New Code
Fashing Cursor

| 2583 | ⊠✓ ABC |

Co	Model ▪	Item	Description	PR	C	Qty	R
UMB	zStripe	Window Shade	48"			6	$24
EGE AL		4'7"X6'7"	1989 ORDER PRICE				$499
EGE AL		6'0"x9'0"	1989 ORDER PRICE				$1,099
EGE AL		8'2"x11'0"	1989 ORDER PRICE				$1,669
GMD		Repl. Cover	Lt Grey			1	$13

.Inventory (DB)

Figure 1-28 New Code at Top of Screen

Adjusting The Inventory   1-33

FIGURE 15–2   *(continued)*

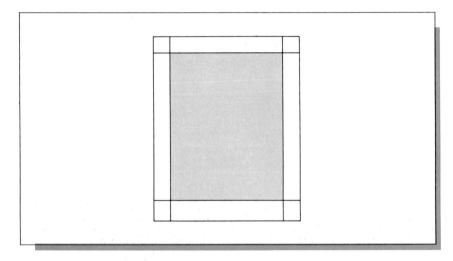

FIGURE 15–3   Communication Area of a Page

the left and right margins, and just as you align all the indented quotations with the special grid lines you establish for them.

When you use a multicolumn grid, you have much more flexibility for arranging your visual material than you do with a single-column grid. The following paragraphs provide some suggestions about ways you can use that flexibility while still maintaining a clear, consistent page design.

How to Align Text and Visual Aids

In multicolumn designs, you can keep your text and visual aids within a single column or you can extend some elements across one or more additional columns. Notice that in the example in the lower right-hand corner of Figure 15–6, the visual elements that cross a gutter go all the way across the next column. In general, when you extend your prose or a visual aid across grid columns, it should occupy the full width of each column.

How to Align Titles and Headings

You can use your grid pattern to establish a visual relationship between your headings and titles and the other elements on your page. To do this, place the headings and titles centered, flush left, or flush right in the appropriate grid columns. A title or heading is *flush left* if it abuts the left-most grid line of the column or columns it labels; it is *flush right* if it abuts the right-most grid line of that column or columns. Figure 15–7 (p. 439) shows headings centered, flush left, and flush right in a single column.

Figure 15–8 shows two page designs in which major titles or headings span more than one column. To see other examples of the placement of head-

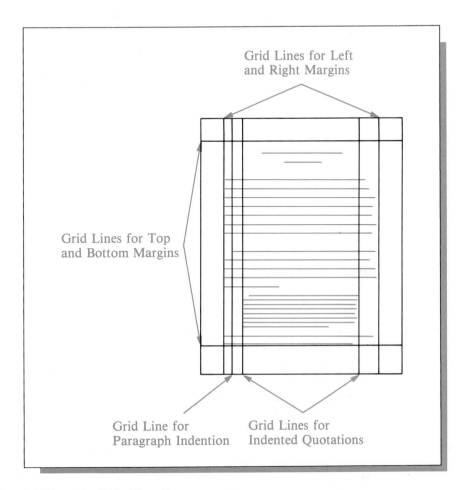

FIGURE 15–4   Grid Lines of a Typical Term Paper

ings and titles, look at the other sample page designs in the figures and exercises of this chapter. Notice that you can use larger sizes of type and special types of lettering for headings and titles, and yet stay within the basic grid framework that holds your pages together visually.

How to Align Related Material in Adjacent Columns

In some communications, you may want to put related material in adjacent columns. For instance, in a troubleshooting guide, you might want to follow the conventional format of devoting the left-hand column to descriptions of problems, the middle column to explanations of the causes of the problems, and the right-hand column to instructions for solving the problems. Similarly, in an instruction manual, you might dedicate the left-hand column to text

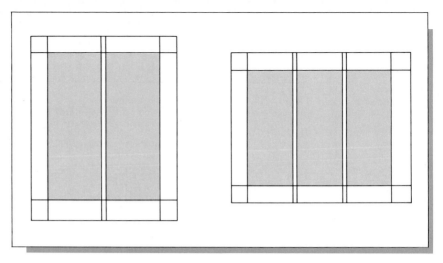

FIGURE 15–5   Gutters Used to Divide the Communication Area into Two and Three Columns

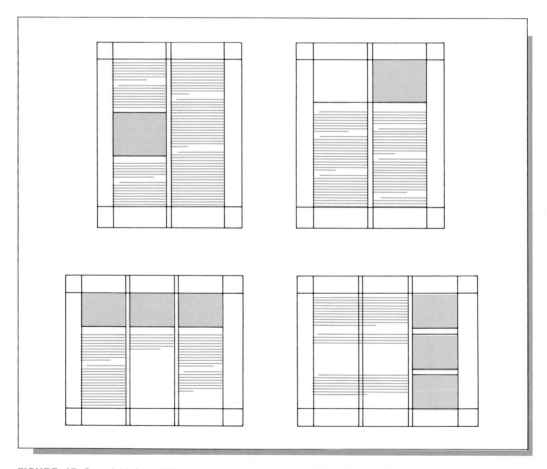

FIGURE 15–6   A Variety of Placements of Text and Visual Aids within Multicolumn Formats

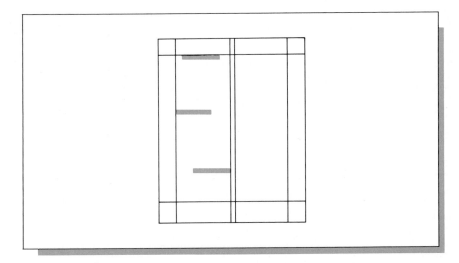

FIGURE 15–7   Alternative Placements for Headings and Titles within a Column

describing each step and the right-hand column to the accompanying illustrations of each step.

In those situations, you want to help your readers quickly match the related material in the adjacent columns—Problem A with its cause and solution, Instruction 17 with its illustration. To do that, you can establish *horizontal gutters*.

Figure 15–9 (page 440) shows two strategies for using horizontal gutters. In each case, the corresponding materials in adjacent columns use the same top

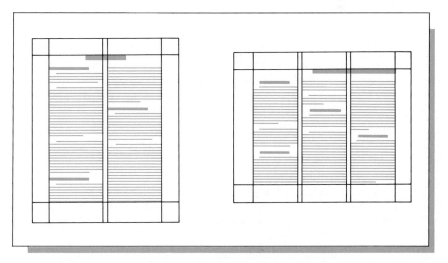

FIGURE 15–8   Some Alternative Placements for Headings and Titles That Span Columns

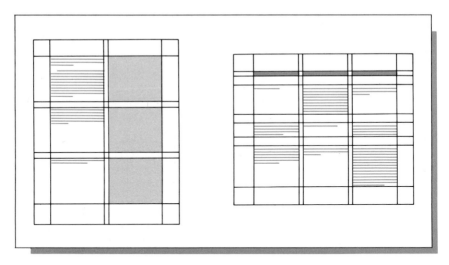

FIGURE 15–9 Two Ways of Coordinating Information by Means of Horizontal Gutters

grid line. You can determine the placement of the grid lines in either of two ways. You may place them at regular intervals, as in the left-hand example in Figure 15–9. Or, you may place them at the end of the longest entry in each cross-column set of associated entries; such placement is shown in the right-hand example in Figure 15–9.

As the examples show, using horizontal grid lines to align associated materials in adjacent columns can leave considerable white space on the page. However, this white space is not wasted space. It helps readers see which text goes with which illustrations. Compare the two page designs shown in Figure 15–10. In their left-hand columns, both pages contain instructions for three steps in a procedure; in their right-hand columns, both contain the three illustrations that accompany those steps. The difference between the two pages is that one uses horizontal gutters to coordinate each step with its accompanying illustration, whereas the other does not. Although the uncoordinated design has the apparent advantage of providing room at the bottom of the left-hand column to print additional steps, it is more difficult for readers to use. That page would be even *more* difficult if the steps added to the bottom of the left-hand column had accompanying illustrations that readers would have to turn the page to see.

It is important for you to note that the use of horizontal gutters applies mainly to communications where each block of text has an associated illustration. In communications where you will include illustrations for some pieces of text but not for others, you may need to consider page designs where horizontal gutters are *not* used to align corresponding text and illustrations. Two examples of such a design are shown in Figure 15–11. The design on the right uses a two-column grid like that in Figure 15–10. Notice that although the

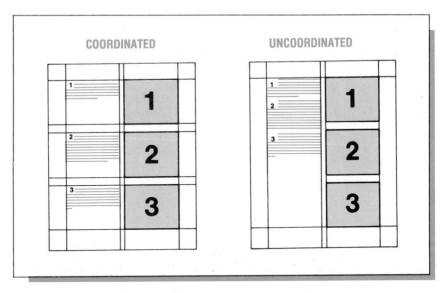

FIGURE 15–10   Comparison of a Page That Coordinates Text and Illustrations by Use of Horizontal Gutters with a Page That Does Not

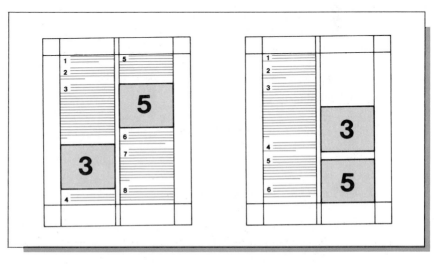

FIGURE 15–11   Two Designs for a Page Where Some Blocks of Prose Do Not Have Associated Illustrations

illustrations are placed across from their associated steps, the associated steps and illustrations do not necessarily use the same top grid line.

The alternative design shown on the left mixes text and illustrations in both columns, placing each illustration immediately after the text that concerns it.

Computer programs for page design make it easy to create grids and then align visual elements with the grid lines. Figure 15–12 shows the computer screen for one such program.

A Note on Creativity

As you can see from looking at the various figures in this chapter, using grids allows you considerable room to creatively design pages that will help you achieve your communication objectives. Experienced graphic designers exercise even greater freedom by occasionally ignoring grid lines with elements they desire to emphasize, while keeping the rest of their design within their

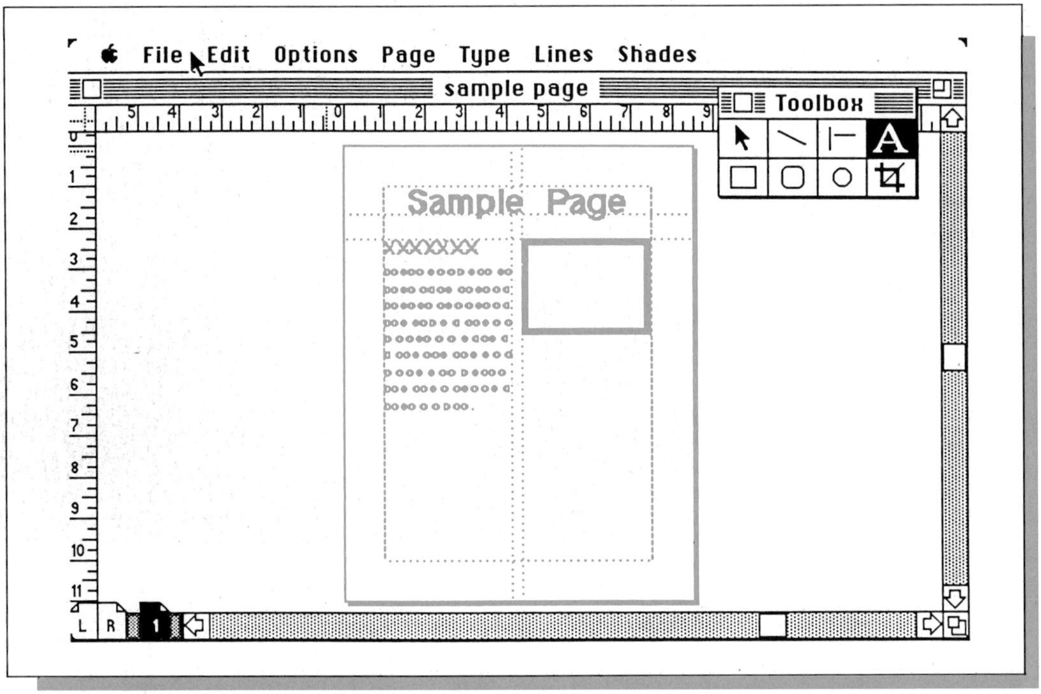

FIGURE 15–12   Computer Program Used for Page Design (Note the alignment of visual elements with the grid lines.)
Used with permission of Apple Computer and Aldus Corporation.

grid framework. They usually do this when making elaborate designs, such as advertisements. Of course, you can experiment in this way too, although for any of the things you are likely to write at work, you will have a perfectly effective design if you rely entirely upon the grid framework.

GUIDELINE
3

USE THE SAME DESIGN FOR ALL PAGES THAT CONTAIN THE SAME TYPES OF INFORMATION

The two guidelines you have just read explain a two-step procedure you can use to design a single page effectively. This concerns multipage communications and suggests that you establish visual consistency among the various pages by using the same page design for all pages that contain the same type of information.

In the visual design of a communication, consistency is both aesthetically pleasing and functional. Pages built upon the same grid pattern make an attractive group because of the visual harmony among them. Further, pages using the same grids help readers read efficiently, a point illustrated in Figure 15–13, which shows two pages from an instruction manual for servicing motorcycles. The student who wrote the manual used this same layout for a sequence of twenty pages. In this layout, readers can find the name of the test in the same places on every page and its purpose in the same places on every page; similarly, readers can find the expected result of the diagnostic activity and also the way to correct problems. Therefore, when turning to a new page readers can readily find the information they want without searching for it: they can simply direct their eyes to the appropriate place on the page.

Some documents contain many different kinds of information. For them, it is not reasonable to use the same page design for *every* page. For example, the motorcycle manual just mentioned has an introduction and also a page of specifications, each needing its own page design. However, even when you use more than one page design in a single communication, you can provide visual unity to your writing by using *related* designs. For example, if you design a major section of your communication so that the long side of the paper (the 11-inch side) is at the top, use the same page orientation throughout. Otherwise, your readers will need to turn the book from side to side as they proceed from section to section. Similarly, if you use a two-column page design in one major section, use a two-column design in all sections.

A special instance of related pages arises in long documents (such as reports, proposals, and instruction manuals) that contain more than one major section. In these documents, writers often treat each section as a new chapter; that is, they begin each section at the top of a new page and they put chapter titles on these pages. Consequently, all first pages of chapters look different from the pages that don't begin chapters—but all the first pages look like one another. See Figure 15–14 for some examples.

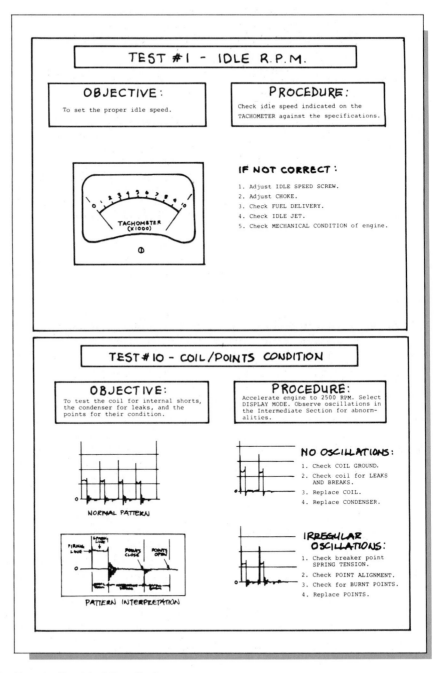

FIGURE 15–13 Use of a Consistent Page Design

Chapter 3
How CHARTER Works

Introduction

The purpose of this chapter is to describe the three activities you complete to produce a graph:

1. Entering the **data** to be graphed
2. Specifying the **format** of the graph
3. **Displaying** the graph on your workstation screen or **printing** it on the 5577 High-Density Matrix Printer

In this chapter you will also learn several ways to use CHARTER in your office.

What is CHARTER?

CHARTER is a powerful software tool that allows you to display and print black and white business graphs **electronically** rather than manually. With electronic drawing, you can **quickly** see what type of graph (line, bar, pie, word) is most effective for a presentation or report because you can **easily** produce the same information in a variety of graph forms without time-consuming hand drawing (much as you can reformat a word processing document without extensive retyping).

A CHARTER Graph

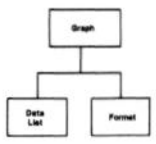

A CHARTER graph consists of two components: a **data list** and a **format.** This is similar to the way a word processing document consists of text and format lines.

The data list is a group of tables that you reference to plot (draw) the data points for any chart in the graph. The format is defined by values that specify the appearance of elements in the chart, for example, chart type, title, border, size, position on the page, type of line, and labels. You specify the format values in fields in the format menus. When you tell CHARTER to draw a graph, the system applies the format values to the data.

How CHARTER Works 3-1

Chapter 6
Advanced Operations

Introduction

The purpose of this chapter is to teach you advanced Wang CHARTER operations. These operations include:

- Producing a line chart, a bar chart, and a pie chart using the **same** data set
- Adding labels, legends, and annotations to charts
- Positioning multiple charts on one page
- Adding sophisticated attributes to charts
- Converting a table in a word processing document into a data set
- Tailoring the format menus so that some fields do not appear on the screen
- Storing copies of your graphs

Remember

The HELP key displays information about a specific menu or field.

The EXIT key lets you exit quickly from any format menu. When you return to the graph, the system displays the menu from which you exited.

Press CANCEL to move back through the menus.

Use GO TO to move from chart to chart, axis to axis, plot to plot, line to line, or bar set to bar set.

To go directly to the X- or Y-axis of a chart, press GO TO (from the X-Y menu, X-Axis menu, or Y-Axis menu), enter the chart number, and type "x" or "y".

Press ERASE or COMMAND and DELETE to remove all characters from the current cursor position.

You do not have to press EXEC after entering data in a menu.

You can display an index quickly by pressing the INDEX key from any Graph, Library, or Volume field.

Advanced Operations 6-1

FIGURE 15–14 Consistent Use of a Single Design for All Pages That Begin Chapters

From Wang Laboratories, *Instruction Manual for Charter Graphics Program* (Lowell, Mass.: Wang Laboratories). Used with permission.

<table>
<tr><td>GUIDELINE
4</td><td>**USE A LIMITED VARIETY OF EASY-TO-READ TYPEFACES**</td></tr>
</table>

If you are preparing your communication on a computer, you may be able to choose among a variety of typefaces and type sizes for your text, titles, and headings. The choices you make can affect greatly your communication's visual impact and readability.

Use Readable Typefaces

One of your most basic choices will be the typeface you use. There are two major kinds: serif and sans serif. The differences between them can be illustrated by comparing the ways they make the letter *A:*

As you can see, the left-hand A has small strokes that extend from the ends of the letter. These strokes are called *serifs* and give the serif typefaces their name. Also, in serif typefaces the major lines vary in thickness; notice that the right leg of the A is thicker than the left leg.

In contrast, the A on the right side has no serifs and its lines are of uniform thickness. Typefaces in this style are called *sans serif* typefaces (*sans* is French for *without*).

There are many varieties of both serif and sans serif typefaces. Here are a few examples:

Serif

This is the Bodoni typeface.

This is the Century Schoolbook typeface.

This is the Palatino typeface.

This is the Times Roman typeface.

Sans Serif

This is the Avante Garde typeface.

This is the Futura typeface.

This is the Gill Sans typeface.

This is the Helvetica typeface.

Some research indicates that people generally prefer reading serif typefaces and that they usually can read them more quickly.[1] This appears to be because the serifs and the differences in line thickness create greater distinctions between similar letters (*a* and *o, n* and *h*) than are found in sans serif typeface. Almost all textbooks, newspapers, and popular magazines use serif typefaces for their prose.

Nevertheless, sans serif typefaces do have their uses. For example, they can work very effectively for headings and labels in figures. In advertising, where prose passages are usually short, they can also be used effectively. But for letters, memos, reports, proposals, and most other communications you write in your career, you can make your writing easier to read by using a serif typeface.

When choosing the type you will use, you should also consider its size. Typesize is measured in units called points. There are 72 points to an inch.

This is 8-point type.

This is 10-point type.

This is 12-point type.

This is 14-point type.

This is 18-point type.

This is 24-point type.

Research has shown that letters 8 points or 12 points high are the easiest to read in prose passages.[2] In most communications written at work, type smaller than 10 point is avoided, and type larger than 12 points is reserved for headings, titles, and the like.

Research has also shown that you should avoid using all capital letters for passages that are longer than a sentence because they are difficult to read in long stretches.[3] On the other hand, for headings, brief warnings, and similar uses, all capital letters can be very effective.

Design Headings to Be Prominent But Not Overwhelming

When designing headings, remember that they should stand out like guideposts but should not dominate a page or draw the readers' eyes away from the prose. You can make your headings stand out by putting them in boldface type and by using a larger typeface for them. However, to prevent your headings from overwhelming the page, keep enlarged headings within two or two and one-half times the size of the type used for the text. For example, if you are using three levels of headings, you might use 12 point type for the lowest level, 14 for the middle level and 18 for the highest level, or 12 for the lowest, 18 for the middle and 24 for the highest. You will need to experiment with the particular pages you are designing to select the best combinations. (Of course, you can use larger type for chapter or section titles or for covers of bound documents.)

Limit the Number of Typefaces You Use

Some computer programs will enable you to use half-a-dozen or more typefaces in a single document. Using more than one typeface can help you create an attractive, easy-to-use communication. However, you should avoid using so many typefaces that you create cluttered, confusing pages. Generally, it's best to stick to two typefaces, one for prose and one for headings. The headings will stand out even more if you use a sans serif typeface for them (presuming that you are using a serif typeface for the prose).

PRACTICAL PROCEDURES FOR DESIGNING PAGES

You do not need any special technical skills to follow the four guidelines for designing your pages. Here are some practical procedures that will help you design pages by using resources available to almost everyone: a pen, a typewriter, press-on letters, scissors, clear tape, and a photocopy machine. The first four steps apply also to people using word processors.

1. **Determine the amounts of text and illustration you will include.** If you are coordinating specific blocks of text or visual aids across adjacent columns by using horizontal gutters, find the longest passage of text that will accompany one illustration. Use that passage to determine the amount of text space you will provide for the prose that accompanies the other illustrations.

2. **Draw thumbnail sketches of a variety of grid patterns.** Figure 15–15 shows nine thumbnail sketches exploring ways of designing a single page of a particular communication. Sketches like these can help you quickly assess the various grid patterns that you might use. Such assessment is important because there is no "correct" or "best" design for all pages.

   Even when making your thumbnail sketches, keep in mind the purpose and readers of your communication, so that you will know what material to emphasize and what relationships to make clear. Think, too, about how much detail your illustrations will include so that you can plan to make your illustrations large enough to be legible. If you are planning to use coordinated columns, keep in mind the size of your illustrations and the

FIGURE 15–15 Nine Thumbnail Page Designs for the Same Material
Courtesy of Professor Joseph L. Cox III.

lengths of your text entries so that you can determine whether or not to place your horizontal grid lines at fixed intervals.

3. **Pick the best thumbnail sketch for your purpose.** Make this judgment by thinking how your readers will react, moment-by-moment, as they read a communication presented with each design.

4. **Make a full-size mock-up page.** A mock-up is a full-scale page that shows what one typical page will look like, though it may not necessarily contain the actual words and visual aids that will be used in the final communication. In fact, the words can be nonsense, or you can represent the lines of type by straight lines of the appropriate length. Your objective is simply to fill in your area for type with about the same density of ink that the type will have on your finished page. For headings, you can use wider, darker lines that indicate the amount of black that the headings will contribute to the final page. Similarly, instead of using actual visual aids, you can fill the spaces for the visual aids with sketches that will have about the same amount of drawing that the final visuals will have. If you are using scissors and tape, you can even fill in the areas for visual aids with colored paper.

Once you have created your full-size mock-up, you will be able to see abstractly how the four elements of your page (text, headings and titles, visual aids, white space), work together. You will then be able to determine whether, for example, the headings need to be larger to stand out properly or whether the visual aids need to be smaller in order to prevent them from dominating the page.

If your full-size mock-up looks attractive and easy for your readers to use, then proceed to the next step. If not, try a grid represented by one of your other thumbnail sketches.

Once you have created a successful mock-up, your procedure will be much different if you are working on a computer than if you are employing scissors and tape. On a computer, you need to begin to enter your actual text and visual aids by following the instructions for the particular program you are using. With scissors and tape, you can follow the steps listed below.

5. **On a blank sheet of paper, carefully draw the grid in dark pen.**

6. **Tape the grid down on a desk or table.**

7. **Type the text of your communication.** Type the text columns exactly as wide as the grid columns.

8. **Cut the typed text.** Cut the columns of text at the appropriate places so that the various text segments will fit into their grid areas.

9. **Tape a blank sheet of paper squarely over the grid.**

10. **Tape the text segments neatly on the blank page.** Use as your guide the grid that shows through from underneath.

11. **Create your illustrations to fit the grid areas.** If you are creating your own drawings, you can draw them to the needed size. If it would be difficult for you to draw that small, draw the illustrations larger and then use a photocopy machine to reduce them to the appropriate size. Often, you can make your page appear crisper if you draw a box around your illustrations; the lines of the box would align directly on the appropriate grid lines. See Figure 15–16 for comparisons of pages with boxed and unboxed figures.

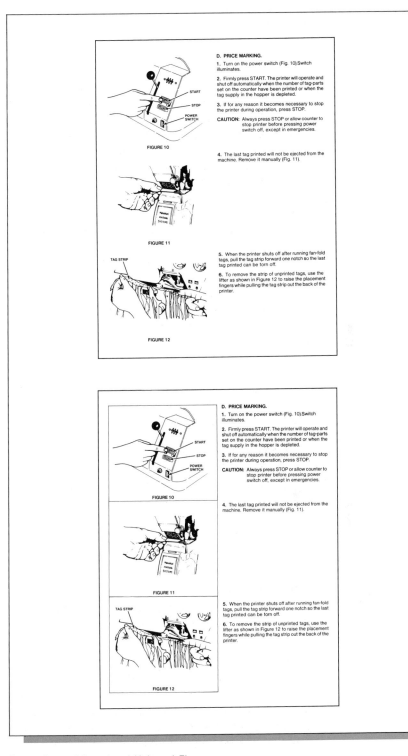

FIGURE 15–16 Comparison of Boxed and Unboxed Figures

From Monarch Marking, *Operating Instructions for Model 100 Dial-A-Price Printer* (Dayton, Ohio: Monarch Marking Systems, Inc.).

12. Cut out and tape down your illustrations as you did your text segments.

13. Use press-on letters to make your headings and titles. Unless you have had practice at using press-on letters, you may want to make the headings and titles on a separate sheet of paper, cut them out, and tape them onto your page in the way that you did with the text and illustrations. Another method of creating headings and titles involves using Kroy lettering machines or other devices that enable you to print strips of letters in large-sized type. A third method of making larger headings is to type the headings on a typewriter, then enlarge the headings as you photocopy them.

14. Photocopy your taped-up pages. If you use a good-quality photocopy machine, the finished product will have much the same appearance as a page printed on an offset press. Sometimes copy machines will record black lines at the edges of taped-down materials. If that happens, use a white masking fluid (available at most copy centers) to cover the lines on the copy and then make a new copy from the touched-up version.

CONCLUSION

Every page you write has some sort of page design. This chapter has explained how you can look at your pages in the way that graphic designers do, thinking about each visual element in terms of the way it affects your readers. By following this chapter's advice, you will be able to create page designs that help your readers read efficiently, emphasize the important contents of your communication, and create a favorable impression.

EXERCISES

1. Describe the page design used in each of the sample pages shown in Figure 15–1. Identify the number of columns; the placement of prose, headings, and visual aids; and other important features of the design.

2. Figure 15–17 shows four package inserts for a prescription medication. All contain the same information, but each was prepared by a different graphic designer. Describe the page designs used in each. Guess at the purpose each designer had in mind. Which design do you think works best? Worst? Why?

3. Look again at Figure 15–14. It contains nine thumbnail sketches, all for the same material. Describe each design in terms of the number of columns, the placement of the title and headings, and the location of the illustration. For each, state what element of the page (title, illustration, headings, and so on) receives primary emphasis.

HYGROTON®
chlorthalidone usp
50 mg. Tablets, 100 mg. Tablets   Oral Antihypertensive-Diuretic

DESCRIPTION
HYGROTON (chlorthalidone) is a monosulfamyl diuretic which differs chemically from thiazide diuretics in that a double-ring system is incorporated in its structure. It is 2-Chlor-5-(1-hydroxy-3-oxo-1-isoindolinyl) benzenesulfonamide. with the following structural formula

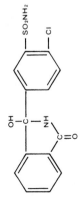

ACTIONS
HYGROTON is an oral diuretic with prolonged action (48-72 hours) and low toxicity. The diuretic effect of the drug occurs within two hours of an oral dose and continues for up to 72 hours.

INDICATIONS
Diuretics such as HYGROTON are indicated in the management of hypertension either as the sole therapeutic agent or to enhance the effect of other antihypertensive drugs in the more severe forms of hypertension and in the control of hypertension of pregnancy.

CONTRAINDICATIONS
Anuria. Hypersensitivity to chlorthalidone. The routine use of diuretics in an otherwise healthy pregnant woman with or without mild edema is contraindicated and possibly hazardous.

WARNINGS
Should be used with caution in severe renal disease. In patients with renal disease, chlorthalidone or related drugs may precipitate azotemia. Cumulative effects of the drug may develop in patients with impaired renal function.

USAGE IN PREGNANCY: Reproduction studies in various animal species at multiples of the human dose showed no significant level of teratogenicity; no fetal or congenital abnormalities were observed.

NURSING MOTHERS. Thiazides cross the placental barrier and appear in cord blood and breast milk.

PRECAUTIONS
Periodic determination of serum electrolytes to detect possible electrolyte imbalance should be performed at appropriate intervals.

Chlorthalidone and related drugs may decrease serum PBI levels without signs of thyroid disturbance

ADVERSE REACTIONS
Gastrointestinal System Reactions:

anorexia	constipation
gastric irritation	jaundice
nausea	(intrahepatic
vomiting	cholestatic
cramping	jaundice)
diarrhea	pancreatitis

Central Nervous System Reactions:

dizziness	headache
vertigo	xanthopsia
paresthesias	

Hematologic Reactions:

leukopenia	thrombocytopenia
agranulocytosis	aplastic anemia

Other Adverse Reactions:

hyperglycemia	muscle spasm
glycosuria	weakness
hyperuricemia	restlessness
impotence	

Whenever adverse reactions are moderate or severe, chlorthalidone dosage should be reduced or therapy withdrawn.

DOSAGE AND ADMINISTRATION
Therapy should be individualized according to patient response. This therapy should be titrated to gain maximal therapeutic response as well as the minimal dose possible to maintain that therapeutic response.

Initiation: Preferably, therapy should be initiated with 50 mg or 100 mg daily. Due to the long action of the drug, therapy may also be initiated in most cases with a dose of 100 mg on alternate days or three times weekly (Monday, Wednesday, Friday). Some patients may require 150 or 200 mg. at these intervals.

Maintenance: Maintenance doses may often be lower than initial doses and should be adjusted according to the individual patient. Effectiveness is well sustained during continued use.

OVERDOSAGE
Symptoms of overdosage include nausea, weakness, dizziness and disturbances of electrolyte balance.

HOW SUPPLIED
HYGROTON (chlorthalidone). White, single-scored tablets of 100 mg. and aqua tablets of 50 mg. in bottles of 100 and 1000; single-dose blister packs, boxes of 500; Paks of 28 tablets, boxes of 6.

CAUTION: Federal law prohibits dispensing without prescription.

ANIMAL PHARMACOLOGY
Biochemical studies in animals have suggested reasons for the prolonged effect of chlorthalidone. Absorption from the gastrointestinal tract is slow, due to its low solubility. After passage to the liver, some of the drug enters the general circulation, while some is excreted in the bile, to be reabsorbed later.

USV PHARMACEUTICAL MFG. CORP.
Manati, P.R. 00701

FIGURE 15-17   Package Inserts for a Prescription Medication
Courtesy of USV Pharmaceuticals

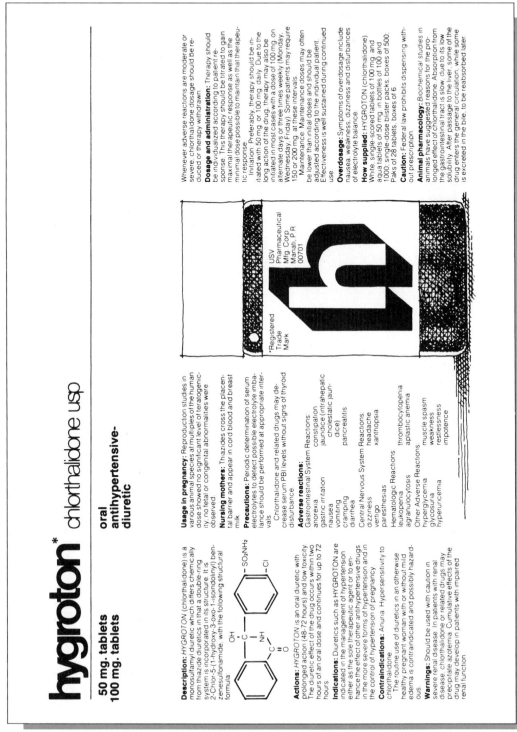

FIGURE 15–17 *(continued)*

B

Hygroton®
chlorthalidone USP

50 mg. Tablets 100 mg. Tablets	Oral **Antihypertensive-Diuretic**	
Description	**Hygroton** (chlorthalidone) is a monosulfamyl diuretic which differs chemically from thiazide diuretics in that a double-ring system is incorporated in its structure. It is 2-Chlor-5-(1-hydroxy-3-oxo-1-isoindolinyl) benzenesulfonamide, with the following structural formula	
Actions	**Hygroton** is an oral diuretic with prolonged action (48-72 hours) and low toxicity. The diuretic effect of the drug occurs within two hours of an oral dose and continues for up to 72 hours	
Indications	Diuretics such as **Hygroton** are indicated in the management of hypertension either as the sole therapeutic agent or to enhance the effect of other	antihypertensive drugs in the more severe forms of hypertension and in the control of hypertension of pregnancy
Contraindications	Anuria Hypersensitivity to chlorthalidone.	The routine use of diuretics in an otherwise healthy pregnant woman with or without mild edema is contraindicated and possibly hazardous
Warnings	Should be used with caution in severe renal disease. In patients with renal disease, chlorthalidone or related drugs may precipitate azotemia	Cumulative effects of the drug may develop in patients with impaired renal function
Usage in Pregnancy:	Reproduction studies in various animal species at multiples of the human dose showed no significant level of teratogenicity; no fetal or congenital abnormalities were observed	
Nursing Mothers:	Thiazides cross the placental barrier and appear in cord blood and breast milk	
Precautions	Periodic determination of serum electrolytes to detect possible electrolyte imbalance should be performed at appropriate intervals	Chlorthalidone and related drugs may decrease serum PBI levels without signs of thyroid disturbance

Adverse Reactions				
Gastrointestinal System Reactions:	anorexia gastric irritation nausea	vomiting cramping diarrhea	constipation jaundice (intrahepatic cholestatic jaundice)	pancreatitis
Central Nervous System Reactions:	dizziness vertigo paresthesias	headache xanthopsia		
Hematologic Reactions:	leukopenia agranulocytosis	thrombocytopenia aplastic anemia		
Other Adverse Reactions:	hyperglycemia glycosuria hyperuricemia muscle spasm	weakness restlessness impotence	Whenever adverse reactions are moderate or severe, chlorthalidone dosage should be reduced or therapy withdrawn	

Dosage and Administration	Therapy should be individualized according to patient response. This therapy should be titrated to gain maximal therapeutic response as well as	the minimal dose possible to maintain that therapeutic response
Initiation:	Preferably therapy should be initiated with 50 mg. or 100 mg. daily. Due to the long action of the drug, therapy may also be initiated in most cases with a dose of 100 mg. on alternate days or	three times weekly (Monday, Wednesday, Friday) Some patients may require 150 or 200 mg. at these intervals
Maintenance:	Maintenance doses may often be lower than initial doses and should be adjusted according to the	individual patient. Effectiveness is well sustained during continued use
Overdosage	Symptoms of overdosage include nausea, weakness, dizziness and disturbances of electrolyte balance	
How Supplied	**Hygroton** (chlorthalidone). White, single scored tablets of 100 mg. and aqua tablets of 50 mg. in	bottles of 100 and 1000; single dose blister packs, boxes of 500. Packs of 28 tablets, boxes of 6
Caution:	Federal law prohibits dispensing without prescription	

C

FIGURE 15–17 *(continued)*

Hygroton®

Chlorthalidone USP

50 mg. Tablets
100 mg. Tablets
Oral Antihypertensive—Diuretic

DESCRIPTION

HYGROTON (chlorthalidone) is a monosulfamyl diuretic which differs chemically from thiazide diuretics in that a double-ring system is incorporated in its structure. It is 2-Chlor-5-(1-hydroxy-3-oxo-1-isoindolinyl) benzenesulfonamide, with the following structural formula:

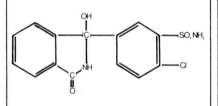

ACTIONS

HYGROTON is an oral diuretic with prolonged action (48-72 hours) and low toxicity. The diuretic effect of the drug occurs within two hours of an oral dose and continues for up to 72 hours.

INDICATIONS

Diuretics such as HYGROTON are indicated in the management of hypertension either as the sole therapeutic agent or to enhance the effect of other antihypertensive drugs in the more severe forms of hypertension and in the control of hypertension of pregnancy.

CONTRAINDICATIONS

Anuria.
Hypersensitivity to chlorthalidone.
The routine use of diuretics in an otherwise healthy pregnant woman with or without mild edema is contraindicated and possibly hazardous.

WARNINGS

Should be used with caution in severe renal disease. In patients with renal disease, chlorthalidone or related drugs may precipitate azotemia. Cumulative effects of the drug may develop in patients with impaired renal function.
Usage in Pregnancy: Reproduction studies in various animal species at multiples of the human dose showed no significant level of teratogenicity; no fetal or congenital abnormalities were observed.
Nursing Mothers: Thiazides cross the placental barrier and appear in cord blood and breast milk.

PRECAUTIONS

Periodic determination of serum electrolytes to detect possible electrolyte imbalance should be performed at appropriate intervals.
Chlorthalidone and related drugs may decrease serum PBI levels without signs of thyroid disturbance.

ADVERSE REACTIONS

Gastrointestinal System Reactions:
　anorexia
　gastric irritation
　nausea
　vomiting
　cramping
　diarrhea
　constipation
　jaundice (intrahepatic cholestatic jaundice)
　pancreatitis
Central Nervous System Reactions:
　dizziness
　vertigo
　paresthesias
　headache
　xanthopsia
Hematologic Reactions:
　leukopenia
　agranulocytosis
　thrombocytopenia
　aplastic anemia
Other Adverse Reactions:
　hyperglycemia
　glycosuria
　hyperuricemia
　muscle spasm
　weakness
　restlessness
　impotence
Whenever adverse reactions are moderate or severe, chlorthalidone dosage should be reduced or therapy withdrawn.

DOSAGE AND ADMINISTRATION

Therapy should be individualized according to patient response. This therapy should be titrated to gain maximal therapeutic response as well as the minimal dose possible to maintain that therapeutic response.
Initiation: Preferably, therapy should be initiated with 50 mg. or 100 mg. daily. Due to the long action of the drug, therapy may also be initiated in most cases with a dose of 100 mg. on alternate days or three times weekly (Monday, Wednesday, Friday). Some patients may require 150 or 200 mg. at these intervals.
Maintenance: Maintenance doses may often be lower than initial doses and should be adjusted according to the individual patient. Effectiveness is well sustained during continued use.

OVERDOSAGE

Symptoms of overdosage include nausea, weakness, dizziness and disturbances of electrolyte balance.

HOW SUPPLIED

HYGROTON (chlorthalidone). White, single-scored tablets of 100 mg. and aqua tablets of 50 mg. in bottles of 100 and 1000; single-dose blister packs, boxes of 500; paks of 28 tablets, boxes of 6.

> **CAUTION: Federal law prohibits dispensing without prescription.**

ANIMAL PHARMACOLOGY

Biochemical studies in animals have suggested reasons for the prolonged effect of chlorthalidone. Absorption from the gastrointestinal tract is slow, due to its low solubility. After passage to the liver, some of the drug enters the general circulation, while some is excreted in the bile, to be reabsorbed later.

**USV PHARMACEUTICAL
MFG. CORP.
Manati, P.R. 00701**

D

FIGURE 15–17 *(continued)*

4. Find three different page designs. (Use no more than one from a popular magazine; look at instructions, insurance policies, leases, company brochures, technical reports, and the like.) If your instructor asks, find all the samples in documents related to your major. Photocopy one page illustrating each design. Then, for each page, describe what you think is the purpose of the document and discuss the specific features of the page design that help or hinder the document from achieving that purpose.

5. This exercise will provide you with practice at evaluating and improving an existing page design.
    a. List the ways in which the design of the page shown in Figure 15–18 could be improved.
    b. Redesign the page by following Steps 1 through 4 of the ''Practical Procedures for Designing Pages.'' Specifically, do the following:
        ● Create three alternative thumbnail sketches for the page.
        ● Create a full-size mock-up of the best design you can create.
    c. Display your mock-up on the wall of your classroom along with the mock-ups prepared by the other students in your class. Decide which mock-ups work most effectively. Discuss the reasons.

6. Find an example of an ineffective page design. If your instructor requests, this might be from a set of instructions or from a document related to your major. Following Steps 1 through 4 of the ''Practical Procedures for Designing Pages,'' draw thumbnail sketches and then make a mock-up of an improved page design for communicating the same material. Write a brief explanation of how you changed the original design and why.

7. Create a multicolumn page design for a course project you are now preparing, or redesign an earlier project using a multicolumn design.

4) Flatten the clay by pounding
it onto the table. Remember,
work the clay thouroughly!!

4)

Slab Roller

The slab roller is a simple but very
efficient mechanical device. A uniform
thickness of clay is guaranteed. The
ease of its operation saves countless
hours of labor over hand rolling techniques.

1) The thickness of the clay is determined by the number of masonite
boards used. There are three ¼" boards and one 1/8" board.
They can be used in any combination to reduce or increase the
thickness of the clay. If
no boards are used, the clay
will be 1" thick. If all   1)
the boards are used, the
clay will be 1/8" thick.
So, for a ½" thickness of
clay, use two ¼" boards.

2) The canvas cloths must be arranged   2)
correctly. They should form a
sort of envelope around the wet
clay. This prevents any clay
from getting on the rollers
and masonite boards.

3) Place the clay flat on the canvas   3)
near the rollers. It may be
necessary to trim some of the
clay since it will spread out as
it passes through the rollers.

4.

FIGURE 15–18   Sample Page for Use with Exercise 5

458

EVALUATING

16

CHECKING

A s you recall from Chapter 1, writing involves five major activities:

- Defining objectives
- Planning
- Drafting
- Evaluating your draft
- Revising

This is the first of three chapters about the fourth activity, evaluating your draft. The chapter begins by discussing the importance and aims of evaluation in the workplace, and then it presents advice for conducting one important evaluation method, checking over your draft yourself. Two other evaluation methods are discussed in Chapters 17 and 18. They are (1) asking others to review your draft and (2) giving your draft to members of your target audience to try out.

AN OVERVIEW OF EVALUATING

When you evaluate, you step back from your draft in order to examine it critically, looking for weaknesses you can correct and for strengths you can make even stronger.

As you first think about evaluating, you might be wondering whether it is really necessary. You may be thinking, "If I have carefully followed the guidelines given in this book for defining objectives, planning, and drafting, won't my communication be fine already? Won't special efforts at evaluating simply waste my time?" Basically, the answer is "No." It's true, of course, that sometimes your evaluation will demonstrate that your communication is just fine as it stands, so it needs no further improvement. However, several kinds of problems can arise in drafts even when you have very conscientiously followed all the advice given elsewhere in this book.

The major reason that problems can arise is that drafting is guesswork. No matter how carefully you have crafted your draft, it really constitutes a guess about how you can achieve the results you desire. You *guess* that by presenting *this* information in *this* manner you will affect your readers in a particular way. It is possible, however, that you have guessed incorrectly, at least in some small (but perhaps crucial) respect. When drafting a report, you may have accidentally left out a piece of information essential to your readers. Or, you may have assumed that the readers are familiar with a situation or concept that they really need to have explained. Or, you unintentionally may have said something that will offend some person or department in your company.

Second, although you may have guessed well about your overall strategy, you may have slipped in carrying it out. Every communicator slips sometimes. The explanation that seemed so plain to us turns out to be murky to our readers. The phrasing we thought would sound conciliatory seems to the readers to be sarcastic.

Evaluation involves a systematic effort to discover such problems *before* you deliver your communication to your readers. It's a form of quality control that helps ensure that your communications will work. Like any quality control procedure, this one will sometimes show that your products are just fine. But at other times, it will help you avoid costly errors. Of course, once you discover a problem in a draft, you will need to find some way of overcoming it. A good evaluation will also help you do that.

In addition, a good evaluation can help you find ways to improve drafts that are already fundamentally sound. Most employers want their employees to write communications that are better than "good enough." They feel that even a basically usable instruction manual should be made more helpful if possible, and even a fundamentally persuasive proposal should be polished enough to win praise from its readers.

Because of the twin goals of avoiding communication failures and of making adequate writing even better, employees spend much time and effort evaluating drafts and making the revisions that evaluation indicates are necessary or desirable.

The Place of Evaluating in the Writing Process

Although evaluating comes after drafting in the list of the five major activities of writing, evaluating is not something that you must wait to do until you have completed a full draft of your communication. Instead, it is an ongoing activity that most writers engage in continuously even as they draft. When writing a long communication, for instance, they may look back over each passage as they finish it, in order to see how well the passage works. When writing a difficult or delicate passage, they may even evaluate each individual sentence as they complete it.

Similarly, evaluating is closely intertwined with the fifth task on the list of writing activities—revising. Although writers sometimes complete their evaluation before making any revisions, they often begin revising as soon as they discover places where improvements can be made.

Generally, however, it's best to postpone your evaluation and revising until you have completed a sizeable chunk of prose (though not necessarily your full communication). This is because drafting and revising involve different mental activities. Drafting is creative: as you develop and elaborate ideas, one idea leads to another. Evaluating is critical: you evaluate by stepping back from your draft to see where it might be improved. If you turn off your drafting efforts after every sentence or paragraph to criticize your results, you can dry up your fountain of ideas and bring on a bad case of writer's block. Consequently, for most people, it's usually best to write long passages, then revise.

How long should the passages be? For a brief communication, a passage might be the entire message, but for a lengthier communication it might be a group of related paragraphs, a section, or a chapter. In general, continue drafting as long as you continue to generate ideas. When your ideas run dry, then revise.

For especially complex, delicate, or important communications, you may need to create many drafts, gradually improving your communication by examining each new draft to discover additional areas requiring improvement. This rhythm of drafting, evaluating, and revising—and then evaluating and revising again—is common on the job.

Evaluating's Relationship to Your Communication Objectives

In addition to being closely intertwined with your work at drafting and revising, evaluating has a close relationship with the objectives you defined when you began work on your communication (and may have revised and refined several times since). Evaluating and your objectives are closely related because when you evaluate something, you measure it against some standard or set of standards. As you evaluate a piece of your writing, one indispensable standard has to be the objectives you set for it. You ask, ''Will my draft do what it is intended to do? Will it enable my readers to perform the tasks they have to perform? Will it affect their attitudes in the way I want it to?''

Thus, the objectives that guide your work at planning and drafting your communications also serve as a standard against which you measure your drafts.

AN INTRODUCTION TO CHECKING

As explained earlier, the rest of this chapter presents advice about checking, which is the evaluation of your drafts that you perform on your own, without asking for help from other people.

The first thing to know about checking is that it is very difficult to find problems in your own writing. Consider Diane's situation. Diane is experiencing one of the most excruciating of all feelings. Ten minutes ago, she mailed an important report to a client. Now, as she puts a photocopy of it in her file, her eye catches an obvious error. How could she have missed it? She read and reread that report before mailing it, looking diligently for errors. And yet she let this one get past her.

Diane's experience is one we have all shared. Of course, we get plenty of practice at checking our own writing. We are expected to do so at both school and work. Still, we let blunders slip by all too often. Why? Some clearly identifiable obstacles hinder our checking. The six guidelines presented below will help you overcome them, so that you can evaluate your writing as effectively as possible.

CHECK FROM YOUR READERS' POINT OF VIEW—AND YOUR EMPLOYER'S

One major obstacle to effective checking is that writers define the purpose of their checking too narrowly. They concentrate solely upon spelling, punctuation, and grammar, forgetting that it is even more important to examine the drafts' effects on their readers and employers.

Take Your Readers' Point of View

The reason for examining your drafts from your readers' point of view has already been stated in this chapter: the ultimate test of your communication is whether it does what you want it to do—namely, affect your particular readers in the specific way you defined when you established your communication objectives.

Consequently, the first step in checking a draft is to refer to your statement of objectives. If you wrote down notes about your objectives either by filling out a Worksheet for Defining Objectives (Figure 3–5) or in some other manner, you should begin your evaluation by pulling out those notes. If you merely noted your objectives mentally, you should go over them now. As you review your objectives, remember to focus on the way you want your communication to affect your readers as they read your message. Consider both the enabling and persuasive elements of purpose. What mental task do you want your communication to help your readers perform? How do you want it to alter their attitudes?

Then, with your objectives fresh in mind, read through your draft, trying to imagine or simulate your readers' attitudes and moment-by-moment responses. If you have written a proposal, for instance, think of the questions and objections your readers will raise. If you have written a procedure, try to follow your own instructions just as your readers would. In every way you can, read in the way your readers will read. By doing so, you will be able to determine whether your communication is likely to achieve its objective of affecting your readers in the way you desire.

Take Your Employer's Point of View

In addition to evaluating from your readers' point of view, you need to evaluate from your employer's. Here, too, you should begin by settling upon appropriate standards for evaluating your particular communication. You will have gathered much of the information about those standards when you followed Chapter 5's advice to investigate the expectations and regulations your employer has that are relevant to your communication. You should now review what you learned. When you do, be sure to consider each of the following concerns:

- **Effect on the organization.** Almost everything you write at work will affect other people and departments in your employer's organization, particularly if you are providing information that will be used to make a decision or if you are writing a recommendation. Ask yourself who these people and departments are and how your communication will affect them. Also, ask what responses they will have to what you say. By anticipating the conflicts and objections that your communication might stir up, you can reduce their severity or, perhaps, avoid them altogether.

- **Commitments.** In many communications, you will make commitments on behalf of your department or employer. When you submit a proposal outside the organization, you are promising that your employer will provide the materials or perform the work you describe. When you move to a managerial position and write policies, you will often be promising that in certain situations your employer will do certain things. Therefore, as you check a draft in which you make such a commitment, you should ask whether you have the authority to do so, whether the commitments are in the best interest of your employer, and whether they are the kind of commitments that your employer wants to make.

- **Compliance.** Determine whether your draft complies with organizational policies governing communications. Be especially careful when you are working on a project your employer considers to be sensitive. For example, for legal reasons your employer may want to restrict when and how you convey information about some aspect of your work. This might happen, for instance, if you are working on a patentable project or a project regulated by a government body, such as a state or federal environmental protection agency. Whatever the reason for your employer's policy, check your draft to be sure the communication complies with it.

Also, be sure to check your communication against whatever written or unwritten regulations your employer has about style and format.

GUIDELINE 2	LET TIME PASS BEFORE YOU CHECK

Guideline 1 suggested that one major obstacle to good checking is that writers sometimes define their objectives too narrowly, looking only for problems in spelling, punctuation, and grammar, while neglecting to consider the communication's likely effects upon readers and employers.

Guideline 2 addresses another major obstacle to checking: the fact that we are too "close" to the things we write ourselves. We have the *disadvantage* of knowing what our page is meant to say. As a result, when we check for errors and mistakes, we often see what we *intended* to write down rather than what is actually there. The word is misspelled, but we see the correct spelling that we intended to use. The paragraph is cloudy, but we clearly see the meaning we intended to convey.

To some extent, you can overcome the difficulties arising from being so close to your own communication by letting time pass between drafting and checking. In this way, you distance yourself from what you have written so you can better spot weaknesses in your draft.

How much time should you let pass? For a brief memo, a few minutes may provide you with all the time you need. Hours and days are better for longer, more complex communications. Of course, you should schedule your work to leave time to set your draft aside for a while.

GUIDELINE 3

READ YOUR DRAFT ALOUD

In addition to letting time pass, another good way to distance yourself from your draft is to read your draft aloud, even if there is no one but you to listen. When you speak your words, you process them mentally in a somewhat different way than you do when reading them silently to yourself. You *hear* your words as well as see them. These small shifts may seem trivial, but they can help you detect problems you might otherwise have overlooked.

Reading aloud can be particularly useful in helping you find sentences and passages that are confusing or graceless. Where you stumble over your own words, your audience is also likely to have difficulties. Reading aloud will also help you determine whether you have succeeded in creating an effective, professional voice (see Guideline 5 in Chapter 5). There is no better way to find out how your voice will ''sound'' to your audience than to hear it read aloud.

GUIDELINE 4

READ YOUR DRAFT MORE THAN ONCE, CHANGING YOUR FOCUS EACH TIME

Another cause for the difficulties we all have checking is summed up in the old adage, ''You can't do two things at once.'' To ''do'' something, according to researchers who study the way humans think, requires ''attention.'' Some activities (like walking) require much less attention than others (like solving algebra problems). We *can* attend to several more-or-less automatic activities at once. However, when we are engaged in any activity that requires us to concentrate, we have difficulty doing anything else well at the same time.[1]

This limitation on attention has important consequences. When you check your drafts, you must attend to many different things, each requiring concentration: spelling, the consistency of your headings, the clarity of your prose, and so on. When you concentrate on one kind of checking, you diminish your ability to concentrate on the others.

You can overcome the limitations of attention by reading through your draft more than once. You should read through it at least twice, once for substantive matters (like clarity and persuasive impact) and once for mechanical ones (like correctness of punctuation and consistency in the format of headings). If you have the time, you might read through it more than twice, sharpening the focus of each reading even more.

To check substantive matters, read your communication the way your readers will read it, whether reading it sequentially or skimming through it. Reflect upon your communication in the way your readers will, considering their reactions to the things you are saying and the way you are saying them.

When checking for mechanical matters, read very slowly, examining each detail of your draft. Pause to check and correct every possible misspelling or ungrammatical sentence. Flip back and forth among your pages as necessary to check things like the consistency of headings.

When checking your draft, you may find it helpful to use a checklist to identify the kinds of problems that you most want to look for, so you don't overlook any. You can group the problems to remind yourself to focus only on related types of problems in each reading.

When creating a checklist, you might use the following sources:

- Any style guide or other directive issued by your employer about how communications should be written.
- The checklist shown in Figure 16–1.
- Notes or recollections of what you were asked to do (if you are working on a communication that someone else asked you to write).

You may also add guidelines given in this book to your checklist. When deciding which guidelines to include, review the comments that your instructor, your classmates, or your supervisor has made on communications you have written recently. Include guidelines for writing activities that give you difficulty, but omit those for activities you have mastered.

| GUIDELINE 5 | **WHEN CHECKING FOR SPELLING, FOCUS ON INDIVIDUAL WORDS** |

When we read, we usually do *not* look closely at each individual word. This may surprise you. Nevertheless, research into the way people read shows that we usually read by the phrase and sentence. Although we identify every word, we do not look closely enough at each one to inspect all its letters.[2] This presents special difficulties for you when you check your written communications for misspellings and typographical errors.

To overcome this difficulty, you must slow down your reading so that you see each letter in each word. Some people can do that by applying their will power. Many cannot. After reading one or two dozen words, they begin

Worksheet

CHECKING YOUR DRAFT

Project _____ Date _____

Check from Your Readers' Point of View

_____ Will your communication be easy for them to use?

_____ Will it persuade them?

Check from Your Employer's Point of View

_____ Consider the effects of your draft on your employer's organization.

     _____ Do you know which people and departments will be affected?

     _____ Do you know *how* they will be affected?

     _____ Do you know what their reactions will be?

_____ Review the commitments you make on behalf of your employer.

     _____ Do you have the authority to make these commitments?

     _____ Are they in your employer's best interest?

     _____ Are they commitments your employer wants to make?

_____ Be sure your communication complies with any special policies your employer has concerning statements on the subjects you discuss.

Check for Mechanical Matters

_____ Are the spelling, grammar, and punctuation correct?

_____ Is your draft consistent in the numbering of figures, appearance of headings, and similar matters?

_____ Does your draft conform to the style your employer prefers? (If your employer has a style guide, did you follow it?)

_____ Is your draft neat and attractive?

FIGURE 16–1 Worksheet for Checking Drafts

reading the *meaning* of their communication, not the letters of their words. If that happens to you, you might try one of the following strategies:

- **Read backwards.** By reading your draft backwards, you see each word out of the context in which it forms a larger meaning. Consequently, you have an easier time concentrating on it and the letters that compose it.
- **Use a peephole card.** In an index card, cut a hole just large enough to see one word at a time through. Like reading backwards, this allows you to concentrate on each word in isolation from the words that surround it.

GUIDELINE 6 — DOUBLE-CHECK EVERYTHING YOU ARE UNSURE ABOUT

When checking a draft, all of us sometimes pause to ask ourselves such questions as: ''Is that word spelled correctly?'' ''Is that number accurate?'' ''Is that precisely what the lab technician told me?'' Since checking drafts is a demanding and difficult task, we may cringe at the thought of getting out the dictionary, looking back through our calculations, or calling up the technician. We may be tempted to answer our questions with, ''Yeah, sure. No need to double-check.'' Unfortunately, whenever we ignore our own doubts, we are ignoring one of the primary signals that there may be a problem. When checking, trust the instinct that made you hesitate in the first place.

Such double-checking can be especially irksome if you are shaky about mechanical matters, such as spelling and punctuation. You may need to look up the same word again and again: *seperate* or *separate? conceivable* or *conceivable?* (The latter in both cases.) What you learn by looking in the dictionary repeatedly is *not* how to spell the word but that *you do not know how* to spell it. Such knowledge is valuable to you *if* you use it to know when to double-check.

GUIDELINE 7 — USE COMPUTER AIDS TO FIND (BUT NOT TO CURE) POSSIBLE PROBLEMS

Some aids to make checking easier are now available in computer programs that can be used with drafts prepared with word-processing software. These aids can be helpful, but you must use them carefully.

Using Spell-Check Programs

If you prepare your communications with a word-processing program, you may also have a spell-check program. Such a program checks each word in a communication against the words in its dictionary and then displays a list of the spellings in the communication that aren't in the dictionary. In addition to misspellings, the list might include many words that are spelled correctly but that don't appear in the program's dictionary, such as proper names and technical words. Further, the list will omit words that are mistyped but that look like other words. For instance, if you mistype *take* as *rake,* the program will consider the word to be spelled correctly.

Despite these shortcomings, computerized spelling checkers enable you to check your entire draft very quickly for possible misspellings. However, you still need to proofread your writing to find words that the program mistakenly thought to be correct.

Using Readability Formulas

Some computer programs evaluate writing according to *readability formulas.* Readability formulas calculate a single number that, according to their creators, represents the difficulty of a piece of writing. While these formulas may indicate some problems, they cannot be trusted very far.

Why? Such formulas typically base their rating on only two factors: sentence length and word length (see Figure 16–2). They assume that bigger words and longer sentences make writing more difficult to read. But that's not necessarily the case. Needlessly long words do make for difficult reading, but many long words (like *excitement*) are much easier to understand than are many short words (like *erg,* a term from physics). Likewise, long sentences can be well written and short ones poorly written.

Furthermore, readability formulas won't identify problems that have nothing to do with sentence length and word length, such as poor organization, poor use of topic sentences, or poor use of headings. Nor do they give you a procedure for curing the problems they do find. George R. Klare, one of the leading authorities on readability formulas, says that trying to cure writing problems simply by using shorter sentences and shorter words is like trying to warm up your house by holding a match under a thermometer.[3] It simply doesn't work. The research in support of his position is overwhelming.[4]

Using Other Computerized Checking Aids

Other computer programs analyze, to a limited extent, other aspects of writing, including grammar. Commonly, these aids look for grammatical errors (such as lack of agreement between subject and verb) and some look for grammatical forms that often are signs of weak prose, such as use of the passive voice instead of the active voice, and the use of *is, are, were,* and other forms of the verb *to be* rather than action verbs.

The limitations of these aids are much like those of readability formulas. Although they can identify passive verbs, for example, they cannot distinguish a good use of the passive voice from a poor use of it (see Chapter 11).

In sum, spelling checkers, readability formulas, and similar aids can help you check your drafts by pointing out places that may have problems. However, the formulas may indicate problems where there are none, they cannot identify certain fundamental kinds of problems, and they do not tell you how to correct the problems they do find.

FLESCH READING EASE SCALE

Reading Ease = 206.835 − .846wl − 1.015sl

where wl = number of syllables per 100 words
sl = average number of words per sentence

The resulting figure will fall on a scale between 0 and 100, with 100 representing the easiest reading.

GUNNING FOG INDEX

Reading Grade Level = .4 (ASL + %PW)

where ASL = average sentence length
(number of words)
%PW = percentage of words with more
than two syllables

The resulting figure will give a number that represents the grade level for which the writing is appropriate (for example, 1 = first grade, 13 = first year of college).

FIGURE 16–2   Two Popular Readability Formulas

The Flesch Reading Ease Scale is from Rudolf Flesch, ''A New Readability Yardstick,'' *Journal of Applied Psychology* 32 (1948): 221–33. The Gunning Fog Index is from Robert Gunning, *The Technique for Clear Writing* (New York: McGraw Hill, 1952 [revised 1968]), 38.

CONCLUSION

This chapter began with an overview of evaluating, the fourth major activity of writing. Essentially, evaluating is a form of quality control in which you step back from your draft to determine if it is likely to succeed in achieving the objectives you set for it.

One indispensable way of evaluating drafts is to check them over yourself, a task that is never easy. This chapter suggested several ways of overcoming some of the major obstacles to effective checking, and it also explained how to use computerized checking aids effectively.

In the next two chapters, you will find advice about two other important ways of evaluating your drafts:

- Reviewing, in which you submit your drafts to someone else for advice.
- Testing, in which you ask one or more people to read and use your draft in the same way your target audience will read and use your finished communication.

EXERCISES

1. Following the advice in this chapter, carefully check a draft you are preparing for your class. Then give your draft to one or more of your classmates to review. Make a list of the problems they find which you overlooked. For three of those problems, explain why you missed them. If possible, pick problems that have different explanations.

2. Make a personal checklist of the aspects of writing that you feel you need to examine most carefully when you check your own drafts. Then, do one of the following:
   a. Compare your list with the lists prepared by other students in your class.
   b. Give one or more of your classmates a draft of an assignment you are working on. After they have reviewed your draft, compare their suggestions with the items on your checklist.

3. The memo shown in Figure 16–3 has 23 misspelled words. Find them by using this procedure:
   a. Read the communication from front to back.
   b. If you don't find them all *on your first check,* look for the rest by using one of the procedures mentioned in the chapter (read backwards or use a card with a peephole).
   c. Unless you found all the words on your first reading (front to back), explain why you missed each of the words you overlooked on that reading.
   d. Based on your performance in this exercise, state in a sentence or two your advice to yourself about how to improve your reading for misspellings and typographical errors.

4. In addition to its many misspellings, the memo shown in Figure 16–3 has several other problems, such as inconsistencies and missing punctuation. Find as many of these problems as you can.

MARTIMUS CORPORATION
Interoffice Memorandum

February 19, 19--

From: T. J. Mueller, Vice President for Developmnet

To: All Staff

RE: PROOFREADING

Its absolutely critical that all members of the staff carefully proofread all communications they write. Last month we learned that a proposal we had submited to the U.S. Department of Transportation was turned down largely becasue it was full of careless errors. One of the referees at the Department commented that, "We could scarcely trust an important contract to a company that cann't proofread it's proposals any better than this. Errors abound.

We received similar comments on final report of the telephone technology project we preformed last year for Boise General.

In response to this widespread problem in Martinus, we are taking the three important steps decsribed below.

I. TRAINING COURSE

We have hired a private consulting firm to conduct a 3-hour training course in editing and proofreading for all staff members. The course will be given l5 times so that class size will be held to twelve participants. Nest week you will be asked to indicates times you can attend. Every effort will be made to accomodate your schehule.

II. ADDITIONAL REVIEWING.

To assure that we never again send a letter, report, memorardum, or or other communication outside the company that will embarrass us with it's carelessness, we are estabishing an additional step in our review proceedures. For each communication that must pass throug the regular review process, an additionnal step will be required. In this step, an appropriate person in each deparmtent will scour the communication for errors of expression, consistancy, and correctness.

III. WRITING PERFORMANCE TO BE EVALUATED

In addition, we are creating new personnel evaluation forms to be used at annual salary reviews. They include a place for evaluating each employees' writing.

III. Conclusion

We at Martimus can overcome this problem only with the full cooperation of every employe. Please help.

FIGURE 16-3   Memo for Use in Exercises 3 and 4.

REVIEWING

DEFINING
OBJECTIVES
PLANNING
DRAFTING
EVALUATING
REVISING

I n the preceding chapter, you learned about the important role that evaluating plays in the writing process. When you evaluate, you examine a draft in order to determine whether any revisions are necessary or desirable and to gain ideas about how to make those revisions effectively.

The preceding chapter also introduced you to the most common of all evaluation methods—checking your drafts over yourself. The chapter you are now reading focuses on another widely used evaluation method, reviewing. In reviewing, you give your drafts to other people so they can read over what you have written and give you advice about it.

REVIEWING IN THE WORKPLACE

Employers consider reviewing to be a very effective evaluation method. In fact, much of the reviewing done on the job is *required* by employers. For instance, consider Jim's situation. Eight months ago, Jim graduated with a degree in biology and began working for a pharmaceutical company. This morning, after days of hard work, he finished drafting a five-page memo reporting on his analysis of a new prescription drug that the company is developing. However, Jim cannot mail his memo yet. Instead, he must meet with his boss to discuss it.

Jim has had similar meetings with his boss concerning every report he has written since he began his job. Usually, he gives his draft to his boss a day or two before their meeting. Before Jim arrives for the meeting, his boss makes many marginal notes on Jim's draft, and sometimes writes additional notes on another sheet of paper. These describe changes that Jim's boss wants Jim to make. Over the months, the changes have involved everything from altering a recommendation in one report to changing a few words in another. After Jim's boss has gone over the notes with Jim, Jim must revise his draft and resubmit it. Only after Jim's boss approves the report can Jim send it to its intended readers.

Jim's situation is typical. At work, you will almost certainly have to submit at least some of your writing for review by others. Even if you are in a management position, you will probably have to do the same. And then, like Jim's boss, you will have the additional responsibility of reviewing things written by other people.

Why Employers Require Reviews

Why is reviewing so often required at work? As Chapter 1 explained, employers consider the things their employees write on the job to be the property of the company. The communications you write will represent your department to the rest of the organization, and they will represent the organization to customers, clients, suppliers, and the rest of the outside world. Furthermore, your employer will have to live with the results of your writing—the commitments you make, the views you state, and the impressions you create. By reviewing your work, your employer will hope to ensure that your communications serve the best interests of the company.

Many companies also have a second reason for requiring reviews. They use reviewing as a way of helping employees—especially new ones—learn to write better. They hope that the writers will be able to sharpen their communication skills by thinking about and acting upon the advice given by their reviewers.

Formal Procedures for Reviewing

Because reviewing is so important to employers, many of them establish formal procedures for this activity. These procedures often require that a communication be reviewed not only by the writer's boss (as in Jim's case) but also by many other people. Depending upon the situation, these other reviewers might include other people in the writer's field, people in other fields who know about the writer's subject, upper-level managers, lawyers, public relations specialists, and even a professional editor.

Some employers who have established formal review procedures use special forms (often called *cover sheets* or *transmittal sheets*) to ensure that all the required reviews are carried out. Each of the designated reviewers must sign the form before the communication can be sent to the intended readers. Often, before signing the form, a reviewer returns a draft to the writer for revision. Occasionally, the writer and reviewer will then negotiate over the reviewer's requests, but often the writer must simply comply with them. Only after the changes have been made to the reviewer's satisfaction will that person sign the approval form. Figure 17–1 shows such a form.

Benefits of Requesting Reviews When They Aren't Required

Even when you are not required to submit your drafts for review, you will often find it to be worthwhile to ask others to look your drafts over. As explained in Chapter 16, it can be very difficult for you to see your writing as your readers will see it. By enlisting the help of other people, you can often gain many valuable insights into ways of improving your communication.

On the other hand, reviewing does take time—both yours and your reviewers'. Therefore, as you consider asking someone to look over your work, you should weigh the time and effort that you and your reviewer would spend on the review against the likelihood that the review will produce some significant improvement in your draft or in your writing ability. The discussion in the chapter on ''Revising'' (Chapter 19) provides a framework for balancing those considerations.

MIAMI UNIVERSITY PROPOSAL TRANSMITTAL FORM

PRELIMINARY BUDGET REVIEW—It is required that your budget be reviewed in advance of final submission of your proposal. **CONTACT:** Budget Office

---

THE PROPOSAL

Principal
Investigator _____ Department _____

Starting Date _____
Title of Project _____ Ending Date _____

New Project ☐
Funding Agency _____ Continuation ☐

Deadline Date for Submission _____ Postmark ☐ Receipt ☐

Human Subjects Yes ____ No ____ Live, Vertebrate Animals Yes ____ No ____ Recombinant DNA Yes ____ No ____

BUDGET DATA

	Funding Agency		Miami U. Cost-Sharing	
	1st year	Total	1st year	Total
Direct Costs	$_____	_____	$_____	_____
Indirect Costs	$_____	_____	$_____	_____
Total Costs	$_____	_____	$_____	_____

APPROVALS

1. Principal Investigator _____ Date

2. Department Chair _____ Date

3. Divisional Dean/Executive Director _____ Date

4. Dean, The Graduate School & Research _____ Date

5. Executive Vice President for Academic Affairs _____ Date

6. Budget Director _____ Date

COMMENTS BY SIGNERS:

FIGURE 17–1 Transmittal Form

RESEARCH OFFICE REVIEW
CHECK RELEVANT ITEM AND COMMENT IF PROPOSAL HAS NOTEWORTHY FEATURE

GENERAL

_____ Scope of the Proposal

_____ Long-Term Commitments

_____ Community Impacts

PERSONNEL

_____ The Principal Investigator

_____ Faculty or Postdoctoral Associates

_____ Graduate or Undergraduate Students

_____ Staff Time/Effort

ACADEMIC RIGHTS

_____ Freedom to Publish

_____ Patent Agreements, Copyrights
or Rights in Data

_____ National Security Restriction

FISCAL

_____ Student Fees

_____ Staff Benefits and Indirect Cost Rates

_____ Cost Sharing

_____ Space

_____ Equipment

COMPLIANCES

_____ Human Subjects

_____ Care of Laboratory Animals

_____ Radiation Hazards

_____ Safety and Health

_____ Biosafety

_____ Computers/Word Processors

RESEARCH OFFICE COMMENTS

Reviewer _____

Date _____

FIGURE 17–1   *(continued)*

REVIEWING IN THE CLASSROOM

Many writing instructors incorporate reviewing into their classes. For example, your instructor may ask to review a draft of an assignment you are writing. Or your instructor may ask you and your classmates to review each other's drafts.

You can benefit in several ways from these reviewing activities. Obviously, you will have a chance to improve your assignments before you turn them in for a grade. More important, you will gain insights into the strengths and weaknesses of your writing, insights you can use in your efforts to learn how to write better. In addition, you will gain practice at encouraging your reviewers to give you good advice and at putting their advice to good use. These skills will be very valuable to you on the job.

Any reviewing you do in your class can also benefit you by helping you learn how to provide good, helpful reviews for other people. Reviewing well requires thought and practice, just as writing well does. The practice you gain by reviewing your classmates' drafts will help you prepare for the many times on the job when you will be asked to look at someone else's drafts. Because reviewing other people's writing is an essential part of a manager's job, your accomplishments in this area can increase your chances of promotion.

IMPORTANCE OF GOOD RELATIONS BETWEEN WRITERS AND REVIEWERS

Before studying this chapter's advice about reviewing, consider for a moment the great importance of the human relationship between a writer and a reviewer. In this relationship, both individuals are deeply influenced by two major factors.

First, both writer and reviewer have professional obligations to fulfill. The writer is expected to seek advice and to use it. The reviewer is expected to provide substantive advice that will improve the communication and increase the writer's writing ability. The writer's and reviewer's senses of their obligations influence the ways these two people relate to one another.

Deeply intertwined with their professional responsibilities are the personal feelings of both the writer and the reviewer. In many cases, the dominant feeling of both persons is insecurity. As writers, we invest much of our personal mental effort and much of our individual creativity in our writing, and we know that we will be judged both professionally and personally on the result. Consequently, when we give our writing to someone else for review, we are likely to be afraid that we will be embarrassed by what we hear from our reviewer.

On the other side, reviewers are fearful also. When we must review someone else's writing, most of us are afraid of making bad suggestions that might reveal our own weaknesses as writers or readers. Also, most of us are fearful of hurting the writer's feelings. If we give in to these fears and withhold our advice, we are letting down the writer, who is depending upon us for assistance.

There is no surefire way to avoid the insecurities that beset writers and reviewers, though these fears will certainly become less troublesome as you gain experience in reviewing situations. Therefore, if you have opportunities to participate in review sessions in your writing course, you should take advantage of them as fully as you can, especially if you have no previous experience with reviewing. This classroom experience will help you significantly as you enter similar situations on the job.

The rest of this chapter presents two sets of guidelines. The first will help you obtain good advice when you have your drafts reviewed. The second set will help you give good advice when you review someone else's drafts.

GUIDELINES FOR HAVING YOUR WRITING REVIEWED

The six guidelines for having your drafts reviewed all share two objectives:

- To encourage your reviewers to think hard about your writing and to share freely their insights.
- To provide your reviewers with information they need in order to be able to review your draft as helpfully as possible.

The first guideline concerns your attitude toward your reviewers. The next two concern the information you should give your reviewers before they begin to look over your draft, and the last three concern the conversation in which your reviewer explains his or her suggestions to you.

GUIDELINE
1

THINK OF YOUR REVIEWERS AS YOUR PARTNERS

Because of your natural insecurity about showing drafts of your writing to other people, you might be tempted to see your reviewers as being ''against'' you, or as obstacles to the completion of your project. However, if you convey those attitudes, your reviewers are likely to feel that you don't really want their help, and consequently they may not give you the full value of their insights.

Thus, to encourage your reviewers to think hard about your draft and to share freely their insights, it is crucial that you communicate to your reviewers that you think they are ''for'' you, that you welcome their help and are grateful for their suggestions. One way to do this is to remind yourself that your reviewers are your partners and that they have the same aims you have of making your communication better and of improving your writing ability. You might say to yourself, ''I have gone as far as I can on my own in preparing a really effective communication. Now my reviewer will give me ideas for making it even better.'' This is an important kind of thing to say to yourself, even if your insecurities make it difficult to say in your heartiest voice.

TELL YOUR REVIEWERS ABOUT YOUR PURPOSE AND READERS

To communicate effectively, you must think constantly about your purpose and readers. For reviewers to work well on your behalf, they too must think constantly about your purpose and audience. Otherwise they cannot judge whether your communication is likely to achieve its purpose, nor can they suggest ways of revising your communication to better accomplish what you want it to. Thus, when you ask your reviewers to ''Please look this over,'' be sure to add, ''It's addressed to such and such an *audience* and has such and such a *purpose*.''

TELL YOUR REVIEWERS WHAT YOU WANT THEM TO DO

In addition to describing your purpose and readers, you should also let your reviewers know what you want them to look at. There are lots of things they *could* look at: organization, selection of material, accuracy, tone, spelling, page design, and so on. Unfortunately, it will be very difficult for them to concentrate on all the aspects of a communication simultaneously. Therefore, you can help your reviewers if you direct their attention to the particular features of the communication that you would most like their advice about.

To determine what you should ask your reviewers to look at, consider such things as the areas of writing that you are most unsure about, the sections of the communication that you had the most trouble drafting, and the sections of the communication you think are most critical to its success.

STIFLE YOUR TENDENCY TO BE DEFENSIVE

For most people, the most difficult part of reviewing is the conversation in which the writer and reviewer discuss the reviewer's suggestions. As writers, our natural reaction is to become defensive. We want to explain why what we wrote is better than what the reviewer suggests.

However, such a defensive response may prompt your reviewers to quit giving you advice, because they feel that their suggestions will inevitably meet not with thanks, but with arguments. Thus, even if you *know* that a particular suggestion is no good, listen to it without argument. Perhaps the next suggestion will be excellent. Don't discourage your reviewer from making it.

On the other hand, in your efforts to stifle your defensiveness, don't suppress all dialogue with your reviewers. When your reviewers' suggestions seem to be based upon a misunderstanding of what you are trying to accomplish, explain your aims. But do so in a way that indicates that you are still open for advice. If your reviewers misunderstood what you were attempting to accomplish, it's likely that you need advice about how to revise anyway.

GUIDELINE 5	ASK YOUR REVIEWERS TO EXPLAIN THE REASONS FOR THEIR SUGGESTIONS

When talking with your reviewers about their suggestions, learn as much as you can about their *reasons* for making the suggestions. Sometimes their explanations of these reasons will be more valuable to you than the suggestions themselves. Perhaps their explanation will stimulate you to think of an even better way of solving the problem they have identified. Or, perhaps your reviewers will have a way of explaining a certain piece of advice that will help you recognize and avoid the same kind of problem in the future. Even when a particular suggestion seems useless to you, you can gain valuable insights by finding out why the reviewers thought a suggestion was needed in the first place. Then, you can devise your own solution to the difficulty that prompted them to suggest a change in the first place.

GUIDELINE 6	TAKE NOTES ON YOUR REVIEWERS' SUGGESTIONS

Finally, you will probably increase the value of your conversations with your reviewers if you take notes during them. Your conversations with your reviewers may head off in many directions as you discuss various aspects of your draft. Consequently, when the conversation is all over, it can be tough to remember what was said. Further, most of us have difficulty remembering the details of conversations in which we feel a little bit on the spot, as most of us do when our writing is being reviewed.

Besides helping you remember what your reviewers have said, notetaking can also help you in another way. If you read your notes back to your reviewers, you can be certain that you have correctly understood all of their points. This check may also remind the reviewers of some additional, important point that they might make.

GUIDELINES FOR REVIEWING DRAFTS WRITTEN BY OTHER PEOPLE

The six guidelines you have just read provide advice about how to get the most out of reviewing when you are the writer. The following eight guidelines will help you when you are the reviewer. The first guideline concerns your general attitude toward the people whose writing you review. The rest lead you from your preparation to review a draft, through your decisions about what to suggest, to your presentation of your suggestions to the writer.

GUIDELINE 1	THINK OF THE WRITERS AS PEOPLE YOU ARE HELPING, NOT PEOPLE YOU ARE JUDGING

Your success as a reviewer will depend largely upon your ability to make writers feel open and receptive to your suggestions. If writers think that you lack respect for their ideas or feelings, they will probably resist your suggestions. If they feel you are genuinely interested in helping them, they will be much more likely to accept your suggestions.

Alternative Roles for Reviewers

A writer's initial reaction to you as a reviewer may depend largely upon the role you have been asked to play by your employer. You may be asked to play the *coach,* a person who merely *suggests* improvements to the writer. The writer is free to take your suggestions or ignore them. On the other hand, if you are asked to play the *gatekeeper,* you will have the authority to *require* the writer to make the revisions you suggest. Only after he or she has done so can the writer deliver the communication to its intended readers.

Whether you are a coach or a gatekeeper, you need to show the same concern for the writer's feelings. This advice may surprise you. It may seem that when you are a coach you must be very considerate of the writer's feelings because you have to persuade the writer to accept your suggestions, but that when you are a gatekeeper you can ignore the writer's feelings because the writer must follow your suggestions no matter how he or she feels about them.

Actually, to be an effective reviewer, you must always be considerate of the writer's feelings. When writers are forced to make changes they don't agree with, they become discouraged and bitter. Instead of trying to write better next time, they may turn in sloppy, thoughtless work, saying to themselves that there is no point in working hard at their writing because whatever they write will be changed arbitrarily anyway. They can also build resentments towards their reviewers that carry over to other parts of their working relationships. Consequently, even when you are a gatekeeper you should work very hard at building a positive, supportive relationship with the writers whose drafts you review.

How to Build a Positive Relationship with Writers

Whether you are acting as a coach or gatekeeper, the most effective strategy for building a good relationship with a writer is to persuade the writer that you genuinely desire to assist him or her. To do this, of course, you must bring a truly helpful attitude to the relationship. That's why it's so important for you to follow Guideline 1's advice: think of your readers as people you want to help, not people you are judging. When writers feel they are being judged, they generally react by trying to protect their communications and their self-images by becoming defensive.

In addition to bringing a positive rather than judgmental attitude to your relationship with writers, you can also encourage them to be open by following the advice given in the remaining seven guidelines for reviewing other people's writing. All seven are directed not only at helping you think of good, substantial advice to provide, but also at helping you establish and maintain a cordial, productive working relationship with the people whose writing you review.

| GUIDELINE 2 | ASK ABOUT THE COMMUNICATION'S PURPOSE AND AUDIENCE |

Your first step as a reviewer should always be to ask the writer to describe the purpose and readers of the communication you are about to examine. By doing so, you indicate to the writer that you are genuinely interested in what he or she is trying to accomplish, that you want to see the communication from his or her point of view. Furthermore, you must have this information to tell where the communication works and where it doesn't, and to formulate good suggestions for improving it.

Your inquiries about purpose and readers will enable you to be especially helpful to a person who usually takes a writer-centered (not a reader-centered) approach to communicating. If you use the information about purpose and readers to explain your suggestions, you will help the writer learn a powerful new strategy for writing, one that can improve his or her effectiveness quickly and dramatically.

| GUIDELINE 3 | ASK WHAT THE WRITER WANTS YOU TO DO |

Before you begin reviewing, you and the writer should agree about what you are to look for. If the writer wants you to proofread only, or to look at one section only, you should find that out, so that you don't waste time on other aspects or other sections of the communication.

Of course, what you look for when reviewing will not always be suggested solely by the writer. When you serve as a gatekeeper, your employer will also tell you what things to be concerned about. Even then, it is important to ask the writer if there is anything special you should consider. A writer may have a good sense of what most needs attention in the draft. Further, by asking the question, you are expressing your personal interest in helping the writer.

| GUIDELINE 4 | DISTINGUISH MATTERS OF SUBSTANCE FROM MATTERS OF PERSONAL TASTE |

Once you know what you should look for in the writer's draft, you can begin your examination of it. As you search for ways to improve the communication, you must be particularly careful to determine whether your ideas are based upon substantial principles of writing or upon your personal taste. We all have our individual ways of expressing ourselves. And things expressed in our style often sound better to us than do things expressed in someone else's style. However, that does not mean that other people should abandon their style and adopt ours. If as a reviewer you suggest changes that would merely replace the writer's preferred way of saying something with your own preferred way, you won't bring about *any* genuine improvement in the writing—and you will almost certainly spark the writer's resentment.

Unfortunately, it can sometimes be difficult for us to determine whether we like an alternative way of saying something because the alternative is genuinely better or because it matches our style more closely. When you are unsure, think about the *reason* you would offer for the change. If all you could say is that, ''My way sounds better,'' you are probably dealing with a matter of taste. Your way of saying it sounds better to you because it is the way you like to say things. On the other hand, if you can offer a more objective reason, for instance an explanation based on one of the guidelines in this book, you are dealing with a matter of substance.

One caution, however. Sometimes your sense that something doesn't sound right is your first clue that there is a problem. For example, as you read through a communication, you may stop at a sentence that doesn't sound right and find, after closer examination, that it has a grammatical error. It is important to follow up your instincts about how something sounds to see whether you can find an objective reason for your dissatisfaction. If so, tell the writer. If not, let the matter drop.

GUIDELINE 5   DETERMINE WHICH REVISIONS WILL BRING THE MOST IMPROVEMENT

After you have identified the revisions that you will suggest to the writer, rank them. By doing so, you help the writer decide which revisions will bring about the greatest improvement in the least time.

A second reason for ranking revisions is that many writers can face only a certain number of suggestions without feeling overwhelmed and defeated. As a reviewer, therefore, you should look over your own list of possible suggestions to see which ones will make the most difference. Convey those to the writer. As for the rest, keep them to yourself, or make it clear that they are less important than the others.

A third reason for ranking possible revisions is to help you use your own time effectively. In many situations you will not have time to work on all possible areas for improvement in a communication. Consequently, you should determine which revisions will make the greatest difference to the communication's effectiveness and then concentrate on those. To help themselves in this effort, some reviewers first scan a draft to see what sorts of suggestions they want to make. Only after they've developed a focus or plan do they begin their closer reading of the draft. The alternative to such planning is simply to proceed through the communication, marking each thing as it occurs to you. Such a strategy can be just as wasteful for a reviewer as it is for the writer who is checking his or her own work.

GUIDELINE 6   BEGIN YOUR SUGGESTIONS BY PRAISING THE STRONG POINTS

Once you have ranked your suggestions, you are ready to present them to the writer. When you do, pay special attention to the way you open the conversation. Like the opening of any communication, the opening of a series of review comments is crucial in establishing the attitude that the person you are addressing will take toward what you have to say. By opening with praise, you let a writer know that you appreciate the good things he or she has done, and you indicate your sensitivity to that person's feelings.

Unfortunately, it is easy to forget to open with praise. After all, the essence of reviewing is to focus on weaknesses that need improvement. You may find it helpful to plan your praise before you meet with a writer so you don't focus too quickly on what is ''wrong'' with the communication.

The praise you give a writer serves other important purposes as well. Often, writers don't know their particular strengths any more than they know their weaknesses. By praising the things they have done well, you encourage them to continue to do those things, and you reduce the chances that they will weaken one of the strong parts of their communication by revising it.

GUIDELINE 7   PHRASE YOUR COMMENTS POSITIVELY

Once you progress from praise to suggestions, phrase them positively. Instead of saying, ''I think you have a problem in the third paragraph'' or ''Your third paragraph doesn't work,'' say ''I have a suggestion about how you can make your third paragraph more understandable or persuasive to your readers.'' Even when you are a gatekeeper who must tell the writer that the draft can't be sent until it is revised, assure the writer that he or she has made a good start and say that you have some suggestions about how to bring about the necessary improvement.

Another way to phrase your comments positively is to explain your suggestions by using positive examples from the writer's own draft. For example, if the writer includes a topic sentence in one paragraph but omits one in another place, you could offer the former paragraph as an example of a way to improve the latter one. By doing this, you indicate that you know that the writer understands the basic writing principle but slipped once in applying it.

In sum, wherever possible, phrase your comments positively to convey to the writer that you see yourself as helping the writer make a good draft better.

GUIDELINE 8   EXPLAIN YOUR SUGGESTIONS FULLY

In addition to phrasing your suggestions positively, you should explain fully your reasons for making them. If you don't, the writer may assume that you are simply expressing a personal preference that makes no substantive difference to his or her communications. As a result, the writer may not take your suggestion, leaving the communication weaker than it might otherwise have been.

Besides persuading a writer to make needed changes, explanations help the writer learn to write better. For example, imagine that you suggest rephrasing a sentence so that the old information is at the beginning and the new information is at the end (see Chapter 11). The writer may agree that your version is better but not know in general how to improve his or her skill at writing sentences unless you explain the *principle* you applied.

Some reviewers withhold explanations because they think that the reasons for the suggestion are obvious. However, suggestions that are obvious to the reviewer are not necessarily obvious to the writer, or the writer probably would have avoided the problem in the first place.

Whenever possible, phrase your comments from the perspective of the intended readers. Instead of saying, "I think you should phrase it like this," say "I think your intended readers will be able to understand your point more clearly if you say it like this." When you explain your suggestions from the target audience's point of view, you help the writer take a reader-centered approach to making improvements based upon your suggestions. Also, this strategy relieves you of presenting yourself as judge and puts you in the role of someone who wants to help the writer convey his or her message to the target audience.

CONCLUSION

Reviewing is a form of evaluation in which one person submits his or her draft to other people for their advice. Employers often require reviews in order to ensure that the communications their employees send will benefit—rather than harm—the organization and to help the employees improve their writing abilities.

Whether you are the writer or the adviser, an important element in any reviewing situation is the human relationship between you and the other person. Both of you will have obligations to fulfill, and both will have personal feelings that color your interaction with one another and, therefore, influence the amount and quality of advice you share.

This chapter has provided two sets of guidelines, one for use when you are having your drafts reviewed and one for use when you are reviewing other people's drafts. Underlying both sets of guidelines are two basic suggestions. First, concentrate on building a positive, productive personal relationship with the other person. Second, do what you can to help the other person work efficiently.

EXERCISES

1. This exercise will give you practice at taking notes on the suggestions your reviewer makes.
    a. Exchange drafts with a classmate.
    b. Make notes on improvements you would like to suggest to your partner, while he or she writes down suggestions for you.
    c. Listen to your partner's suggestions, taking notes on them. (Your partner should not show you what he or she has written, but may refer to those notes while talking with you.)
    d. Compare the notes you took with the notes from which your partner spoke. Do the two sets of notes correspond closely? Did additional suggestions emerge from your discussion with your partner?
    e. Switch roles, so that you are the reviewer and your partner is the writer.

2. This exercise will help you develop your skills at delivering your suggestions to a writer in a constructive way. (See the chapter's advice about reviewing the writing of other people.)

    a. Exchange drafts with a classmate. Also exchange information about the purpose and audience for the drafts.

    b. Carefully read your partner's draft, playing the role of a reviewer. Your partner should read your draft in the same way.

    c. Make your comments to your partner, following the suggestions for reviewing the drafts of others.

    d. Evaluate your success in delivering your comments to the writer in a way that makes the writer feel comfortable while still providing the writer with substantive, understandable advice. Do this by writing down three specific things you think you did well and three ways in which you think you could improve. At the same time, your partner should be making a similar list of observations about your delivery. Both of you should focus on such things as how you opened the discussion, how you phrased your various comments, and how you explained them.

    e. Talk over with your partner the observations that each of you made.

    f. Repeat steps *c* through *e*, but have your partner give you his or her comments on your draft.

3. In this exercise, you are to prepare review comments for a writer. Depending upon what your instructor asks you to do, you may present these comments orally or you may convey them in a memo to the writer. In either case, be sure to follow *all* the suggestions for reviewing other people's writing. Remember that *how* you present your comments to the writer can be as important as *what* you suggest. For the sake of this exercise, imagine that you are a coach (not a gatekeeper), so that the writer has the choice of following or not following your suggestions.

    The memo you are to review is shown in Figure 17–2. Your review comments are to be directed to the writer of the memo, S. Benjamin Bradstreet, who works for a firm that builds manufacturing plants in other countries. Yesterday, Ben received a call from Dick Saunders, who has been in the Philippines for the past two years. Dick has been overseeing the construction of a factory for manufacturing roof shingles. The shingles are made from bagasse, the fibers that remain after the sugar has been squeezed out of sugar cane.

    Dick reached Ben while trying to call his own boss, Tom Wiley, who wasn't in. Consequently, Dick asked Ben to take notes on their conversation and to forward them to Tom.

    The audience for Ben's note will include not only Tom, who will be interested in all the information it contains, but also several other people, who will want only certain pieces of that information. These other people include Tom's assistant, who will be responsible for sending the things Dick requests, and other men and women in the sales department, who will need to talk with Tom on the phone or in person while he is in the States.

INTERNATIONAL MANUFACTURING, INCORPORATED
New York, New York

Interoffice Correspondence

DATE: 7 July 19–

FROM: S. Benjamin Bradstreet, Marketing Department

TO: Tom Wiley, New Factory Development Department

Your department's representative Dick Saunders called us on Tuesday, 5 July, from Manila, requesting some product information, an old formula, and information about the status of shipments to him. He also told us his projected return schedule.

First of all, he plans on returning to the States on 12 July, arriving in Chicago and then going directly to North Hampton, Massachusetts. He will stay at the Colonial–Hilton, (413) 586–1211, with Mr. Rossini of the Appliance Factories Group (13, 14, and 15 July).

A Victory luncheon is planned in Manila for Friday, 8 July, based on the anticipated success of our roofing factory there. General Tobias, the equivalent of our Secretary of Health and Human Services, is pleased with the factory and sees the work done this past week as the start of a production that will continue for some time.

Dick requested some additional information that is to be answered by means of a cable. This was to include when and how the Osprey drawings and the Quikmold release agent were shipped. He also wanted to know the exact cost of the release agent because it would be paid for with a check that he will carry back with him.

Dick said that Snyder and Leigh, of the American consulate, have been replaced by Wilson F. Brady. In future correspondence, we can use his name as a contact.

Dick said that they were having considerable success in making the shingles at the new roofing plant and that the 361

FIGURE 17–2   Memorandum for Use in Exercise 3

Filipinos will be able to take over full operation of the plant in a few days. These are the results that please General Tobias so.

Dick will return to New York on 18 July.

He said that the hard—rubber roofs installed last fall look good and that people are living in the houses. The only problem is some holes where nails were inadvertently pounded through in the wrong place.

Dick is also interested in a fire—retardant formula for the dry-blend phenolic system, and specifications for the Slobent 37, so he can find some locally available substitute before he leaves. He was aware of the retardant having a specific gravity of 2.4. Please send him the answers to these questions by cable.

FIGURE 17–2   *(continued)*

18

TESTING

DEFINING
OBJECTIVES
PLANNING
DRAFTING
EVALUATING
REVISING

An employee of an electronics company, Imogen, has written a set of instructions for installing a car radio and tape deck made by her employer. She has checked her instructions carefully, and she has asked her boss and two engineers to review a draft of them. Through these efforts, she found many ways to improve her instructions, which she now feels pretty good about.

However, Imogen still doesn't know one crucial thing: how well her instructions will *really* work for their intended readers, ordinary consumers who buy the radio and tape deck and want to install it themselves. When she checked her draft over, she *guessed* about how useful it would be to these readers. When her boss and the engineers reviewed her draft, they guessed as well.

One way Imogen can find out how good her guesses have been is to print up her instructions so that her employer can begin packing them with the radio and tape decks. If it turns out that her guesses were good, that strategy will work well. Both the customers and her employer will be happy. However, if her guesses were not good, she won't find out until after customers have difficulty using the instructions and begin complaining. In contrast, if she tests a draft of her instructions with members of her target audience before they are printed in final form, Imogen can discover and correct any problems before the instructions reach customers' hands.

This chapter presents six guidelines that Imogen—and you—can follow in order to create helpful, informative tests.

THE BASIC STRATEGY OF TESTING

When you test a communication, you try it out in a way that will enable you to predict what will happen when your intended readers read it. To do this, you ask a small group of people, called *test readers,* to read and use a draft of your communication in the way your intended readers will read and use the final draft. Then, by means of various techniques, you gather information from your test readers that will serve as the basis for your prediction. If your test is well-constructed, you can predict with some confidence that your intended readers will respond to your communication in much the same way as did your test readers.

For your prediction to be good, your test must resemble the real situation as much as possible. Your test readers and test draft must closely resemble your intended readers and final draft. Also, you must have your test readers read in the same way your intended readers will read. Deviations between your test and the actual reading situation will reduce the likelihood that your test results will predict accurately what will happen when your intended readers use your communication.

By following the guidelines presented in this chapter, you will be able to make your tests resemble actual reading situations as closely as possible.

THE TWO MAJOR QUESTIONS THAT YOU CAN ANSWER THROUGH TESTING

The prediction you base upon your test readers' responses to your draft can help you answer two key questions:

- Does your communication work well enough to send to your intended readers?
- How can you improve your communication?

The following paragraphs discuss the ways that testing addresses these two questions.

Does Your Communication Work Well Enough?

One way to determine if your communication works well enough is to see if it meets some predetermined standard in the test of its effectiveness. Therefore, when writers test their drafts, they often establish such standards before giving their drafts to their test readers.

For instance, before testing her instructions, Imogen decided that they would be good enough if a typical reader could use them to successfully install the radio and tape deck within two hours. Other types of standards are appropriate for other types of communications. For example, when communications are intended to teach, minimum test scores are often expressed in terms of percentage of correct answers. Thus, a person who wrote a description of the electromagnetic fields that surround the earth might say that the description achieved a satisfactory score if readers could answer correctly at least 75% of questions on a test based on it. (Remember that the tests described in this chapter evaluate the communication, not its readers—a point you should emphasize to your test readers before they begin their reading.)

Sometimes, you may decide to test your communications in ways that do not produce numerical results. For instance, you might decide to give your communication to your test readers and then talk with them about the communications after they have finished reading. To interpret the results, you need to use good judgment and common sense. Did your readers seem to be able to use the communication easily? Were their problems, if any, minor ones?

In fact, you will often need to use good judgment and common sense to interpret the outcome of tests that produce numerical results. Suppose, for instance, that Imogen found that her readers needed three hours, not two, to install the radio and tape deck. If she observed that her readers worked steadily throughout without having any difficulties understanding or using her instructions, she might conclude that three hours were needed because of the difficulty of the task, not because the instructions were written poorly. On the other hand, if she observed that the readers stumbled through the procedure because they had trouble understanding and using the instructions, she might conclude that her instructions were not yet clear enough.

How Can You Improve Your Communication?

The second question a test can help you answer is, "How can I improve my communication?" For many tests this is the *primary* question to be addressed. That's because by the time writers have worked on their communications long enough to prepare them for testing, the communications will usually work satisfactorily. The purpose of these tests is diagnostic—to determine ways to make a satisfactory communication even better. Of course, if the test shows that the communication really isn't satisfactory, the writer will still want ideas for improvement. Because of the great emphasis on gaining ideas for improvement in almost all tests, this chapter emphasizes diagnostic testing.

WHEN SHOULD YOU TEST A COMMUNICATION?

Before starting to read this chapter, you may never have thought that you might want to test the communications you will write at work. Indeed, of the three methods of evaluation described in this book—checking, reviewing, and testing—testing is used least often. At work, writers almost always check their own work, and a great many communications are reviewed before they are delivered to their intended readers. In contrast, only a small proportion of the communications written at work are tested.

Testing is used less often than the other two methods of evaluation because testing a communication usually requires substantial effort over and above the work already spent on checking and reviewing it. For many kinds of communications, such as technical reports and routine memos, people rarely even consider the possibility of testing. They believe that the improvement that testing might bring would not be great enough to compensate for the time and effort testing requires. In contrast, step-by-step instructions are often tested, especially for consumer products. That's because tests are often (though not always) fairly easy to conduct and because the writers and their employers can foresee that hidden problems with the instructions can seriously harm the company, by reducing sales of the product, for instance, or by making the company vulnerable to product liability lawsuits. If testing can uncover these hidden problems before the final instructions are issued, the company has much to gain.

Nevertheless, at present, many communications that could benefit immensely from testing are not being tested, a fact that a slowly growing number of organizations are beginning to realize. In your future job, you may be in a position to persuade your employer of the value of testing some of its communications. When should you urge that a communication be tested? The more important the communication and the larger the number of people who will read it, the more reason there is to test it.

THE THREE ELEMENTS OF A TEST

The guidelines in this chapter are organized into three groups, corresponding to the three basic elements of a test:

- The *draft* you will test—what it should be like (Guideline 1)
- The *people* who will read your communication in your test—what kinds of people you should ask to be the readers in your test, and how many you should have (Guidelines 2 and 3)
- The *test activities*—what you will ask your readers to do and the way you will gather information about their response to the draft (Guidelines 4 through 6)

GUIDELINE 1

TEST A DRAFT THAT IS AS CLOSE AS POSSIBLE TO A FINAL DRAFT

For your test to provide the basis for an accurate prediction about how your readers will respond to your finished communication, you must give your test readers a draft that very closely resembles what you have planned for your final draft. All differences between the draft you test and the final draft you plan will increase the uncertainty of your test results.

Suppose, for instance, that you are testing a draft that is missing one of the figures you plan to include in the finished communication. Then, you might find yourself asking, "Did my test readers have trouble understanding this important point from my communication because I didn't include the drawing that helps to illustrate it or because my prose needs to be rewritten?" Or, imagine that you are testing a draft in which you use a different page design from the one you plan to employ in the finished communication. Then you might find yourself asking, "Did my test readers take so long using my instructions because I need to rethink the way I have written them or because my test draft didn't have the same page design that I plan for the final draft, where I will place each figure on the same page as the prose it illustrates?"

Because the accuracy of your test results depends upon a close correspondence between the draft you test and the final draft you plan, it's best to test a *pilot* draft, one that is identical to the planned final draft in every feasible respect. For that reason, testing should follow both checking and reviewing. And the draft you test should *look* as much as possible like the planned communication. It should be neat, it should have all the drawings and figures, and it should have the same margins, number of columns of print per page, arrangement of headings, and so on.

In some situations, however, it is reasonable to allow some differences between the test draft and the planned final draft. Sometimes, for example, preparing a test draft identical to the planned final draft can require too much time, effort, and money, especially in light of the fact that the test may reveal that substantial changes need to be made. At other times, when you are writing a long instruction manual for instance, you may want to test parts of your communication early. Doing so will enable you to see if your basic communication strategies are working; if they are not, you can change them before you write the rest of the communication.

In situations where you have decided to allow differences between the test and planned final draft, focus your tests on the aspects of your communication that you believe will be most crucial to its success. Thus, if you would find it to be too expensive to prepare final drafts of your illustrations, include sketches of those figures that show all the critical features of the final drafts. Similarly, if you are going to test only one part of a report or booklet in which some points are much more important than others, test the parts that deal with the key points. Thus, even if your test draft isn't almost identical to your planned final draft, be sure that it lets you test the key features of the finished communication.

| GUIDELINE 2 | **PICK YOUR TEST READERS FROM YOUR TARGET AUDIENCE** |

You should select your test readers in much the same way that you prepare the draft you will test: be sure that the readers who try out your test draft closely resemble the target readers for your finished communication. If you are writing to plumbers, have plumbers test the communication. If you are writing to adults suffering from asthma, have adults with asthma serve as your subjects. In short, wherever possible, pick your test readers from your target audience.

It's important to use members of your target audience because different kinds of people will read the same communication differently. They bring different needs and expectations to it. They bring different attitudes and different levels of knowledge and interest in the subject. Consequently, the best way to test a draft is to present it to the kind of people it is written for.

If this is impossible, then seek individuals who are as similar as possible to the members of your target audience. In your class, for instance, you may be writing a report to engineers who design robots. If you can't obtain such engineers for your test readers, you might instead use engineering seniors who have had coursework in robotics.

| GUIDELINE 3 | **USE ENOUGH TEST READERS TO DETERMINE TYPICAL RESPONSES** |

As you know, we all read differently. A reaction or problem that one reader has with a particular communication may not bother the next hundred readers. For that reason, besides choosing test readers from your target audience, you should also be sure to use enough test readers to feel confident that the results you obtain are *typical,* not just the reflections of personal idiosyncrasies.

How many readers are enough? This depends on several factors. Generally, there is little variation from reader to reader in tests where you ask people to follow step-by-step instructions. If you get two or three members of your target audience to serve as test readers, you may have enough. More variation among readers occurs when you are testing people's understanding of the ideas or arguments presented in a communication. Common practice, based upon practical experience, suggests that about a dozen readers is a good number for such tests. Another factor that determines the number of test readers is your own knowledge of your audience. If you know it well, you will be able to use your personal experience to distinguish responses that probably are typical from those that probably are not.

Guidelines 1 through 3 have stressed the importance of testing a draft that closely resembles your planned final communication or using test readers who closely resemble your target readers, or both. The remaining guidelines in this chapter shift your focus to tasks you ask your test reader to perform and the ways you gather information about your draft's success.

GUIDELINE 4	**ASK YOUR TEST READERS TO USE YOUR DRAFT THE SAME WAYS YOUR TARGET READERS WILL**

When you draft, you create a communication designed to achieve a particular purpose. Use that purpose to design your test. How? If you have followed the guidelines in Chapter 3, you defined your communication's purpose in terms of:

- The tasks you want your communication to enable your readers to perform while reading (the enabling element of purpose)
- The way you want it to alter your readers' attitudes (the persuasive element of purpose)

The guideline you are now reading concerns testing related to your readers' tasks, and the next guideline provides advice about testing related to changes in your readers' attitudes.

At work, tests usually focus on the readers' efforts to carry out one or more of the following tasks:

- **Perform a procedure,** as when reading instructions
- **Understand content,** as when trying to learn about something through reading
- **Locate information,** as when looking for a certain fact in a reference manual

For some communications, all three tasks are important. For example, the owner's manuals for some personal computers open with a section describing how computers work (the readers' chief task is to *understand*), proceed to step-by-step instructions (the readers' chief task is to *perform* a procedure), and conclude with a reference section (where the readers must *locate* the particular pieces of information they need).

You should test your communication's effectiveness at enabling your readers to perform each and every task crucial to its success. Although this may mean that you test a single communication for all three types of tasks, you should test each type in a different way. The following paragraphs describe appropriate tests for each task.

Performance Tests

To test a communication's effectiveness at enabling readers to perform a procedure, give the communication to a few test readers and watch them use it. In addition to assigning your test readers the same tasks your target readers will perform, have them work in the same kind of setting. That way, you can learn how the communication will work for readers who have the same resources (for example, work space, lighting, tools) that your target readers will have, not any more nor any less. Also, as you watch your test readers, avoid intruding upon their work, either by getting in their way or by providing them with assistance that your target readers will not have.

An Example Test To see how you might design and conduct a performance test, consider what Imogen did to test her instructions for the car radio and tape deck. First, she asked two friends, Rob and Janice, to use her instructions to install the equipment in their own cars. (Her company gave them the equipment, considering that to be part of the expense of developing good instructions.) Imogen asked each of them to work alone because she imagined that most purchasers of the radio and tape deck would install it without help. She had Rob work at his home and Janice in the parking lot of her apartment building, where each would have the kind of work area and tools that would be available to most purchasers.

To gather information about how the instructions worked, Imogen watched Rob and Janice throughout their work, taking notes about places where they had difficulties. She also asked Rob and Janice about the instructions after they were done installing the equipment. Imogen had prepared her questions in advance, and others emerged from her discussion with them.

While watching Rob and Janice, Imogen was careful not to intrude upon their work. When Rob began his work, he several times asked Imogen about various instructions before trying them. Instead of answering, Imogen urged him to do his best without her help. If she had instead begun to give oral instructions, she would no longer have been testing her writing. She did help both Rob and Janice at one point when it became clear that they could not understand the instructions at all. If she hadn't helped them at that point, they would have had to stop work, so that Imogen couldn't have found out how well the remaining instructions worked.

Performance Test Laboratories   To provide test situations that are realistic and permit extensive observation that does not interrupt the reader, some companies have built special test facilities. Typically, these include two adjacent rooms. In the first room, the test readers work with the instructions. It is furnished to resemble the setting in which the target audience will use the instructions. For instance, if the instructions tell how to use a piece of office equipment, such as a typewriter or word-processing system, the room is outfitted like an office. If the instructions tell service personnel how to repair something, the room is furnished like the kind of shop that service personnel work in.

The second room is an observation room. From it, the people conducting the test watch the readers and record information. They might watch through one-way mirrors or with the help of television cameras. When readers are stumped by the instructions they are trying to follow, they can use a telephone to call the observation room for help, as if they were calling the manufacturer's customer service number.

Although such test facilities are relatively rare, their existence helps to underscore the importance of designing tests in which you do the following:

- Have your test readers work in a situation that resembles as closely as possible the situation in which your readers will work.
- Gather information without interfering with the readers' activity.

Adjusting Your Tests to Circumstances   Sometimes it may be impossible for you to ask the test readers to use your communication in exactly the way your target readers would. Imagine, for instance, that Imogen's instructions included a troubleshooting section designed to enable consumers to make some repairs to their radios and tape decks on their own. Target readers would use that section only when their radios and tape decks quit working properly. The instructions would tell them what symptoms to look for. Based upon the symptoms they found, readers would determine what the problem is and then fix it.

To test her troubleshooting section in the ideal way, Imogen would need several broken radios and tape decks, each with different symptoms. But it might be difficult for her to persuade her company to spend the time and money to make such broken equipment. In that case, she might prepare *written descriptions* of malfunctioning radios and tape decks. She would ask her test readers to diagnose the problem and describe the repair they would make, based on the information contained in her descriptions.

At work, practical circumstances may require you to adjust your test procedures in similar, creative ways.

Understandability Tests

Sometimes, you may want to see if your readers understand a communication accurately. You can test your communication's understandability by having some people read your communication and then asking them questions about its subject matter.

For example, Norman has been assigned by his employer, an insurance company, to write a new, more comprehensible version of its automobile insurance policies. Before the company begins using Norman's version, however, the company wants to test it. Through a newspaper ad, Norman has arranged to have two dozen people read a draft and answer questions about it.

Understandability tests can be used to test understandability alone or to test memorability in addition. To test understandability alone, Norman would let his test readers refer to the insurance policy while answering the questions—as if they were taking an open-book test. To test memorability also, he would have them read the entire policy, set it aside, and then answer the questions without looking at it again. He might even introduce a long interval between their reading of the policy and their work on the questions.

To decide whether or not to test memorability as well as understandability, you must think about how your target readers will normally use your communication. Norman determined that most people read their insurance policies only when they have specific questions about their coverage and that they do not try to memorize the policy's contents. For that reason, he will not test for memorability. There are situations, however, for which a memorability test is very important. For example, Jason is writing a booklet to teach cardiopulmonary resuscitation (CPR), a method for keeping people alive when they have stopped breathing and their hearts have stopped beating. Memorability would be very important because most people do not have their CPR instruction booklets with them when an emergency arises. Therefore, Jason will ask his test questions after his test readers finish reading the booklet and give it back to him.

The sections below describe three ways you can construct questions to test understandability alone or both understandability and memorability. The sample questions all refer to the following paragraph, which is from the liability portion of the automobile insurance policy that Norman drafted:

> In return for your insurance payments, we will pay damages for bodily injuries or property loss for which you or any other person covered by this policy becomes legally responsible because of an automobile accident. If we think it appropriate, we may defend you in court at our expense against a claim for damages rather than pay the person making the claim. Our duty to pay or defend ends when we have given the person making a claim against you the full amount of money for which this policy insures you.

Ask Your Test Readers to Recognize a Correct Paraphrase of Your Information   When you test your readers' ability to recognize a paraphrase of your communication, you see whether or not they have understood your communication well enough to recognize essentially the same meaning as expressed in different words. In your test questions, you should present paraphrases rather than direct quotations of your communication because readers may recognize the quotation through a simple act of memory, without having understood its meaning.

To test a reader's ability to recognize a paraphrase, you can use true-false questions. Here is such a question that Norman might use:

True or False:

We will defend you in court against a claim when we feel that is the best thing to do.

Or you can ask multiple-choice questions:

When will we defend you in court? (A) When you ask us to. (B) When the claim against you exceeds $20,000. (C) When we think that is the best thing to do.

Ask Your Test Readers to Provide a Paraphrase of Your Communication   When you ask your test readers to paraphrase your communication, you can find out whether the readers understand the material well enough to explain it in their own words to someone else. Here is a sample question:

In your own words, tell when we will defend you in court.

Presumably, if the readers provide a correct paraphrase your communication is understandable, and if they don't, it isn't. However, you must use your judgment to determine whether an unclear paraphrase results from ineffective writing on your part or from the test readers' lack of skill at explaining things. For this reason, if you are going to use paraphrase questions, it is important for you to have more than one test reader. Then you can identify readers who typically have difficulty expressing their thoughts.

Ask Your Test Readers to Apply Your Information   A third way to test your communication's understandability is to ask your readers to apply the information you provide. By doing so, you can see if they can understand your communication well enough to figure out how they should use your information in a particular situation.

To test your readers' ability to apply information, you need to construct a fictional situation in which the information can be used. Here is a situation that

Norman devised to test his readers' ability to understand one portion of the insurance policy (not the section quoted above).

> You own a sports car. Your best friend asks to borrow it so he can attend his sister's wedding in another city. You agree. Before leaving for the wedding, he takes his girlfriend on a ride through a park, where he loses control of the car and hits a hot dog stand. No one is injured, but the stand is damaged. Is that damage covered by your insurance? Explain why or why not.

Location Tests

At work, you will sometimes write things in which your readers will want to find specific pieces of information without reading (or rereading) the rest of your communication.

To find the information, the readers may use many of your communication's features, including its headings, topic sentences, table of contents, and index. Even the way you have arranged your prose and visual aids on the page can affect the ease with which your readers can find what they are looking for. By conducting a location test, you can determine how effective all these features are at helping searching readers.

A location test is similar to a performance test: give your communication to the test readers and ask them to find specific pieces of information as rapidly as possible. As in a performance test, you should have the test readers work in a situation that is as close as possible to the one in which the target readers will read, and you should gather information about the test readers' performance without disrupting their natural reading activity.

When devising a location test, you must decide what kind of assignment you want to give your test readers. The most straightforward way is to ask them to find the pages on which specific pieces of information are presented. For instance, if Norman were to construct a location test for his insurance policy, one of his questions might be:

> On which page do you find information about the liability
> insurance that is included in this policy?

Alternatively, you can start your test readers on their search with a problem-solving question. To do this, describe a situation that they can solve only by finding certain information. Such questions are most helpful to you as a writer if they present problems similar to those your target readers will use your communication to solve. Here is a problem-solving question that Norman might ask:

> You are making preparations for a party. Because you are running
> late, you ask a friend to use your car to pick up some things at a

bakery so you can finish putting up the decorations. On the way back, your friend hits a parked car while trying to avoid a child who has run into the street. Find the page that tells you whether or not we will pay for the damage to the other car.

Location tests can easily be combined with an understandability test. Simply ask the test readers to apply the information they find. For example, Norman could make an understandability question out of the sample just given by rewriting the final sentence to say, "Will we pay for damage to the other car?"

Similarly, you can easily combine a location question with a performance question. Imagine, for instance, that Imogen wanted to test the success with which readers could use her instructions to solve problems with their car radio and tape deck. She could do this by presenting her readers with her instructions and a broken car radio. Her assignment to the test readers would be to find the place in her instructions that tells how to fix the problem and then to do what it says.

In sum, to test your readers' ability to use the communications you write on the job, you can use one or more of the following: a performance test, an understandability test, and a location test.

<table>
<tr><td>GUIDELINE
5</td><td>LEARN HOW YOUR DRAFT AFFECTS YOUR TEST READERS' ATTITUDES</td></tr>
</table>

In the preceding guideline, you learned about ways to find out how well your communication is likely to succeed at enabling your readers to perform their reading tasks. This guideline looks at the second element of purpose by helping you predict the success your communication is likely to have in altering your readers' attitudes in the way you desire.

At work, tests usually focus on the readers' attitudes towards one or both of the following:

- The *subject matter* of the communication. For example, does this pamphlet by the National Cancer Institute change the readers' attitudes about prohibiting smoking in public places?
- The *quality* of the communication. For example, do readers think this instruction manual is easy to use, complete, and accurate?

Compare Your Readers' Attitudes before and after Reading

To determine how your communication affects your test readers' attitudes, you must learn about their attitudes both before and after reading it. Consider, for example, the test that Janet might give to determine the effects of a brochure she wrote concerning the hazards of smoking. Suppose that one of the things she learned from her test was that after reading her brochure her test readers were moderately in favor of banning smoking in public places. Unless she knew what her readers' attitudes were *before* they read her brochure, she could not conclude that their attitude about a smoking ban was caused by what she wrote. Maybe her test readers felt exactly the same way before reading, and her brochure made no difference. Or maybe they were strongly in favor of such a ban, and her brochure made them question the desirability of one.

One way to gather information about your readers' attitudes before and after reading is to ask them—at *both* times—the same set of questions about their attitudes. Another way is to ask the readers after reading how their present attitudes differ from those they held before reading. There is a danger, however, in asking readers to report their earlier attitudes: readers sometimes misremember their earlier attitudes.

In some situations, you may reasonably assume that you know what your readers' attitudes are before reading, without making a special effort to find out about them. That assumption can simplify your work at testing. However, it is usually safe to make assumptions about your readers' attitudes before reading only when you can be reasonably sure that your readers were neutral. That happens most often when the readers have not read or thought about your subject before reading your communication.

The following paragraphs provide specific advice about two techniques for gathering information about reader attitudes: interviewing and asking your readers questions constructed around a scale.

Gather Information about Your Readers' Attitudes through Interviewing

When you interview your test readers about their attitudes, focus on the specific aspects of your subject or communication most crucial to you. "Do you think that the rights of smokers would be unfairly ignored if smoking were banned in public places?" "Do you feel that these instructions are thorough? Clear? Helpful?"

At the end of a series of focused questions, you may want to ask a more open question, such as, "What else would you like to say about this subject (or this communication)?" An open question can uncover additional insights into your readers' attitudes and their reactions to your communication.

Instead of presenting such interview-type questions orally, you may present them in writing. In either case, prepare your questions in advance, so that you are sure you get the information you want.

Gather Information about Your Readers' Attitudes by Asking Them Questions Constructed around a Scale

When you ask your readers to answer questions on a scale, you are asking them to assign a number to their feelings:

Circle the number that most closely corresponds to your feelings about this communication.

It is complete.   Disagree   1   2   3   4   5   6   7   Agree

It is easy to use.   Disagree   1   2   3   4   5   6   7   Agree

Questions of this sort are especially useful when you want to gather precise information about your readers' attitudes both before and after they read your communication. To determine the effect of your communication, you can simply compare the number you receive before with the number you receive afterwards. Here, for example, are some questions Janet might use to test the effectiveness of her brochure about smoking:

Smoke in a room can   Disagree   1   2   3   4   5   6   7   Agree
harm the nonsmokers
who are there.

The rights of   Disagree   1   2   3   4   5   6   7   Agree
smokers would be
unfairly ignored if
smoking were
banned in theaters,
stores, offices, and
other public places.

A formal analysis of results obtained in this way requires the use of statistics, something outside the scope of this book. However, if you use ten or more readers you can gain at least a general idea about the effectiveness of your communication by comparing the average score before reading and the average score after reading. Don't be discouraged, by the way, if the change you see is small. Especially in situations when you are writing on topics that are personally important to your readers, you cannot expect that reading your communication once will shift their attitudes greatly. A small change is a reasonable goal and a commendable accomplishment.

GUIDELINE
6

LEARN HOW YOUR TEST READERS INTERACT WITH YOUR DRAFT WHILE READING

As emphasized at the beginning of this chapter, you should use your tests to learn as much as possible about how to *improve* your drafts. For this reason, you should find out not only about the outcome of your test readers' reading but also about your test readers' moment-by-moment interaction with the text. The following paragraphs describe four ways you can do this.

Sit Beside Your Test Readers during the Test

One simple, direct method of gathering information about your readers' progress through your draft is to pull up a chair next to your readers and watch them read. You need no special equipment, no special preparation. This method of observation is particularly useful for performance and location tests.

While watching, you should try to remain as unobtrusive as possible. Simply by being there you have already changed your test readers' reading situation from the situation in which your target readers will read. You will magnify this difference if you begin to answer questions from your test readers or help them in any way. Postpone discussions until after the readers are done reading.

Videotape Your Test Readers

You can also observe test readers by using videotape equipment. Depending upon the placement of the camera and the nature of the readers' task, you can learn a great deal from such tapes.

Videotaping has several advantages over direct observation. Most important, it removes you from the setting. Second, videotaping provides a detailed record that you and each test reader can look at together. This is important because either one of you might have difficulty remembering the details of the reader's progress through a draft. Third, by using more than one videotape camera at a time, you can see the reader from several angles of view simultaneously. This can be especially helpful when the reader is following step-by-step instructions and therefore moves from one place to another (from a desk to a machine, for instance).

Ask Your Test Readers to Provide Reading-Aloud Protocols

In a reading-aloud protocol, the test readers don't read your draft aloud but rather report continuously upon their thoughts and feelings as they proceed through it. The special advantage of such protocols is the insight they give into the readers' ''invisible'' mental processes. For example, without a protocol you might see (in person or on tape) that a reader is looking for something, but you would not know for certain what the person was trying to find—or why—unless the person told you.

Protocols can also help you discover good strategies for overcoming problems in your draft. For instance, in an experiment in which they collected reading-aloud protocols, Flower, Hayes, and Swarts found that when readers tried to understand difficult, impersonal prose from some government regulations, the readers reconstructed the statements in terms of stories in which particular people performed specific actions. Here, for example, is one part of the regulations the researchers used in their experiment. It concerns loans made by the Small Business Administration (SBA):

Advance payments may be approved for a Section 8(a) business concern when the following conditions are found by SBA to exist:

(1) A Section 8(a) business concern does not have adequate working capital to perform a specific Section 8(a) subcontract; and . . .

When trying to determine the meaning of those sentences, readers would transform them into something like the following:

If the owner of a small business needs the money to fulfill a contract, the SBA will lend it. However, if the owner already has the needed money on hand, the SBA won't make her a loan so that she can use the money on hand to buy land or things like that.

By using read-aloud protocols to learn about the ways that readers rework such statements, Flower, Hayes, and Swarts learned a great deal about how to write the regulations more effectively.[1]

The simplest method of collecting reading-aloud protocols is to have the readers speak into a tape recorder. Or, if you are videotaping the test readers, you can create a sound track at the same time.

Interview Your Test Readers after They Have Finished Reading

In interviews, you can ask your test readers for detailed information about their efforts to use your communication and their reactions to it. Usually, you will want to combine this technique for gathering information with one of the other three you have just read about. In interviews, you should focus on problems your readers have had, because those problems may indicate areas in which you can improve your draft.

There are several ways to be sure that you discuss each of the places that caused difficulties for the test reader. For instance, you can ask the readers to make marginal marks as they are reading at every point where they have difficulty. Alternatively, you can make a list of problem-causing spots by watching your readers read and marking a copy of your draft at every point where you see a reader make a mistake, flip through pages, or act in other ways that signal a problem. If you have your test readers make reading-aloud protocols, you can listen to each reader's tape with that person, stopping to talk at each mention of a problem.

During the interviews, you should also ask your test readers to tell you how they tried to overcome the problems they encountered. Their answers can be as helpful as reading-aloud protocols for suggesting ways you can improve troublesome parts of your draft. In addition, ask your test readers what they think you should do when you revise your communication. Test readers sometimes make excellent suggestions.

During interviews, you can also ask your test readers about their attitudes while reading. When inquiring about attitudes, you should ask not only what those attitudes are but about how they were shaped, moment by moment, as the readers read.

CONCLUSION

Like all the other evaluating that you do, testing should enable you to answer two important questions about something you are writing: (1) is your communication good enough to send to your intended readers and (2) how can you improve your communication?

For your test to enable you to make reliable predictions about how your target readers will respond to your finished communication, your test draft must closely resemble the final communication you have planned. Also, because different people read in different ways, your test readers should be from your target audience; if this isn't possible, the test readers should at least resemble the target readers in as many significant ways as possible. Ask these test readers to use your communication in the same way your target audience will. As they do, carefully gather information that indicates how successful your communication is in achieving the objectives you have identified for it.

EXERCISES

1. Explain how you would test each of the following communications:
    a. A display in a national park that is intended to explain to the general public how the park's extensive limestone caves were formed.
    b. Instructions that tell home owners how to design and construct a home patio. Assume that you must test the instructions without having your test readers actually build a patio.
2. Following the guidelines in this chapter, test one of the communications that you are preparing in your writing class.

VI
REVISING

19

REVISING

DEFINING
OBJECTIVES
PLANNING
DRAFTING
EVALUATING
REVISING

I n this chapter, you will find advice about the fifth major activity of writ-ing—revising. When you are revising, you act upon the insights that your evaluation—through checking, reviewing, or testing—provides about ways to improve your communication.

THE COMPLEXITIES OF REVISING

It might seem that revising involves nothing more than making, one by one, all the changes suggested by your evaluation. In fact, this is often the case, especially with short and routine communications. However, revising can also be much more complicated than that. Consider Wayne's situation.

Wayne has just returned to his office. On top of his desk, which is already crowded with work, he drops four binders. Each contains a copy of the current draft of the first major report he has written at work. In this lengthy report, Wayne makes several far-reaching recommendations, so his draft has been ex-amined by several reviewers—a co-worker, his boss, and his boss's boss, and the manager of another department. Their review comments are contained in the binders Wayne has just placed on his desk.

Wayne sits down and begins to thumb through the four binders. He is surprised at what he finds. Together, his reviewers have made well over one hundred comments that range from a suggestion about rephrasing particular sentences to a request that he substantially reorganize one section. Wayne looks at his calendar. In only three days, he must finish the report. During those same days, he has one other project to complete, several others to work on, and a staff meeting to prepare for and attend. Clearly, he needs to improve his communication, but he cannot possibly make all the revisions his reviewers have suggested.

To complicate matters, Wayne soon discovers that in places his reviewers have identified problems with his draft but have not given him any ideas about how to solve those difficulties. And even worse, his reviewers sometimes con-tradict one another in the advice they give and the changes they request.

YOUR VARIOUS ROLES WHEN REVISING

Wayne's situation resembles ones that you are likely to face frequently on the job. You, too, are likely to discover that you should make more changes than you have time for, that you have ideas about where problems arise but no immediate solutions for them, and that the advice you receive from some re-viewers contradicts the advice that you have received from other reviewers. Because of these sorts of complexities, you will sometimes need to play a variety of roles when revising. Depending upon the situation, you may need to be an investor and a diplomat. Always, of course, you will also need to be a writer.

- **Investor.** When revising, the resources you will have to invest are your energy, talent, and time. After evaluating a communication, you must consider the entire situation in order to determine how much of your resources you want to invest in revising your communication rather than in working on your other projects and responsibilities.

- **Diplomat.** When revising, you may have to negotiate with others. This could be necessary, for instance, if your boss or another powerful person requests a change that you think is dead wrong. When trying to persuade that person to change his or her mind, you need to behave very diplomatically so that you remain in the person's good graces. Moreover, when you receive conflicting advice or demands from two or more reviewers, you will need to negotiate with those individuals to settle upon the strategy that you will follow, and you must conduct these negotiations so that you maintain everybody's good will toward you.

- **Writer.** Despite the pressures and distractions you encounter from tight deadlines, other job responsibilities, and the demands of reviewers, you must remember that you are a writer. As a writer, you must work within the constraints imposed by time and reviewers to create a communication that your readers will find understandable, useful, and persuasive.

This chapter presents four guidelines that will help you play these three roles effectively.

GUIDELINE 1 ADJUST YOUR EFFORT TO THE SITUATION

Taken together, this guideline and the next suggest that you begin your task of revising by deciding how you will proceed. This guideline focuses on revising from the point of view of your responsibilities as an investor. It suggests that you begin revising by planning the amount of time, talent, and energy you will devote to it.

Often, it's very easy to plan the amount of time you will spend revising. Through your evaluation, you may find that your draft needs no revision at all, or so little that you make all the modifications very quickly. Also, when you receive comments from reviewers who have the authority to require you to make the changes they request, you will simply do what they say, no matter how long it takes. And, unfortunately, you may sometimes find that your deadline is short enough and the problems uncovered by your evaluation are large enough that you obviously must devote every available moment to revision. No special decision about your investment is necessary.

However, in many situations, you may have more difficulty deciding how long to continue revising your communication and when to tell yourself that you have revised enough and should stop. In those situations, you should consider three things: how good your communication needs to be, how much effort (if any) is needed to bring your draft up to that level, and how much additional effort you can reasonably spend making your communication *better* than ''good enough.''

How Good Does Your Communication Need to Be?

Some employees think it peculiar to ask how good a communication needs to be. ''After all,'' they observe, ''shouldn't we try to make everything we write as good as it can be?''

Of course, all the communications you write at work will have to achieve some minimum level of quality: they must convey essential information and they must be understandable. Beyond that, however, different communications need different levels of polish, a point you will recall reading in Chapter 1. For example, a co-worker who has asked you to provide a small piece of information is not expecting highly polished prose and might be very surprised to receive it. If you provide the key fact clearly, your memo will probably succeed no matter how rough your prose is. In contrast, in a proposal to a prospective client, you are probably writing to someone who expects your prose to be very polished. In fact, though your message might be valid, if your writing isn't polished, the client may infer that you and your organization are not capable of doing high-quality work on the proposed project.

To determine how good a particular communication needs to be, you can do four things: think about your purpose, consider a few general expectations that people at work hold about quality, look at similar communications, and ask somebody.

Think about Your Purpose   Of course, the most important factor determining how polished a communication needs to be is its purpose. How carefully does it have to be crafted to help your readers perform their mental tasks? How polished does it have to be to affect your readers' attitudes in the way you desire? Be sure to think about your purpose when deciding about the level of quality needed for your communication.

Consider General Expectations about Quality   When determining how polished your communication needs to be, you should also consider the general expectations of business people. Most agree that you need to polish your communications more in the following situations:

- **When you try to gain something** rather than give something. For example, you need to prepare proposals more carefully than reports.
- **When you address people outside of your organization** rather than those inside
- **When you address people at a higher level in your organization** than those who are at or below your own level

These general expectations can be translated into a list that ranks some typical communications according to the level of polish they usually need (Figure 19–1).

Look at Similar Communications   Another way to gain information about the level of quality needed is to look at similar communications. Often the level

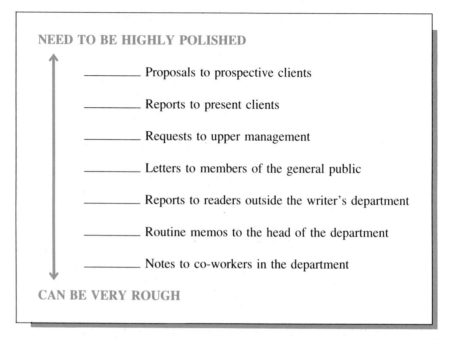

NEED TO BE HIGHLY POLISHED

_____ Proposals to prospective clients

_____ Reports to present clients

_____ Requests to upper management

_____ Letters to members of the general public

_____ Reports to readers outside the writer's department

_____ Routine memos to the head of the department

_____ Notes to co-workers in the department

CAN BE VERY ROUGH

FIGURE 19–1   Amount of Polish Typically Needed by Several Kinds of Communication

of quality needed will be determined in large part by what's customary. If you achieve the customary level of quality, you will do fine. If not, the shortcoming will be noticed and will undermine your communication's effectiveness. By looking at communications similar to the one you are preparing, you can see what level of quality has succeeded in the past.

Ask Someone   While the advice just given can help you determine the quality generally needed for the *type* of communication you are preparing, you will be best off if you obtain specific information about the *particular* communication you are preparing. One obvious way to get this information is to ask someone, such as your boss, co-workers, or members of your target audience. When you do, be sure to ask focused questions, because it can be difficult for anyone to define abstractly the level of quality expected. You are most likely to get a useful answer if you ask such questions as, "Does this need to be as polished as the report I wrote last week?" or "Can I show you a preliminary draft so you can tell me what else I need to do?"

How Much Effort Is Needed to Make Your Communication "Good Enough"?

Once you have determined how good your communication needs to be, compare that level of quality against the level your evaluation has shown you that your draft has achieved. The amount of effort needed to eliminate that difference is the *minimum* you should spend revising your communication. However, as the next section explains, it does not necessarily define the maximum effort you should invest.

How Much Additional Effort Can You Reasonably Spend?

From the discussion so far, it may seem that the amount of time you should spend revising a communication depends solely upon the amount of work needed to raise your communication to the minimum quality required for it. In many situations, however, you will have good reasons for wanting to surpass the minimum level.

First, your employer will probably encourage you to do so, particularly when you are addressing clients and customers. Second, you will want to enjoy the satisfaction of doing the best job you can do. Third, by making the extra effort you can favorably impress your readers, so that they say not only, ''I picked up a lot of useful information from your report,'' but also ''And you wrote it so well!'' Such comments are very pleasant to hear. They indicate that your work has commanded respect and may be increasing your status and influence. And they can lead to a favorable impression of you among those who make decisions about promotions and raises.

For all these reasons, you will often want to prepare communications that are better than just ''good enough.'' On the other hand, you will certainly want to avoid spending too much time on a communication, robbing time from other projects that deserve it more.

This leaves you facing the sometimes difficult question, ''How much better than 'good enough' should I make my communications?''

To answer this question, you must weigh the following two considerations simultaneously:

- **How much additional time will you need to invest to obtain a** *noticeable* **improvement?** For example, if it will take you only a short time to convert your choppy memo into a smooth one, it could be worthwhile to make the effort. However, if it will take you a long time, you should be more cautious about spending any time on that effort if the memo already possesses the necessary level of quality.

- **How important are your** *other* **projects?** As explained above, the amount of time and effort you should invest on a particular communication depends largely upon the benefit of investing it there compared to investing it in your other duties and responsibilities.

Two Cautions

Finally, as you decide how much time and effort to invest in revising a particular communication, you need to observe two cautions. First, when you are new on the job, people who are not involved with your work from day to day may base their opinions of your ability solely upon your writing. For this reason, when you begin work you should be especially diligent about revising your written communications carefully.

Second, communications prepared at work often have a wider audience than the communicator assumes. As you read above, notes to co-workers and to your immediate manager can often be fairly rough. Sometimes, however, those individuals may pass your communications along to upper management or to people in other departments, for whom you would want to prepare more polished communications. Therefore, when determining what level of quality you must reach to achieve your objectives, be sure to consider all the possible audiences and uses of your communication.

This discussion of Guideline 1 has touched upon several factors that you should consider when deciding how much time and energy to devote to revising a communication. The basic point for you to remember is that at work you will be an investor who will have to be sure that all the effort you spend on revising brings you sufficient benefit in return.

<table>
<tr><td>GUIDELINE
2</td><td>**MAKE THE MOST SIGNIFICANT REVISIONS FIRST**</td></tr>
</table>

When you follow Guideline 1, you have the first element in a solid plan for carrying out your revising efforts: you know how much time and energy you are going to spend at it. By following Guideline 2, you create the second part of such a plan by deciding which revisions you will tackle first, which second, and so on.

Why concern yourself with the order in which you make your revisions? When revising, your overall goal is to bring about the greatest improvement as quickly as possible. To avoid squandering your time and effort, begin with those revisions that will bring about the most significant improvement. Start with any that are needed to bring the draft up to the minimum level of quality, and then (if you decide to make the additional investment) work on the revisions that will most increase the effectiveness of your communication.

Because the advice to make the most important revisions first seems so sensible, you might wonder whether anyone ever proceeds in a different way. In fact, many people do. They begin revising by looking at the first paragraph, making whatever improvements they see there, and then continuing through the communication in the same way until time runs out. That procedure will work fine—if the writer is lucky enough to have the most serious shortcomings in the first paragraphs. But the most serious shortcomings sometimes come later. And some sorts of problems—for instance inconsistencies or organizational flaws—often are not evident until later portions of the communication are evaluated.

To avoid such an ill-planned effort, you should generally revise in three stages: evaluate the entire draft (or part of a draft) at once, rank the improvements you see, and then—and only then—begin making changes. In some situations the stages don't need to be quite that distinct, especially in shorter communications where big problems can't hide for long. With longer communications, however, you should make sure that you have an overall sense of what needs to be done and of what the priorities are before you invest any significant time making changes.

How to Rank Revisions

There is no single ranked list of revisions that you can apply to all communications. The importance of most kinds of revision depends very much on the situation. For one thing, the relative importance of a certain kind of revision is determined by the other kinds of revisions that are needed or desirable. Consider, for instance, the place of ''supplying topic sentences'' in the ranked list (Figure 19–2) that Wayne created after surveying the comments he received from the four reviewers of his report (mentioned at the beginning of this chapter). In comparison with the other items in the list, supplying topic sentences is *relatively* unimportant. However, in another communication in which Wayne had handled effectively all of the other aspects of writing, providing topic sentences might bring more improvement than any other revision he could make.

Another factor affecting the relative importance of a revision is its location in a communication. In long reports, for instance, a summary addressed to decision-makers contributes more than the appendixes to the overall

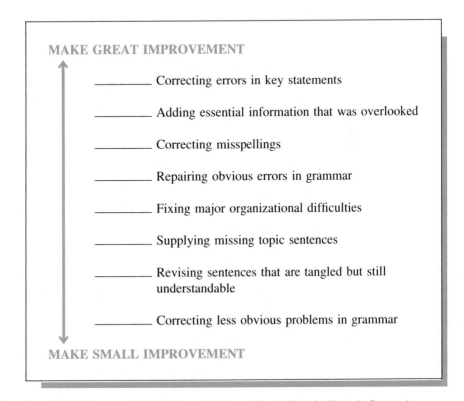

MAKE GREAT IMPROVEMENT

_____ Correcting errors in key statements

_____ Adding essential information that was overlooked

_____ Correcting misspellings

_____ Repairing obvious errors in grammar

_____ Fixing major organizational difficulties

_____ Supplying missing topic sentences

_____ Revising sentences that are tangled but still understandable

_____ Correcting less obvious problems in grammar

MAKE SMALL IMPROVEMENT

FIGURE 19–2   Amount of Improvement That Various Revisions Would Make in Wayne's Proposal

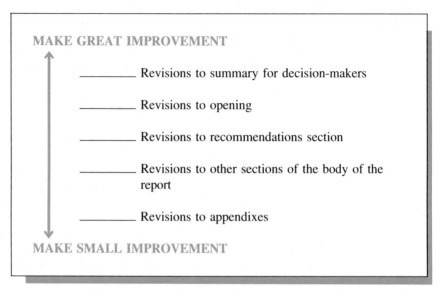

FIGURE 19–3   Amount of Improvement That Comparable Revisions Would Make in Various Parts of a Communication

effectiveness of a communication. The same principle applies within sections: some parts are more crucial than others. Generally, the beginning and the end are more important than the middle. Parts that make key points are more important than those that provide background information or explanations. Figure 19–3 shows the relative amounts of improvement that can be made by comparable revisions in various parts of a typical communication.

A third factor can come into play when you are ranking revisions requested by reviewers: the importance of the person making the request. Sometimes, for obvious and practical reasons, you will want to give first attention to revisions requested by people in positions of authority, even if you would not otherwise have given those same revisions high priority.

If you have only a limited amount of time to revise, some of the revisions on your ranked list may never be made. By making such a list, however, you ensure that the revisions you do complete are the ones that will contribute most to making your communication work.

Be Sure to Correct Mechanical Problems

No matter how you rank the other possible improvements, be sure to place the correction of mechanical problems at the top of your list of revisions. The following list describes the major mechanical issues that you should focus on.

● **Correctness.** Be sure that you are correct in matters where there is a clear right and wrong (as with spelling, grammar, and punctuation).

- **Consistency.** Where two or more items should have the same form, be sure that they do. For instance, be sure that parallel tables look the same. Be sure that the figure after Figure 3 is correctly numbered as Figure 4. Similarly, if you tell your readers to look on Page 10 for a particular drawing, be sure that the drawing is there, not on Page 11.
- **Conformity.** Be sure that you have followed all of your employer's policies concerning the mechanical aspects of writing, such as the width of margins, the use of abbreviations, and the contents of title pages. This will be relatively easy for you if your employer has put these policies in writing. If your employer hasn't, remember that your communication must still conform to your employer's informal policies and expectations.
- **Attractiveness.** Be sure your communication looks appealing and neat.

Why give such a high priority to fixing mechanical problems? The working world is very intolerant of them, particularly of errors in spelling and basic grammar. In fact, the working world is much harsher in judging lapses in these areas than are most college teachers (including most college writing teachers). On the job, a typographical error can seriously affect the readers' attitudes toward the writing—even though the error doesn't prevent the readers from knowing what the writer meant. Consequently, when you plan your revisions, always leave time for a careful, thorough check of mechanical matters.

Usually, it is best to leave this check for mechanical matters until the end. There is no point in investing time in correcting passages that may later be deleted or replaced. At the same time, if you will ask others to help you evaluate your draft, either by reviewing it for you or by testing it, be sure to clear up mechanical problems in the draft you give them. If the errors remain, your readers may become distracted by them, thereby missing the more substantive issues with which you most need their help.

GUIDELINE 3	**TO REVISE WELL, FOLLOW THE GUIDELINES FOR WRITING WELL**

Guidelines 1 and 2 concern the planning efforts that you will often find helpful to undertake before you begin making any specific improvements in your drafts. In contrast, Guideline 3 advises you about what to do as you modify your communications. It suggests, quite simply, that when revising you should follow the same guidelines that you followed when drafting. If you want to change the material you include, for instance, follow the guidelines in Chapters 4 and 5, ''Planning to Meet Your Readers' Informational Needs'' and ''Planning Your Persuasive Strategies.'' Similarly, if you want to polish your paragraphs, sentences, or visual aids, follow the guidelines in the chapters that pertain to them.

''Is that right?'' you may be asking yourself. ''Shouldn't a chapter on revising tell me how to rewrite a sentence, restructure a paragraph, and so on?'' Not necessarily, at least not if you can find the most important advice about those very matters elsewhere in the same book, as you can in this one. Whether you are drafting or revising, the principles and guidelines for good communication remain the same. Where your evaluation shows that you might improve your communication, you need to apply the guidelines in a different way—not search for alternative guidelines.

| GUIDELINE 4 | **REVISE TO LEARN** |

This chapter's first three guidelines have focused on the ways that revising can improve the effectiveness of your communications. Revising can also help you to learn to write better.

When revising, you will often work at applying writing skills that you have yet to master fully. By consciously practicing those skills, you can improve them. Then you will be able to use them when you *draft*. For example, if you spend much of your revising time supplying topic sentences for your paragraphs, you probably do so because you have not yet learned to write topic sentences in your first drafts. Through the practice you receive when revising, you can master the use of topic sentences and learn to supply them automatically in your first drafts.

Because you can learn so much from revising, it can often be worthwhile to invest *extra* time in revising. Even if your communication is enough to satisfy the needs and expectations of your readers and employer, consider revising it further for the sake of what you can learn by doing so.

CONCLUSION

This chapter has presented four guidelines for revising your drafts. Because of the complexities of working situations, you will sometimes need to play several roles while revising. As an investor, you will need to decide how much time to devote to revising a communication. As a diplomat, you will have to negotiate with reviewers about the changes they have requested or demanded. And, as a writer, you will have to keep in mind that the primary goal of your writing is to create a communication that (despite deadlines and despite the demands of other people) will be understandable, useful, and persuasive to your readers.

This chapter has presented four guidelines for revising. By following the first two, you will be able to create an advance plan that tells how much time you should invest in revising and also describes what you should do with that time to obtain as much improvement as possible from it.

The third guideline reminds you that strategies you use to write well are the same ones you should use to revise well.

Finally, the fourth guideline points out that revising can be educational. Whether you are on the job or in school, revising gives you practice at skills you have not completely mastered. If the situation permits, spend a little extra time revising for the sake of what it can teach you.

SUPERSTRUCTURES

20

GENERAL REPORTS

A t work you will often be called upon to convey and interpret information for other people. That's because you will possess a great deal of important knowledge that others will want to use. Sometimes, this will be knowledge you have gained in your college courses. At other times, it will be knowledge you have gained through special research, perhaps in the library or the laboratory. And at still other times, it will be knowledge you have gained through general, thoughtful observation of the things that go on around you on the job. Whatever the source of that knowledge, you will often respond to people's requests for it by writing a report.

This chapter is the first of four that will help you prepare effective reports on the job. In it, you will learn about the basic superstructure for reports. The advice given here will be useful almost every time you compose a report at work.

The following three chapters describe variations of the basic superstructure. Each represents an adaptation of the basic superstructure to meet the needs of readers in a particular type of reporting situation. The superstructures described in those chapters are for empirical research reports (Chapter 21), feasibility reports (Chapter 22) and progress reports (Chapter 23). Whenever you write one of those types of reports, you will find those chapters to be helpful *supplements* to this one. You should not, however, use them as *substitutes* for this chapter because this one contains considerable advice that is not repeated in the others.

HOW TO USE THE ADVICE IN CHAPTERS 20 THROUGH 23

Whether you read only about the basic superstructure or also about the variations of it, keep in mind that no two reporting situations are exactly alike. You will have to use your imagination and creativity to adapt the superstructures you find here to the particular reports you are preparing. Before reading farther in this chapter, you may want to reread the discussion of superstructures and their uses that appears in Chapter 4 (pp. 121–23).

Remember also that whatever superstructure you use for a report, you will have to present the report in some format, usually one of the three discussed in Appendix A: letter, memo, or book. Therefore, as you prepare your report, look also at the discussion of the appropriate format.

VARIETIES OF REPORT-WRITING SITUATIONS

Reports come in many varieties. The following examples will give you some idea of their diversity:

● A one-hundred-page report on a seven-month project to test a special method of venting high-speed engines for use in space vehicles

- A twelve-page report based on library research to determine which long-distance telephone company provides the most reliable service
- A two-paragraph report based upon a manufacturing engineer's visit to a new plant that is about to be put into service
- A two-hundred-page report addressed to the general public concerning the environmental impact of mining certain portions of public land in Utah

As these examples suggest, there are many ways in which the reports you write may differ from one another and from the reports written by other people:

- Sources of Your Information. You may base your report on information gathered from one or more of a wide variety of sources, including your own research, reading, and interviews.
- Amount of Time You Spent Gathering Your Information. This may vary from a few minutes to many years.
- Number of Readers. Your report may have many readers or only one.
- Kinds of Readers. Your readers may be people employed in your own organization or they may be employed in other organizations. In some situations, you may address the general public.

YOUR READERS WANT TO USE THE INFORMATION YOU PROVIDE

Despite these and the many other differences among them, almost all report-writing situations have one factor in common: your readers will want to put the information you provide to some professional or practical use. The precise kind of use will vary, of course, from situation to situation. For example, your readers may want to use your information to solve an organizational problem (where typical goals are to increase efficiency and profit), a social problem (where typical goals are to improve the general health and welfare of groups of people), or a personal problem (where typical goals are to satisfy individual preferences and values). Regardless of these differences, however, the readers' desire to use your information will almost always be a key factor when you prepare reports, a factor that you should take into account when planning and drafting every part of your communication.

Many other times in this book, you have read about the importance of considering the way your readers will use your communications. The point deserves repetition—and special emphasis—here because so many people forget their readers when they prepare reports. They persuade themselves that their purpose is to "tell what I know" or "tell what I have done" rather than to provide their readers with information that the readers can use.

THE QUESTIONS THAT READERS ASK MOST OFTEN

When trying to use the information they find in reports, readers usually ask the same basic questions. The general superstructure for reports is a pattern that writers and readers have found to be successful for answering those basic questions. Therefore, by thinking about your readers' questions, you will prepare yourself to understand the superstructure and to use it effectively. The readers' six basic questions are as follows:

- **What will we gain from your report?** Most people at work want to read only those communications that are directly useful to them. Therefore, you need to explain your communication's relevance to the readers' interests, responsibilities, and concerns.
- **Are your facts reliable?** Readers want to be certain that the facts you supply will provide a sound basis for their decisions or actions.
- **What do you know that is useful to us?** Readers don't want you to tell them everything you know about your subject; they want you to tell them only those facts they must know to do the job that lies before them. (Example: "The most important sales figures for this quarter are as follows: . . .")
- **How do you interpret those facts from our point of view?** Facts alone are meaningless. To give facts meaning, people must interpret them by pointing out relationships or patterns among them. (Example: "The sales figures show a rising demand for two products but not for two others.") Usually, your readers will want you to make those interpretations rather than leave that work to them.
- **How are those facts significant to us?** Readers generally want you to go beyond an interpretation of the facts to explain what the facts mean in terms of the readers' responsibilities, interests, or goals. (Example: "The demand for one product falls during this season every year, though not quite this sharply. The falling demand for the other may signal that the product is no longer competitive.")
- **What do you think we should do?** Because you will have studied the facts in detail, your readers will often want you to tell them what action you think they should take. (Example: "You should continue to produce the first product, but monitor its future sales closely. You should find a way to improve the second product or else quit producing it.")

Of course, those six questions are very general. For large reports, people need to take hundreds, even thousands, of pages to answer them. That's because business people seek answers to these basic questions by asking a multitude of more specific, subsidiary questions.

GENERAL SUPERSTRUCTURE FOR REPORTS

The general superstructure for reports contains six elements, one for each of the six basic questions you just read about: introduction, method of obtaining facts, facts, discussion, conclusions, and recommendations.

Figure 20–1 shows how the six elements relate to the readers' basic questions. Of course, each element of the superstructure may serve important purposes in addition to answering the general question identified with it. Also, the six elements may be arranged in many ways, and one or more of them may be omitted if circumstances warrant. In some brief reports, for example, the writers begin with a recommendation, move to a paragraph in which the facts and conclusions are treated together, and state the sources of their facts in a concluding, single-sentence paragraph. Also, people sometimes present two or more of the six elements under a single heading. For instance, they may include in their introduction information about how they obtained their facts, and they frequently present and interpret their facts in a single section of their report.

The following paragraphs briefly describe each of the six elements of the general superstructure for reports.

Introduction

In the introduction of a report, you answer your readers' question, "What will we gain by reading your report?" In some reports, you can answer the question in a sentence or less. Consider, for instance, the first sentence of a report

Report Element	Readers' Question
Introduction	What will we gain by reading your report?
Method of obtaining facts	Are your facts reliable?
Facts	What do you know that is useful to us?
Discussion	How do you interpret those facts from our point of view?
Conclusions	How are those facts significant to us?
Recommendations	What do you think we should do?

FIGURE 20–1   Elements of a Report and Their Relationship to the Basic Questions Readers Ask

written by Lisa, an employee of a university's fund-raising office, who was asked to investigate the university's facilities and programs in horseback riding. Because her reader, Matt, had assigned her to prepare the report, she could tell him what he would gain from it simply be reminding him why he had requested it:

> In this report I present the information you wanted to have before deciding whether to place new university stables on next year's list of major funding drives.

In longer reports, your explanation of the relevance of your report to your readers may take many pages, in which you tell such things as (1) what problem your report will help solve, (2) what activities you performed toward solving that problem, and (3) how your audience can apply your information in their own efforts toward solving the problem. In the discussion of Guideline 1 in Chapter 9, you will find detailed advice about how to tell readers why they should pay attention to your report.

Besides telling your readers what your communication offers them, your introduction may serve many other functions. The most important of these is to tell your main points. In most reports, your main point will be your major conclusions and recommendations. Although you should save a full discussion of these topics for the sections devoted to them at the end of your report, your readers will usually appreciate a brief summary of them—perhaps in a sentence or two—in your introduction. Lisa provided such a summary in the second, third, and fourth sentences of her horseback-riding report:

Summary of conclusions	Overall, it seems that the stables would make a good fund-raising project because of the strength of the current programs offered there, the condition of the current facilities, and the existence of a loyal core of alumni who used the facilities while undergraduates.
Summary of recommendations	The fund-raising should focus on the construction of a new barn, costing $125,000. An additional $150,000 could be sought for a much-needed arena and classroom, but I recommend that this construction be saved for a future fund-raising drive.

In brief reports (for example, one-page memos), a statement of your main points may even replace the conclusions and recommendations that would otherwise appear at the end. For an additional discussion of the value of putting your main points in your introduction, see Guideline 2 in Chapter 9.

Four other important functions that an introduction may serve are to tell how the report is organized, outline its scope, encourage openness to your message, and provide background information the readers will need in order to understand the rest of the report. You will find a detailed discussion of these functions in Chapter 9.

Method of Obtaining Facts

In a report, your discussion of your method of obtaining your facts can serve a wide variety of purposes. Report readers want to assess the reliability of the facts you present: your discussion of your method tells them how and where you got your facts. It also suggests to your readers how they can gain additional information on the same subject. If you obtained your information through reading, for example, you direct your readers to those sources. If you obtained your information through an experiment, survey, or other special technique, your account of your method may help others design similar projects.

In her investigation of the university stables, Lisa gathered her information through interviews. She reported her method in the following way:

> I obtained the information given below from Peter Troivinen, Stable Manager. Also, at last month's Alumni Weekend, I spoke with a half-dozen alumni interested in the riding programs. Information about construction costs comes from Roland Taberski, whose construction firm is experienced in the kind of facility that would be involved.

Facts

Your facts are the individual pieces of evidence that underlie and support your conclusions and recommendations. If your report, like Lisa's, is based upon interviews, your facts are the things people told you. If your report is based upon laboratory, field, or library research, your facts are the verifiable pieces of information that you gathered: the laboratory data you obtained, the survey responses you recorded, or the knowledge you assembled from printed sources. If your report is based upon your efforts to design a new product, procedure or system, your facts are the various aspects of the thing you designed or created. In sum, your facts are the separate pieces of information you present as objectively verifiable.

You may present your facts in a section of their own or you may combine your presentation of your facts with your discussion of them, as explained next.

Discussion

Taken alone, facts mean nothing. They are a table of data, a series of isolated observations or pieces of information without meaning. Therefore, an essential element of every report you prepare will be your discussion of the facts, in which you interpret the facts in a way significant to your readers.

Sometimes, writers have trouble distinguishing between a presentation of the facts and a discussion of them. The following example may help to make the distinction clear. Imagine that you observed that when the temperature on the floor of your factory is 65°F, workers produce 3 percent rejected parts; when it is 70°F, they produce 3 percent rejected parts; when it is 75°F, they produce 4.5 percent rejected parts, and when it is 80°F, they produce 7 percent rejected parts. Those would be your facts. If you were to say, "As the temperature rises above 70°F, so too does the percentage of rejected parts," you would be interpreting those facts. Of course, in many reports you will be dealing with much larger and more complicated sets of facts that require much more sophisticated and extended interpretation. But the basic point remains the same: when you begin to make general statements based upon your facts, you are interpreting them for your readers. You are discussing them.

In many of the communications you write, you will weave your discussion of the facts together with your presentation of them. In such situations, the interpretations often serve as the topic sentences for paragraphs. Here, for example, is a paragraph in which Lisa mixes facts and discussion:

<table>
<tr><td>Interpretation</td><td>The university's horseback riding courses have grown substantially in recent years, due largely to the enthusiastic and effective leadership of Mr. Troivinen, who took over as Stable Manager five years ago.</td></tr>
<tr><td>Facts</td><td>When Mr. Troivinen arrived, the university offered three courses: beginning, intermediate, and advanced riding. Since then two new courses have been added, one in mounted instruction and one in the training of horses.</td></tr>
</table>

Whether you integrate your presentation and discussion of the facts or treat the two separately, it is important for you to remember that your readers count upon you not only to select the facts that are relevant to them, but also to discuss those facts in a way that is meaningful to them.

Conclusions

Like interpretations, conclusions are general statements based on your facts. However, conclusions focus not simply on interpreting the facts but on answering the readers' question, "How are those facts significant to us?" In her report, for instance, Lisa provided many paragraphs of information about the university riding programs, the state of the current stable facilities, and the likely interest among alumni in contributing money for new stable facilities. After reading her presentation and discussion of those facts, Lisa's reader, Matt, might ask, "But what, exactly, does all that mean in terms of my decision about whether to start a fund-raising project for the stables?" To anticipate and answer this question, Lisa provided the following conclusions:

In conclusion, my investigation indicates that the university's riding programs could benefit substantially from a fund-raising effort. However, the appeal of such a program will be limited primarily to the very supportive alumni who used the university stables while students.

Such brief, explicit statements of conclusions are almost always desired and welcomed by report readers.

Recommendations

Just as conclusions grow out of interpretations of the facts, recommendations grow out of conclusions. They answer the readers' question, "If your conclusions are valid, what should we do?" Depending on many factors, including the number and complexity of the things you are recommending, you may state your recommendations in a single sentence or in many pages.

As mentioned above, you can help your readers immensely by stating your major recommendations at the beginning of your report. In short reports where you can state your recommendations in a few words or sentences, this may be the only place you need to present them. On the other hand, if your communication is long or if a full discussion of your recommendations requires much space, you can summarize your recommendations generally at the beginning of your report and then treat them more extensively at the end. This is what Lisa did when she summarized her recommendations in two sentences in the first paragraph of her introduction and presented and explained the recommendations in three paragraphs at the end of her report. To be sure that her readers could readily find this fuller discussion, she placed it under the heading "Conclusions and Recommendations," and she began the first paragraph of her recommendations with the words, *I recommend*.

Although readers usually want recommendations in reports, you may encounter some situations in which you will not want to include them. That might happen, for instance, in either of the two following situations:

- The decision being made is clearly beyond your competence and you have been asked to provide only a small part of the information your readers need to make the decision.
- You are working in a situation where the responsibility for making recommendations belongs to your boss or other people.

Nevertheless, in the usual situation your recommendations will be expected, or at least welcomed. If you are uncertain about whether to provide them, ask your boss or the person who asked you to report. Don't omit recommendations out of shyness or because you are guessing about what is wanted.

A NOTE ABOUT SUMMARIES

The preceding discussion concentrates on the elements found in most reports written on the job. Many longer reports share another feature: they are preceded by a separate summary of the report overall. (This summary is distinct from the summary of conclusions and recommendations that appears in the introduction.) Such summaries are often called *executive summaries* because they usually are addressed to decision-makers. Executive summaries devote special attention to conclusions and recommendations, but they also provide a general overview of the report, introducing it briefly, summarizing the communicator's method of obtaining facts, and citing and discussing (perhaps only in a sentence or two) the key findings and their interpretation.

Although such summaries are often associated with reports that have covers and are bound like books, they are also appropriate at the beginning of reports in the letter and memo formats that are more than a page long. You will find detailed advice about how to write report summaries in Appendix A, "Formats for Letters, Memos, and Books."

SAMPLE OUTLINES

The six elements of the general superstructure for reports can be incorporated into many organizational patterns. Figure 20–2 illustrates one of the possibilities. It contains the outline for a report written by Brian, who works in the Human Resource Development Department of Centre Corporation. Much to the credit of Brian's department, the company's executive management recently decided to make it a major corporate goal to develop the full potential of all of the company's employees. As a first step in that effort, the executive management asked Brian's department to study employee morale. Brian has organized his report on the study into six sections that correspond directly to the six elements of the general superstructure.

Figure 20–3 shows the much different outline for Lisa's report to Matt about a university fund-raising campaign on behalf of the stables. Like Brian's report, Lisa's has six sections. However, those sections do not bear a one-to-one relationship with the six elements of the general superstructure—although all six elements are present in her report. In her opening section, she explains what the reader will gain from her communication and describes her methods for gathering the facts she presents in the report. In each of the next four sections, she mingles her presentation of the facts with her discussion of them; each of these four sections is built around one of the four major focuses of her investigation. In the last section Lisa combines her conclusions and her recommendations.

As these two outlines suggest, when writing a report you should be certain that your communication contains all six elements of the superstructure and you should present them in an organizational pattern suited to your particular purpose and readers.

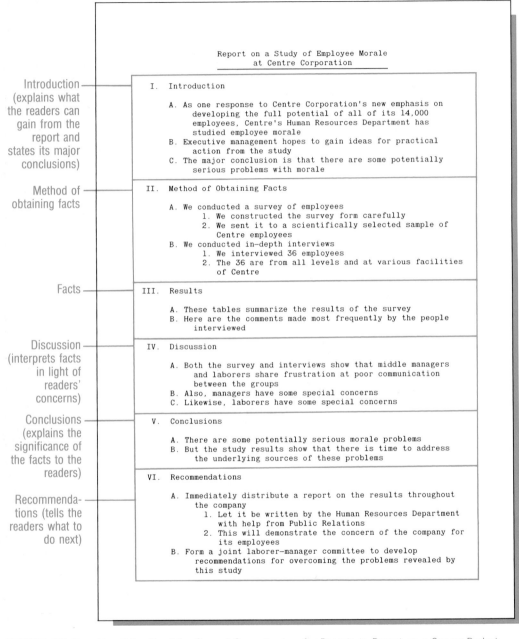

Introduction (explains what the readers can gain from the report and states its major conclusions)

Method of obtaining facts

Facts

Discussion (interprets facts in light of readers' concerns)

Conclusions (explains the significance of the facts to the readers)

Recommendations (tells the readers what to do next)

Report on a Study of Employee Morale
at Centre Corporation

I. Introduction

   A. As one response to Centre Corporation's new emphasis on
      developing the full potential of all of its 14,000
      employees, Centre's Human Resources Department has
      studied employee morale
   B. Executive management hopes to gain ideas for practical
      action from the study
   C. The major conclusion is that there are some potentially
      serious problems with morale

II. Method of Obtaining Facts

   A. We conducted a survey of employees
      1. We constructed the survey form carefully
      2. We sent it to a scientifically selected sample of
         Centre employees
   B. We conducted in-depth interviews
      1. We interviewed 36 employees
      2. The 36 are from all levels and at various facilities
         of Centre

III. Results

   A. These tables summarize the results of the survey
   B. Here are the comments made most frequently by the people
      interviewed

IV. Discussion

   A. Both the survey and interviews show that middle managers
      and laborers share frustration at poor communication
      between the groups
   B. Also, managers have some special concerns
   C. Likewise, laborers have some special concerns

V. Conclusions

   A. There are some potentially serious morale problems
   B. But the study results show that there is time to address
      the underlying sources of these problems

VI. Recommendations

   A. Immediately distribute a report on the results throughout
      the company
      1. Let it be written by the Human Resources Department
         with help from Public Relations
      2. This will demonstrate the concern of the company for
         its employees
   B. Form a joint laborer-manager committee to develop
      recommendations for overcoming the problems revealed by
      this study

FIGURE 20–2   How Brian Used the General Superstructure for Reports to Report on a Survey Project

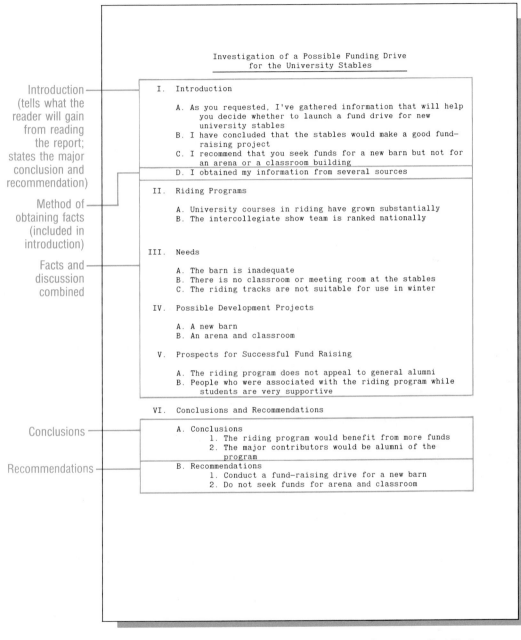

Introduction (tells what the reader will gain from reading the report; states the major conclusion and recommendation)

Method of obtaining facts (included in introduction)

Facts and discussion combined

Conclusions

Recommendations

Investigation of a Possible Funding Drive
for the University Stables

I. Introduction

   A. As you requested, I've gathered information that will help
      you decide whether to launch a fund drive for new
      university stables
   B. I have concluded that the stables would make a good fund-
      raising project
   C. I recommend that you seek funds for a new barn but not for
      an arena or a classroom building
   D. I obtained my information from several sources

II. Riding Programs

   A. University courses in riding have grown substantially
   B. The intercollegiate show team is ranked nationally

III. Needs

   A. The barn is inadequate
   B. There is no classroom or meeting room at the stables
   C. The riding tracks are not suitable for use in winter

IV. Possible Development Projects

   A. A new barn
   B. An arena and classroom

V. Prospects for Successful Fund Raising

   A. The riding program does not appeal to general alumni
   B. People who were associated with the riding program while
      students are very supportive

VI. Conclusions and Recommendations

   A. Conclusions
      1. The riding program would benefit from more funds
      2. The major contributors would be alumni of the
         program
   B. Recommendations
      1. Conduct a fund-raising drive for a new barn
      2. Do not seek funds for arena and classroom

FIGURE 20–3 How Lisa Used the General Superstructure for Reports to Report on a Fact-Finding Project

PLANNING GUIDE

Figure 20–4 shows a worksheet that may help you plan your reports. As you fill out the worksheet, remember that it is intended merely to stimulate your thinking. To achieve your objectives, you may need to say many things that you don't write down on it, and you may discover that some of the worksheet's questions are irrelevant to the particular report you are writing. Also, remember that the superstructure on p. 2 of Figure 20–4 does not necessarily provide the outline for your report.

SAMPLE REPORTS

Figure 20–5 (pp. 539–43) shows Lisa's full report. The next two chapters also contain samples that will help you understand the general superstructure for reports. While those samples show specific varieties of reports, those varieties are really just specialized versions of the general report superstructure. You might be interested in comparing the sample feasibility report (Chapter 22), which uses the book format, with Lisa's report shown in Figure 20–5 which uses the memo format.

WRITING ASSIGNMENTS

Assignments that involve writing reports are included in Appendix C.

Planning Guide

REPORT

Subject _____ **Due Date** _____

Overall Purpose

Reader Profile

Who are your primary readers?

Who else might read your report?

What are your readers' titles and professional responsibilities? How will they influence what your readers want from your report?

How well do your readers understand the technical, scientific, or other specialized terms and concepts you might use?

Do your readers have any communication preferences you should take into account? If so, what are they?

Are there any other considerations you should keep in mind when addressing these readers?

Readers' Informational Needs

What will your readers' key questions be?

Readers' Attitudes

What are your readers' present attitudes toward your *subject?* What do you want them to be?

What are your readers' present attitudes toward *you?* What do you want them to be?

FIGURE 20–4   Planning Guide for Reports

Superstructure

Introduction
>   What reason can you give your readers to read your report?
>
>   What is your main point in this report?
>
>   What background information do your readers need about your subject?
>
>   Can you help your readers by forecasting the rest of your report?

Method
>   Do you need to tell your readers anything about your method to persuade them that your facts are reliable? If so, what?

Facts
>   What data, information, and ideas do you possess or did you discover that are relevant to your readers?

Discussion
>   What generalizations do you draw from those facts? (Remember to make your generalizations from your readers' point of view.)

Conclusions
>   What is the significance of your facts to your readers?

Recommendations
>   Based upon your conclusions, what do you think your readers should do?

Organization

What organization would your readers find useful and persuasive? (Provide a brief sketch or outline to identify the major sections and subsections.)

Visual aids

What visual aids would your readers find helpful or persuasive?

FIGURE 20–4 *(continued)*

Lisa's
introduction
explains the
significance of
her report to
her reader and
summarizes her
conclusions and
recommendations

Lisa explains
her method of
obtaining facts

Lisa begins the
first of four
sections that
present and
discuss her
findings

MEMORANDUM

Central University
Development Office

FROM: Lisa Beech   February 26, 19—

TO: Matt Fordyce, Director of Funding Drives

SUBJECT: POSSIBLE FUNDING DRIVE FOR UNIVERSITY STABLES

   In this report I present the information you wanted to have
before deciding whether to place new university stables on next
year's list of major funding drives. Overall, it seems that the
stables would make a good fund–raising project because of the
strength of the current programs offered there, the condition of the
current facilities, and the existence of a loyal core of alumni who
used the facilities while undergraduates. The fund–raising should
focus on the construction of a new barn, costing $125,000. An
additional $150,000 could be sought for a much–needed arena and
classroom, but I recommend that this construction be saved for a
future fund–raising drive.

   I obtained the information given below from Peter Troivinen,
Stable Manager. Also, at last month's Alumni Weekend, I spoke with
a half–dozen alumni interested in the riding programs. Information
about construction costs comes from Roland Taberski, whose
construction firm is experienced in building the kind of facility
that would be involved.

RIDING PROGRAMS

   Begun in 1936, Central University's riding programs fall into two
categories: regular university courses offered through the stable,
and the university's horse show team, which competes nationally
through the Intercollegiate Horse Show Association.

University Courses

   The university's horseback riding courses have grown
substantially in recent years, due largely to the enthusiastic and
effective leadership of Mr. Troivinen, who took over as Stable
Manager five years ago. When Mr. Troivinen arrived, the university
offered three courses: beginning, intermediate, and advanced riding.
Since then two new courses have been added, one in mounted
instruction and one in the training of horses. In all riding
courses, students may choose either English or Western style.

FIGURE 20–5   Report That Uses the General Superstructure

February 26, 19—   Page 2

   In the past five years, the number of students taking a horseback
riding course for college credit has more than doubled from 310 to
725. That has placed a tremendous burden on both the stable and the
horses.

Intercollegiate <u>Show</u> <u>Team</u>

   When Mr. Troivinen began working at the stable, he helped
students form an intercollegiate show team, which originally had six
members. It now has over one hundred, making it one of the largest
in the country. Furthermore, the team has become competitive both
regionally and nationally. Three years ago it was first in the
region; two years ago it was third in the nation in English; and
last year it was first in the nation in Western.

<u>NEED</u>

   Mr. Troivinen and the alumni identify three major problems facing
the equestrian programs: poor barn facilities, lack of a classroom
and meeting room, and lack of an all-weather riding area.

<u>Barn</u>

   Clearly, the most pressing problem facing the equestrian programs
is the size of the barn. The barn does not provide adequate
stabling for the increased number of horses Mr. Troivinen has
purchased in response to the rising number of riders in classes and
on the horse show team. To provide at least some space for these
horses, he has made some modifications to the barn. When originally
built, the barn had six box stalls and two tie stalls. Box stalls
are the best housing for most horses because they allow the horses
to move around. Tie stalls have many uses, including providing a
place for horses between classes and while they are being curried.
However, tie stalls are not desirable for permanent housing because
the horses' heads are tied at all times. Mr. Troivinen has had to
convert all but one of the box stalls to tie stalls in order to make
room for the additional horses. Also, some horses are now kept in
the aisles.

   The barn is now full, so that no additional horses can be
purchased to meet the student demand for riding courses. As a
result, many of the horses work seven hours a day instead of the
more reasonable four or five that is standard for the type of horse
used for riding. With a seven-hour workload, the chances of health
problems increase—and the workload of a sick horse must be shifted
to other horses. Also, with this workload, the horses grow tired by
the afternoon classes and they are no longer willing to cooperate
with their riders. Consequently, the riders (especially beginning

Lisa begins the
second section
that reports and
discusses her
findings

FIGURE 20–5   *(continued)*

February 26, 19--   Page 3

riders) have trouble getting their horses to perform. However, the only way to increase the number of horses now would be to tie them up in enclosures along a fence behind the barn. Mr. Troivinen is reluctant to do that because horses kept there would have no shelter from the weather.

Other problems with the barn include the following:

- <u>Storage space for feed is too small</u>. Hay, grain, and the special feeds required for some of the horses are stored in the aisle at one end of the barn. Horses that get loose during the night sometimes eat the feed until they get sick.

- <u>Lack of storage place for equipment</u>. Saddles and bridles are kept along the aisles, where they are exposed to damage from bites or kicks of passing horses. Moreover, other necessary equipment can't be purchased because there is no place to store it.

- <u>Lack of a wash area</u>. Washing has to be done in the courtyard, where drainage is poor and there is no place to tie the horses. As a result, the horses rarely get a bath, rendering them more susceptible to fungus and skin disease.

<u>Classroom and Meeting Room</u>

At present, there is no classroom for riding classes. In good weather the classes meet outdoors and in bad weather they meet in the aisles of the barn. For the basic and intermediate riding classes, that is not a great problem. It is a problem, however, for the advanced riding class and the elementary training class, where lectures, films, and other indoor instruction are appropriate.

Also, there is no meeting room for the riding club, which in addition to its regular business meetings sometimes has guest speakers and films. The nearest rooms they can use for these purposes are classrooms in Haddock Hall, a quarter mile from the barn.

<u>Riding Tracks</u>

Riding is done on four outdoor cinder tracks. Therefore, student interest is very low during winter sessions, when the weather is cold and the riding tracks are often muddy or even covered with snow and ice. Snow and ice create additional problems. They greatly increase the chances that a horse will slip, injuring itself and its rider. Also, snow builds up on the horses' feet, impairing their

FIGURE 20–5   *(continued)*

February 26, 19——   Page 4

movements. When that happens, riding is restricted to the slower
gaits of walk and trot, so students don't receive the full content
of the course they are taking. Finally, the cold air of winter is
hard on the horses' lungs.

<u>POSSIBLE DEVELOPMENT PROJECTS</u>

Lisa begins the third section on her findings

 Mr. Troivinen has identified two subjects for a possible fund
drive: a new barn to replace the old one and a riding arena with a
classroom and viewing area attached.

<u>Barn</u>

 According to both Mr. Troivinen and the alumni I spoke with, an
adequate barn would be a 70 × 170-foot steel building. The building
would contain 32 box stalls, 27 tie stalls, 3 storage rooms for
equipment, another storage area for feed, and a washing area for
bathing the horses. Like the present barn, it would also include an
office for the Stable Manager.

 The stall area and aisles would have dirt floors to provide good
drainage. The other areas would have regular concrete floors,
except the washing area, which would need a special nonslip concrete
floor. The building could be faced with wood siding and roofed with
wood shingles to help it blend in with its setting.

 The cost of such a barn would be about $125,000. Small savings
could be realized by replacing some of the box stalls with tie
stalls, which cost about one-third as much, but as explained above
the tie stalls are not as desirable.

<u>Arena and Classroom</u>

 An indoor riding facility could be provided by building an 80 ×
200-foot arena. It would be used for riding instruction, team
practice, and horse shows. Like the barn, this could be a steel
building covered on the outside with appropriate siding and roofing.
It would have a dirt floor.

 The arena would have no seating area, under the assumption that
a classroom would be attached to its side. The classroom would
extend for 40 feet along one wall of the arena, and would be 30 feet
deep. Along the common wall, it would have a 30-foot window through
which students could watch demonstrations and spectators could watch
horse shows. The room would be equipped with a movie screen and
restroom facilities. It would also provide a meeting room for the

FIGURE 20–5   *(continued)*

February 26, 19—   Page 5

horse show club. The arena and classroom together would cost about
$150,000.

<u>PROSPECTS FOR SUCCESSFUL FUND RAISING</u>

When I spoke with alumni at last month's alumni weekend, I found
that a fund-raising drive for the stables would not be widely
appealing to Central's general alumni. Many alumni perceive that
contributions to such a drive would benefit only a relatively small
number of students, primarily those on the riding team.

However, I also found that alumni who have been involved in the
riding programs over the years are very supportive, especially
alumni from more than ten years ago, when students who owned their
own horses could board them at the barn. These alumni are
particularly impressed with the success of the intercollegiate show
team. In the past few months, one alumnus has pledged $4000 toward
the new barn, and the parents of a current member of the riding club
have donated a new treadmill, valued at over $5000, which will be
used to train young horses. Mr. Troivinen has a long list of alumni
who have expressed support for the program.

<u>CONCLUSIONS AND RECOMMENDATIONS</u>

In conclusion, my investigation indicates that the university's
riding programs could benefit substantially from a fund-raising
effort. However, the appeal of such a program will be limited
primarily to the very supportive alumni who used the university
stables while students.

Therefore, I recommend that we conduct a fund-raising drive
focused on the barn alone, leaving the arena and classroom for a
future program. The drive should be announced to all alumni in one
of the brief sketches in the brochure we send to all alumni each
year to describe our development plants. The description should
emphasize the classes taught and the need to provide better housing
for the horses.

A more extensive description of the project should be prepared
for alumni known to be interested in the riding programs. In
addition to the information provided to all alumni, the description
should emphasize the success of the horse show team and appeal to
the desire of the alumni to have a top-quality equestrian program at
their alma mater.

Though Mr. Troivinen will be disappointed with a decision not to
seek funds for the arena and classroom next year, I am sure that he
will work with us enthusiastically and effectively in the fund-
raising drive I have outlined.

This is the last section in which Lisa presents and discusses her findings

Lisa explicitly states her conclusions

Lisa presents her detailed recommendations

FIGURE 20–5   *(continued)*

21

EMPIRICAL RESEARCH REPORTS

Vernon has just completed a major research project—not library research but an experiment. He has tested two new, lightweight metal alloys to see which one will work best in the blades of jet engines. Working with a team of engineers, he placed fan blades made of each alloy in engines, which he ran through the same fifty-hour-long sequence of operating conditions. Throughout that time, he and the others monitored the performance of the blades, and afterwards they analyzed the blades to determine how well they were holding up.

Elaine is just finishing another research project, a survey to learn what kinds of outdoor recreation activities elderly people prefer and what factors influence how much they participate in those activities. With the help of six assistants, she has gathered responses to eighty carefully written questions from a scientifically selected group of two hundred elderly individuals, each interviewed at home. Then, she spent weeks performing statistical analyses of the responses.

Research of the kind performed by Vernon and Elaine is called *empirical research*. In it, the researcher gathers information by means of carefully planned, systematic observations or measurements. Although your major field may be much different from those of Vernon and Elaine, you will probably conduct some kind of empirical research regularly in your career. In this chapter, you will learn how to use the conventional superstructure for reporting empirical research.

TYPICAL WRITING SITUATIONS

You will be able to use the superstructure for empirical research reports most successfully if you understand the purposes of the research discussed in them. Basically, empirical research has two distinct purposes. Most of it aims to help people make practical decisions. For instance, the results of Vernon's experiment will be used by the engineers who design engines for his employer. The results of Elaine's survey will be used by the state agency in charge of outdoor recreation as it decides what sorts of services and facilities it must provide to meet the needs of older citizens. Similarly, when you write about your empirical research, you may be writing to people who want to use your information to make practical decisions; these people may work in your own organization or in a client's.

A smaller amount of the empirical research has a different purpose: to extend general human knowledge. The researchers set out to learn how fish remember, what the molten core of the earth is like, or why people fall in love. This research is carried out even though it has no immediate practical value, and it is usually reported in scholarly journals, such as the *Journal of Chemical Thermodynamics,* the *Journal of Cell Biology,* and the *Journal of Social Psychology,* whose readers are concerned not so much with making practical, business decisions as with extending the frontiers of human understanding and knowledge.

In some situations, these two aims of research overlap. Some organizations sponsor basic research, usually in the hope that what is learned can later be turned to some practical use. Likewise some practical research turns up results that are of interest to those who desire to learn more about the world in general.

THE QUESTIONS READERS ASK MOST OFTEN

Whether it aims to support practical decisions, extend human knowledge, or achieve some combination of the two purposes, almost all empirical research is customarily reported in the same superstructure. That's largely because the readers of all types of empirical research have the same seven general questions about it:

- **Why is your research important to us?** Readers concerned with solving specific practical problems want to know what problems your research will help them address. Readers concerned with extending human knowledge want to know how you think your research contributes to what humans know.

- **What were you trying to find out?** A key part of an empirical research project is the careful formulation of the research questions that the project will try to answer. Readers want to know what those questions are so they can determine whether they are significant questions.

- **Was your research method sound?** Unless your method is appropriate to your research questions and unless it is intellectually sound, your readers will not place any faith in your results or in the conclusions and recommendations you base upon them.

- **What results did your research produce?** Naturally, your readers will want to find out what results you obtained.

- **How do you interpret those results?** Your readers will want you to interpret your results in ways that are meaningful to them.

- **What is the significance of those results?** What answers do your results imply for your research questions, and how do your results relate to the problems your research was to help solve or to the area of knowledge your research set out to expand?

- **What do you think we should do?** Readers concerned with practical problems want to know what you advise them to do. Readers concerned with extending human knowledge want to know what you think your results imply for future research.

observed

SUPERSTRUCTURE FOR EMPIRICAL RESEARCH REPORTS

To answer the readers' typical questions about empirical research projects, writers use a superstructure that has the following elements: introduction, objectives of the research, method, results, discussion, conclusions, and recommendations. As Figure 21–1 shows, each of these elements addresses one of the readers' seven questions.

The rest of this chapter explains how you can develop each element of an empirical research report. If you have not already done so, you should also read Chapter 20, which describes the _general_ superstructure for reports. It provides information not repeated here that will help you write empirical research reports effectively.

Much of the advice provided in this chapter is illustrated through the use of two sample reports. The first is presented in full at the end of this chapter. Its aim is practical. It was written by engineers who are developing a satellite communication system that will permit companies with large fleets of trucks to communicate directly with their drivers at any time.[1] In the report, the writers tell decision-makers and other engineers in their organization about the first operational test of the system, in which they sought to answer several practical engineering questions.

overview results
(RESults up front)
Context
factom
Data

Report Element	Readers' Question
Introduction	Why is your research important to us?
Objectives of the research	What were you trying to find out?
Method	Was your research method sound?
Results	What results did your research produce?
Discussion	How do you interpret those results?
Conclusions	What is the significance of those results?
Recommendations	What do you think we should do?

FIGURE 21–1   Elements of an Empirical Research Report and Their Relationship to the Questions Readers Ask

In contrast, the aim of the second sample report is to extend human knowledge. The researcher tells about his study of the ways people develop friendships. Selected passages from this report are quoted throughout the rest of this chapter; if you wish to read the entire report you will find it in the April 1985 issue of the *Journal of Personality and Social Psychology*.[2]

Introduction

In the introduction to an empirical research report, you should seek to answer the readers' question, "Why is this research important to us?" Typically, writers answer that question in two steps: they announce the topic of their research and then explain the importance of the topic to their readers.

Announcing the Topic

You can often announce the topic of your research simply by including that topic as the key phrase in the opening sentence of your report. For example, consider the first sentence of the report on the satellite communication system:

Topic of report

> For the past eighteen months, the Satellite Products Laboratory has been developing a system that will permit companies with large, nationwide fleets of trucks to communicate directly to their drivers at any time through a satellite link.

Here is the first sentence from the report on the way that people develop friendships:

Topic of report

> Social psychologists know very little about the way real friendships develop in their natural settings.

Explaining the Importance of the Research

To explain the importance of your research to your readers, you can use either or both of the following methods: state the relevance of the research to your organization's goals, and review the previously published literature on the subject.

Relevance to Organization Goals
In reports written to readers in organizations (whether your own or a client's), you can explain the relevance of your research by relating it to some organizational goal or problem. Sometimes, in fact, the importance of your research to the organization's needs will be so

obvious to your readers that merely naming your topic will be sufficient. At other times, you will need to discuss at length the relevance of your research to the organization. In the first paragraph of the satellite report, for instance, the writers mention the potential market for the satellite communication system they are developing. That is, they explain the importance of their research by saying that it can lead to a profit. For detailed advice about how to explain the importance of your research to readers in organizations, see the advice given in Guideline 1 of Chapter 9, ''Writing the Beginning of a Communication.''

Literature Reviews   A second way to establish the importance of your research is to review the existing knowledge on your subject. Writers usually do this by reviewing the previously published literature. Generally, you can arrange a literature review in two parts. First, present the main pieces of knowledge communicated in the literature. Then, identify some significant gap in this knowledge—the very gap your own research will fill. In this way, you establish the special contribution that your research will make.

The following passage from the opening of the report on the friendship study shows this two-stage pattern of development. Notice that the writer uses author–year citations, as described in Appendix B.

The writer tells what is known on his topic

A great deal of research in social psychology has focused on variables influencing an individual's attraction to another at an initial encounter, usually in laboratory settings (Bergscheid and Walster, 1978; Bryne, 1971; Huston and Levinger, 1978), yet very little data exists on the processes by which individuals in the real world move beyond initial attraction to develop a friendship; even less is known about the way developing friendships are maintained and how they evolve over time (Huston and Burgess, 1979; Levinger, 1980).

The writer identifies the gaps in knowledge that his research will fill

The writer continues this discussion of previous research for three paragraphs. Each follows the same pattern: it identifies an area of research, tells what is known about that area, and identifies gaps in the knowledge—gaps that will be filled by the research that the writer has conducted. These paragraphs serve an important additional function also performed by many literature reviews: they introduce the established facts and theories that are relevant to the writer's work and necessary to the understanding of the report.

Writers almost always include literature reviews in the reports they write for professional journals. In contrast, they often omit reviews when writing to readers inside an organization. That's because such reviews are often unnecessary when addressing organizational readers. Organizational readers judge the importance of a report in terms of its relevance to the organization's goals and problems, not in terms of its relation to the general pool of human knowledge.

For example, the typical readers of the truck-and-satellite communication report were interested in the report because they wanted to learn how well their company's system would work. To them, a general survey of the literature on satellite communication would have seemed irrelevant—and perhaps even annoying.

A second reason that writers often omit literature reviews when addressing readers in organizations is that such reviews rarely help such readers understand the reports. That's because the research projects undertaken within organizations usually focus so sharply on a particular, local question that published literature on the subject is beside the point. For example, a review of previously published literature on satellite communications would not have helped readers understand the truck-and-satellite report.

Sometimes, of course, literature reviews do appear in reports written to organizational readers. Often, they say something like this: "In a published article, one of our competitors claims to have saved large amounts of money by trying a new technique. The purpose of the research described in this report is to determine whether or not we could enjoy similar results." Of course, the final standard for judging whether you should include a literature review in your report is your understanding of your purpose and readers. In some way or another, however, the introductions to all your empirical research reports should answer your readers' question, "Why is this research important to us?"

Objectives of the Research

Every empirical research project has carefully constructed objectives. These objectives define the focus of your project, influence the choice of research method, and shape the way you interpret your results. Thus, readers of empirical research reports want and need to know what the objectives are.

The following example from the satellite report shows one way you can tell your readers about your objectives:

> In particular, we wanted to test whether we could achieve accurate
> data transmissions and good-quality voice transmissions in the variety
> of terrains typically encountered in long-haul trucking. We wanted
> also to see what factors might affect the quality of transmissions.

When reporting on research that involves the use of statistics, you can usually state your objectives by stating the hypotheses you tested. Where appropriate, you can explain these hypotheses in terms of existing theory, again citing previous publications on the subject. The following passage shows how the writer who studied friendship explains some of his hypotheses. Notice how the author begins with a statement of the overall goal of the research:

Overall goal

First objective
(hypothesis)
Second objective
(hypothesis)
Third objective
(hypothesis)

> The goal of the study was to identify characteristic behavioral and attitudinal changes that occurred within interpersonal relationships as they progressed from initial acquaintance to close friendship. With regard to relationship benefits and costs, it was predicted that both benefits and costs would increase as the friendship developed, and that the ratings of both costs and benefits would be positively correlated with ratings of friendship intensity. In addition, the types of benefits listed by the subjects were expected to change as the friendships developed. In accord with Levinger and Snoek's (1972) model of dyadic relatedness, benefits listed at initial stages of friendship were hypothesized to be more activity centered and to reflect individual self-interest (e.g., companionship, information) than benefits at later stages, which were expected to be more personal and reciprocal (e.g., emotional support, self-esteem).

Method

When reading reports of your empirical research, people will look for precise details concerning your method. Those details serve three purposes. First, they let your readers assess the soundness of your research design and its appropriateness for the problems you are investigating. Second, the details enable your readers to determine the limitations that your method might place upon the conclusions you can draw. Third, the description of your method provides information that will help your readers repeat your experiment if they wish to verify your results or conduct similar research projects of their own.

The nature of the information you should provide about your method depends upon the nature of your research. For instance, the writer studying friendship began his description of his research methods in this way:

> At the beginning of their first term at the university, college freshmen selected two individuals whom they had just met and completed a series of questionnaires regarding their relationships with those individuals at 3-week intervals through the school term.

In the rest of that paragraph, the writer explains that the questionnaires asked the freshmen to tell about such things as their attitudes toward each of the other two individuals and the specific things they did with each of the other two. However, that paragraph is just a small part of the researcher's account of his method. He then provides a 1200-word discussion of the students he studied and of the questionnaires and procedures he used.

The writers of the satellite report likewise provided detailed information about their procedures: three paragraphs and two tables explaining their equipment (truck radios and satellite), two paragraphs and one map describing the eleven-state region covered by the trucks, and two paragraphs describing their data analysis.

How can you decide what details to include? The most obvious way is to follow the general reporting practices in your field. Find some research reports that use a method similar to yours and see what they report. You can check the scope of your description in the following ways:

- List every aspect of your procedure that you made a decision about when planning your research.
- Identify every aspect of your method that your readers might ask about.
- Ask yourself what aspects of your procedure might limit the conclusions you can draw from your results.
- Identify every procedure that other researchers would need to understand in order to design a similar study.

Results

The results of empirical research are the data you obtain. Although your results are the heart of your empirical research report, they may take up a very small portion of it. Generally, results are presented in one of two ways:

- **Tables and Graphs.** The satellite report, for instance, uses two tables. The report on friendship uses four tables and eleven graphs. For information on using tables, graphs, and other visual aids, see Chapters 13 and 14, "Using Visual Aids" and "Creating Twelve Types of Visual Aids."
- **Sentences.** When placed in sentences, results are often woven into a discussion that combines data and interpretation, as explained in the next paragraphs.

Discussion

Sometimes writers briefly present all their results in one section and then discuss them in a separate section. Sometimes they combine the two in a single, integrated section. Whichever method you use, your discussion must link your interpretative comments with the specific results you are interpreting.

One useful way of making that link is to refer to the key results shown in a table or other visual aid and then comment on them as appropriate. The following passage shows how the writers of the satellite report did that for some of the results that they presented in one of their tables:

Writers emphasize
a key result shown
in a table

Writers draw
attention to other
important results

Writers interpret
those results

> As Table 3 shows, 91% of the data transmissions were successful. These data are reported according to the region in which the trucks were driving at the time of transmission. The most important difference to note is the one between the rate of successful transmissions in the Southern Piedmont region and the rates in all the other regions. In the Southern Piedmont area, we had the truck drive slightly outside the ATS-6 footprint so that we could see if successful transmissions could be made there. When the truck left the footprint, the percentage of successful data transmissions dropped abruptly to 43%.

When you present your results in prose only (rather than in tables and graphs), you can weave those results together with your discussion by beginning your paragraphs with general statements that are really interpretations of your data. Then, cite the relevant results as evidence in support of the interpretation. Here is an example from the friendship report:

General
interpretation

Specific results
presented as
support for the
interpretation

> Intercorrelations among the subjects' friendship intensity ratings at the various assessment points showed that friendship attitudes became increasingly stable over time. For example, the correlation between friendship intensity ratings at 3 weeks and 6 weeks was .55; between 6 weeks and 9 weeks, .78; between 9 weeks and 12 weeks, .88 (all $p < .001$).

In a single report, you may use both of these methods of combining the presentation and discussion of your results.

Conclusions

Besides interpreting the results of your research, you need to explain what your results mean in terms of the original research questions and the general problem you set out to investigate. Your explanations of these matters are your conclusions.

If your research project is sharply focused (for example) on only a single hypothesis, your conclusion can be very brief, perhaps only a restatement of your chief results. However, if your research has many threads, your conclusion should draw those strands together.

In either case, the presentation of your conclusions should correspond very closely to the objectives for your research that you identified toward the beginning of your report. Consider, for instance, the correspondence between objectives and conclusions in the satellite study. The first objective was to determine whether accurate data transmissions and good-quality voice transmissions could be obtained in the variety of terrains typically encountered in long-haul trucking. The first of the conclusions addresses that objective:

> The Satellite Product Laboratory's system produces good-quality data and voice transmissions throughout the eleven-state region covered by the satellite's broadcast footprint.

The second objective was to determine what factors affect the quality of transmissions, and the second and third conclusions relate to it:

> The most important factor limiting the success of transmissions is movement outside the satellite's broadcast footprint, which accurately defines the satellite's area of effective coverage.

> The system is sensitive to interference from certain kinds of objects in the line of sight between the satellite and the truck. These include trees, mountains and hills, overpasses, and buildings.

The satellite research concerns a practical question. Hence its objectives and conclusions address practical concerns of particular individuals—in this case, the engineers and managers in the company that is developing the satellite system. In contrast, research that aims primarily to extend human knowledge often has both objectives and conclusions that focus on theoretical issues.

For example, at the beginning of the friendship report, the researcher identifies several questions that his research investigated, and he tells what answers he predicted his research would produce. In his conclusion, then, he systematically addresses those same questions, talking about them in terms of the results his research produced. Here is a summary of some of his objectives and conclusions. (Notice how he uses the technical terminology commonly employed by his readers.)

Objective	Conclusion
As they develop friendships, do people follow the kind of pattern theorized by Guttman, in which initial contacts are relatively superficial and later contacts are more intimate?	Yes. "The initial interactions of friends . . . correspond to a Guttman-like progression from superficial interaction to increasingly intimate levels of behavior."

Do *both* the costs (or unpleasant aspects) and the benefits of personal relationships increase as friendships develop?	Yes. "The findings show that personal dissatisfactions are inescapable aspects of personal relationships and so, to some degree, may become immaterial. The critical factor in friendships appears to be the amount of benefits received. If a relationship offers enough desirable benefits, individuals seem willing to put up with the accompanying costs."
Are there substantial differences between the friendships women develop with one another and the kind men develop with one another?	Apparently not. "These findings suggest that—at least for this sample of friendships—the sex differences were more stylistic than substantial. The bonds of male friendship and female friendship may be equally strong, but the sexes may differ in their manner of expressing that bond. Females may be more inclined to express close friendships through physical or verbal affection; males may express their closeness through the types of companionate activities they share with their friends."

Typically, in a discussion of the conclusions of an empirical research project directed primarily at increasing human knowledge, writers will also discuss the relationship of their findings to the findings of other researchers and to various theories that have been advanced concerning their subject. The writer of the article on friendship did that. The table you have just read presents only a few snippets from his overall discussion, which is several thousand words long and is full of thoughts about the relationship of his results to the results and theories of others.

In the discussion sections of their empirical research reports, writers sometimes discuss any flaws in their research method or limitations on the generalizability of their conclusions. For example, the writer on friendship points out that his subjects were all college students and most lived in dormitories. It is possible, he cautions, that the things he found while studying this group may not be true for other groups.

Recommendations

The readers of some empirical research reports want to know what, based on the research, the writer thinks should be done. This is especially true in situations where the research is directed at solving a practical problem. Consequently, research reports usually include recommendations. For example, the satellite report contains three. The first is the general recommendation that work on the project be continued. The other two involve specific actions that the writers think should be taken: design a special antenna for the trucks and develop a plan that tells what satellites would be needed to provide coverage throughout the 48 contiguous states, Alaska, and Southern Canada. As is common in research addressed to readers in organizations, these recommendations concern practical business and engineering decisions.

Even in reports directed toward extending human knowledge, writers sometimes have recommendations. These usually concern their ideas about future studies that should be made, adjustments in methodology that seem to be called for, and the like. In the last paragraph of the friendship report, for instance, the writer suggests that additional research be conducted that studies different groups (not just students) in different settings (not just college) to establish a more comprehensive understanding of how friendships develop.

AN IMPORTANT NOTE ABOUT HEADINGS

In the discussion you have just read, you learned about the seven elements of the conventional superstructure for reports on empirical research. Such reports may be divided into seven sections that correspond directly to those seven elements. That is especially likely to happen in long reports written to readers in organizations (as distinct from the readers of professional journals). However, in some reports some of the elements are combined under a single heading. For example, in the satellite report, the objectives are included at the end of the section entitled "Introduction," and the results and discussion are combined under the single heading "Results and Discussion." Even when only a few major headings are used, almost all empirical research reports include all seven of the elements described above. You should determine which headings to use in your reports by considering your purpose and your readers' expectations.

PLANNING GUIDE

Figure 21–2 (pp. 558–59) shows a worksheet that may help you plan your empirical research reports. As you fill out the worksheet, remember that it is intended merely to stimulate your thinking. To achieve your objectives, you may need to say many things that you don't write down on it, and you may discover that some of the worksheet's questions are irrelevant to the particular report you are writing. Also, remember that the superstructure on p. 2 of Figure 21–2 does not necessarily provide the outline for your report.

SAMPLE RESEARCH REPORT

Figure 21–3 (pp. 560–77) shows the full report on the truck-and-satellite communication system, which is addressed to readers within the writers' own organization. To see examples of empirical research reports presented as journal articles, consult journals in your field.

WRITING ASSIGNMENT

An assignment that involves writing an empirical research report appears in Appendix C.

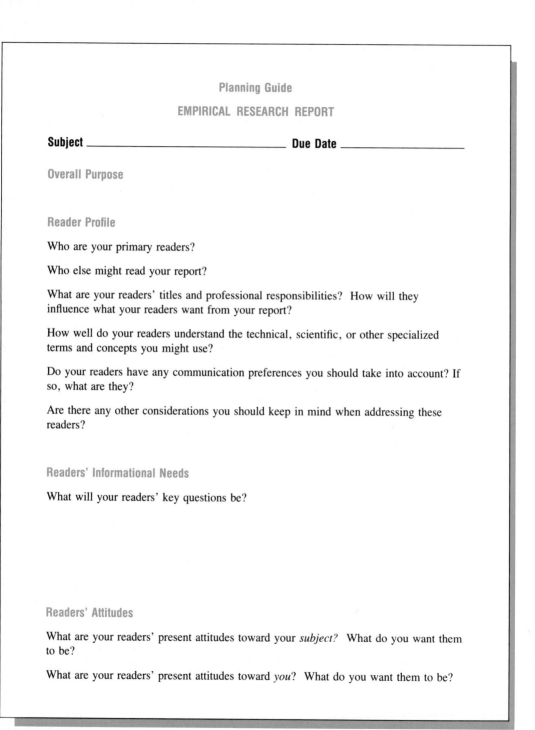

Planning Guide

EMPIRICAL RESEARCH REPORT

Subject _____ **Due Date** _____

Overall Purpose

Reader Profile

Who are your primary readers?

Who else might read your report?

What are your readers' titles and professional responsibilities? How will they influence what your readers want from your report?

How well do your readers understand the technical, scientific, or other specialized terms and concepts you might use?

Do your readers have any communication preferences you should take into account? If so, what are they?

Are there any other considerations you should keep in mind when addressing these readers?

Readers' Informational Needs

What will your readers' key questions be?

Readers' Attitudes

What are your readers' present attitudes toward your *subject?* What do you want them to be?

What are your readers' present attitudes toward *you*? What do you want them to be?

FIGURE 21–2 Planning Guide for Empirical Research Reports

Superstructure

Introduction
   Why is your research important to your readers?

   What is your main point in this report?

   What background information do your readers need about your subject?

   Can you help your readers by forecasting the rest of your report?

Objectives
   What were you trying to find out?

Method
   What, if anything, will your readers want to know about your method?

   What additional things should you tell your readers about your method to persuade them that your method was sound?

Results
   What results did your research produce?

Discussion
   What generalizations do you draw from the results?

Conclusions
   What is the significance of these results?

Recommendations
   Based on your conclusions, what do you think your readers should do?

Visual Aids

What visual aids would your readers find helpful or persuasive?

Organization

What organization would your readers find useful and persuasive? (On a separate sheet provide a brief sketch or outline to identify the major sections and subsections.)

FIGURE 21–2   *(continued)*

Cover

ELECTRONICS CORPORATION OF AMERICA

Truck-to-Satellite Communication System: First Operational Test

September 30, 19--

Internal Technical Report

Number TR-SPL-0931

FIGURE 21–3   Empirical Research Report

Title Page

ELECTRONICS CORPORATION OF AMERICA

Truck–to–Satellite Communication System:
First Operational Test

September 30, 19--

Research Team

Margaret C. Barnett

Erin Sanderson

L. Victor Sorrentino

Raymond E. Wu

Internal Technical Report
Number TR–SPL–0931

Read and Approved:

Cynthia Robinson   _10-23_
Laboratory Director   Date

FIGURE 21–3   *(continued)*

<u>EXECUTIVE SUMMARY</u>

Topic of report

Method

Results and
discussion

Conclusions

Recommendations

For the past eighteen months, the Satellite Products Laboratory
has been developing a system that will permit companies with large,
nationwide fleets of trucks to communicate directly to their drivers
at any time through a satellite link. During the week of May 18, we
tested our concepts for the first time, using the ATS-6 satellite
and five trucks that were driven over an eleven-state region with
our prototype mobile radios.

More than 91% of the 2500 data transmissions were successful and
more than 91% of the voice transmissions were judged to be of
commercial quality. The most important factor limiting the success
of transmissions was movement outside the satellite's broadcast
footprint. Other factors include the obstruction of the line of
sight between the truck and the satellite by highway overpasses,
mountains and hills, trees, and buildings.

Overall, the test demonstrated the soundness of the prototype
design. Work on it should continue as rapidly as possible. We
recommend the following actions:

- Develop a new antenna designed specifically for use in
  communications between satellites and mobile radios.

- Explore the configuration of satellites needed to provide
  thorough footprint coverage for the 48 contiguous states,
  Alaska, and Southern Canada at an elevation of 25° or more.

1

FIGURE 21–3   *(continued)*

Cross-reference
to related
reports

Acknowledgments

<u>FOREWORD</u>

Previous technical reports describing work on this project are TR–SPL–0785, TR–SPL–0795 through TR–SPL–0798, TR–SPL–0823, and TR–SPL–0862.

We could not have conducted this test without the gratifying cooperation of Smithson Moving Company and, in particular, the drivers and observers in the five trucks that participated. We are also grateful for the cooperation of the United States National Aeronautics and Space Administration in the use of the ATS–6 satellite.

The test described in this report was supported through internal product development funds.

ii

FIGURE 21–3   *(continued)*

Table of
contents

iii

FIGURE 21–3   *(continued)*

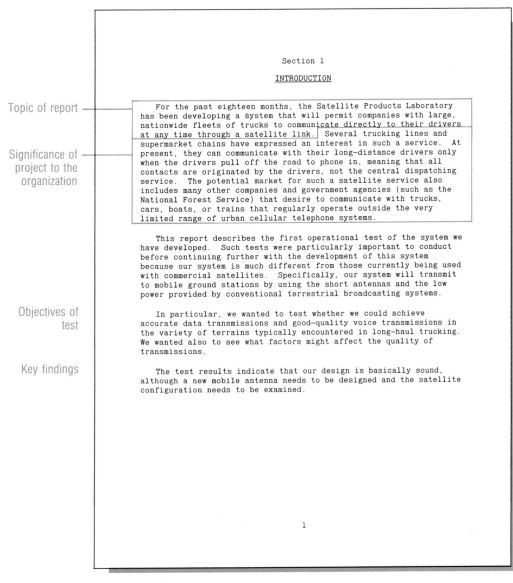

Section 1

INTRODUCTION

Topic of report —

    For the past eighteen months, the Satellite Products Laboratory
has been developing a system that will permit companies with large,
nationwide fleets of trucks to communicate directly to their drivers
at any time through a satellite link. Several trucking lines and

Significance of
project to the
organization
supermarket chains have expressed an interest in such a service. At
present, they can communicate with their long-distance drivers only
when the drivers pull off the road to phone in, meaning that all
contacts are originated by the drivers, not the central dispatching
service. The potential market for such a satellite service also
includes many other companies and government agencies (such as the
National Forest Service) that desire to communicate with trucks,
cars, boats, or trains that regularly operate outside the very
limited range of urban cellular telephone systems.

    This report describes the first operational test of the system we
have developed. Such tests were particularly important to conduct
before continuing further with the development of this system
because our system is much different from those currently being used
with commercial satellites. Specifically, our system will transmit
to mobile ground stations by using the short antennas and the low
power provided by conventional terrestrial broadcasting systems.

Objectives of
test
    In particular, we wanted to test whether we could achieve
accurate data transmissions and good-quality voice transmissions in
the variety of terrains typically encountered in long-haul trucking.
We wanted also to see what factors might affect the quality of
transmissions.

Key findings
    The test results indicate that our design is basically sound,
although a new mobile antenna needs to be designed and the satellite
configuration needs to be examined.

1

FIGURE 21–3   *(continued)*

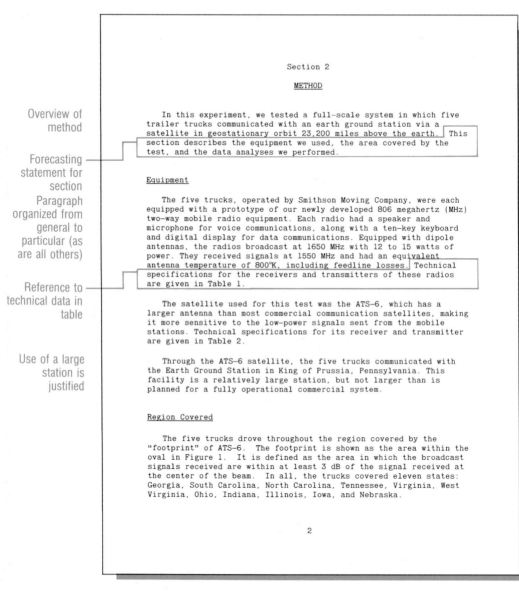

Overview of method

Forecasting statement for section

Paragraph organized from general to particular (as are all others)

Reference to technical data in table

Use of a large station is justified

Section 2

METHOD

In this experiment, we tested a full-scale system in which five trailer trucks communicated with an earth ground station via a satellite in geostationary orbit 23,200 miles above the earth. This section describes the equipment we used, the area covered by the test, and the data analyses we performed.

Equipment

The five trucks, operated by Smithson Moving Company, were each equipped with a prototype of our newly developed 806 megahertz (MHz) two-way mobile radio equipment. Each radio had a speaker and microphone for voice communications, along with a ten-key keyboard and digital display for data communications. Equipped with dipole antennas, the radios broadcast at 1650 MHz with 12 to 15 watts of power. They received signals at 1550 MHz and had an equivalent antenna temperature of 800°K, including feedline losses. Technical specifications for the receivers and transmitters of these radios are given in Table 1.

The satellite used for this test was the ATS-6, which has a larger antenna than most commercial communication satellites, making it more sensitive to the low-power signals sent from the mobile stations. Technical specifications for its receiver and transmitter are given in Table 2.

Through the ATS-6 satellite, the five trucks communicated with the Earth Ground Station in King of Prussia, Pennsylvania. This facility is a relatively large station, but not larger than is planned for a fully operational commercial system.

Region Covered

The five trucks drove throughout the region covered by the "footprint" of ATS-6. The footprint is shown as the area within the oval in Figure 1. It is defined as the area in which the broadcast signals received are within at least 3 dB of the signal received at the center of the beam. In all, the trucks covered eleven states: Georgia, South Carolina, North Carolina, Tennessee, Virginia, West Virginia, Ohio, Indiana, Illinois, Iowa, and Nebraska.

2

FIGURE 21–3 *(continued)*

Table 1
Specifications for Satellite–Aided Mobile Radio

Transmitter

Frequency	1655.050 MHz
Power Output	16 watts nominal
	12 watts minimum
Frequency Stability	±0.0002% (−30° to +60°C)
Modulation	$16F_3$ Adjustable from 0 to ±5 kHz swing FM with instantaneous modulation limiting
Audio Frequency Response	Within +1 dB and −3 dB of a 6 dB/octave pre–emphasis from 300 to 3000 HZ per EIA standards
Duty Cycle	EIA 20% Intermittent
Maximum Frequency Spread	±6 MHz with center tuning
RF Output Impedance	50 ohms

Receiver

Frequency	1552.000 MHz
Frequency Stability	±0.0002% (−30° to +60°C)
Noise Figure	2.6 dB referenced to transceiver antenna jack
Equivalent Receiver Noise Temperature	238° Kelvin
Selectivity	−75 dB by EIA Two–Signal Method
Audio Output	5 watts at less than 5% distortion
Frequency Response	Within +1 and −8 dB of a standard 6 dB per octave deemphasis curve from 300 to 3000 Hz
Modulation Acceptance	±7 kHz
RF Input Impedance	50 ohms

3

FIGURE 21–3   *(continued)*

Table 2
Performance of ATS–6 Spacecraft L–Band Frequency
Translation Mode

<u>Receive</u>

Receiver Noise Figure (dB)	6.5
Equivalent Receiver Noise Temperature (°K)	1005
Antenna Temperature Pointed at Earth (°K)	290
Receiver System Temperature (°K)	1295
Antenna Gain, peak (dB)	38.4
Spacecraft G/T, peak (dB/°K)	7.3
Half Power Beamwidth (degrees)	1.3
Gain over Field of View (dB)	35.4
Spacecraft G/T over Field of View (dB/°K)	4.3

<u>Transmit</u>

Transmit Power (dBw)	15.3
Antenna Gain, peak (dB)	37.7
Effective Radiated Power, peak (dBw)	53.0
Half Power Beamwidth (degrees)	1.4
Gain over Field of View (dB)	34.8
Effective Radiated Power over Field of View (dBw)	50.1

4

FIGURE 21–3   *(continued)*

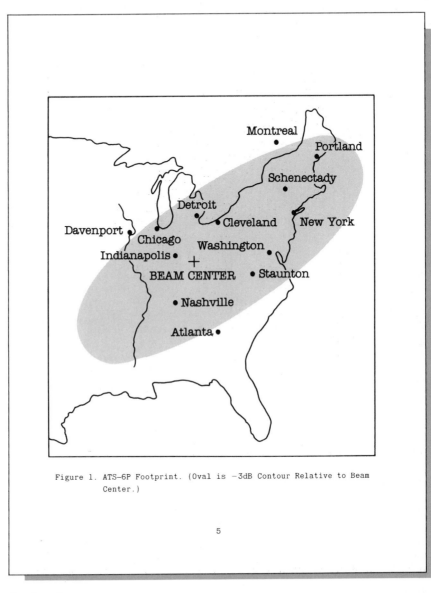

Figure 1. ATS-6P Footprint. (Oval is −3dB Contour Relative to Beam Center.)

5

FIGURE 21–3   *(continued)*

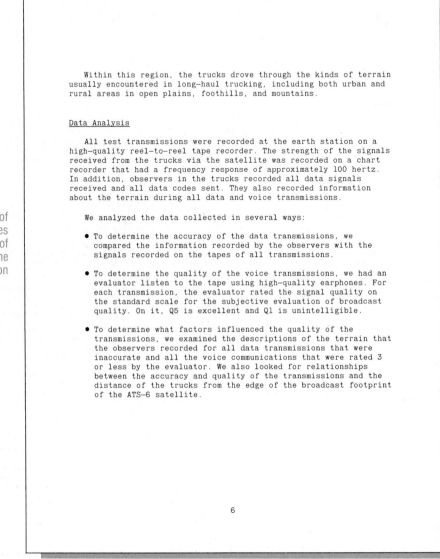

Within this region, the trucks drove through the kinds of terrain usually encountered in long—haul trucking, including both urban and rural areas in open plains, foothills, and mountains.

<u>Data Analysis</u>

All test transmissions were recorded at the earth station on a high—quality reel—to—reel tape recorder. The strength of the signals received from the trucks via the satellite was recorded on a chart recorder that had a frequency response of approximately 100 hertz. In addition, observers in the trucks recorded all data signals received and all data codes sent. They also recorded information about the terrain during all data and voice transmissions.

We analyzed the data collected in several ways:

- To determine the accuracy of the data transmissions, we compared the information recorded by the observers with the signals recorded on the tapes of all transmissions.

- To determine the quality of the voice transmissions, we had an evaluator listen to the tape using high—quality earphones. For each transmission, the evaluator rated the signal quality on the standard scale for the subjective evaluation of broadcast quality. On it, Q5 is excellent and Q1 is unintelligible.

- To determine what factors influenced the quality of the transmissions, we examined the descriptions of the terrain that the observers recorded for all data transmissions that were inaccurate and all the voice communications that were rated 3 or less by the evaluator. We also looked for relationships between the accuracy and quality of the transmissions and the distance of the trucks from the edge of the broadcast footprint of the ATS—6 satellite.

6

Presentation of data analyses parallels list of objectives in the introduction

FIGURE 21–3 *(continued)*

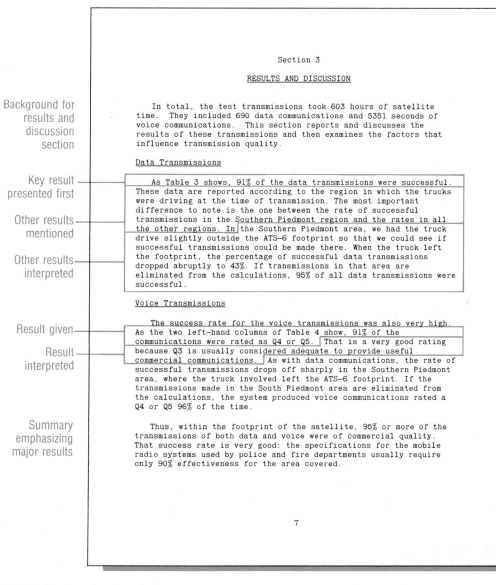

Background for
results and
discussion
section

Key result
presented first

Other results
mentioned

Other results
interpreted

Result given

Result
interpreted

Summary
emphasizing
major results

Section 3

RESULTS AND DISCUSSION

In total, the test transmissions took 603 hours of satellite
time. They included 690 data communications and 5351 seconds of
voice communications. This section reports and discusses the
results of these transmissions and then examines the factors that
influence transmission quality.

Data Transmissions

As Table 3 shows, 91% of the data transmissions were successful.
These data are reported according to the region in which the trucks
were driving at the time of transmission. The most important
difference to note is the one between the rate of successful
transmissions in the Southern Piedmont region and the rates in all
the other regions. In the Southern Piedmont area, we had the truck
drive slightly outside the ATS-6 footprint so that we could see if
successful transmissions could be made there. When the truck left
the footprint, the percentage of successful data transmissions
dropped abruptly to 43%. If transmissions in that area are
eliminated from the calculations, 95% of all data transmissions were
successful.

Voice Transmissions

The success rate for the voice transmissions was also very high.
As the two left-hand columns of Table 4 show, 91% of the
communications were rated as Q4 or Q5. That is a very good rating
because Q3 is usually considered adequate to provide useful
commercial communications. As with data communications, the rate of
successful transmissions drops off sharply in the Southern Piedmont
area, where the truck involved left the ATS-6 footprint. If the
transmissions made in the South Piedmont area are eliminated from
the calculations, the system produced voice communications rated a
Q4 or Q5 96% of the time.

Thus, within the footprint of the satellite, 95% or more of the
transmissions of both data and voice were of commercial quality.
That success rate is very good: the specifications for the mobile
radio systems used by police and fire departments usually require
only 90% effectiveness for the area covered.

7

FIGURE 21–3   *(continued)*

Table 3
Success in Decoding DTMF Automatic Transmitter

REGION	STATES	ATS TRANSMISSIONS		ELEVATION
		Sent by Vehicle	Received and Decoded Correctly	Angle to Satellite (°)
Open Plains	Indiana, Ohio, Nebraska, Illinois, Iowa	284	283 (100%)	17–26
Western Appalachian Foothills	Ohio, Tennessee	55	53 (96%)	15–19
Appalachian Mountains	West Virginia	112	93 (83%)	15–17
Piedmont	Virginia, North Carolina	190	178 (97%)	11–16
Southern Piedmont	Georgia, South Carolina	49	21 (43%)	17–18
TOTAL		690	628 (91%)	

8

FIGURE 21–3   *(continued)*

Table 4
Quality of Voice Communication Signal[1]

Area	Transmission Time (Seconds)	No Blockage Time (Secs)		Trees			Mountains and Hills			Overpasses (Momentary Dropouts)			Buildings		
		Q5	Q4	Q3	Q2	Q1	Q3	Q2	Q1	Q3	Q2	Q1	Q3	Q2	Q1
Open Plains	2481	2334	73	1	1	0	4	1	3	13	2	20	15	10	4
Western Foothills	344	322	2	0	0	0	6	12	13	0	0	0	0	0	0
Appalachian Mountains	1037	614	267	42	17	0	20	31	37	0	2	7	0	0	0
Piedmont	1219	481	623	5	15	19	25	16	10	4	1	4	8	4	1
Southern Piedmont	270	0	149	109	3	12	0	0	0	0	0	2	0	0	0
TOTAL TIME (seconds)	5351	3751	1114	157	36	31	55	60	63	17	5	33	23	14	5

[1]Total times in seconds for each quality of received signals. Q5 is excellent; Q1 is unintelligible.

9

FIGURE 21–3 *(continued)*

<u>Factors Affecting the Quality of Transmissions</u>

The factor having the largest effect on transmission quality is the location of the truck within the footprint of ATS-6. The quality of transmissions is even and uniformly good throughout the footprint, but almost immediately outside of it the quality drops well below acceptable levels.

Several factors were found to disrupt transmissions even when the trucks were in the satellite's footprint. The four right-hand columns in Table 4 show what these factors are for the 4% of the transmissions in the footprint that were not of commercial quality. In all cases the cause is some object passing in the line-of-sight between the satellite and the truck.

Trees caused 45% of the disruptions, more than any other source. At the frequencies used for broadcasting in this system, trees and other foliage have a high and very sharp absorption. Of course, the tree must be immediately beside the road and also tall enough to intrude between the truck and the satellite, which is at an average elevation of 17° above the horizon. And the disruption caused by a single tree will create only a very brief and usually insignificant dropout of one second or less. Only driving past a group of trees will cause a significant loss of signal. Yet this happened often in the terrain of the Appalachian Mountains and the South Piedmont.

We believe we could eliminate many of the disruptions caused by trees if we developed an antenna specifically for use in communications between satellites and mobile radios. In the test, we used a standard dipole antenna. Instead, we might devise a Wheeler-type antenna that is omnidirectional in azimuth and with gain in the vertical direction to minimize ground reflections.

Mountains and hills caused 36% of the disruptions. That happened mostly in areas where the satellite's elevation above the horizon was very low. Otherwise a hill or mountain would have to be very steep to block out a signal. For example, if a satellite were only 17° above the horizon, the hill would have to rise over 1500 feet per mile to interfere with a transmission—and the elevation would have to be precisely in a line between the truck and the satellite.

Most of the disruptions caused by mountains and hills can be eliminated by using satellites that have an elevation of at least 25°. That would place them above all but the very steepest slopes.

10

FIGURE 21–3   *(continued)*

Most important factor presented first

Reference to detailed data in table

Key result presented and explained

    Highway overpasses accounted for 11% of the disruptions, but
these disruptions had little effect on the overall quality of the
broadcasts. As one of the test trucks drove on an open stretch of
interstate highway, the signal was strong and steady, with fading
less than 2 decibels peak-to-peak. About two seconds before the
truck entered the overpass, there was detectable but not severe
multipath interference. The only serious disruption was a one-
second dropout while the truck was directly under the overpass.
This one-second dropout was so brief that it did not cause a
significant loss of intelligibility in voice communications. Only a
series of overpasses, such as those found where interstate highways
pass through some cities, cause a significant problem.

    Finally, buildings and similar structures accounted for about 8%
of the disruptions. These were experienced mainly in large cities,
and isolated buildings usually caused only brief disruptions.
However, when the trucks were driving down city streets lined
with tall buildings, they were unable to obtain satisfactory
communications until they were driven to other streets.

                                  ·11

FIGURE 21–3   *(continued)*

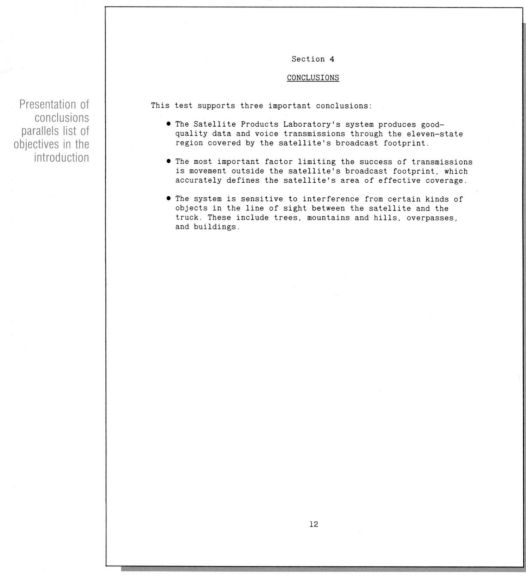

Presentation of
conclusions
parallels list of
objectives in the
introduction

Section 4

CONCLUSIONS

This test supports three important conclusions:

- The Satellite Products Laboratory's system produces good-quality data and voice transmissions through the eleven-state region covered by the satellite's broadcast footprint.

- The most important factor limiting the success of transmissions is movement outside the satellite's broadcast footprint, which accurately defines the satellite's area of effective coverage.

- The system is sensitive to interference from certain kinds of objects in the line of sight between the satellite and the truck. These include trees, mountains and hills, overpasses, and buildings.

12

FIGURE 21–3   *(continued)*

Section 5

<u>RECOMMENDATIONS</u>

Based on this test, we believe that work should proceed as rapidly as possible to complete an operational system. In that work, the Satellite Products Laboratory should do the following things:

1. <u>Develop a new antenna designed specifically for use in communications between satellites and mobile radios</u>. Such an antenna
   would probably eliminate many of the disruptions caused by trees,
   the most common cause of poor transmissions.

2. <u>Define the configuration of satellites needed to provide service
   throughout our planned service area</u>. We are now ready to
   determine the number and placement in orbit of the satellites we
   will need to launch in order to provide service to our planned
   service area (48 contiguous states, Alaska, Southern Canada).
   Because locations outside of the broadcast footprint of a
   satellite probably cannot be given satisfactory service, our
   satellites will have to provide thorough footprint coverage
   throughout all of this territory. Also, we should plan the
   satellites so that each will be at least 25° above the horizon
   throughout the area it serves; in that way we can almost entirely
   eliminate poor transmissions due to interference from mountains
   and hills.

13

FIGURE 21–3   *(continued)*

22

FEASIBILITY REPORTS

W hen you perform a feasibility study, you evaluate the practicality and desirability of pursuing some course of action. Imagine, for instance, that you work for a company that designs and builds sailboats. The company thinks it might reduce manufacturing costs without hurting sales if it uses high-strength plastics to manufacture some parts traditionally made from metal. Before making such a change, however, the company wants you to answer many questions. Would plastic parts be as strong, durable, and attractive as metal ones? Is there a supplier who would make the plastic parts? Would the plastic parts really be less expensive than the metal ones? Would boat buyers accept the change? The information and analyses you provide in your report will be the company's primary basis for deciding whether to pursue this course of action. In this chapter, you will learn how to prepare effective reports about feasibility studies.

TYPICAL WRITING SITUATION

All feasibility reports share one essential characteristic: they are written to help decision-makers choose between two or more courses of action. Even when a feasibility report seems to focus primarily on one course of action, the readers are always considering a second course: to leave things the way they are—for example, to continue to use metal parts rather than plastic parts in the sailboats. In many situations, however, your readers will already have decided that some change is necessary and will be choosing between two or more alternatives to the status quo.

THE QUESTIONS READERS ASK MOST OFTEN

As they think about the choice they must make, decision-makers ask many questions. From situation to situation, these basic questions remain the same. That's what makes it possible for one superstructure to be useful across nearly the full range of situations in which people prepare feasibility reports.

The basic questions usually asked by people when they read feasibility reports are the following:

- **Why is it important for us to consider these alternatives?** Decision-makers ask this question because they want to know why they have to make any choice in the first place. Your readers may need a detailed explanation of a problem to appreciate the importance of considering alternative courses of action. On the other hand, if the readers are familiar with

Importance

the problem, they may see the importance of considering the alternatives if you simply remind them of the situation.

- Are your criteria reasonable and appropriate? To help your readers choose between the alternative courses of action, you must evaluate the alternatives in terms of specific criteria. At work, people want these criteria to reflect the needs and aims of their organization. And they want you to tell them explicitly what the criteria are so they can judge them.

- Are your facts reliable? Decision-makers want to be sure that your facts are reliable before they take any action based on those facts.

- What are the important features of the alternatives? So that they can understand your detailed discussion of the alternatives, readers want you to present an overview that highlights the key features of each alternative.

- How do the alternatives stack up against your criteria? The heart of a feasibility study is your evaluation of the alternatives in terms of your criteria. Your readers want to know the results.

- What overall conclusions do you draw about the alternatives? Based upon your detailed evaluation of the alternatives, you will reach some general conclusions about the merits of each. Decision-makers need to know your conclusions because these overall judgments form the basis for decision-making.

- What do you think we should do? In the end, your readers must choose one of the alternative courses of action. Because of your expertise on the subject, they want you to help them by telling what you recommend.

SUPERSTRUCTURE FOR FEASIBILITY REPORTS

To answer your readers' questions about your feasibility studies, you can use a superstructure that has the following seven elements: introduction, criteria, method of obtaining facts, overview of alternatives, evaluation, conclusions, and recommendations. As Figure 22–1 shows, each element corresponds to one of the decision-makers' basic questions described above. Of course, you may combine the elements in many different ways, depending on your situation. For instance, you may integrate your conclusions into your evaluation, or you may omit a separate discussion of criteria if they need no special explanation. But when preparing any feasibility report, you should consciously determine whether to include each of the seven elements, based upon your understanding of your purpose, audience, and situation.

The rest of this chapter explains how you can develop each of the seven elements to create an effective feasibility report. If you have not already done so, you should also read Chapter 20, which describes the general superstructure for reports.

Report Element	Readers' Question
Introduction	Why is it important for us to consider these alternatives?
Criteria	Are your criteria reasonable and appropriate?
Method of Obtaining Facts	Are your facts reliable?
Overview of Alternatives	What are the important features of the alternatives?
Evaluation	How do the alternatives stack up against your criteria?
Conclusions	What overall conclusions do you draw about the alternatives?
Recommendations	What do you think we should do?

FIGURE 22–1   Elements of a Feasibility Study and Their Relationship to the Questions Readers Ask

Introduction

In the introduction to a feasibility report you should answer your readers' question, "Why is it important for us to consider these alternatives?" The most persuasive way to answer this question is to identify the problem your feasibility study will help your readers solve or the goal it will help them achieve. Be sure to identify a problem or goal that is significant from the point of view of your employer or client: to reduce the number of rejected parts, to increase productivity, and so on. Beyond identifying the problem or goal that your study addresses, your introduction should announce the alternative courses of action you studied and tell generally what you did to investigate them.

Consider, for example, the way Phil wrote the introduction of a feasibility report he prepared. (Phil's entire report is presented at the end of this chapter.) A process engineer in a paper mill, Phil was asked to evaluate the feasibility of substituting one ingredient for another in the furnish for one of the papers it produces (*furnish* is the combination of ingredients used to make the pulp for paper):

Problem

Possible solution

What the writer did
to investigate the
possible solution

At present we rely on the titanium dioxide (TiO_2) in our furnish to provide the high brightness and opacity we desire in our paper. However . . . the price of TiO_2 has been rising steadily and rapidly for several years. We now pay roughly $1400 per ton for TiO_2, or about 70¢ per pound.

Some mills are now replacing some of the TiO_2 in their furnish with silicate extenders. Because the average price for silicate extenders is only $500 per ton, well under half the cost of TiO_2, the savings are very great.

To determine whether we could enjoy a similar savings for our 30-pound book paper, I have studied the physical properties, material handling requirements, and cost of two silicate extenders, Tri-Sil 606 and Zenolux 26 T.

Generally, the introduction to a long feasibility report (and most short ones) should also include a preview of the main conclusions and, perhaps, the major recommendations. Phil included his major conclusion:

I conclude that one of the silicate extenders, Zenolux 26 T, looks promising enough to be tested in a mill run.

As another example, consider the way Ellen wrote the introduction of a feasibility report she prepared for the board of directors of the bank that employs her. Ellen was asked to evaluate the feasibility of opening a new branch in a particular suburban community. She began by announcing the topic of her report:

This report discusses the feasibility of opening a branch office of Orchard Bank in Rolling Knolls, Tennessee.

Then, after giving a sentence of background information about the source of the bank's interest in exploring this possibility, Ellen emphasized the importance of such a feasibility study:

In the past, Orchard Bank has approached the opening of new branches with great care, which is undoubtedly a major reason that in the twelve years since its founding it has become one of Tennessee's most successful small, privately owned financial institutions.

Ellen also included her major conclusions:

Overall, the Rolling Knolls location offers an enticing opportunity, but would present Orchard Bank with some challenges it has not faced before.

She finished by briefly summarizing her major recommendations:

> We should proceed carefully.

The introduction of a feasibility report is often combined with one or more of the other six elements, such as a description of the criteria, a discussion of the method of obtaining facts, or an overview of the alternatives.

Of course, the opening of a feasibility report may also include various kinds of background, explanatory, and forecasting information that may be found in the beginning of any technical communication (see Chapter 9).

Criteria

Criteria are the standards that you apply in a feasibility study to evaluate the alternative courses of action that you are considering. For instance, to assess the feasibility of opening the new branch office, Ellen used many criteria, including the existence of a large enough market, a good possibility of attracting depositors away from the competition, the likelihood that profits on the deposits at the branch will exceed the expenses of operating it, and the reasonableness of the financial outlay required to open the office. If she had found that the proposed branch failed to meet any of those criteria, she would have determined that the branch office was not feasible. Likewise, Phil evaluated the two silicate extenders in terms of several specific properties.

Two Ways of Presenting Criteria

There are two common ways of telling your readers what your criteria are:

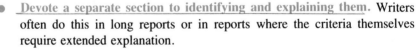

- Devote a separate section to identifying and explaining them. Writers often do this in long reports or in reports where the criteria themselves require extended explanation.
- Integrate your presentation of them into other elements of the report. Phil did this in the following sentence from the third paragraph of his introduction:

Criteria named

> To determine whether we could enjoy a similar savings for our 30-pound book paper, I have studied the physical properties, material handling requirements, and cost of two silicate extenders, Tri-Sil 606 and Zenolux 26 T.

For each of the general criteria named in the quoted sentence, Phil had some more specific criteria. These he described when he discussed his methods and results. For instance, at the beginning of his discussion of the physical properties of the two extenders, he named the three properties he evaluated.

Importance of Presenting Criteria Early

Whether you present your criteria in a separate section or integrate them into other sections, you should introduce them early in your report. There are three good reasons for doing this. First, your readers know that the validity of your conclusions depends on the criteria you use to evaluate the alternatives, and they want to evaluate the criteria themselves. They will ask, for instance, ''Did you take into account all of the considerations relevant to this decision?'' and ''Are the standards that you are applying reasonable in this circumstance?''

Second, your discussion of the criteria tells your readers a great deal about the scope of your study. Did you restrict yourself to technical questions, for instance, or did you also consider relevant organizational issues such as profitability and management strategies?

The third reason for presenting your criteria early in your report is that your discussion of the alternative courses of action will make much more sense to your readers if they know in advance the criteria by which you evaluated the alternatives.

Sources of Your Criteria

You may wonder how to come up with the criteria you will use in your study and report. Often, the person who asks you to undertake a study will simply tell you what criteria to apply. In other situations, particularly when you are conducting a feasibility study that requires technical knowledge that you have but your readers don't, your readers may expect you to identify the relevant criteria for them.

In either case, you are likely to refine your sense of the criteria as you conduct your study. The writing process itself can help you refine your criteria because as you compose you must think in detail about the information you have obtained and decide how best to evaluate it.

Four Common Types of Criteria

As you develop your criteria, you may find it helpful to know that at work, criteria often address one or more of the following questions:

- **Will the course of action really do what's wanted?** This question is especially common when the problem is a technical one: Will this reorganization of the department really improve the speed with which we can process loan applications? Will the new type of programming really reduce the computer time?

- **Can we implement it?** Even though a particular course of action may work technically, it may not be practical. Maybe it requires overly extensive changes in existing operations, or equipment or materials that are not readily available, or special skills that employees do not possess.

- **Can we afford it?** Cost can be treated in several ways. You may seek an alternative that costs less than some fixed amount, or one that will save enough to pay for itself in a fixed period (for example, two years). Or, you may simply be asked to determine whether the costs are "reasonable."

- **Is it desirable?** Sometimes, a solution must be more than effective, implementable, and affordable. Many otherwise feasible courses of action are rejected because they create undesirable side effects. For example, a company might reject a plan for increasing productivity because the plan would have the undesirable side effect of decreasing employee morale. Similarly, the citizens of a region might feel that although a new power plant is technically feasible and would bring many benefits, the land it would occupy should be used for other purposes, such as farming.

Of course, your selection of criteria for a particular feasibility study will depend upon the problem you address and the professional responsibilities, goals, and values of the people who will use your report. In some instances, you will need to deal only with criteria related to the question, "Does it work?" At other times, you might need to deal with all the criteria mentioned above, plus others. No matter what your criteria, however, announce them to your readers before you discuss your evaluation.

Method of Obtaining Facts

When you tell how you obtained your facts, you answer your readers' question, "Are your facts reliable?" That is, by showing that you used reliable methods, you assure your readers that your facts form a sound basis for decision-making.

The source of your facts will depend upon the nature of your study—library research, calls to manufacturers, interviews, meetings with other experts in your organization, surveys, laboratory research, and the like.

How much detail must you provide about your method of gathering facts? The amount of detail you need to supply depends upon your readers and situation, but in each case your goal is to say enough to satisfy your readers that your information is trustworthy. For example, Ellen used some fairly technical procedures to estimate the amount of deposits that Orchard Bank could expect from a new branch in Rolling Knolls, Tennessee. However, those procedures are standard in the banking industry and are well-known to her readers. For that reason, she did not need to provide a detailed explanation of them.

In contrast, Calvin needed to provide a great deal of information about his methods in a feasibility study that he prepared for his employer, a manufacturer of packaged mixes for such foods as cakes, breads, and cookies. Calvin's assignment was to determine the feasibility of making a new, low-gluten muffin mix for a new line of products for people with special health conditions. To evaluate the alternative muffin mixes, Calvin used several test procedures common in the food industry. However, to persuade his very demanding readers that his results were valid, he had to explain in detail how he had obtained them. Readers in another organization might have required less detail to accept the validity of his findings.

Where is the best place to describe your methods? Sometimes the answer depends partly upon how many different techniques you used. If you used one or two techniques—say library research and interviews—you might explain each in a separate paragraph or section near the beginning of your report. You could name the books read and individuals interviewed there or, if the list is long, in an appendix. On the other hand, if you used several different techniques, each pertaining to a different part of your analysis, you might mention each of them at the point at which you discuss the results you obtained.

Of course, if your method is obvious, you may not need to describe it at all. You must always be sure, however, that your readers know enough about your methods to trust that your facts are reliable.

Overview of Alternatives

Before you begin your detailed evaluation of the alternatives, you must be sure that your readers understand what the alternatives are. Sometimes you need to devote only a few words to that task. Imagine, for instance, that you work for a chain of convenience stores that has asked you to investigate the feasibility of increasing starting salaries for store managers as a way of attracting stronger applicants for job openings. Surely, your readers will not require any special explanation to understand the course of action you are assessing.

However, you may sometimes need to provide extensive background information or otherwise explain the alternative to your readers. George needed to do so when he wrote a report on the feasibility of replacing his employer's company-owned, building-wide telephone system. That's because the alternative systems he considered are complex and differ from one another in many ways. By providing an initial overview of the alternative phone systems he was considering, George helped his readers piece together the more detailed comments he later made in his point-by-point evaluation of the systems.

Of course, if George's readers had already understood the alternative systems, he would not have needed to explain them no matter how complex the alternatives were. Your job is to ensure that your readers understand the alternatives before you begin evaluating them, regardless of whether you must take a few words or many pages of explanation to provide that understanding.

Evaluation

The heart of a feasibility report is the detailed evaluation of the course or courses of action you studied. In most feasibility studies, writers organize their evaluation sections around their criteria. For example, in her study of the feasibility of opening a new branch office of a bank, Ellen devoted one section to the size of the market, another to the competition, a third to prospective income and expenses, and so on. Similarly, in his report on the various telephone systems, George organized his evaluation around his criteria: ability to handle voice communications, ability to handle data communication, and so on.

The following sections discuss three pieces of advice that will help you present your evaluations clearly and effectively: choose carefully between the alternating and divided patterns, dismiss obviously unsuitable alternatives, and put your most important points first.

Choose Carefully between the Alternating and Divided Patterns

In reports where you compare two or more alternatives, you can organize your evaluation according to either the divided pattern or the alternating pattern (see Chapter 5). Choose the pattern that will enable your readers to read, understand, and use your information most efficiently. For example, George knew that readers of his report on the phone systems would want to compare the systems directly in terms of each of their specific criteria. Because many of the comparisons involved complex details, George felt that he could help his readers by using the alternating pattern, which enabled him to present each set of relevant details in one place.

On the other hand, Calvin used the divided pattern in his evaluation of the alternative recipes for low-gluten muffins. Each individual test on each muffin produced a single, specific result that could be communicated in only a few words. He judged that his readers would have difficulty building a coherent understanding of each muffin's strengths and weaknesses if he used the alternating pattern. By using the divided pattern, he presented all the results for a particular muffin in one fairly brief passage. Then, in his conclusions, Calvin found himself able to compare the muffins quickly in terms of the most significant differences among them.

When organizing your evaluation section, you should imitate George and Calvin by using your knowledge of your readers to choose between the alternating and divided patterns.

Dismiss Obviously Unsuitable Alternatives

Sometimes you will want to mention several alternatives but treat only one or a few thoroughly. Perhaps your investigation showed that the other alternatives failed to meet one or more of the critical criteria so that they should not be considered seriously.

Usually, it makes no sense to discuss obviously unqualified solutions at length. One good way to handle them is to explain briefly the alternatives that you have dismissed and tell why you dismissed them. This entire discussion might take only a sentence or a paragraph. You might include it in the introduction (when you are talking about the scope of your report) or in your overview of the alternatives.

It is usually not wise to postpone the discussion of unqualified solutions until the end of a report. If you do this, throughout the time they are reading or listening to the earlier parts of your report your readers may keep asking, "But why didn't you consider so and so?" That question may distract their attention from important material that you are presenting.

Put Your Most Important Points First

The final piece of advice for presenting your evaluation of the alternatives is to remember the guideline in Chapter 7, "Drafting Paragraphs, Sections, and Chapters," that urged you to begin each portion of your communications with the most important information. This advice applies to the entire section in which you evaluate the alternatives, as well as to its parts. By presenting important information early, you save your readers the trouble of trying to figure out what generalizations they should draw from the details you are presenting. For instance, in her report to the bank Ellen begins one part of her evaluation of the Rolling Knolls location by saying:

> The proposed Rolling Knolls branch office would be profitable after
> four years if Orchard Bank successfully develops the type of two-
> pronged marketing strategy outlined in the preceding section.

Ellen then spends two pages discussing the estimates of deposits, income, and expenses that support her overall assessment.

Similarly, Phil begins one part of his evaluation of the silicate extenders in this way:

> With respect to material handling, I found no basis for choosing
> between Zenolux and Tri-Sil.

Phil then spends two paragraphs reporting the facts he has gathered about the physical handling of the two extenders.

Conclusions

Your conclusions are your overall assessment of the feasibility of the courses of action you studied. You might present your conclusions in two or three places in your report. You should certainly mention them in summary form near the beginning. If your report is long (say several pages), you might also remind your readers of your overall conclusion at the beginning of your evaluation section. Finally, you should usually provide the detailed discussion of your conclusions in a separate section that follows your evaluation of the alternatives.

Recommendations

It is customary to end a feasibility report by answering the decision-maker's question, "What do you think we should do?" Because you have investigated and thought about the alternatives so thoroughly, your readers will place special value on your recommendations. Depending on the situation, you might need to take only a single sentence or else many pages to present your recommendations.

Sometimes your recommendations will pertain directly to the course of action you studied: "Do this" or "Don't do it." At other times you may perform preliminary feasibility studies to determine whether the course of action you are studying is promising enough to warrant a more thorough investigation. In that case, your recommendation would focus not on whether to pursue a certain course of action but on whether to continue studying it. Ellen's report about opening a new bank office is of that type. She determined that there is a substantial possibility of making a profit with the branch, but felt the need for expert assistance before making a final decision. Consequently, she recommended that the bank hire a marketing agency to evaluate the prospects.

Sometimes, too, you may discover that you were unable to gather all the information you needed to make a firm recommendation. Perhaps your deadline was too short or your funds too limited. Perhaps you uncovered an unexpected question that needs further investigation. In all these situations, you should point out the limitations of your report and let your audience know what else they should find out so that they can be confident that they are making a well-informed decision.

SAMPLE OUTLINES

The seven elements of the superstructure for feasibility reports can be incorporated into many different organizational patterns. Figures 22–2 through 22–4 illustrate some of the possibilities by showing you the outlines for the reports by George, Calvin, and Ellen.

In his report on the telephone systems (Figure 22–2), George uses seven sections that correspond directly with the seven elements of a feasibility report (Figure 22–1). In contrast, Calvin uses only four sections in his report on the muffins (Figure 22–3). He does this by combining some of the superstructure's elements: in the opening section, he presents both the introduction and overview; in the next section he describes both his method of obtaining facts and his criteria; and in the final section, he intertwines his conclusions and recommendations.

In her report on the new branch office for the bank (Figure 22–4, p. 593), Ellen uses six sections but she devotes three of them to her evaluation of her alternatives. In contrast, both George and Calvin devote only a single section to evaluation.

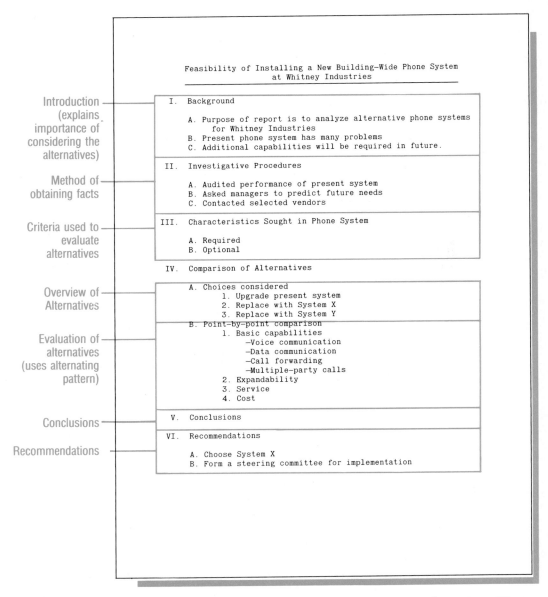

Feasibility of Installing a New Building—Wide Phone System
at Whitney Industries

Introduction
(explains
importance of
considering the
alternatives)

Method of
obtaining facts

Criteria used to
evaluate
alternatives

Overview of
Alternatives

Evaluation of
alternatives
(uses alternating
pattern)

Conclusions

Recommendations

I.   Background

   A. Purpose of report is to analyze alternative phone systems
        for Whitney Industries
   B. Present phone system has many problems
   C. Additional capabilities will be required in future.

II.   Investigative Procedures

   A. Audited performance of present system
   B. Asked managers to predict future needs
   C. Contacted selected vendors

III. Characteristics Sought in Phone System

   A. Required
   B. Optional

IV.   Comparison of Alternatives

   A. Choices considered
        1. Upgrade present system
        2. Replace with System X
        3. Replace with System Y
   B. Point—by—point comparison
        1. Basic capabilities
             —Voice communication
             —Data communication
             —Call forwarding
             —Multiple—party calls
        2. Expandability
        3. Service
        4. Cost

V.   Conclusions

VI.   Recommendations

   A. Choose System X
   B. Form a steering committee for implementation

FIGURE 22–2   How George Used the Superstructure for Feasibility Reports in His Comparison of Three
Phone Systems

If you examine these three outlines further, you will find many additional
differences among them. The most important points for you to note are (1) that
all three reports contain all seven elements of a feasibility report and (2) that
the superstructure for the feasibility report can be applied in many different
ways to suit the purpose and readers of a particular communication.

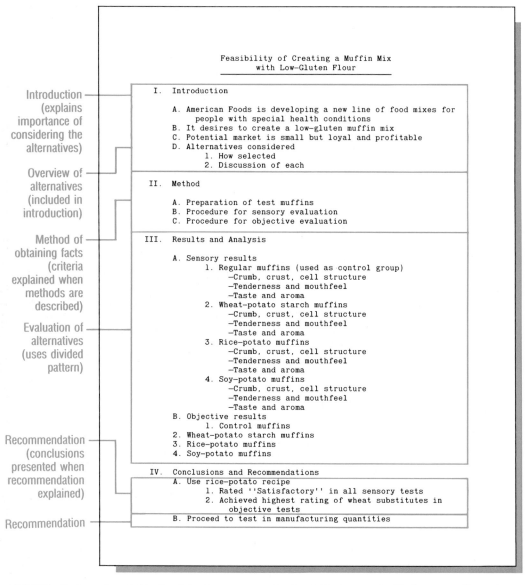

Introduction (explains importance of considering the alternatives)

Overview of alternatives (included in introduction)

Method of obtaining facts (criteria explained when methods are described)

Evaluation of alternatives (uses divided pattern)

Recommendation (conclusions presented when recommendation explained)

Recommendation

```
              Feasibility of Creating a Muffin Mix
                     with Low-Gluten Flour

     I.   Introduction

          A. American Foods is developing a new line of food mixes for
             people with special health conditions
          B. It desires to create a low-gluten muffin mix
          C. Potential market is small but loyal and profitable
          D. Alternatives considered
               1. How selected
               2. Discussion of each

     II.  Method

          A. Preparation of test muffins
          B. Procedure for sensory evaluation
          C. Procedure for objective evaluation

     III. Results and Analysis

          A. Sensory results
               1. Regular muffins (used as control group)
                    —Crumb, crust, cell structure
                    —Tenderness and mouthfeel
                    —Taste and aroma
               2. Wheat—potato starch muffins
                    —Crumb, crust, cell structure
                    —Tenderness and mouthfeel
                    —Taste and aroma
               3. Rice—potato muffins
                    —Crumb, crust, cell structure
                    —Tenderness and mouthfeel
                    —Taste and aroma
               4. Soy—potato muffins
                    —Crumb, crust, cell structure
                    —Tenderness and mouthfeel
                    —Taste and aroma
          B. Objective results
               1. Control muffins
               2. Wheat—potato starch muffins
               3. Rice—potato muffins
               4. Soy—potato muffins

     IV.  Conclusions and Recommendations
          A. Use rice—potato recipe
               1. Rated ''Satisfactory'' in all sensory tests
               2. Achieved highest rating of wheat substitutes in
                  objective tests
          B. Proceed to test in manufacturing quantities
```

FIGURE 22–3   How Calvin Used the Superstructure for Feasibility Reports in His Report on Alternative Ways of Making a Food Product

PLANNING GUIDE

Figure 22–5 shows a worksheet that may help you plan your feasibility reports. As you fill out the worksheet, remember that it is intended merely to stimulate your thinking. To achieve your objectives, you may need to say many things

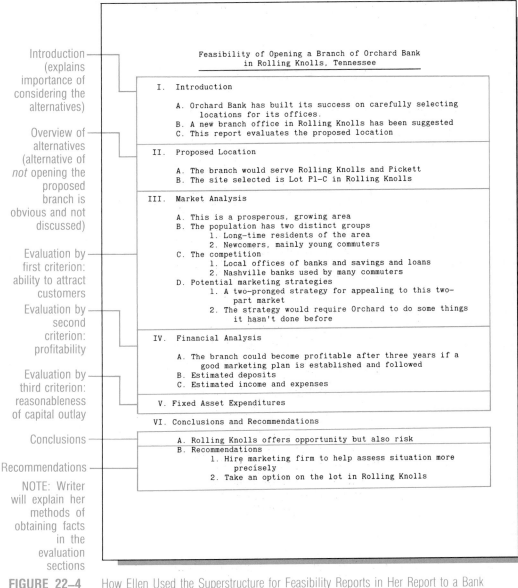

Introduction
(explains
importance of
considering the
alternatives)

Overview of
alternatives
(alternative of
not opening the
proposed
branch is
obvious and not
discussed)

Evaluation by
first criterion:
ability to attract
customers

Evaluation by
second
criterion:
profitability

Evaluation by
third criterion:
reasonableness
of capital outlay

Conclusions

Recommendations

NOTE: Writer
will explain her
methods of
obtaining facts
in the
evaluation
sections

Feasibility of Opening a Branch of Orchard Bank
in Rolling Knolls, Tennessee

I.   Introduction

   A. Orchard Bank has built its success on carefully selecting
      locations for its offices.
   B. A new branch office in Rolling Knolls has been suggested
   C. This report evaluates the proposed location

II. Proposed Location

   A. The branch would serve Rolling Knolls and Pickett
   B. The site selected is Lot Pl-C in Rolling Knolls

III. Market Analysis

   A. This is a prosperous, growing area
   B. The population has two distinct groups
         1. Long-time residents of the area
         2. Newcomers, mainly young commuters
   C. The competition
         1. Local offices of banks and savings and loans
         2. Nashville banks used by many commuters
   D. Potential marketing strategies
         1. A two-pronged strategy for appealing to this two-
            part market
         2. The strategy would require Orchard to do some things
            it hasn't done before

IV. Financial Analysis

   A. The branch could become profitable after three years if a
      good marketing plan is established and followed
   B. Estimated deposits
   C. Estimated income and expenses

V. Fixed Asset Expenditures

VI. Conclusions and Recommendations

   A. Rolling Knolls offers opportunity but also risk
   B. Recommendations
         1. Hire marketing firm to help assess situation more
            precisely
         2. Take an option on the lot in Rolling Knolls

FIGURE 22–4   How Ellen Used the Superstructure for Feasibility Reports in Her Report to a Bank

that you don't write down on it, and you may discover that some of the work-sheet's questions are irrelevant to the particular report you are writing. Also, remember that the superstructure on page 2 of Figure 22–5 does not necessarily provide the outline for your report.

Planning Guide

FEASIBILITY REPORT

Subject _____ **Due Date** _____

Overall Purpose

Reader Profile

Who are your primary readers?

Who else might read your report?

What are your readers' titles and professional responsibilities? How will they influence what your readers want from your report?

How well do your readers understand the technical, scientific, or other specialized terms and concepts you might use?

Do your readers have any communication preferences you should take into account? If so, what are they?

Are there any other considerations you should keep in mind when addressing these readers?

Readers' Informational Needs

What will your readers' key questions be?

Readers' Attitudes

What are your readers' present attitudes toward your *subject?* What do you want them to be?

What are your readers' present attitudes toward *you?* What do you want them to be?

FIGURE 22–5   Planning Worksheet for Feasibility Reports

Superstructure

Introduction
>   Why is it important for your readers to consider the alternative(s) you discuss?
>
>   What is your main point in this report?
>
>   What background information do your readers need about your subject?
>
>   Can you help your readers by forecasting the rest of your report?

Criteria
>   What criteria will you apply in your evaluation of the alternatives?

Method
>   What, if anything, will your readers want to know about your method?
>
>   What additional things should you tell your readers about your method to persuade them you have done a good job?

Overview of Alternatives
>   What are the important features of each alternative?

Evaluation
>   Based on the information you gathered, how do these alternatives stack up against one another? (Remember to use your criteria as the basis for this evaluation.)

Conclusions
>   What overall conclusions do you draw about the alternatives?

Recommendations
>   Based on your conclusions, what do you think your readers should do?

Visual Aids

What visual aids would your readers find helpful or persuasive?

Organization

What organization would your readers find useful and persuasive? (On a separate sheet provide a brief sketch or outline to identify the major sections and subsections.)

FIGURE 22–5   *(continued)*

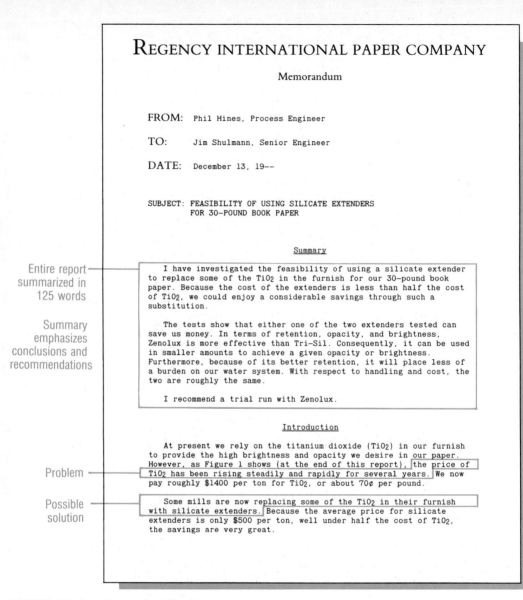

Entire report summarized in 125 words

Summary emphasizes conclusions and recommendations

Problem

Possible solution

REGENCY INTERNATIONAL PAPER COMPANY

Memorandum

FROM: Phil Hines, Process Engineer

TO: Jim Shulmann, Senior Engineer

DATE: December 13, 19—

SUBJECT: FEASIBILITY OF USING SILICATE EXTENDERS FOR 30-POUND BOOK PAPER

Summary

I have investigated the feasibility of using a silicate extender to replace some of the TiO_2 in the furnish for our 30-pound book paper. Because the cost of the extenders is less than half the cost of TiO_2, we could enjoy a considerable savings through such a substitution.

The tests show that either one of the two extenders tested can save us money. In terms of retention, opacity, and brightness, Zenolux is more effective than Tri-Sil. Consequently, it can be used in smaller amounts to achieve a given opacity or brightness. Furthermore, because of its better retention, it will place less of a burden on our water system. With respect to handling and cost, the two are roughly the same.

I recommend a trial run with Zenolux.

Introduction

At present we rely on the titanium dioxide (TiO_2) in our furnish to provide the high brightness and opacity we desire in our paper. However, as Figure 1 shows (at the end of this report), the price of TiO_2 has been rising steadily and rapidly for several years. We now pay roughly $1400 per ton for TiO_2, or about 70¢ per pound.

Some mills are now replacing some of the TiO_2 in their furnish with silicate extenders. Because the average price for silicate extenders is only $500 per ton, well under half the cost of TiO_2, the savings are very great.

FIGURE 22–6   Sample Feasibility Report

SAMPLE FEASIBILITY REPORT

Figure 22–6 shows the feasibility report written by Phil concerning the silicate extenders.

WRITING ASSIGNMENT

A writing assignment that involves a feasibility report is included in Appendix C.

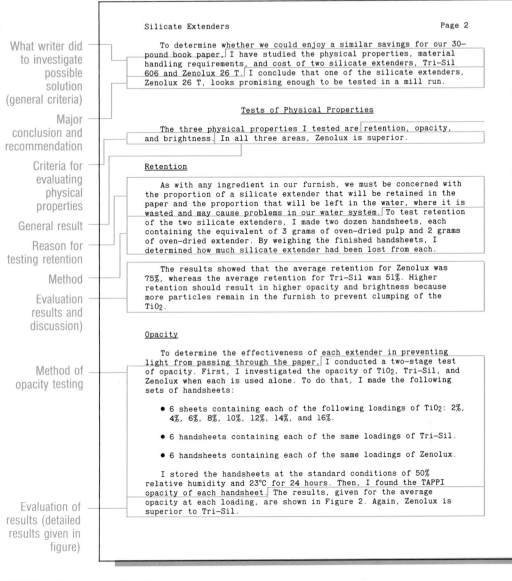

What writer did to investigate possible solution (general criteria)

Major conclusion and recommendation

Criteria for evaluating physical properties

General result

Reason for testing retention

Method

Evaluation results and discussion)

Method of opacity testing

Evaluation of results (detailed results given in figure)

Silicate Extenders   Page 2

    To determine whether we could enjoy a similar savings for our 30-pound book paper, I have studied the physical properties, material handling requirements, and cost of two silicate extenders, Tri-Sil 606 and Zenolux 26 T. I conclude that one of the silicate extenders, Zenolux 26 T, looks promising enough to be tested in a mill run.

Tests of Physical Properties

    The three physical properties I tested are retention, opacity, and brightness. In all three areas, Zenolux is superior.

Retention

    As with any ingredient in our furnish, we must be concerned with the proportion of a silicate extender that will be retained in the paper and the proportion that will be left in the water, where it is wasted and may cause problems in our water system. To test retention of the two silicate extenders, I made two dozen handsheets, each containing the equivalent of 3 grams of oven-dried pulp and 2 grams of oven-dried extender. By weighing the finished handsheets, I determined how much silicate extender had been lost from each.

    The results showed that the average retention for Zenolux was 75%, whereas the average retention for Tri-Sil was 51%. Higher retention should result in higher opacity and brightness because more particles remain in the furnish to prevent clumping of the TiO_2.

Opacity

    To determine the effectiveness of each extender in preventing light from passing through the paper, I conducted a two-stage test of opacity. First, I investigated the opacity of TiO_2, Tri-Sil, and Zenolux when each is used alone. To do that, I made the following sets of handsheets:

- 6 sheets containing each of the following loadings of TiO_2: 2%, 4%, 6%, 8%, 10%, 12%, 14%, and 16%.

- 6 handsheets containing each of the same loadings of Tri-Sil.

- 6 handsheets containing each of the same loadings of Zenolux.

    I stored the handsheets at the standard conditions of 50% relative humidity and 23°C for 24 hours. Then, I found the TAPPI opacity of each handsheet. The results, given for the average opacity at each loading, are shown in Figure 2. Again, Zenolux is superior to Tri-Sil.

FIGURE 22-6   *(continued)*

Silicate Extenders   Page 3

In the second stage of the opacity test, I made two additional sets of handsheets:

- 6 handsheets with each of the following pigment loadings: 100% TiO_2 and 0% Tri–Sil; 75% TiO_2 and 25% Tri–Sil; 50% TiO_2 and 50% Tri–Sil; 25% TiO_2 and 75% Tri–Sil; and 0% TiO_2 and 100% Tri–Sil.

- 6 handsheets with each of the same proportions of TiO_2 and Zenolux.

As Figure 3 shows, Zenolux is again superior.

<u>Brightness</u>

Using the three sets of handsheets employed in the first–stage opacity tests, I calculated the GE brightness achieved by each of the three pigments. Figure 4 shows the results. Although not as bright as TiO_2, Zenolux is brighter than Tri–Sil.

Using the two sets of handsheets employed in the second–stage opacity tests, I examined the brightness of each of the ratios of TiO_2 to the extenders. The results, shown in Figure 5, indicate that, as expected, TiO_2 with Zenolux is brighter than TiO_2 with Tri–Sil.

<u>Material Handling</u>

With respect to material handling, I found no basis for choosing between Zenolux and Tri–Sil. Both are available from suppliers in Chicago. Both are available in dry form in bags and as a slurry in bulk hopper cars. Zenolux slurry is also available in 20,000–gallon tank cars, which provide a small savings, but we do not have the storage facility needed to receive it in this way.

Similarly, both silicate extenders are quite safe. Both are 100% pigment and both are chemically stable. Neither poses a hazard of fire or explosion and neither is hazardous when mixed with any other substance, whether liquid, solid, or gas. Both create the same effects in the event of overexposure: dehydration of the respiratory tract, eyes, and skin. These effects can be avoided by purchasing the extenders in slurry rather than dry form.

<u>Cost</u>

Either silicate extender could save us money, and they both cost about the same: $421 per ton for Zenolux and $391 per ton for Tri–

Overall evaluation of material handling

Results concerning material handling (method not mentioned because it is obvious)

FIGURE 22–6   *(continued)*

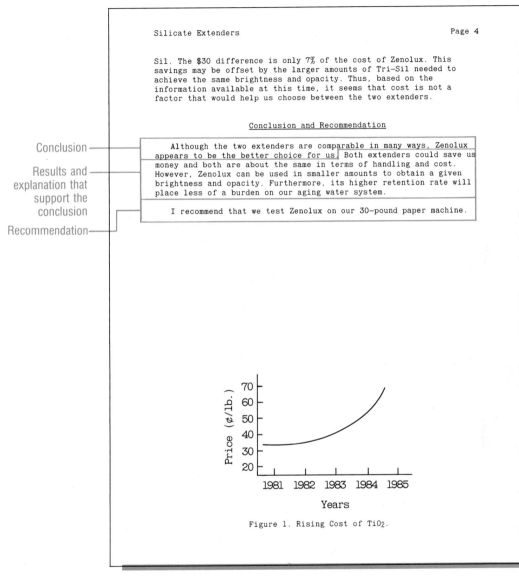

Silicate Extenders   Page 4

Sil. The $30 difference is only 7% of the cost of Zenolux. This
savings may be offset by the larger amounts of Tri-Sil needed to
achieve the same brightness and opacity. Thus, based on the
information available at this time, it seems that cost is not a
factor that would help us choose between the two extenders.

                    Conclusion and Recommendation

Conclusion ————   Although the two extenders are comparable in many ways, Zenolux
              appears to be the better choice for us. Both extenders could save us
Results and —— money and both are about the same in terms of handling and cost.
explanation that However, Zenolux can be used in smaller amounts to obtain a given
support the   brightness and opacity. Furthermore, its higher retention rate will
conclusion   place less of a burden on our aging water system.

Recommendation ——   I recommend that we test Zenolux on our 30-pound paper machine.

Figure 1. Rising Cost of TiO_2.

FIGURE 22–6   *(continued)*

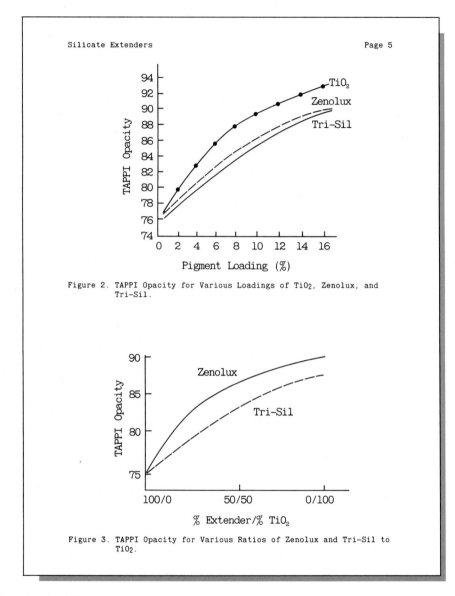

Silicate Extenders   Page 5

Figure 2. TAPPI Opacity for Various Loadings of TiO2, Zenolux, and Tri-Sil.

Figure 3. TAPPI Opacity for Various Ratios of Zenolux and Tri-Sil to TiO2.

FIGURE 22–6   *(continued)*

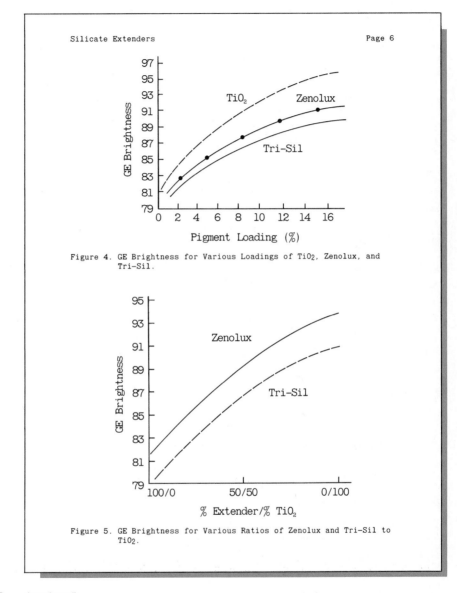

Silicate Extenders   Page 6

Figure 4. GE Brightness for Various Loadings of TiO$_2$, Zenolux, and Tri-Sil.

Figure 5. GE Brightness for Various Ratios of Zenolux and Tri-Sil to TiO$_2$.

FIGURE 22–6   *(continued)*

23

PROGRESS REPORTS

I n a progress report, you tell about work that you have begun but not yet completed. The typical progress report is one of a series submitted at regular intervals, such as every week or month.

TYPICAL WRITING SITUATIONS

Progress reports are prepared in two types of situations. In the first, you tell your readers about your progress on one *particular* project. As a geologist employed by an engineering consulting firm, Lee must do this. His employer has assigned him to study the site that a large city would like to use for a civic center and large office building. The city is worried that the site might not be geologically suited for such construction. Every two weeks, Lee must submit a progress report to his supervisor and to the city engineer. Lee's supervisor uses the progress report to be sure that Lee is conducting the study in a rapid and technically sound manner. The city engineer uses the report to see that Lee's study is proceeding according to the tight schedule planned for it. She also uses it to look for preliminary indications about the likely outcome of the study. Other work could be speeded up or halted as a result of these preliminary findings.

In the second type of situation, you prepare progress reports that tell about your work on *all* your projects. Many employers require their workers to report on their activities at regular intervals all year round, year in and year out. Jacqueline is a person who must write such progress reports (often called periodic reports). She works in the research division of a large manufacturer of consumer products, where she manages a department that is responsible for improving the formulas for the company's laundry detergents—making them clean and smell better, making them less expensive to manufacture, and making them safer for the environment. At any one time, Jacqueline's staff is working on between ten and twenty different projects.

As part of her regular responsibilities, Jacqueline must write a report every two weeks to summarize the progress on each of the projects. These reports have many readers, including the following people: her immediate superiors, who want to be sure that her department's work is proceeding satisfactorily; researchers in other departments, who want to see whether her staff has made discoveries they can use in the products they are responsible for (for example, dishwashing detergents); and corporate planners, who want to anticipate changes in formulas that will require alterations in production lines, advertising, and so on.

As the examples of Lee and Jacqueline indicate, progress reports can vary from one another in many ways: they may cover one project or many; they may be addressed to people inside the writer's organization or outside it; and they may be used by people with a variety of reasons for reading them, such as learning things they need to know to manage and to make decisions.

THE READERS' CONCERN WITH THE FUTURE

Despite their diversity, however, almost all progress reports have this in common: their readers are primarily concerned with the *future*. That is, even though most progress reports talk primarily about what has happened in the past, their readers usually want that information so that they can plan for and act in the future.

Why? Consider the responsibilities that your readers will be fulfilling by reading your progress reports. From your report, they may be trying to learn the things they need to know to manage *your* project. They will want to know, for instance, what they should do (if anything) to keep your project going smoothly or to get it back on track. The reports written by Lee and Jacqueline are used for this purpose by some of their readers.

From your report, some people may also be trying to learn things they need to know to manage *other* projects. This is because almost all projects in an organization are interdependent with other projects. For example, other people and departments may need the results of your project as they work on their own projects. Maybe you are conducting a marketing survey whose results they need so that they can design an advertising campaign, or maybe your company is building a piece of equipment that must be installed before your customer can install other equipment. If your project is going to be late, the schedules of those other projects will have to be adjusted accordingly. Similarly, if your project costs more than expected, money and resources will have to be taken away from other activities to compensate. Because of interdependencies like these, your readers need information about the past accomplishments and problems in your project so that they can make plans for the future.

Similarly, your readers will often be interested in learning the preliminary results of your work. Suppose, for instance, that you complete one part of a research project before you complete the others. Your audience may very well be able to use the result of that part immediately. The city engineer who reads Lee's reports about the possible building site is especially interested in making this use of the information Lee provides.

THE QUESTIONS READERS ASK MOST OFTEN

The readers' concern with the implications of your progress for their future work and decisions leads them to want you to answer the following questions in your progress reports. If your report describes more than one project, your readers will ask these questions about each of them.

● **What work does your report cover?** To be able to understand anything else in a progress report, readers must know what project or projects and what time period the report covers.

- **What is the purpose of the work?** Readers need to know the purpose of your work to see how your work relates to their responsibilities and to the other work, present and future, of the organization.
- **Is your work progressing as planned or expected?** Your readers will want to determine if adjustments are needed in the schedule, budget, or number of people assigned to the project or projects you are working on.
- **What results did you produce?** The results you produce in one reporting period may influence the shape of work in future periods. Also, even when you are still in the midst of a project, readers will want to know about any results they can use in other projects now, before you finish your overall work.
- **What progress do you expect during the next reporting period?** Again, your readers' interests will focus on such management concerns as schedule and budget and on the kinds of results they can expect.
- **How do things stand overall?** This question arises especially in long reports. Readers want to know what the overall status of your work is, something they may not be able to tell readily from all the details you provide.
- **What do you think we should do?** If you are experiencing or expecting problems, your readers will want your recommendations about what should be done. If you have other ideas about how the project could be improved, they too will probably be welcomed.

SUPERSTRUCTURE FOR PROGRESS REPORTS

To answer your readers' questions, you can use the conventional superstructure for writing progress reports, which has the following elements: introduction, facts and discussion, conclusions, and recommendations.

Figure 23–1 shows the relationship between these elements and the general questions readers ask. The rest of this chapter explains how you can develop each of those four elements to create an effective progress report. If you have not already done so, you should also read Chapter 20, which describes the general superstructure for reports.

Introduction

In the introduction to a progress report, you should answer the first two questions shown in Figure 23–1. You can usually answer the question, "What work does your report cover?" by opening with a sentence that tells what project or projects your report concerns and what time period it covers.

Sometimes you will not need to answer the second question—"What is the purpose of the work?"—because all your readers will already be quite

Report Element	Readers' Question
Introduction	What work does your report cover? What is the purpose of the work?
Facts and Discussion	
Past work	Is your work progressing as planned or expected? What results did you produce?
Future work	What progress do you expect during the next reporting period?
Conclusions	(In long reports: How do things stand overall?)
Recommendations	What do you think we should do?

FIGURE 23–1 Elements of a Progress Report and Their Relationship to the General Questions Readers Ask

familiar with your work's purpose. At other times, however, it will be crucial for you to tell your work's purpose because your readers will include people who don't know or may have forgotten it. You are especially likely to have such readers when your written report will be widely circulated in your own organization or when you are writing to another organization that has hired your employer to do the work you describe. You can usually explain purpose most helpfully by describing the problem that your project will help your readers solve.

The following sentences show how one manager answered the readers' first two questions:

Project and period covered —— This report covers the work done on the Focus Project from July 1 through September 1. Sponsored by the U.S. Department of Energy, the aim of the Focus Project is to overcome the technical difficulties encountered in manufacturing photovoltaic cells that can be used to generate commercial amounts of electricity.

Purpose of project ——

Of course, in your introduction you should also provide your readers with any background information they will need in order to understand the rest of your report.

Facts and Discussion

In the discussion section of your progress report, you should answer these questions from your readers: ''Is your work progressing as planned or expected?'' ''What results did you produce?'' and, ''What progress do you expect during the next reporting period?''

Answering Your Readers' Questions

In many situations, the work for each reporting period is planned in advance. In such cases you can easily tell about your progress by comparing what happened with what was planned. Where there are significant discrepancies between the two, your readers will want to know why. The information you provide about the causes of problems will help your readers decide how to remedy them. It will also help you explain any recommendations you make later in your report.

When you are discussing preliminary results that your readers might use, be sure to explain them in terms that allow your readers to see their significance. In research projects, preliminary results are often tentative. If this is the case for you, let your readers know how certain—or uncertain—the results are. This information will help your readers decide how to use the results.

Providing the Appropriate Amount of Information

When preparing progress reports, people often wonder how much information they should include. Generally, progress reports are brief because readers want them that way. While you need to provide your readers with specific information about your work, don't include details except when the details will help your readers decide how to manage your project or when you believe that your readers can make some immediate use of them. In many projects, you will learn lots of little things and you will have lots of little setbacks and triumphs along the way. Avoid talking about these matters. No matter how interesting they may be to you, they are not likely to be interesting to your readers. Stick to the information your readers can use.

Organizing the Discussion

You can organize your discussion section in many ways. One is to arrange your material around time periods:

   I. What happened during the most recent time period.
  II. What's expected to happen during the next time period.

You will find that this organization is especially well-suited for reports in which you discuss a single project that has distinct and separate stages, so that you work on only one task at a time. However, you can also expand this structure for reports that cover either several projects or one project in which several tasks are performed simultaneously:

I. What happened during the most recent time period.
   A. Project A (or Task A)
   B. Project B

II. What's expected to happen during the next time period.
   A. Project A
   B. Project B

When you prepare reports that cover more than one project or more than one task, you might also consider organizing around those projects or tasks:

I. Work on Project A (or Task A)
   A. What happened during the last time period.
   B. What's expected to happen during the next time period.

II. Work on Project B
   A. What happened during the last time period.
   B. What's expected to happen during the next time period.

This organization works very well in reports that are more than a few paragraphs long because it keeps all the information on each project together, making the report easy for readers to follow.

Emphasizing Important Findings and Problems

As mentioned, your findings and problems are important to your readers. Your findings are important because they may involve information that can be used right away by others. The problems you encounter are important because they may require your readers to change their plans.

Because your findings and problems can be so important to your readers, be sure that you devote enough discussion to them to satisfy your readers' needs and desires for information. Also, place these discussions prominently and mark them with headings or other appropriate devices so that they are easy to find.

Conclusions

Your conclusions are your overall views on the progress of your work. In short progress reports, there may be no need to include them, but if your report covers many projects or tasks, a conclusion may help your readers understand the general state of your progress.

Recommendations

If you have any ideas about how to improve the project or increase the value of its results, your readers will probably want you to include them. Your recommendations might be directed at overcoming some difficulty that you have experienced or anticipate encountering in the future. Or they might be directed at refocusing or otherwise altering the project.

A Note on the Location of Conclusions and Recommendations

For most of your readers, your conclusions and recommendations are the most important information in your progress report. Therefore, you should generally include them at the beginning, either in the introduction or at the head of your discussion section. This may be the only place you need to state them if your conclusions and recommendations are brief. If they are long or if your readers will be able to understand them only after reading or hearing your discussion section, you can present your conclusions and recommendations at the end of your report, while still including a summary of them at the beginning.

TONE IN PROGRESS REPORTS

You may wonder what tone to use in the progress reports you prepare. In them, you will generally aim to persuade your readers that you are doing a good job. The pressure you feel to make your readers feel satisfied can be especially great when you are new on the job and when your readers might discontinue a project if they feel that it isn't progressing satisfactorily.

Because of this strong persuasive element in progress reports, some people are tempted to use an inflated or highly optimistic tone. This sort of tone, however, can lead to difficulties. It might lead you to make statements that sound unbusinesslike—more like advertising copy than a professional communication. Such a tone is more likely to make your readers suspicious than agreeable. Also, if you present overly optimistic accounts of what can be expected, you risk creating an unnecessary disappointment if things don't turn out the way you seem to be promising. And if you consistently turn in overly optimistic progress reports, your credibility with your readers will quickly diminish.

In progress reports, it's best to be straightforward about problems so that the readers can take appropriate measures to overcome them and so that they can adjust their expectations realistically. It is best to sound pleased and proud of your accomplishments without seeming to puff them up.

SAMPLE OUTLINES

By comparing the outlines of two progress reports prepared by Margo and Lloyd, you can see how the six elements of the superstructure for progress reports can be adapted to different situations. Margo is reporting to her technical writing instructor about her progress on a course project (Figure 23–2), and Lloyd is submitting a weekly update on a project he heads to introduce a new line of high-fashion women's clothes (Figure 23–3, p. 613).

One of the most striking differences between the two reports is that Margo organizes around time periods, devoting two sections to her past work and a separate section to her future work. In contrast, Lloyd organizes around major areas of activity: manufacturing, marketing, and so on. Within the section on manufacturing, he talks about past work (emphasizing a problem) and about future work; he even makes a recommendation there, rather than at the most traditional place at the end of a memo. Incidentally, Margo's progress report contains no recommendation because she has no action to suggest to her instructor; she's the one who must do something to get her work back on schedule.

When using the superstructure to guide your preparation of progress reports, you will have to be as responsive to your purpose and readers as Margo and Lloyd have been to theirs.

PLANNING GUIDE

Figure 23–4 (p. 614) shows a worksheet that may help you plan your progress reports. As you fill out the worksheet, remember that it is intended merely to stimulate your thinking. To achieve your objectives, you may need to say many things that you don't write down on it, and you may discover that some of the worksheet's questions are irrelevant to the particular report you are writing. Also, remember that the superstructure on page 2 of Figure 23–4 does not necessarily provide the outline for your report.

SAMPLE PROGRESS REPORT

Figure 23–5 (p. 616) shows the full progress report that Margo used to tell her instructor about her work on a project in her technical writing class.

EXERCISE

1. **a.** Recommend ways that the writer could improve the progress report printed in Figure 23–6 (p. 619). Assume that his readers are busy managers who frequently express a desire to receive memos they can read

as efficiently as possible. In your recommendations, focus on his selection and organization of information. The report concerns work done at a company that is developing a product that construction companies can use to make large numbers of styrofoam panels at their building sites. The product is to consist of a liquid and a powder that are poured into heated metal molds, where they will foam up to fill the mold and take the mold's shape.

b. Rewrite the memo by following the recommendations you developed for the writer.

WRITING ASSIGNMENT

An assignment that involves writing a progress report is given in Appendix C.

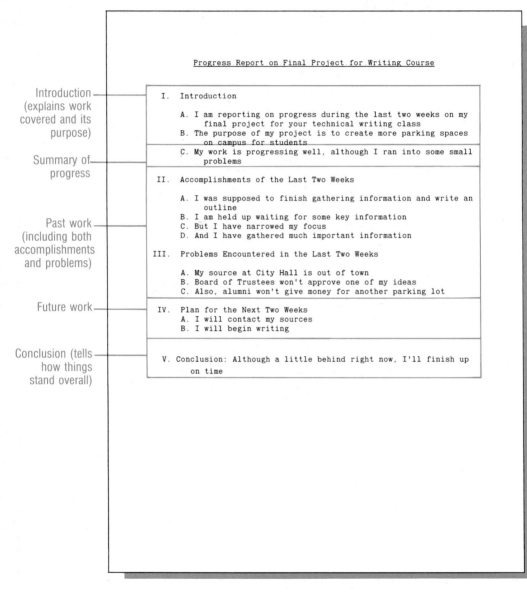

Introduction (explains work covered and its purpose)

Summary of progress

Past work (including both accomplishments and problems)

Future work

Conclusion (tells how things stand overall)

Progress Report on Final Project for Writing Course

I. Introduction

    A. I am reporting on progress during the last two weeks on my final project for your technical writing class
    B. The purpose of my project is to create more parking spaces on campus for students
    C. My work is progressing well, although I ran into some small problems

II. Accomplishments of the Last Two Weeks

    A. I was supposed to finish gathering information and write an outline
    B. I am held up waiting for some key information
    C. But I have narrowed my focus
    D. And I have gathered much important information

III. Problems Encountered in the Last Two Weeks

    A. My source at City Hall is out of town
    B. Board of Trustees won't approve one of my ideas
    C. Also, alumni won't give money for another parking lot

IV. Plan for the Next Two Weeks
    A. I will contact my sources
    B. I will begin writing

V. Conclusion: Although a little behind right now, I'll finish up on time

FIGURE 23–2 Outline of Margo's Progress Report on a Writing Project

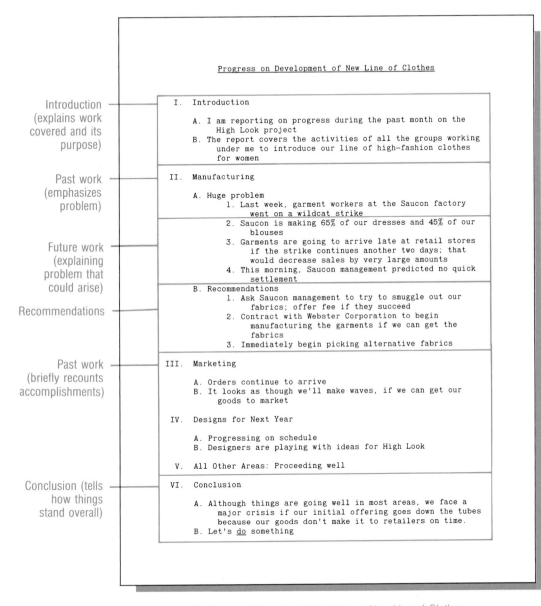

Progress on Development of New Line of Clothes

Introduction
(explains work
covered and its
purpose)

I.   Introduction

     A. I am reporting on progress during the past month on the
        High Look project
     B. The report covers the activities of all the groups working
        under me to introduce our line of high-fashion clothes
        for women

Past work
(emphasizes
problem)

II. Manufacturing

     A. Huge problem
             1. Last week, garment workers at the Saucon factory
                went on a wildcat strike

Future work
(explaining
problem that
could arise)

             2. Saucon is making 65% of our dresses and 45% of our
                blouses
             3. Garments are going to arrive late at retail stores
                if the strike continues another two days; that
                would decrease sales by very large amounts
             4. This morning, Saucon management predicted no quick
                settlement

Recommendations

     B. Recommendations
             1. Ask Saucon management to try to smuggle out our
                fabrics; offer fee if they succeed
             2. Contract with Webster Corporation to begin
                manufacturing the garments if we can get the
                fabrics
             3. Immediately begin picking alternative fabrics

Past work
(briefly recounts
accomplishments)

III. Marketing

     A. Orders continue to arrive
     B. It looks as though we'll make waves, if we can get our
        goods to market

IV. Designs for Next Year

     A. Progressing on schedule
     B. Designers are playing with ideas for High Look

 V. All Other Areas: Proceeding well

Conclusion (tells
how things
stand overall)

VI. Conclusion

     A. Although things are going well in most areas, we face a
        major crisis if our initial offering goes down the tubes
        because our goods don't make it to retailers on time.
     B. Let's do something

FIGURE 23-3   Outline of Lloyd's Progress Report on Introducing a New Line of Clothes

Planning Guide

PROGRESS REPORT

Subject _____ **Due Date** _____

Overall Purpose

Reader Profile

Who are your primary readers?

Who else might read your report?

What are your readers' titles and professional responsibilities? How will they influence what your readers want from your report?

How much do your readers know about your projects?

How well do your readers understand the technical, scientific, or other specialized terms and concepts you might use?

Do your readers have any communication preferences you should take into account? If so, what are they?

Are there other considerations you should keep in mind when addressing these readers?

Readers' Informational Needs

What will your readers' key questions be?

Readers' Attitudes

What are your readers' present attitudes toward your progress on your projects? What do you want them to be?

What are your readers' present attitudes toward _you?_ What do you want them to be?

FIGURE 23–4   Planning Guide for Progress Reports

Superstructure

Introduction
What work does your report cover?

What is the purpose of the work?

What period does your report cover?

Can you help your reader by forecasting the rest of your report?

Facts and Discussion
What should you tell your readers about your past work?
- Are you on schedule?
- What have you accomplished?
- What work that was planned is not completed?
- Have you obtained any results that your readers would like to know about immediately?
- Have you encountered any significant problems that your readers should know about?

Future Work
- What progress do you expect to make during the next reporting period?
- Do you forsee any problems that your readers should know about?

Conclusions
Do you need to tell your readers how things stand overall?

Recommendations
What action, if any, do you think your readers should take? (Be sure to answer this question if your project is behind schedule or has encountered significant problems.)

Visual Aids

What visual aids would your readers find helpful or persuasive?

Organization

What organization would your readers find useful and persuasive? (On a separate sheet, provide a brief sketch or outline to identify the major sections and subsections.)

FIGURE 23–4   *(continued)*

Project and
period covered

Purpose of
project

Important
background
information

Summary of
conclusions

Discussion
compares work
planned with
work
accomplished

Important result
emphasized

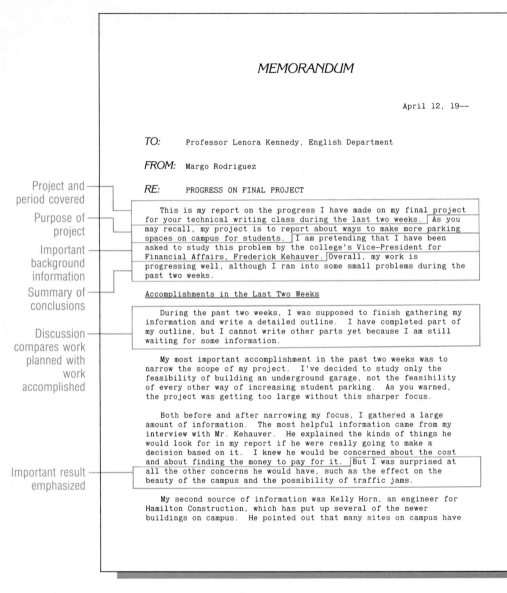

MEMORANDUM

April 12, 19--

TO: Professor Lenora Kennedy, English Department

FROM: Margo Rodriguez

RE: PROGRESS ON FINAL PROJECT

   This is my report on the progress I have made on my final project
for your technical writing class during the last two weeks. As you
may recall, my project is to report about ways to make more parking
spaces on campus for students. I am pretending that I have been
asked to study this problem by the college's Vice-President for
Financial Affairs, Frederick Kehauver. Overall, my work is
progressing well, although I ran into some small problems during the
past two weeks.

Accomplishments in the Last Two Weeks

   During the past two weeks, I was supposed to finish gathering my
information and write a detailed outline. I have completed part of
my outline, but I cannot write other parts yet because I am still
waiting for some information.

   My most important accomplishment in the past two weeks was to
narrow the scope of my project. I've decided to study only the
feasibility of building an underground garage, not the feasibility
of every other way of increasing student parking. As you warned,
the project was getting too large without this sharper focus.

   Both before and after narrowing my focus, I gathered a large
amount of information. The most helpful information came from my
interview with Mr. Kehauver. He explained the kinds of things he
would look for in my report if he were really going to make a
decision based on it. I knew he would be concerned about the cost
and about finding the money to pay for it. But I was surprised at
all the other concerns he would have, such as the effect on the
beauty of the campus and the possibility of traffic jams.

   My second source of information was Kelly Horn, an engineer for
Hamilton Construction, which has put up several of the newer
buildings on campus. He pointed out that many sites on campus have

FIGURE 23–5   Sample Progress Report

Professor Kennedy   April 12, 19--
Page 2

underground springs or unstable soil that make them unsuitable for
building an underground garage. He's promised to send me a map that
identifies these spots. Best of all, though, he's going to send me
some cost figures--so many dollars per parking space and the like.
He will include different sets of figures for the different kinds of
locations that I might choose.

Third, I've talked with many students to get their views on a
parking garage. They all like it, but some say that they wouldn't
pay to park there, and some are very concerned about where it might
be put. They don't want it to be too far from their classes.

Finally, I have talked on the phone with several administrators
at Green University, which has recently built an underground garage
similar to the one I am investigating. From them I learned a lot
about how to evaluate alternative plans for such a structure and I
learned some of the arguments people can make for (and against)
building an underground garage here.

Problems Encountered in the Last Two Weeks

Separate section devoted to problems

I have run into a couple of problems getting the information I
need to complete my outline. The biggest one is that Cecilia
Norton, the Assistant City Planner, has been on vacation and won't
be back until next week. City Hall has told me she's the person I
must depend upon for information about the city's regulations and
concerns.

Also, I found out from Ron Thiemann, who is a special aide to the
university president, that the university's board of trustees almost
certainly wouldn't approve an addition to student tuition to pay for
the lot. Also, the Alumni Office says that it would be impossible
to give money for a parking garage. Alumni want to pay for
classroom buildings and scholarships, not garages. A tuition
surcharge and alumni contributions were my best ideas about funding
the lot, so I will have to see if there are other ways to pay the
bill.

Overall, during the past two weeks I've gotten lots of
information and started my detailed outline, but I still need more
information before I can finish it.

Plan for the Next Two Weeks

Expected progress during next reporting period

Like everyone else in the class, I am rushing so that I can
complete my rough draft April 29, the day we will have our drafts

FIGURE 23–5   *(continued)*

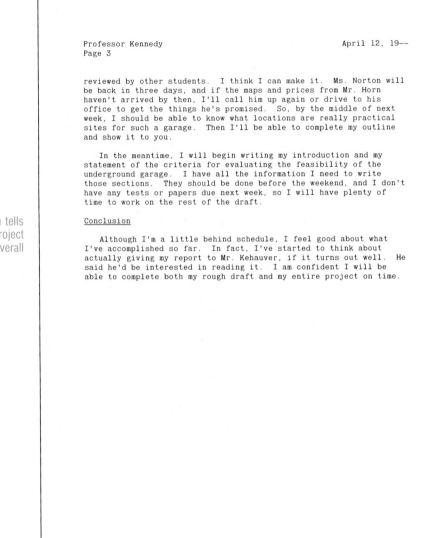

Professor Kennedy   April 12, 19—
Page 3

reviewed by other students. I think I can make it. Ms. Norton will
be back in three days, and if the maps and prices from Mr. Horn
haven't arrived by then, I'll call him up again or drive to his
office to get the things he's promised. So, by the middle of next
week, I should be able to know what locations are really practical
sites for such a garage. Then I'll be able to complete my outline
and show it to you.

In the meantime, I will begin writing my introduction and my
statement of the criteria for evaluating the feasibility of the
underground garage. I have all the information I need to write
those sections. They should be done before the weekend, and I don't
have any tests or papers due next week, so I will have plenty of
time to work on the rest of the draft.

Conclusion

Although I'm a little behind schedule, I feel good about what
I've accomplished so far. In fact, I've started to think about
actually giving my report to Mr. Kehauver, if it turns out well. He
said he'd be interested in reading it. I am confident I will be
able to complete both my rough draft and my entire project on time.

Conclusion tells
how the project
stands overall

FIGURE 23–5   *(continued)*

FROM: Buddy McCormick
TO: Craig Skelton
DATE: January 29, 19—
RE: PROGRESS REPORT ON INSTANT STYROFOAM MOLDS

Over the past month, we have experimenting with designs for the molds for the new ''Instant Styrofoam'' product being developed by the Chemical Products Division. We started by looking for devices that could clamp our standard mold shut once these new ingredients were poured into it and began to foam up. The only appropriate devices found to clamp the standard mold and hold it closed during the foaming process were C-clamps (12) around all the edges. Obviously, this type of clamping required considerable time for closing and opening, but a variety of quick-action type clamping devices (all designed to take the temperatures and pressures expected) proved unsatisfactory in earlier experiments.

Whereas the clamps did hold the mold closed, this resulted in some secondary damage to the molds, specifically warpage of the aluminum plates. This was due to the internally generated pressure during the foaming of the panels. This pressure reached a maximum about 10 minutes after placing the mold into the oven. This resulted in cleaning problems of both the mold and oven, and produced an unusable part.

Having determined that the aluminum could not withstand the pressure without deflection, we decided to build a steel mold using 1.25 cm (1/2 inch) thick steel and to bolt it together. To reduce the time required to close the mold, the lid was split two-thirds and one-third, allowing two-thirds of the mold to be bolted shut prior to the adding of the foam ingredients. Once the foam ingredients were added, only about a dozen bolts needed to be inserted. Such a mold was fabricated and the first attempt at molding a part resulted in blowing the hinges off the end of the mold. Replacing the hinges and doubling up on the bolts with the hinges on, we made a second molding that actually caused a deflection of the 1.25 cm (1/2 inch) thick steel.

Thus, we have been unsuccessful in our efforts to design a mold that can accommodate the high pressure created during foaming by the higher density foams. I recommend that during the next month you schedule us to work exclusively on this important problem.

FIGURE 23-6   Memo for Exercise 1

24

PROPOSALS

I n a proposal, you make an offer. And you try to persuade your readers to accept that offer. You say that, in exchange for money or time or some other form of support from your readers, you will give them something they want, make something they desire, or do something they wish to have done.

Throughout your career you will have many occasions to make such offers. You may think up a new product that you could develop—if your employer will give you the time and funds to do so. Or, you may devise a plan for increasing your employer's profits, but you must obtain your employer's authorization to put it into effect. Perhaps, you will work for one of the many companies that sell their products and services through proposals (rather than through advertising), in which case you may offer to provide those products and services to your employer's prospective clients and customers.

In this chapter, you will learn about the conventional superstructure for writing proposals, one that will help you persuade people to accept your offer and invest their money, time, and trust in your ideas and in your employer's products and services.

THE VARIETY OF PROPOSAL-WRITING SITUATIONS

As the second paragraph of this chapter suggests, you may write proposals in a wide variety of situations. Some of the major ways in which these situations vary from one another are as follows:

- Your readers may be employed in your own organization, or they may be employed in other organizations.

- Your readers may have asked you to submit a proposal, or you may submit it to them on your own initiative.

- Your proposal may be in competition against others, or it may be considered on its own merits alone.

- Your proposal may need to be approved by various people in your organization before you submit it to your readers, or you may submit it directly yourself.

- You may have to follow regulations concerning the content, structure, and format of your proposal, or you may be free to write your proposal entirely as you think best.

- Once you have delivered your proposal to your readers, they may follow any of a wide variety of methods for evaluating it.

To illustrate these variables in actual proposal-writing situations, the following paragraphs describe the circumstances in which two successful proposals were written. The information provided about these two proposals will also be useful to you later in this chapter, where several pieces of advice are explained through the example of these proposals.

Example Situation 1

Helen wanted permission to undertake a special project. She thought her employer should develop a computer program that employees could use to reserve conference rooms. She concluded that her company needed such a program after several instances in which she arrived for a meeting at a conference room she had reserved only to find that someone else had reserved it also. Because she is employed to write computer programs, she is well qualified to write this one. However, her work is assigned to her by her boss, and she cannot write the scheduling program without permission from him and from his boss. Consequently, she wrote a proposal to them.

As she wrote, Helen had to think about only two readers, because her boss and her boss's boss would decide about her proposed project without consulting other people. Because her employer has no specific guidelines for the way such internal proposals should be written, she could have used whatever content, structure, and format she thought would be most effective. Furthermore, she did not need anyone's approval to submit this proposal to people within her own department, although she would need approval before sending a proposal to another department.

Finally, Helen did not have to worry about competition from other proposals, because hers would be considered on its own merits. However, her readers would approve her project only if they were persuaded that the time she would spend writing the scheduling program would not be better spent on her regular duties. (Helen's entire proposal is presented at the end of this chapter.)

Example Situation 2

The second proposal was written under much different circumstances than was Helen's. To begin with, three people, not just one, wrote it. The writers were a producer, a scriptwriter, and a business manager at a public television station that seeks funds from nonprofit corporations and the federal government to produce television programs. These writers learned that the U.S. Department of Education (DOE) was interested in sponsoring programs about environmental concerns. To learn more about what the DOE wanted, the writers obtained copies of the DOE's "request for proposals" (RFP). After studying the RFP, the group decided to propose to develop materials that high school teachers and community leaders could use to teach about hazardous wastes.

In their proposal, the writers addressed an audience much different from Helen's. The DOE receives about four proposals for every one it can fund. To evaluate these proposals, it follows a procedure widely used by government agencies. It sends batches of the proposals to appropriate experts around the country. These experts, called reviewers, rate and comment on each proposal. Every proposal is seen by several reviewers. Then, the reviews for each proposal are gathered and interpreted by staff members at the DOE. Those proposals that have received the best response from the reviewers are funded. To secure funding, then, the writers needed to persuade their reviewers that their proposed project came closer to meeting the announced objectives of the DOE than did at least three-quarters of the other proposed projects.

Before the writers could even mail their proposal to the DOE, they had to obtain approval for it from several administrators at the station. That's because the proposal, if accepted, would become a contract between the station and the DOE. By means of its approval process, the station assures itself that all the contracts it makes are beneficial to it.

PROPOSAL READERS ARE INVESTORS

The descriptions of the proposals written by Helen and by the three employees of the public television station illustrate some of the many differences that exist in various proposal-writing situations. Despite these differences, however, almost all proposal-writing situations have two important features in common—features that profoundly affect the way you should write your proposals:

- In your proposals, you ask decision-makers to invest some resource, such as time or money, so that the thing you propose can be done.
- Your readers will make their investment decisions *cautiously*. They will be acutely aware that their resources are limited, that if they decide to invest in the purchase or project you propose, those resources will not be available for other uses. For example, to let Helen spend two weeks creating the new scheduling program, her bosses had to decide that Helen would not spend that time on the other projects the department had. To spend money on the education project proposed by the three writers, the DOE had to turn down some other projects that were seeking those same dollars. These readers, like all proposal readers, wanted to be sure they invested their resources wisely.

THE QUESTIONS READERS ASK MOST OFTEN

As cautious investors, proposal readers ask many questions about the purchases, projects, and other things proposed to them. From situation to situation, these questions remain basically the same. Furthermore, the kinds of answers that people at work find satisfying and persuasive are—at a very basic level—the same also. That's what makes it possible for one superstructure to be useful across nearly the full range of proposal-writing situations.

The questions asked by decision-makers when they consider a proposal generally concern the following three topics:

- **Problem.** Your readers will want to know why you are making your proposal and why they should be interested in it. What problem, need, or goal does your proposal address—and why is that problem, need, or goal important to them?

- **Solution.** Your readers will want to know exactly what you propose to make or do and how it relates to the problem you described. Therefore, they will ask, "What kinds of things will a successful solution to this problem have to do?" and "How do you propose to do those things?" They will examine carefully your responses, trying to determine whether it is likely that your overall strategy and your specific plans will work.

- **Costs.** What will your proposed product or activity cost your readers—and is it worth the cost to them?

In addition, if you are proposing to perform some work (rather than supply a ready-made product), your readers will want the answer to this question:

- **Capability.** If your readers pay or authorize you to perform this work, how do they know they can depend upon you to deliver what you promise?

STRATEGY OF THE CONVENTIONAL SUPERSTRUCTURE FOR PROPOSALS

The conventional superstructure for proposals is a framework for answering those questions—one that has been found successful in repeated use in the kinds of situations you will encounter on the job.

When you follow this superstructure, you provide information on up to ten topics, which are listed in Figure 24–1. In some cases, you will include information on all ten topics, but in others, you will cover only some of them. Even in the briefest of proposals, however, you will probably need to treat the following four topics if you are to succeed: introduction, problem, solution, costs.

When you provide information on these topics, however, you should do much more than supply data. You should also try to make persuasive points. In its right-hand column, Figure 24–1 identifies the overall persuasive point you should try to make when discussing each of the ten topics of the conventional superstructure.

Topic	Your Readers' Question	Your Persuasive Point
*Introduction	What is this communication about?	Briefly, I propose to do the following.
*Problem	Why is the proposed project needed?	The proposed action addresses a problem that is important to you.
Objectives	What features will a solution to this problem need in order to be successful?	A successful solution must achieve the following objectives.
*Product	How do you propose to do those things?	Here's what I plan to produce and how it will work effectively at achieving the objectives.
Method Resources Schedule Qualifications Management	Are you going to be able to deliver what you describe here?	Yes, because I have a good plan of action (method); the necessary facilities, equipment, and other resources; a workable schedule; appropriate qualifications; and a sound management plan.
*Costs	What will it cost?	The cost is reasonable.

*Topics marked with an asterisk are important in almost every proposal, whereas the others are needed only in certain ones.

FIGURE 24–1   Relationship of the Standard Topics in a Proposal to Your Readers' Questions and the Persuasive Points You Need to Make

As you write, you will find it helpful to see the relationships among the ten topics. Think of them as a sequence in which you lead your readers through the following progression of thoughts:

1. The readers learn generally what you want to do. (Introduction)

2. The readers are persuaded that there is a problem, need, or goal that is important to them. (Problem)

3. The readers are persuaded that the proposed action will be effective in solving the problem, meeting the need, or achieving the goal that the readers now agree is important. (Objectives, Product)

4. The readers are persuaded that you are capable of planning and managing the proposed solution. (Method, Resources, Schedule, Qualifications, Management)

5. The readers are persuaded that the cost of the proposed action is reasonable in light of the benefits the action will bring. (Costs)

There is no guarantee, of course, that your readers will actually read your proposal from front to back or concentrate on every word you write. Consider how readers often approach long proposals. These proposals usually include a summary or abstract at the beginning. Instead of reading these long proposals straight through, many people will read the summary, perhaps the first few pages of the body, and then skip around through the other sections.

In fact, in some competitive situations, reviewers of proposals are *prohibited* from reading the entire proposal. For instance, when companies compete for huge contracts to build major parts of space shuttles for the National Aeronautics and Space Administration (NASA), they submit proposals in three volumes: one explaining the problem and their proposed solution, one detailing their management plan, and one analyzing their costs. Each volume is evaluated by a separate set of experts: technical experts for the first volume, management experts for the second, and budget experts for the third.

Even when readers will not read your proposal straight through, the account given above of the relationships among the parts can help you write a tightly focused proposal in which all the parts support one another effectively.

VARIOUS LENGTHS FOR PROPOSALS

The preceding discussion mentions proposals that are several volumes long. Such proposals can run to hundreds or even thousands of pages. On the other hand, some proposals are less than one page. How will you know how long your written proposals should be? There is no simple answer. In each case you need to determine how much you must say to win your point with your readers.

Sometimes you can be brief and still be persuasive. Often you will need to touch upon only a few of the ten topics listed in Figure 24–1. For instance, Helen's proposed project involved only one person: Helen. Consequently, she didn't need any management plan. Similarly, because her readers were already familiar with her qualifications as a writer of programs, she didn't have to say anything about them, except perhaps to point out the experience she had had in preparing the particular kind of program she proposed to write. And because she was asking only for two weeks of time to spend on the project she proposed, she didn't have to present a detailed budget, though she did need to justify her proposed schedule.

In other situations, like those involving the proposals to NASA mentioned above, you may need to write very lengthy proposals. Those will be long because you will need to address all ten topics, and your discussion of each of those topics must answer fully all the questions your readers will have. In the end, then, to decide how long a proposal needs to be, you must think about your readers, anticipating their questions and their reactions to what you are writing.

SUPERSTRUCTURE FOR PROPOSALS

The rest of this chapter describes in detail each of the ten topics that form the conventional superstructure for proposals. As you read this information, keep in mind that the conventional superstructure represents only a general plan. You must use your imagination and creativity to adapt it effectively to your particular situation.

In addition, as you plan and write your proposal, remember that the ten topics identify kinds of information you need to provide, not necessarily the titles of the sections you will include. In brief proposals, some parts take only a sentence or a paragraph, so that several are grouped together. For instance, writers often combine their announcement of their proposal, their discussion of the problem, and their explanation of their objectives under a single heading, which might be "Introduction," "Problem," or "Need."

Also, remember that the conventional superstructure may be used with any of the three common formats: letter, memo, and book. While writing your proposal, you should consult Appendix A for information about the particular format you are going to use.

The sections that follow take up the ten topics in the order in which they appear in Figure 24–1: introduction, problem, objectives, product, method, resources, schedule, qualifications, management, and costs.

Introduction

At the beginning of a proposal you want to do the same thing that you do at the beginning of anything else you write on the job: tell your readers what you are writing about. In a proposal, this means announcing what you are proposing.

How long and detailed should this introductory announcement be? In proposals, introductions vary considerably in length but they are almost always relatively brief. By custom, writers reserve the full description of what they propose until later, after they have discussed the problem that their proposal will help to solve.

You may be able to introduce your proposal in a single sentence. Helen did this in her proposal:

> I request permission to spend two weeks writing, testing, and implementing a program for scheduling conference rooms.

CHAPTER 24 PROPOSALS **629**

When you propose something more complex than a two-week project, you may need more words to introduce it. In addition, sometimes you may need to provide background information to help your readers understand what you have in mind. Here, for example, is the introduction from the proposal written by the public television station to the DOE:

> Chemicals are used to protect, prolong, and enhance our lives in numerous ways. Recently, however, society has discovered that some chemicals also present serious hazards to human health and the environment. In the coming years, citizens will have to make many difficult decisions to solve the problems created by these hazardous substances and to prevent future problems from occurring.
>
> To provide citizens with the information and skills they will need to decide wisely, WPET Television proposes to develop two educational packages entitled ''Hazardous Substances: Handle with Care!'' One package, for high school students, will include five fifteen-minute videotape programs and a teacher's guide. The other package, for communities, will consist of a thirty-minute color film and a discussion leader's guide.

Problem

Once you've announced what you are proposing, you must persuade your readers that your proposed action will address some problem significant to *them*. Your description of the problem is crucial to the success of your proposal. Although you might persuade your readers that your proposed project will achieve its objectives and that your project's costs are reasonable, you cannot hope to win approval unless you show that the project is worth doing in the first place—from your readers' point of view. You must not only identify a problem, but also make that problem seem important to your readers; not only describe a need, but also make it seem significant to your readers; not only define a goal, but also make its achievement seem worthwhile to your readers.

To do this requires both creativity and research. However, the precise nature of your effort to describe the problem addressed by your proposed project will depend upon your proposal-writing situation. The following paragraphs offer advice that applies to each of three situations you are likely to encounter on the job: when your readers define the problem for you, when your readers provide you with a general statement of the problem, and when you must define the problem yourself.

When Your Readers Define the Problem for You

You need to do the least research about the problem when your readers define it for you. This can happen when you are writing a proposal that your readers have asked you to submit. For instance, the readers might issue an RFP that explains in complete detail some technical problem that they would like your firm (or one of your competitors) to solve. In such situations, your primary objective in describing the problem will be to show your readers that you thoroughly understand what they want.

When Your Readers Provide a General Statement of the Problem

At other times, even when you have an RFP, you will need to devote considerable research and creativity to describing the problem. This happened to the three people who wrote the proposal concerning environmental education. In its RFP, the DOE provided only the general statement that it wanted to support projects that would "develop educational practices and resources dealing with the relation of various aspects of the natural and man-made environment to the total human environment." Proposal writers were then left to identify the particular kind of practice or resource they wished to develop and to devise their own arguments that those practices or resources were worthy of financial support from the DOE.

When you are in a similar situation, you should find out what sort of problem your readers will consider important. For instance, the writers at the public television station discovered that the DOE made the following statement elsewhere in the RFP: "Thus the environmental education process is multifaceted, multidisciplinary, and issue- or problem-oriented." This statement suggested to the writers that they should describe the problem they would address as an issue- or problem-oriented one that would require multidisciplinary knowledge to solve.

Accordingly, the writers decided to say that the materials they would prepare on hazardous substances would help citizens make decisions about the handling and regulating of hazardous substances in their own communities. Further, they described the citizens' problem in making these decisions as one that required them to think about hazardous wastes from many points of view: health, economics, technology, and so on. Once the writers framed their general approach, they found appropriate facts and articles to explain and support this need. They also investigated to prove that materials addressed to this need did not already exist. Finally, they identified two groups of people in particular need of this information: high school students, who would soon have to use these skills as adult, voting citizens, and residents who already face decisions about hazardous wastes because industries in their communities produce them or because such wastes are disposed of in their region. In the end, the writers included in their proposal a four-page discussion of the need for the particular kind of educational materials they would develop—a need they presented in terms they knew would be persuasive to their readers.

When You Must Define the Problem Yourself

In other situations, you may not have the aid of explicit statements from your readers to help you formulate the problem. This is most likely to happen when you are preparing a proposal on your own initiative, without being asked by someone else to submit it. The challenge of describing the problem in such situations can be particularly great because the argument that will be persuasive to your readers might be entirely distinct from your own reasons for writing the proposal.

Think about Helen's situation, for instance. She originally came up with the idea of writing the program for scheduling conference rooms because she felt frustrated and angry on the many occasions that she went to a meeting room that she had reserved only to find that someone else had reserved it too. Her bosses, however, are not likely to approve the scheduling project simply to help Helen avoid frustrations. For the project to appeal to them, it must be couched in different terms, terms that are allied to their own responsibilities and professional interests.

In such situations, you can pursue two strategies to define the problem. The first is to think about how you can make your proposed project important to your readers. What goals or responsibilities do your readers have that your proposal will help them achieve? What kinds of concerns do they typically express that your proposal could help them address? A good place to begin is to think about some of the standard concerns of organizations: efficiency and profit.

When she did this, Helen realized that from her employer's point of view, the problems involved with scheduling conference rooms were creating great inefficiencies. Time was wasted as people looked around for another place to meet. Further, the time wasted was not just of one person, but of all the people involved with the displaced meeting.

The second strategy for defining the problem is to speak with one or more of the people to whom you will send your proposal. This conversation can have two advantages. It will let you know whether or not your proposal has at least some chance of succeeding. If it doesn't, you might as well find out before you invest a large amount of time writing it. Second, by talking with someone who will later be a reader of your proposal, you can find out how the problem appears to him or her.

When Helen spoke to her boss, she discovered another aspect of the conference room scheduling that she had not thought of. Sometimes the rooms are used for meetings with customers. When customers see confusions arising over something as simple as meeting rooms, they are not only annoyed but also prompted to wonder if the company has similar problems with other parts of its operation—problems that would affect the company's products and services.

Objectives

When using the conventional superstructure for proposals, you discuss in two stages your ideas for solving the problem you have identified:

1. You answer your readers' question, "What will a solution to this problem need in order to be successful?" Your answer will be a statement of the objectives of your proposed solution.

2. You answer your readers' question, "How do you propose to do those things?" In your answer, you describe the product of your proposed action—the thing you are asking your readers to authorize or support. You must do so in a way that persuades your readers that your proposed action will in fact achieve the objectives and thereby solve the problem.

As you can see, your statement of objectives plays a crucial role in the logical development of your proposal: it links your proposed action to the problem by telling how the action will solve the problem. To make that link tight, you must formulate each of your objectives so that it grows directly out of some aspect of the problem you describe. Here, for example, are three of the objectives that the writers devised for their proposed environmental education program. To help you see how these objectives grew out of the writers' statement of the problem, each objective is followed by the point from their problem statement that serves as the basis for it. (The writers did not include the bracketed sentences in the objectives section of their proposal.)

1. To teach high school students the definitions, facts, and concepts necessary to understand both the benefits and risks of society's heavy reliance upon hazardous substances. [This objective is based on the evidence presented in the problem section that high school students do not have that knowledge.]

2. To employ an interdisciplinary approach that will allow students to use the information, concepts, and skills from their science, economics, government, and other courses to understand the complex issues involved with our use of hazardous substances. [This objective is based on the writers' argument in the problem section that people need to understand the use of hazardous substances from many points of view to be able to make wise decisions about them.]

3. To teach a technique for identifying and weighing the risks and benefits of various possible courses of action. [This objective is based on the writers' argument in the problem section that people must be able to weigh the risks and benefits of the use of hazardous substances if they are to make sound decisions about that use.]

The writers created similar lists of objectives for the other parts of their proposed project, each likewise based upon specific points made in their discussion of the problem.

Like the writers at the television station, Helen carefully based her objectives upon her description of the problem. The exception is that she based her second objective not upon the problem that existed when she wrote the proposal but upon a problem that would have been created if her employer sought to solve the scheduling problem through another strategy, which Helen rejected. Here are two of Helen's three objectives (which she presented in paragraph form):

1. To maintain a completely up-to-date schedule of reservations that can be viewed by salaried personnel in every department. [This objective corresponds to her argument that the problem arises partly because there is no convenient way for people to find out what reservations have been made.]

2. To allow designated persons in every department to add, change, or cancel reservations through their terminals. [This objective relates to Helen's observation that making everyone follow the present regulations will create too much work for the people who would have to take everybody's calls about reservations.]

In proposals, writers usually describe the objectives of their proposed solution without describing the solution itself at all. Consequently, their readers can imagine many solutions that might achieve those objectives. The television writers, for example, wrote their objectives so that their readers could imagine achieving them through the creation of a textbook, an educational movie, a series of lectures, or a slide–tape show—as well as through the creation of the videotape programs the instructors proposed. Similarly, Helen described objectives that might be achieved by many kinds of computer programs. She withheld her ideas about the structure and strategies of her program's internal design until the next section of her proposal.

The purpose of separating the writer's objectives from the description of the solution is not to keep readers in suspense. In a proposal's opening paragraph, writers generally tell what they are proposing so readers know at least generally what is being proposed before reading the objectives. Rather, the separation of objectives from solution enables readers to evaluate the aims of the project separately from the writers' particular strategies for achieving those aims. Consequently, you must write your objectives so they will seem positive and desirable to your readers. Unless your readers feel that your objectives are desirable, they are unlikely to feel that your proposed way of achieving those objectives is worth their support. The most important way to make them seem desirable, of course, is to base them upon your description of the problem that your project will solve.

As the examples quoted from the television proposal suggest, writers often present objectives in lists. Furthermore, even when they don't, they usually describe their objectives very briefly. For instance, Helen used only a paragraph to present her objectives. The television group presented all of their objectives in 3 pages of their 98-page proposal. The challenge presented to you as a proposal writer is to make your objectives clear and appealing while still making them brief.

Product

When you describe the product that your proposed project will produce, you explain your plan for achieving the objectives you told your readers about. For example, Helen described the various parts of the computer program she would write, talking both about how they would be built and how they would be used. The television writers described in detail each of the four components of their environmental education package. Scientists seeking money for cancer research would describe the experiments they wish to conduct.

To describe your product persuasively, you need to do three things. First, be sure to let your readers know how you will achieve each of your objectives. For instance, to be sure that the description of their proposed educational program matched their objectives, the writers included detailed descriptions of the following: the definitions, facts, and concepts their videotape program would teach (see Objective 1, above), the strategy they would use to help students take an interdisciplinary approach to the question of hazardous waste (Objective 2), and the technique for weighing costs and benefits that they would help students learn (Objective 3).

The second thing you must do when describing your product is provide enough detail to satisfy your readers that you have planned it carefully and thoughtfully. To do this, you will have to begin actual work on the project you are proposing. For instance, to provide enough detail to her boss about the computer program she would write, Helen had to design most of its aspects. Similarly, the television group had to design many aspects of the proposed educational materials. For instance, they developed an outline for each of the five videotape programs and another outline for the teacher's guide that would accompany them.

The third thing you must do is to explain, where appropriate, the desirability of the product of your project. For example, the television writers planned to use a case-study approach in their materials, and in their proposal they offer this explanation of the advantages of that approach:

> Because case studies represent real-world problems and solutions, they are effective tools for illustrating the way that politics, economics, and social and environmental interests all play parts in hazardous substance problems. In addition, they can show the outcome of the ways that various problems have been solved— successfully or unsuccessfully—in the past. In this way, case studies provide students and communities with an opportunity to learn from past mistakes and successes.

Of course, you should include such statements only where they won't be perfectly obvious to your readers. In her proposal, Helen did not include any because she planned to use standard practices whose advantages would be perfectly evident to her readers, both of whom had been promoted to their present positions after several years of doing exactly the kind of work Helen is doing now.

A Note about the Relationships among Problem, Objectives, and Product

The first three elements of a proposal are closely related to one another. In fact, you can increase the likelihood of succeeding in your proposal by ensuring that the three parts fit tightly together. Be sure that the objectives grow directly out of your statement of the problem, and that the project's product will address those objectives.

As a result of the close interrelationship among these parts, your work on any one of them has implications for the others. For example, your definition of the problem will shape the product you propose. When the television writers discovered the importance the DOE placed on issue-oriented and multidisciplinary approaches, they knew they would have to describe a problem, define an objective, and design their product (instructional materials) in issue-oriented, multidisciplinary terms. In a similar fashion, Helen discovered that, from the company's point of view, the impression that the present system creates on customers is a major problem. Therefore, she emphasized that her proposed computer program would include some method of creating priorities among potential users of the conference rooms so that users meeting with customers would be assured of having conference rooms to use even if other groups wanted them also.

It's crucial that you realize that the parts of your proposal must be well integrated and that your work on one topic will naturally affect what you will say about the other topics.

Method

The decision-makers who act on proposals sometimes need to be assured that you can, in fact, produce the results that you promise. That happens especially in situations where you are offering something that takes special expertise—something to be customized or created only if your proposal is approved.

To assure themselves that you can produce what you promise, your readers will look for information about several aspects of your project: your method or plan of action for producing the result; the facilities, equipment, and other resources you plan to use; your schedule; your qualifications; and your plan for managing the project. Method is described in this section; the other topics are discussed in the sections that follow.

To determine how to explain your proposed method, imagine that your readers have asked you, "How will you bring about the result you have described?"

In some cases, you will not need to answer that question. For example, Helen did not talk at all about the programming techniques she planned to use because her readers were already familiar with them. On the other hand, her readers did not know before reading her proposal how she planned to train people to use the program she would create. Therefore, in her proposal she explained her plans for training.

In contrast, the television writers had an elaborate plan for creating their educational materials. An important part of their plan, for instance, was to use three review teams: one to assess the accuracy of the materials they drafted, another to advise about the effectiveness of the videotapes, and a third to advise about the effectiveness of the community film and the discussion leader's guide. In their proposal, the writers described these review teams in great detail.

In addition, the writers described each phase of their project to show that they would conduct all phases in a way that would lead to success. These phases include research, scripting, review, revision, production, field-testing, revision, final production, and distribution. In a similar fashion, when you write proposals you may have to supply detailed descriptions of your method of creating your product.

Resources

By discussing the facilities, equipment, and other resources to be used for your proposed project, you assure your readers that you will use whatever special equipment is required to do the job properly. If part of your proposal is to request that equipment, tell your readers what you need to acquire and why.

Of course, when you propose something that requires no special resources, you do not need to include such a section. Helen did not need to include one. In contrast, the television writers needed many kinds of resources. In their proposal, they described the excellent library facilities that were available for their research. Similarly, to persuade their readers that they could produce high-quality programs, they described the videotaping facilities they would use.

Schedule

People who read or listen to your proposals have several reasons for wanting to know the schedule for your proposed project. First, they want to know when they can enjoy the final result. Second, they want to know how the work will be structured so they can be sure the schedule is reasonable and is a sound way of organizing the work. In addition, they may want to plan other work around the project: When will this project have to coordinate with others? When will it take people's attention from other work? When will other work be disrupted and for how long? Finally, proposal readers want a schedule they can use once the project has begun so they can determine if the project is proceeding according to plan.

The most common way to provide a schedule is to use a schedule chart. You can find detailed instructions for creating schedule charts in Chapter 14, "Creating Twelve Types of Visual Aids." Figure 14–30 shows a sample schedule chart.

Qualifications

When they are thinking about investing in a project, proposal readers want to be sure that the proposers have the experience and capabilities to carry out the project successfully. For that reason, a discussion of the qualifications of the personnel involved with a project is a standard part of most proposals. For example, in their proposal, the television writers discussed their qualifications in two places. First, in a section entitled "Qualifications," they presented the chief qualifications of each of the eight key people who will work on the project. In addition, they included a detailed resume for each in an appendix.

In other situations, much less information might be needed. For instance, Helen's qualifications as a programmer were evident to her readers because they were employing her as one. If that experience alone were enough to persuade her readers that she could carry out the project successfully, Helen would not have needed to include any section on qualifications. However, her readers might have wondered whether she was qualified to undertake the particular program she proposed, because different kinds of programs require different knowledge and skills. Therefore, Helen wrote the following:

> As you know, although I usually work with our IBM system, I am also familiar with the Hewlett-Packard computer on which the schedule will be placed. In addition, as an undergraduate I took a course in scheduling and transportation problems, which will help me here.

In some situations, your readers will want to know not only about the qualifications of the personnel who will work on the proposed project, but also the qualifications of the organization for which you all work.

Management

When you propose a project that will involve more than about four people, you increase the persuasiveness of your proposal by describing the management structure of your group. That's because proposal readers know that even well-qualified people cannot work successfully on a project if their activities aren't coordinated and overseen in an effective manner. In projects with relatively few people, you can describe the management structure by first identifying the person or persons who will have management responsibilities and then telling what their duties will be. In larger projects, you might need to provide a full organizational chart for the project (see Chapter 14 for information about creating organizational charts) and a detailed description of the management techniques and tools that will be used.

Because her project involved only one person, Helen did not establish or describe any special management structure. However, the television writers did. Because they had a complex project involving many parts and several workers, they set up a project planning and development committee to oversee the activities of the principal workers. In their proposal, they described the make-up and functions of this committee, and in the section on qualifications, they also described the credentials of the committee members.

Costs

As emphasized throughout this chapter, when you propose something, you are asking your readers to invest resources, usually money and time. Naturally, then, you need to tell them how much your proposed project will cost.

One way to discuss costs is to include a budget statement. Sometimes, a budget statement needs to be accompanied by a prose explanation that persuades your readers that any unusual expenses are justified and that each of the items in the budget is calculated in a reasonable fashion. (See Chapter 14 for information about presenting budget statements.)

In proposals where dollars are not involved, information about the costs of required resources may be provided elsewhere. For instance, in her discussion of the schedule for her project, Helen explained the number of hours she would spend, the time that others would spend, and so on.

In some proposals, you may demonstrate the reasonableness of the costs of your proposal by also calculating the savings that will result from your project.

PLANNING GUIDE

Figure 24–2 shows a worksheet that may help you plan your proposals. As you fill out the worksheet, remember that it is intended merely to stimulate your thinking. To achieve your objectives, you may need to say many things that you don't write down on it, and you may discover that some of the worksheet's questions are irrelevant to the particular proposal you are writing. Also, remember that the superstructure on pages 2 and 3 of Figure 24–2 does not necessarily provide the outline for your proposal.

SAMPLE PROPOSAL

Figure 24–3 shows Helen's proposal.

WRITING ASSIGNMENT

An assignment that involves writing a proposal is included in Appendix C.

Planning Guide
PROPOSAL

Subject _____ **Due Date** _____

Overall Purpose

Reader Profile

Who are your primary readers?

Who else might read your proposal?

What are your readers' titles and professional responsibilities? How will they influence what your readers look for in your proposal?

How well do your readers understand the technical, scientific, or other specialized terms and concepts you might use?

Do your readers have any communication preferences you should take into account? If so, what are they?

Are there any other considerations you should keep in mind when addressing these readers?

Readers' Informational Needs

What will your readers' key questions be?

Readers' Attitudes

What are your readers' present attitudes toward your proposal? What do you want them to be?

What are your readers' present attitudes toward *you*? What do you want them to be?

What will be your readers' initial responses to your proposal?

FIGURE 24–2   Planning Guide for Proposals

Superstructure

Introduction
> Briefly, what do you propose to do?

> Can you help your readers by forecasting the rest of your report?

Problem
> Why is your proposed project needed? (Remember to answer this question from your readers' point of view.)

> Do your readers already perceive the need for your proposed project, or will you have to persuade them of the need?

> What background information do your readers need about your subject?

Objectives
> What features will a solution to this problem need in order to be successful (e.g., low cost, durability, etc.)?

Product
> What do you plan to produce?

> How will your product work effectively at achieving the objectives you identified?

Method
> How will you go about producing your product?

> What do you need to say about your method to persuade your readers that it will be successful?

Resources
> What resources do you have or need to produce your product?

FIGURE 24–2 *(continued)*

Schedule
    When will your product be ready?

    When will you perform the major steps involved with creating your product?

Qualifications
    If your readers will want to know your qualifications, what can you say to persuade the readers that you are qualified?

Management
    If your project is large, how will you organize the people working on it to ensure that they work productively?

Costs
    Are there any parts of the cost that your readers will want you to explain?

    What will it cost?

    Do you need to say anything to persuade your readers that the costs are reasonable?

Counterarguments

Are your readers likely to raise counterarguments to anything you state in your proposal? If so, what will the counterarguments be and how will you address them?

Visual Aids

What visual aids would your readers find helpful or persuasive?

Organization

What organization would your readers find useful and persuasive? (On a separate sheet provide a brief sketch or outline to identify the major sections and subsections.)

FIGURE 24–2   *(continued)*

PARKER   MANUFACTURING   COMPANY
Memorandum

TO: Floyd Mohr and Marcia Valdez

FROM: Helen Constantino

DATE: July 14, 19--

RE: Proposal to Write a Program for Scheduling Conference
         Rooms

Helen tells what is proposed and what it will cost →

I request permission to spend two weeks writing, testing, and
implementing a program for scheduling conference rooms in the plant.
This program will eliminate several problems with conference room
schedules that have become acute in the last six months.

Present System

Background information

At present, the chief means of coordinating room reservations is the
monthly "Reservations Calendar" distributed by Peter Svenson of the
Personnel Department. Throughout each month, Peter collects notes
and phone messages from people who plan to use one of the conference
rooms sometime in the next month. He stores these notes in a folder
until the fourth week of the month, when he takes them out to create
the next month's calendar. If he notices two meetings scheduled for
the same room, he contacts the people who made the reservations so
they can decide which of them will use one of the other seven
conference rooms in the new and old buildings. He then prints the
calendar and distributes it to the heads of all seventeen
departments. The department heads usually give the calendars to
their secretaries.

Someone who wants to schedule a meeting during the current month
usually checks with the department secretary to see if a particular
room has been reserved on the monthly calendar. If not, the person
asks the secretary to note his or her reservation on the
department's copy of the calendar. The secretary is supposed to call
the reservation in to Peter, who will see if anyone else has called
about using that room at that time.

Problems with the Present System

The present system worked adequately until about six months ago,
when two important changes occurred:

- The new building was opened, bringing nine departments here
  from the old Knoll Boulevard plant.

FIGURE 24–3   Sample Proposal in the Memo Format

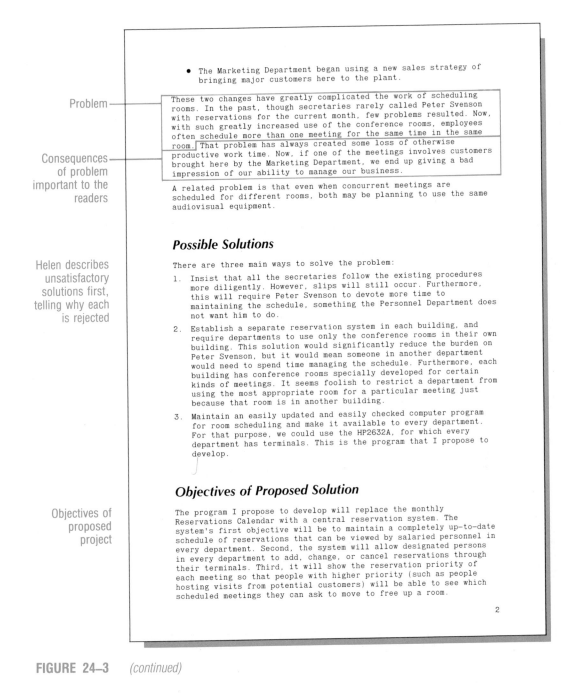

- The Marketing Department began using a new sales strategy of bringing major customers here to the plant.

Problem —

These two changes have greatly complicated the work of scheduling rooms. In the past, though secretaries rarely called Peter Svenson with reservations for the current month, few problems resulted. Now, with such greatly increased use of the conference rooms, employees often schedule more than one meeting for the same time in the same room. That problem has always created some loss of otherwise productive work time. Now, if one of the meetings involves customers brought here by the Marketing Department, we end up giving a bad impression of our ability to manage our business.

Consequences of problem important to the readers

A related problem is that even when concurrent meetings are scheduled for different rooms, both may be planning to use the same audiovisual equipment.

Possible Solutions

There are three main ways to solve the problem:

Helen describes unsatisfactory solutions first, telling why each is rejected

1. Insist that all the secretaries follow the existing procedures more diligently. However, slips will still occur. Furthermore, this will require Peter Svenson to devote more time to maintaining the schedule, something the Personnel Department does not want him to do.

2. Establish a separate reservation system in each building, and require departments to use only the conference rooms in their own building. This solution would significantly reduce the burden on Peter Svenson, but it would mean someone in another department would need to spend time managing the schedule. Furthermore, each building has conference rooms specially developed for certain kinds of meetings. It seems foolish to restrict a department from using the most appropriate room for a particular meeting just because that room is in another building.

3. Maintain an easily updated and easily checked computer program for room scheduling and make it available to every department. For that purpose, we could use the HP2632A, for which every department has terminals. This is the program that I propose to develop.

Objectives of Proposed Solution

Objectives of proposed project

The program I propose to develop will replace the monthly Reservations Calendar with a central reservation system. The system's first objective will be to maintain a completely up-to-date schedule of reservations that can be viewed by salaried personnel in every department. Second, the system will allow designated persons in every department to add, change, or cancel reservations through their terminals. Third, it will show the reservation priority of each meeting so that people with higher priority (such as people hosting visits from potential customers) will be able to see which scheduled meetings they can ask to move to free up a room.

2

FIGURE 24–3 *(continued)*

Overview of proposed project

Detailed description of product of proposed work

Helen has already planned the testing

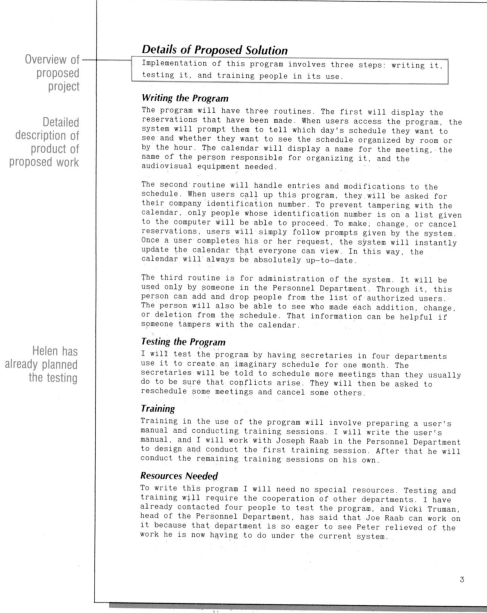

Details of Proposed Solution

Implementation of this program involves three steps: writing it, testing it, and training people in its use.

Writing the Program

The program will have three routines. The first will display the reservations that have been made. When users access the program, the system will prompt them to tell which day's schedule they want to see and whether they want to see the schedule organized by room or by the hour. The calendar will display a name for the meeting, the name of the person responsible for organizing it, and the audiovisual equipment needed.

The second routine will handle entries and modifications to the schedule. When users call up this program, they will be asked for their company identification number. To prevent tampering with the calendar, only people whose identification number is on a list given to the computer will be able to proceed. To make, change, or cancel reservations, users will simply follow prompts given by the system. Once a user completes his or her request, the system will instantly update the calendar that everyone can view. In this way, the calendar will always be absolutely up-to-date.

The third routine is for administration of the system. It will be used only by someone in the Personnel Department. Through it, this person can add and drop people from the list of authorized users. The person will also be able to see who made each addition, change, or deletion from the schedule. That information can be helpful if someone tampers with the calendar.

Testing the Program

I will test the program by having secretaries in four departments use it to create an imaginary schedule for one month. The secretaries will be told to schedule more meetings than they usually do to be sure that conflicts arise. They will then be asked to reschedule some meetings and cancel some others.

Training

Training in the use of the program will involve preparing a user's manual and conducting training sessions. I will write the user's manual, and I will work with Joseph Raab in the Personnel Department to design and conduct the first training session. After that he will conduct the remaining training sessions on his own.

Resources Needed

To write this program I will need no special resources. Testing and training will require the cooperation of other departments. I have already contacted four people to test the program, and Vicki Truman, head of the Personnel Department, has said that Joe Raab can work on it because that department is so eager to see Peter relieved of the work he is now having to do under the current system.

3

FIGURE 24–3   *(continued)*

Schedule also
tells the cost (in
hours of work)

Qualifications

Schedule

I can write, test, and train in eight eight-hour days.

Task	Hours
Designing Program	12
Coding	24
Testing	8
Writing User's Manual	12
Training First Group of Users	8
Total	64

The eight hours estimated for training includes the time needed both to prepare the session and to conduct it one time.

Qualifications

As you know, although I usually work with our IBM system, I am also familiar with the Hewlett-Packard computer on which the schedule will be placed. In addition, as an undergraduate I took a course in scheduling and transportation problems, which will help me here.

Conclusion

I am enthusiastic about the possibility of creating this much-needed program for scheduling conference rooms and hope that you are able to let me work on it.

4

FIGURE 24–3   *(continued)*

25

INSTRUCTIONS

Y ou may often write instructions on the job. When doing so, you will act as your readers' guide and coach. Perhaps your readers will be people who have just purchased a product made by your employer. Through your instructions, you will tell them how to use that product. Perhaps your readers will be new employees in your organization. Through your instructions, you will teach them how to perform their basic duties. Perhaps your readers will be experienced engineers or scientists. Through your instructions, you will tell them how to perform a special procedure that you have developed.

Whoever your readers are, they will be counting on you to guide them quickly and safely toward the successful completion of their task. In this chapter, you will learn the conventional superstructure for writing instructions, which will help you guide your readers effectively.

THE VARIETY OF INSTRUCTIONS

If you were to look at a sampling of the various kinds of instructions written at work, you would see that instructions vary greatly in length and complexity. The simplest and shortest are only a few sentences long. Consider, for example, the instructions that the state of Ohio prints on the back of the 1×1-inch registration stickers that Ohio citizens must buy and affix to their automobile license plates each year:

Application Instructions

1. Position sticker on clean, dry surface in lower right-hand corner of rear plate (truck tractor front plate). If plate has a previous sticker, place new sticker to cover old sticker.

2. Rub edges down firmly.

Note: Do not moisten or apply at temperatures less than 0° F.

Other instructions are hundreds—or even thousands—of pages long. Examples of these long and highly complex instructions are those written by General Electric, Rolls Royce, and McDonnell Douglas for servicing the airplane engines they manufacture. Other examples are the manuals that IBM, Control Data, and NCR write to accompany their large mainframe computers.

This chapter describes the superstructure for instructions in a way that will enable you to use the patterns for any instructions you write at work, whether long or short.

THREE IMPORTANT POINTS TO REMEMBER

When writing instructions, you should keep in mind three points: instructions shape attitudes, good visual design is essential, and testing is often indispensable. Each of these points is discussed briefly in the following paragraphs.

Instructions Shape Attitudes

All the communications you write at work have a double aim: to help your readers perform some task and to affect your readers' attitudes in some way. However, many writers of instructions focus their attention so sharply on the task they want to help their readers perform that they forget about their readers' attitudes. To write effective instructions, you must not commit this oversight.

The most important attitude with which you should concern yourself is that of your readers toward the instructions themselves. Most people dislike using instructions. When faced with the work of reading, interpreting, and following a set of instructions, they are often tempted to toss the instructions aside and try to do the job using common sense. However, you and your employer will often have good reasons for wanting people to use the instructions you write. Maybe the job you are describing is dangerous if it isn't done a certain way, or maybe the product or equipment involved can be damaged. Maybe you know that failure to follow instructions will lead many readers to an unsatisfactory outcome, which they might then blame on your employer. For these reasons, it is often very important for you to persuade your readers that they should use your instructions.

In addition, as an instruction writer, you may want to shape your readers' attitudes toward your company and its products. If your readers feel that the product is reliable and that the company thoroughly backs it with complete support (including good instructions), they will be more likely to buy other products from your employer and to recommend those products to other people.

Good Visual Design Is Essential

To create instructions that will help your readers and also shape their attitudes in the ways you want, you must pay special attention to the instructions' visual design, including both the page design and the design of the drawings, charts, flow diagrams, and other visual aids you might use.

Page Design

In instructions, good page design is important for several reasons. First, readers almost invariably use instructions by alternating between reading and acting. They read a step and then do the step, read the next step and then do that step. By designing your pages effectively, you can help your readers easily find the instructions for the next step each time they turn their eyes back to your page. This may seem a trivial concern, but readers quickly become frustrated if they have to search through a page or a paragraph to find their places. When readers are frustrated by a set of instructions, they may quit trying to use them.

Second, through good page design you can help your readers grasp quickly the connections between related blocks of material in your instructions, such as the connection between an instruction and the drawing or other visual aid that accompanies it.

Third, the appearance of instructions influences readers to use or not use them. If the instructions appear dense and difficult to follow, or if they appear careless and unattractive, readers may decide not even to try them.

For advice about creating effective page designs, see Chapter 15, "Designing Pages."

Visual Aids

You can increase the effectiveness of most instructions by including visual aids. Well-designed visual aids are much more economical than words in showing readers where the parts of a machine are located or what the result of a procedure should look like. On the other hand, visual aids that are poorly planned and prepared can be just as confusing and frustrating for readers as poorly written prose.

For general advice about creating effective visual aids, see Chapter 13, "Using Visual Aids." For specific advice about preparing twelve of the most commonly used types of visual aids, see Chapter 14, "Creating Twelve Types of Visual Aids."

Testing Is Often Indispensable

It may seem that instructions are among the easiest of all communications to write and therefore among those that least need to be tested. After all, when you write instructions, you usually describe a procedure you know very well—and your objective is simply to tell your readers as clearly and directly as possible what to do, one little step at a time. Actually, instructions present a considerable challenge to you as a writer. You will find that it is often difficult to find the words that will tell your readers what to do in a way that they will understand quickly and clearly. Also, because you know the procedure so well, it will be easy for you to accidentally leave out some critical information because you don't realize that your readers may need to be told it.

The consequences of even relatively small slips in writing—even in only a few of the directions in a set of instructions—can be very great. Every step contributes to the successful completion of the task, and the difficulties the readers have with any step can prevent them from completing the task satisfactorily. Even if the readers can eventually figure out how to perform all the steps, their initial confusion with one or two can greatly increase the time it takes them to complete the procedure. Furthermore, in steps that are potentially dangerous, one little mistake can create tremendous problems.

For these reasons, it's often absolutely necessary to determine for certain if your instructions will work for your intended audience. And the only way to find this out for sure is to give a draft to representatives of your audience and ask them to try the instructions. For detailed advice about designing and creating tests of your instructions, see Chapter 18, "Testing."

CONVENTIONAL SUPERSTRUCTURE FOR INSTRUCTIONS

The conventional superstructure for instructions contains six elements:

- Introduction
- Description of the equipment (if the instructions are for running a piece of equipment)
- Theory of operation
- List of materials and equipment
- Directions
- Guide to troubleshooting

The simplest instructions contain only the directions. More complex instructions contain some or all of the other five elements, the selection depending upon the aims of the writer and the needs of the readers.

Many instructions also include elements often found in longer communications such as reports and proposals. Among these elements are a cover, title page, table of contents, appendixes, list of references, glossary, list of symbols, and index. Because these elements are not peculiar to instructions, they are not described here but rather in the discussion of the book format in Appendix A.

One good way to use the chapter you are now reading is to do the following. After you have carefully defined your purpose and studied your readers (see Chapter 3), read through the following sections, determining which elements of the conventional pattern for writing instructions will help you write effectively. Then, reread the sections for the elements that do seem relevant, thinking about ways to apply the advice given there to your particular situation.

That last step is important. No two writing situations are exactly alike. You cannot write successful instructions if you blindly follow the conventional superstructure described here. You must adapt that pattern to your particular readers, purpose, and circumstances by using your imagination and creativity and by following the guidelines given throughout this book.

Introduction

As mentioned above, some instructions contain only directions, and no introduction. Often, however, readers find an introduction to be helpful—or even necessary. You will find general guidelines for writing introductions in Chapter 9, "Beginning a Communication." The following paragraphs will help you see how to apply that general advice when you are writing instructions.

In the conventional superstructure for instructions, an introduction tells some or all of the following things about the instructions:

- Subject
- Aim (purpose or outcome of the procedure described)
- Intended readers
- Scope
- Organization
- Usage (advice about how to use the instructions most effectively)
- Motivation (reasons why readers should use rather than ignore the instructions)
- Background (information the readers will find helpful or necessary)

The following paragraphs discuss ways you can handle each of these six topics in your introduction.

Subject

Writers usually announce the subject of their instructions in the first sentence. Here is the first sentence from the operator's manual for a ten-ton machine used at the end of assembly lines that make automobile and truck tires:

> This manual tells you how to operate the Tire Uniformity Optimizer (TUO).

Here is the first sentence from the owner's manual for a small, lightweight personal computer:

> This manual introduces you to the Apple Macintosh™ Computer.

Aim

From the beginning, readers want to know the answer to the question, "What can we achieve by doing the things this communication instructs us to do?"

With some of the instructions you write, the purpose or outcome of the procedure described will be obvious. For example, most people who buy computers know many of the things that can be done with them. For that reason, a statement about what computers can do would be unnecessary in the Macintosh instructions, which in fact contain none.

However, other instructions do have to answer the readers' questions about the aim of the instructions. In instructions for operating pieces of equipment, for example, writers often answer the readers' inquiry about what the procedure will achieve by telling the capabilities of the equipment. Here, for instance, is the second sentence of the manual for the Tire Uniformity Optimizer:

Depending upon the options on your machine, it may do any or all of the following jobs:

- Test tires
- Find irregularities in tires
- Grind to correct the irregularities, if possible
- Grade tires
- Mark tires according to grade
- Sort tires by grade

Intended Readers

Many readers will ask themselves, ''Are these instructions written for us—or for people who differ from us in interests, responsibilities, level of knowledge, and so on?''

Often, readers will know the answer to that question without being told explicitly. For instance, the operator's manual for the Tire Uniformity Optimizer is obviously addressed to people hired to operate that machine.

In contrast, people who pick up a computer manual often wonder whether the manual will assume that they know more (or less) about computers than they do. In such situations, it is most appropriate for you to answer that question. Here is the third sentence of the Macintosh manual:

> You don't need to know anything about Macintosh or any other computer to use this manual.

Scope

Information about the scope of the instructions answers the readers' question, ''What kinds of things will we learn to do in these instructions—and what things won't we learn?'' The writers of the manual for the Tire Uniformity Optimizer answer that question in its third and fourth sentences:

> This manual explains all the tasks you are likely to perform in a normal shift. It covers all of the options your machine might have.

The writers of the Macintosh manual answer the same question in this way:

> This manual tells you how to:
>
> - use the mouse and keyboard to control your Macintosh (Chapter 1)
> - get started with your own work, make changes to it, and save it (Chapter 1)

☐ find out more about Macintosh concepts and how to use your new techniques to establish a daily working routine (Chapter 2)

☐ organize your documents on the Macintosh (Chapters 2 and 3)

☐ get the most out of your Macintosh system by adding other products to it (Chapter 5)

☐ care for your Macintosh (Chapter 6)

☐ do simple troubleshooting and find further help (Chapter 6)

Organization

By describing the organization of their instructions, writers answer the readers' question, "How is the information given here put together?" Your readers may want to know the answer so they can look for specific pieces of information. Or, they may want to know about the overall organization simply because they can then understand the instructions more rapidly and thoroughly than they could without such an overview.

The writers of the Macintosh manual announce its organization at the same time that they tell the manual's scope. They do this by citing the appropriate chapter number when describing the manual's scope (see above).

The writers of the manual for the Tire Uniformity Optimizer (TUO) explain its contents in a different way (notice that this information fills out the readers' understanding of the scope of the manual):

> The rest of this chapter introduces you to the major parts of the TUO and its basic operation. Chapter 2 tells you step-by-step how to prepare the TUO when you change the type or size of tire you are testing. Chapter 3 tells you how to perform routine servicing, and Chapter 4 tells you how to troubleshoot problems you can probably handle without needing to ask for help from someone else. Chapter 5 contains a convenient checklist of the tasks described in Chapters 3 and 4.

Usage

As they begin to use a set of instructions, readers often ask themselves, "How can we get the information we need as quickly and easily as possible?" Sometimes, the answer is obvious. If the readers' job is simply to follow the instructions from beginning to end or to look for a certain set of steps and then follow them, you don't need to say anything about how to use the instructions. The manual for the Tire Uniformity Optimizer is used in just such a straightforward way, so it contains no special advice about how readers should use it.

In contrast, in some of the instructions you write, you may be able to help your readers considerably by providing advice about how to use your communication. For that reason, the writers of the Macintosh manual provide this advice on the first page, under the heading "How to Use This Manual":

Read Chapter 1 to learn the basics and to get started using one of the application programs you probably purchased with your Macintosh. Then continue on with this manual or go to the manual that came with the application you're going to use. Return to Chapter 3 of this manual when you want to know more about organizing your work. Use Chapter 4 for reference. Read Chapter 6 soon after you get your Macintosh to learn how to care for it.

Motivation

As pointed out above, when people are faced with the work of using a set of instructions, they often are tempted to toss the instructions aside and try to do the job using common sense. You can do several things to persuade your readers not to ignore your instructions. For instance, you can use an inviting and supportive tone and an attractive appearance, such as are used in the Macintosh manual. You may also include sentences that tell the readers directly why it is important for them to use the instructions. The three examples that follow illustrate some of the kinds of statements that writers sometimes provide.

From the Operating Instructions for a Typewriter

To take advantage of the automatic features of the IBM 60 you need to take the time to do the training exercises offered in this manual.

From a Service Manual for Electric Motors

If, through proper installation and maintenance, we can keep our customers' motors in trouble free operation, we have satisfied customers. Everyone needs ''satisfied customers'' because . . . .
''OUR CUSTOMERS ARE OUR EMPLOYERS''

From an Operating Manual for an Office Photocopy Machine

Please read this manual thoroughly to ensure correct operation.

Background

You may recall that Chapter 9 on ''Beginning a Communication'' advised you to include in your introduction any background information that would help your readers understand and use the rest of your communication. That advice applies to instructions. The particular pieces of background information your readers need vary from one set of instructions to the next. However, for instructions that involve machines or other equipment, two kinds of background information—a description of the equipment and an explanation of the theory of operation of the equipment—are so often helpful that they are discussed separately in the next two sections.

So that you can see what their parts look like when put together, look at the introductions to the Tire Uniformity Optimizer and the Macintosh manuals shown in Figures 25–1 and 25–2. You will find the introduction to another manual in Figure 9–5.

As you look at these figures, notice that only the manual for the Tire Uniformity Optimizer uses the word *Introduction*. The introduction to the Ma-

Chapter 1 — Introduction

This manual tells you how to operate the Tire Uniformity Optimizer (TUO) and its controller, the Tire Quality Computer (TQC). The TUO has many options. Depending upon the options on your machine, it may do any or all of the following jobs:

- Tests tires
- Finds irregularities in tires
- Grinds to correct the irregularities, if possible
- Grades tires
- Marks tires according to grade
- Sorts tires by grade

This manual explains all the tasks you are likely to perform in a normal shift. It covers all of the options your machine might have.

The rest of this chapter introduces you to the major parts of the TUO and its basic operation. Chapter 2 tells you step-by-step how to prepare the TUO when you change the type or size of tire you are testing. Chapter 3 tells you how to perform routine servicing, and Chapter 4 tells you how to trouble-shoot problems with the TUO. Chapter 5 contains a convenient checklist of the tasks described in Chapter 3.

Major Parts of the TUO

You can find the major parts of the TUO by looking at Figure 1-1. To operate the TUO, you will use the Operator's Control Panel and the Computer Panel.

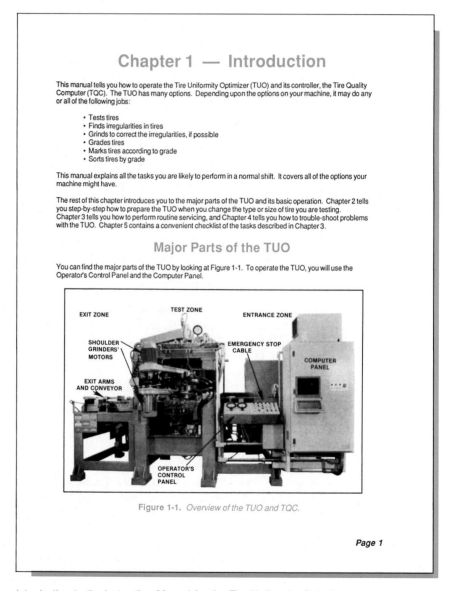

Figure 1-1. *Overview of the TUO and TQC.*

Page 1

FIGURE 25–1 Introduction to the Instruction Manual for the Tire Uniformity Optimizer.

From Akron Standard, *Operator's Manual for the Tire Uniformity Optimizer* (Akron, Ohio: Akron Standard, 1986), 1. Used with permission of Eagle-Picher Industries, Inc., Akron Standard Division.

About This Manual

This manual introduces you to the Apple® Macintosh™ computer. Use it now to learn the basic Macintosh skills, and pick it up again later to use as a reference. You don't need to know anything about Macintosh or any other computer to use this manual. And you won't have to keep learning new ways of doing things. Once you've mastered a few new techniques, you'll use them whenever you use your Macintosh.

You can also take a guided tour of Macintosh by listening to the cassette tape (use it in any cassette player). In the guided tour, your Macintosh demonstrates itself, introducing—in a different way—the same skills this manual teaches.

This manual tells you how to:

☐ use the mouse and keyboard to control your Macintosh (Chapter 1)

☐ get started with your own work, make changes to it, and save it (Chapter 1)

☐ find out more about Macintosh concepts and how to use your new techniques to establish a daily working routine (Chapter 2)

☐ organize your documents on the Macintosh (Chapters 2 and 3)

☐ get the most out of your Macintosh system by adding other products to it (Chapter 5)

☐ care for your Macintosh (Chapter 6)

☐ do simple troubleshooting and find further help (Chapter 6)

How to Use This Manual

Read Chapter 1 to learn the basics and to get started using one of the **application programs** you probably purchased along with your Macintosh. Then continue on with this manual or go to the manual that came with the application you're going to use. Return to Chapter 3 of this manual when you want to know more about organizing your work. Use Chapter 4 for reference. Read Chapter 6 soon after you get your Macintosh to learn how to care for it.

The appendixes contain technical information. A glossary of Macintosh terms and an index are also included.

Now turn to the first chapter and get started.

5   ABOUT THIS MANUAL

FIGURE 25–2   Introduction to the Manual for the Macintosh Personal Computer

From Apple Computer, *Macintosh* (Cupertino, Cal.: Apple Computer, 1986), 5. Reprinted by permission of Apple Computer, Inc.

cintosh manual is called ''About This Manual'' and the introduction to the Detroit Diesel 53 (Figure 9–5) is called ''General Information.'' Indeed, the material that this chapter refers to as the introduction is called many other names in other instructions. Do not be distracted by the variety of names used for it. The material itself, whatever it is called, is a customary part of the conventional pattern for instructions, and it can help your readers greatly when you include it in appropriate situations.

Description of the Equipment

Many sets of instructions concern the operation or repair of equipment: cars, computers, manufacturing machines, and laboratory instruments, for example. To be able to operate or repair the equipment, readers need to know the location and function of its parts. For that reason, a description of the equipment is an important section of many sets of instructions. The manual for the Tire Uniformity Optimizer, for instance, includes on the first page a photograph of the overall machine, with its major parts labelled. In many instructions, the drawings are accompanied by written explanations of the equipment and its parts.

Theory of Operation

An explanation of the way a piece of equipment operates can be extremely useful to readers when they must do something more than merely follow a step-by-step procedure. If your readers will use the material in your instructions to design something (such as a computer program or an experiment), or if they must figure out what is wrong with a malfunctioning piece of equipment, they will be able to work much more effectively if they understand the theory of operation of the tools or equipment they use.

Writers often place such information near the beginning of their instructions, especially if the instructions are for servicing or operating a piece of equipment. No such information is provided in the operator's manual for the Tire Uniformity Optimizer, probably because the operators are supposed to stick strictly to the jobs explicitly laid out for them. However, much general background information is provided at the beginnings of other manuals for this same machine. For example, the manual for setting up the machine provides an overview of the basic features of the special computer language used to program the machine. Without that overview, computer programmers would have a very difficult time writing programs to instruct the machine to do what they wanted done.

In the Macintosh manual, much of the overview material is presented in the chapter on ''Finding Out More About Your Macintosh.'' There, for instance, readers find the answer to the question, ''Where Does Your Information Go?'' Knowledge of how information is stored by the computer enables the readers to figure out and remember some of the sequences of steps that are critical to their effective use of the computer.

List of Materials and Equipment

Some instructions describe processes for which the readers will need to use materials or equipment that they wouldn't normally have at hand. When you are writing in such a situation, you can help your readers greatly by inserting a list of the things needed *before* you give the step-by-step directions. By doing so, you save the readers from the unpleasant surprise that they cannot go on to the next step until they have gone to the shop, supply room, or store to get something they didn't realize they would need. The instructions shown in Figure 25–3 include such a list under the heading "What You Need."

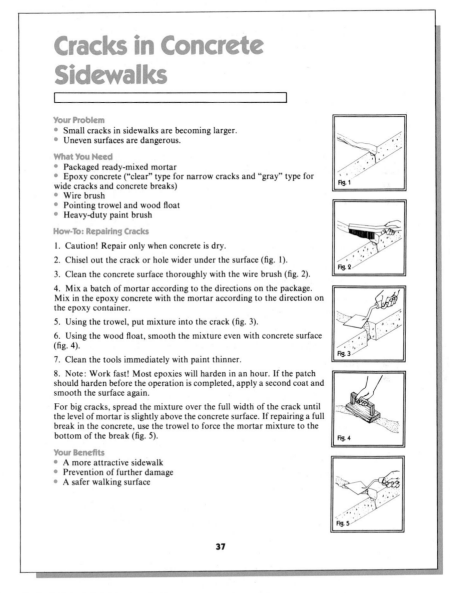

Cracks in Concrete Sidewalks

Your Problem
- Small cracks in sidewalks are becoming larger.
- Uneven surfaces are dangerous.

What You Need
- Packaged ready-mixed mortar
- Epoxy concrete ("clear" type for narrow cracks and "gray" type for wide cracks and concrete breaks)
- Wire brush
- Pointing trowel and wood float
- Heavy-duty paint brush

How-To: Repairing Cracks

1. Caution! Repair only when concrete is dry.

2. Chisel out the crack or hole wider under the surface (fig. 1).

3. Clean the concrete surface thoroughly with the wire brush (fig. 2).

4. Mix a batch of mortar according to the directions on the package. Mix in the epoxy concrete with the mortar according to the direction on the epoxy container.

5. Using the trowel, put mixture into the crack (fig. 3).

6. Using the wood float, smooth the mixture even with concrete surface (fig. 4).

7. Clean the tools immediately with paint thinner.

8. Note: Work fast! Most epoxies will harden in an hour. If the patch should harden before the operation is completed, apply a second coat and smooth the surface again.

For big cracks, spread the mixture over the full width of the crack until the level of mortar is slightly above the concrete surface. If repairing a full break in the concrete, use the trowel to force the mortar mixture to the bottom of the break (fig. 5).

Your Benefits
- A more attractive sidewalk
- Prevention of further damage
- A safer walking surface

Fig. 1

Fig. 2

Fig. 3

Fig. 4

Fig. 5

37

FIGURE 25–3   Typical Set of Brief Instructions

From United States Department of Agriculture Extension Service, *Simple House Repairs . . . Outside* (Washington, D.C.: U.S. Government Printing Office, 1978).

Directions

The heart of a set of instructions is the directions for performing the procedure they describe. The conventional pattern for writing effective directions is undoubtedly familiar to you. Its most obvious characteristics are the use of numbered steps and the use of illustrations. In the ''How-To'' section of the instructions in Figure 25–3, you will see many features of this pattern illustrated. The following are guidelines for writing directions that use that pattern.

Present the Steps in a List

As explained above, people usually use instructions by reading about one step, performing that step, reading the next step, performing it, and so on. If you run your directions together within a paragraph, you make it difficult for your readers to find their place each time they look back at your instructions. In contrast, by presenting your directions in a list, you help your readers find their place each time. You can help your reader even more by placing some distinctive mark at the beginning of each item in your list, perhaps a number or a solid circle (called a *bullet*).

In Your List, Give One Step Per Entry

If you present more than one step per number, you are clumping your directions into the kinds of paragraphs that the list is intended to avoid. At times you may want to present substeps. You can do this by using short, indented lists within your larger list:

14. Drain the cannister.

   ● Release the latch that locks the cannister's drain cap.
   ● Unscrew the cap.

Use Headings and Titles to Indicate the Overall Structure of the Task

For instance, if you were writing a manual for operating a 35-mm camera, you might use the following headings and subheadings:

Preparing for use
   Installing the batteries
   Checking battery power
   Loading film
   Setting the film speed

Operating the camera
   Setting the shutter speed
   Setting the aperture

Focusing
>Regular photography
>Infrared photography
Making the exposure
Advancing the film
Unloading the film

By using headings, subheadings, and (in the book format) chapter titles to show the overall structure of the procedure you are describing, you help your readers *learn* the procedure so that they will be able to perform it without instructions in the future. You also help them find the directions for the specific parts of the procedure that they need assistance with.

Use the Active Voice and the Imperative Mood

Active, imperative verbs give commands: "*Stop* the engine." They allow you to speak directly to your readers, telling them as briefly as possible what to do:

>*Set* the dial to seven. (Much simpler than, "The operator then sets the dial to seven.")

>*Clean* the parts with oil. (Much simpler than, "The parts should be cleaned in oil.")

Use Illustrations

Drawings, photographs, and similar illustrations often provide the clearest and simplest means of telling your reader such important things as:

- Where things are. For instance, Figure 25–4 tells the readers of an instruction manual where to find two control switches.
- How to perform steps. For instance, by showing someone's hand performing a step, you provide your readers with a model to follow as they interpret the words that tell them what to do (see Figure 25–5).
- What should result. By showing readers what should result from performing a step, you help them understand what they are trying to accomplish. You also help them determine whether or not they have performed a step correctly.

Place Warnings Where Readers Will See Them before Performing the Steps to Which They Apply

Your readers will depend upon you to warn them about actions that would

- Endanger them or others
- Damage equipment they are using
- Ruin their results

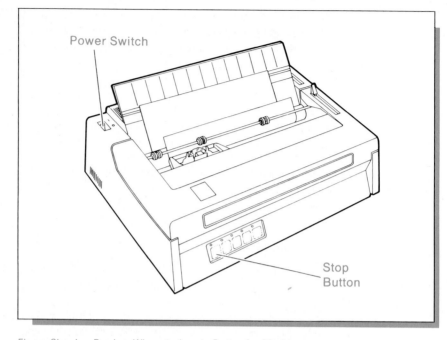

FIGURE 25–4 Figure Showing Readers Where to Locate Parts of a Machine

From International Business Machines, *Wheelprinter E: Guide to Operations* (Lexington, Ky: International Business Machines, 1985), 3–26. © 1985 by International Business Machines Corporation.

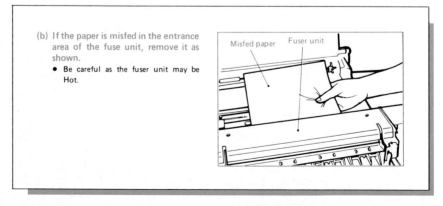

FIGURE 25–5 Illustration Showing Readers How to Perform a Step

From Toshiba Corporation, *Operator's Manual: Plain Paper Copier BD-7816* (Tokyo, Japan: Toshiba Corporation, 1984), 27. Courtesy of Toshiba Corporation.

You can ensure that your readers will see these warnings in time to benefit from them by placing the warnings *before* the steps to which they apply. Otherwise, your readers may look away to perform the step before reading the warning:

> WARNING: Before performing any of the following calibrations,
> follow the initial setup procedures described in sections 6.1 and 6.2.
> If you fail to do so, you could damage the chisel blades or the
> devices upon which they are mounted.

If the sentence describing the step is short, you can also place the warning immediately *after* that sentence. If you do this, you may want to use capital letters or some other device to ensure that your readers read the sentence containing the warning before performing the step described in the preceding sentence:

> 8. Rinse the reservoir. CAUTION: Do not use detergents to clean it.
>   They contain chemicals that will damage the seals.

Both of these sample warnings explain why readers should follow them. Sometimes the reasons will be obvious. When the reasons aren't, readers may ignore or forget the warnings unless you explain the consequences of doing so.

Tell Your Readers What to Do in the Case of a Mistake or Unexpected Result

You should anticipate the places where readers might make mistakes in the procedure. If it will not be obvious to them how to correct or compensate for a mistake, you should tell them how to do so. Similarly, you should tell your readers what to do in places where a correct action by them might not produce the expected result:

> 5. Depress and release the START, RESET, and RUN switches on
>   the operator's panel. NOTE: If the machine stops immediately
>   and the FAULT light illuminates, reposition the second reel and
>   repeat step 5.

Where Alternative Steps May Be Taken, Help Your Readers Quickly Find the One They Want

Sometimes you may write instructions where your readers may choose alternative courses of action, depending upon such things as the equipment or other resources they have or the results they desire. In these situations, be sure to make clear to your readers that they have a choice and then arrange your material so that they can quickly locate the alternative they want. Here is a sample from a manual written by IBM.[1]

You're ready to duplicate your new DPPX/SP system throughout the network. If you plan to:

First
alternative
- Install remote sites with skilled personnel using SLU, repeat the central site installation procedures (Chapters 2 through 4). However, consider the note on CFE/IPF catalogs in Chapter 5 under the heading "Preparing the Distribution Sites for the SLU."

Second
alternative
- Install the DPPX/SP Service Level Update again at each distributed site and customize each system *from the central site,* follow the steps in Chapter 5.

Third
alternative
- Duplicate the central system on tape or diskettes, and restore the contents of the tape or diskettes at the distributed sites, follow the steps in Chapter 6.

Fourth
alternative
- Send the customized central system to distributed sites with DSX, follow the steps in Chapter 7.

Provide Enough Detail for Your Readers to Do Everything They Must Do

One of the questions you ask yourself as you prepare a set of directions is, "How much detail should I give when telling my readers what to do?" The answer, of course, is that you must tell your readers how to do everything your readers don't already know how to do.

To identify places where you may not have included enough detail, you can use the following strategy, which should sound familiar to you: think about your readers in the act of reading. Specifically, imagine whether or not, as they read each of the steps in your procedure, your readers will ask, "How do we do that?"

When you come to some steps, you will know immediately that your readers will not ask such a question. Here is an example:

Set the toggle switch to the ON position.

Because you know that your reader will not ask for detailed information about how to work a toggle switch, you do *not* need to include the following details about the substeps involved:

Extend the index finger of one of your hands.

Place the tip of your finger under the end of the toggle switch.

Push up on the switch until it snaps to the ON position.

With other steps, however, you may discover that your readers are likely to need additional information. Here are examples:

Calibrate the scales. ("But *how* do I calibrate the scales?")

Set the potentiometers to the proper setting. ("*How* do I adjust the potentiometers?" "How can I find out what the proper setting *is*?")

In both of these examples, the writer should provide further information to answer the readers' questions.

Troubleshooting

When they read instructions, people want to learn what to do if things don't work out as they expect—if the equipment they are using fails to work properly or if they don't get the result they want, for instance. You can provide such information in a troubleshooting section.

Usually, you will be able to provide troubleshooting information most helpfully if you use a table format. Figure 25–6 shows the format used in the troubleshooting section of the manual for the Tire Uniformity Optimizer, and Figure 25–7 shows the format used in the troubleshooting section of the Macintosh manual. Other formats are possible, but they share these characteristics: the left-hand column lists the problems that might arise and the right-hand column lists the action to be taken. Information about the probable cause of the problem is often provided either in a middle column or in the right-hand column.

PLANNING GUIDE

Figure 25–8 shows a worksheet that may help you plan your instructions. As you fill out the worksheet, remember that it is intended merely to stimulate your thinking. To achieve your objectives, you may need to say many things that you don't write down on it, and you may discover that some of the worksheet's questions are irrelevant to the particular instructions you are writing. Also, remember that the superstructure on page 2 does not necessarily provide the outline for your instructions.

SAMPLE INSTRUCTIONS

Figure 25–9 shows a set of instructions written by a student.

EXERCISE

1. Find and photocopy a short set of instructions (five pages or less). Attach to the photocopy your analysis of the instructions, noting how the writers have handled each of the elements of the conventional superstructure. If the writers have omitted certain elements, explain why you think they did so. Be sure to comment on the page design and visual aids (if any) used in the instructions.

   Then evaluate the instructions. Tell what you think works best about them, and identify ways you think they can be improved.

Chapter 4 — Trouble-shooting

This chapter tells you what to check when trouble-shooting the TQC. It lists the problems that may occur, the probable causes, and the remedies.

The first list in this chapter consists of the error messages that appear on the CRT when a problem occurs. Next to the error messages are the causes of the problem and the possible remedies. A list of all the error messages can be found in Appendix B. The second list consists of observable phenomena that are listed in order of normal TQC operation.

One easily-solved problem is caused by entering entries too quickly to the TQC through the keyboard. If the operator does not wait for the TQC to respond to one request before entering another, errors and inaccurate data will result. Make sure you allow sufficient time for the TQC to respond to your input before you press another key.

Warning

EXTERNAL TEST EQUIPMENT CAN DAMAGE THE TQC. If you use external equipment to trouble-shoot the TQC, make sure that it does not introduce undesired ground circuits or AC leakage currents.

Trouble-shooting with Error Messages

Power-Up Error Messages

Error Message	Probable Cause	Remedy
BACKUP BATTERY IS LOW	1. Battery on Processor Support PCB.	1. Replace the battery on the Processor Support PCB.
CONTROLLER ERROR	1. PC Interface PCB. 2. Processor Support PCB.	1. Swap the PC Interface PCB. 2. Swap the Processor Support PCB.
EPROM CHECKSUM ERROR	1. Configuration tables. 2. Analog Processor PCB.	1. Check the configuration tables. 2. Swap the Analog Processor PCB 88/40.
KEYBOARD MALFUNCTION: PORT	1. Keyboard or keyboard cable. 2. Processor Support PCB.	1. Check the keyboard and cable. 2. Swap the Processor Support PCB.
RAM FAILURE AT 0000:	1. Main Processor 86/30.	1. Swap the 86/30.
RAM FAILURE AT 1000:	1. Main Processor 86/30.	1. Swap the 86/30.
TIGRE PROGRAM CHECKSUM ERROR	1. TIGRE program.	1. Renter the TIGRE program or debug the program.

Table 4-1. *Power-up error messages.*

Page 59

FIGURE 25–6   Troubleshooting Section from the Manual for the Tire Uniformity Optimizer

From Akron Standard, *Operator's Manual for the Tire Uniformity Optimizer,* (Akron, Ohio: Akron Standard, 1986), 57. Used with permission of Eagle-Picher Industries, Inc., Akron Standard Division.

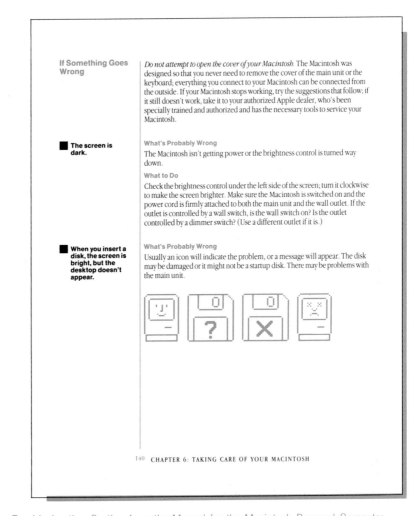

FIGURE 25–7   Troubleshooting Section from the Manual for the Macintosh Personal Computer

From Apple Computer, *Macintosh* (Cupertino, Cal.: Apple Computer, 1986), 140. Reprinted by permission of Apple Computer, Inc.

Planning Guide

INSTRUCTIONS

Subject _____ **Due Date** _____

Overall Purpose

Reader Profile

Who are your primary readers?

Who else might read your instructions?

What will your readers be trying to accomplish when using your instructions?

What are your readers' titles and professional responsibilities?

How much do your readers know about the procedure described in your instructions?

How great is your readers' knowledge of any technical subjects related to your topic?

Do your readers have any communication preferences you should take into account? If so, name them.

Are there any other considerations you should keep in mind when addressing these readers?

Readers' Informational Needs

What will your readers' key questions be?

Readers' Attitudes

What will be your readers' initial attitudes toward the procedure and your instructions? What do you want them to be?

FIGURE 25–8   Planning Worksheet for Instructions

Superstructure

Introduction

What, if anything, will you need to tell your readers about the following topics?
- Aim, purpose, or desired outcome of the procedure
- Intended readers for the instructions (their knowledge level, job descriptions, etc.)
- Scope of the instructions
- Organization of the instructions
- Ways to use the instructions effectively
- Reasons for following the instructions (rather than using some other procedures)
- Background information the readers might find helpful

Description of the Equipment

If your instructions concern the operation or repair of complex equipment, what should you tell your readers about the location and function of the equipment's key parts?

Theory of Operation

What, if anything, will they need to know about the theory of operation of that equipment?

List of Materials and Equipment

What, if any, materials and equipment should you tell your readers to gather in advance of starting the procedure?

Directions

What are the major groups of steps in the procedure?

Troubleshooting

What problems might your readers encounter and what will you tell them about how to overcome those problems?

Visual Aids

What visual aids would your readers find helpful or persuasive?

Organization

What organization would your readers find useful and persuasive? (On a separate sheet provide a brief sketch or outline to identify the major sections and subsections.)

FIGURE 25–8   *(continued)*

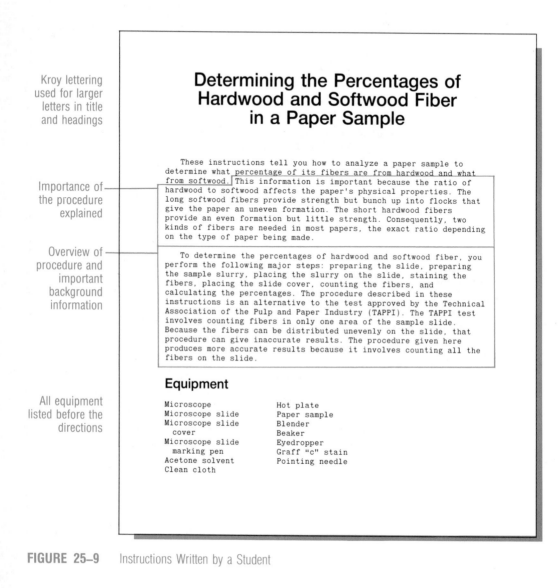

Kroy lettering used for larger letters in title and headings

Determining the Percentages of Hardwood and Softwood Fiber in a Paper Sample

Importance of the procedure explained

These instructions tell you how to analyze a paper sample to determine what percentage of its fibers are from hardwood and what from softwood. This information is important because the ratio of hardwood to softwood affects the paper's physical properties. The long softwood fibers provide strength but bunch up into flocks that give the paper an uneven formation. The short hardwood fibers provide an even formation but little strength. Consequently, two kinds of fibers are needed in most papers, the exact ratio depending on the type of paper being made.

Overview of procedure and important background information

To determine the percentages of hardwood and softwood fiber, you perform the following major steps: preparing the slide, preparing the sample slurry, placing the slurry on the slide, staining the fibers, placing the slide cover, counting the fibers, and calculating the percentages. The procedure described in these instructions is an alternative to the test approved by the Technical Association of the Pulp and Paper Industry (TAPPI). The TAPPI test involves counting fibers in only one area of the sample slide. Because the fibers can be distributed unevenly on the slide, that procedure can give inaccurate results. The procedure given here produces more accurate results because it involves counting all the fibers on the slide.

Equipment

All equipment listed before the directions

Microscope	Hot plate
Microscope slide	Paper sample
Microscope slide	Blender
cover	Beaker
Microscope slide	Eyedropper
marking pen	Graff "c" stain
Acetone solvent	Pointing needle
Clean cloth	

FIGURE 25–9   Instructions Written by a Student

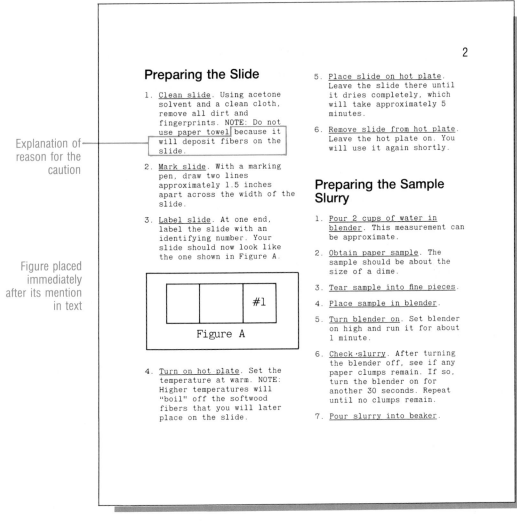

2

Preparing the Slide

1. <u>Clean slide</u>. Using acetone solvent and a clean cloth, remove all dirt and fingerprints. NOTE: Do not use paper towel because it will deposit fibers on the slide.

2. <u>Mark slide</u>. With a marking pen, draw two lines approximately 1.5 inches apart across the width of the slide.

3. <u>Label slide</u>. At one end, label the slide with an identifying number. Your slide should now look like the one shown in Figure A.

#1

Figure A

4. <u>Turn on hot plate</u>. Set the temperature at warm. NOTE: Higher temperatures will "boil" off the softwood fibers that you will later place on the slide.

5. <u>Place slide on hot plate</u>. Leave the slide there until it dries completely, which will take approximately 5 minutes.

6. <u>Remove slide from hot plate</u>. Leave the hot plate on. You will use it again shortly.

Preparing the Sample Slurry

1. <u>Pour 2 cups of water in blender</u>. This measurement can be approximate.

2. <u>Obtain paper sample</u>. The sample should be about the size of a dime.

3. <u>Tear sample into fine pieces</u>.

4. <u>Place sample in blender</u>.

5. <u>Turn blender on</u>. Set blender on high and run it for about 1 minute.

6. <u>Check slurry</u>. After turning the blender off, see if any paper clumps remain. If so, turn the blender on for another 30 seconds. Repeat until no clumps remain.

7. <u>Pour slurry into beaker</u>.

Explanation of reason for the caution

Figure placed immediately after its mention in text

FIGURE 25–9 *(continued)*

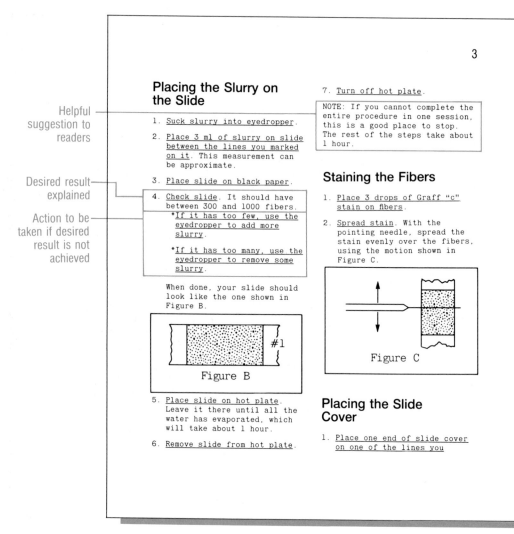

The following annotations point to parts of the figure:

- Helpful suggestion to readers
- Desired result explained
- Action to be taken if desired result is not achieved

3

Placing the Slurry on the Slide

1. <u>Suck slurry into eyedropper</u>.

2. <u>Place 3 ml of slurry on slide between the lines you marked on it</u>. This measurement can be approximate.

3. <u>Place slide on black paper</u>.

4. <u>Check slide</u>. It should have between 300 and 1000 fibers.
   *<u>If it has too few, use the eyedropper to add more slurry</u>.
   *<u>If it has too many, use the eyedropper to remove some slurry</u>.

When done, your slide should look like the one shown in Figure B.

```
        ┌────────────┐
        │ ▒▒▒▒▒▒▒ #1 │
        │ ▒▒▒▒▒▒▒    │
        └────────────┘
          Figure B
```

5. <u>Place slide on hot plate</u>. Leave it there until all the water has evaporated, which will take about 1 hour.

6. <u>Remove slide from hot plate</u>.

7. <u>Turn off hot plate</u>.

NOTE: If you cannot complete the entire procedure in one session, this is a good place to stop. The rest of the steps take about 1 hour.

Staining the Fibers

1. <u>Place 3 drops of Graff "c" stain on fibers</u>.

2. <u>Spread stain</u>. With the pointing needle, spread the stain evenly over the fibers, using the motion shown in Figure C.

```
          Figure C
```

Placing the Slide Cover

1. <u>Place one end of slide cover on one of the lines you</u>

FIGURE 25–9   *(continued)*

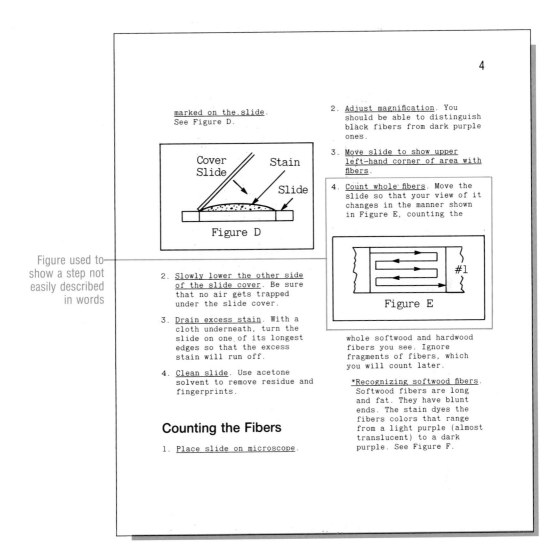

4

marked on the slide.
See Figure D.

Figure used to
show a step not
easily described
in words

Cover
Slide   Stain

Slide

Figure D

2. Slowly lower the other side
   of the slide cover. Be sure
   that no air gets trapped
   under the slide cover.

3. Drain excess stain. With a
   cloth underneath, turn the
   slide on one of its longest
   edges so that the excess
   stain will run off.

4. Clean slide. Use acetone
   solvent to remove residue and
   fingerprints.

Counting the Fibers

1. Place slide on microscope.

2. Adjust magnification. You
   should be able to distinguish
   black fibers from dark purple
   ones.

3. Move slide to show upper
   left-hand corner of area with
   fibers.

4. Count whole fibers. Move the
   slide so that your view of it
   changes in the manner shown
   in Figure E, counting the

#1

Figure E

whole softwood and hardwood
fibers you see. Ignore
fragments of fibers, which
you will count later.

*Recognizing softwood fibers.
Softwood fibers are long
and fat. They have blunt
ends. The stain dyes the
fibers colors that range
from a light purple (almost
translucent) to a dark
purple. See Figure F.

FIGURE 25–9   *(continued)*

5

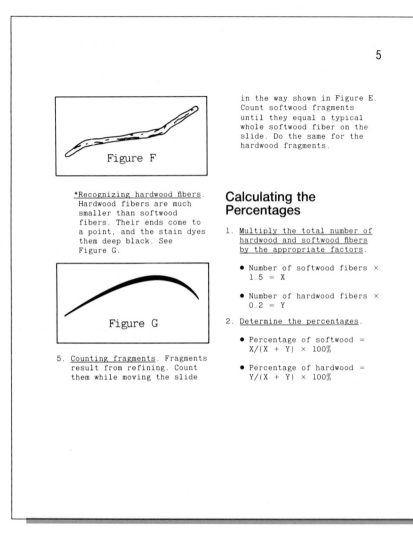

Figure F

in the way shown in Figure E.
Count softwood fragments
until they equal a typical
whole softwood fiber on the
slide. Do the same for the
hardwood fragments.

*Recognizing hardwood fibers.
Hardwood fibers are much
smaller than softwood
fibers. Their ends come to
a point, and the stain dyes
them deep black. See
Figure G.

Figure G

5. Counting fragments. Fragments
result from refining. Count
them while moving the slide

Calculating the Percentages

1. Multiply the total number of
   hardwood and softwood fibers
   by the appropriate factors.

   - Number of softwood fibers ×
     1.5 = X

   - Number of hardwood fibers ×
     0.2 = Y

2. Determine the percentages.

   - Percentage of softwood =
     X/(X + Y) × 100%

   - Percentage of hardwood =
     Y/(X + Y) × 100%

FIGURE 25–9 *(continued)*

SPECIAL ACTIVITIES

26

WRITING COLLABORATIVELY

W ith satisfaction and pride, Jack sets his copy of the report down on his desk. For the past three months, he and four co-workers have labored together to evaluate three possible sites where their employer might locate a new manufacturing plant. Together they researched, together they evaluated, and together they wrote the sixty-page report with which Jack is now so pleased. Such group writing efforts are very common at work. The site evaluation report was Jack's third collaborative writing project in the past twelve months. For Jack, as for most professionals, the ability to work effectively on writing teams is an indispensable skill.

Employees work together on writing teams for several reasons. Many projects require the expertise of people specializing in several fields. Other projects are too large to be completed on time without the combined efforts of several people. Just as important, experience has shown that groups can generate a larger and more creative pool of ideas than can one person working alone. For this last reason, management theory and practice increasingly emphasize teamwork over individual efforts in all areas, including writing.

Despite these good reasons for establishing writing teams, the thought of working on one can create anxiety. Jack experienced such apprehensions the first time his boss assigned him to a writing team. Jack hadn't gained any experience with writing groups in school, so he didn't know what to expect. He feared that others in the group might overrule his good ideas or that his ideas might be accepted but blended in with the others so that he didn't receive the recognition he deserved. He also feared that he might be left with a huge portion of the work because others would not feel as committed to the project as he did. In fact, Jack has encountered some of these problems while working on teams, but overall he has found teamwork to be quite satisfying. And he has earned a reputation for being an effective writing-team member, a quality he hadn't previously realized his employer valued or he possessed.

SPECIAL CHALLENGES OF GROUP WRITING

All of this book's advice about writing applies just as validly to communications you co-author as it does to those you prepare by yourself. That's because, at one level, writing collaboratively involves exactly the same process as writing independently. You need to define your objectives, plan a strategy for meeting those objectives, and then draft, evaluate, and revise the communication in which you carry out that strategy.

However, when you join with others to form a writing team, you and your teammates have some additional tasks to perform. Imagine that your team is preparing a brief communication that can be completed in a single meeting. You and your teammates might discuss the issue until a consensus is reached about what to say. Then, one person could draft what he or she thinks the group has decided. The team then edits the draft and adjourns. Even when a team works on such brief communications, the writing skills required include those necessary to run the meeting efficiently, obtain good contributions from all team members, negotiate conflicting ideas about the communication's content, and draft something that all present will find acceptable.

Other writing projects, like Jack's, are too large to be completed in a single meeting. In fact, some are huge. The proposals from companies that wanted to build the space shuttle for the National Aeronautics and Space Administration were thousands of pages long, required contributions from hundreds of people, and cost millions of dollars to prepare. When working on large projects, a team will typically alternate between work done as a group and work done independently by individual team members. Consequently, in addition to needing all the special skills needed for team writing on brief communications, the team members face the additional writing challenges of figuring out how to coordinate group efforts and individual work and how to create a single, unified communication from sections drafted by different people, each with a distinctive personal writing style.

This chapter's four guidelines will help you and your teammates meet the special challenges presented by group writing.

GUIDELINE 1	ENCOURAGE DEBATE AND DIVERSITY OF IDEAS

As mentioned above, one of the chief benefits of group work is that many people bring their expertise and creativity to a project. To take full advantage of that benefit, all team members must offer their ideas freely—even if the ideas conflict with one another. In fact, debate and disagreement can be very useful if carried out in a courteous and nonthreatening way. Debate ensures that you aren't settling for the first or most obvious suggestion. It allows you to explore different possibilities and to select the ones that truly make the most sense.

Encouraging debate and diversity of ideas can be difficult. Some people are naturally shy of speaking, and many shy away from disagreeing, especially if they fear that their ideas will be treated with hostility rather than openness and politeness. To promote healthy debate and the consideration of a rich diversity of ideas, you and your teammates can use these four strategies: invite everyone to speak, listen with respect and interest, be considerate when discussing drafts, and treat drafts as group property, not individual property.

Invite Everyone to Speak

The quiet members of a team often have good ideas but are timid about offering them. They may need to be invited before they will speak up. You might simply say, "What do you think of that idea, Keith?" or "What have you been thinking about this topic, Jessica?" If some team members are very talkative, you may have to create a place for the quiet person in the team's conversation by interrupting the more talkative people. You might say, "I've liked the things Lynn has had to say, but I wonder what Pat's ideas are,'' or ''Let's take a minute to hear what Terry feels about the ideas we've been discussing.'' Another way of inviting everyone to speak is to establish an operating procedure that gives each person an equal amount of time to talk about every major new topic before the floor is opened to a more general discussion.

Listen with Interest and Respect

When team members do speak, it is critical that others respond with interest and respect—even if they disagree with what the person is saying. If someone feels that his or her ideas have been treated rudely or harshly, that person is unlikely to contribute openly again.

One simple way to indicate that you welcome someone's contribution is to make remarks that show you are paying attention. If you agree with the person, say such things as ''That's interesting,'' ''I hadn't thought of that,'' or even simply ''Ah-huh.''

If you disagree, indicate that you welcome the person's contribution by showing that you want to understand the person's position fully. Ask questions if you are unclear about anything. Or, paraphrase the person's position, perhaps after saying something like, ''What I think I hear you saying is . . .'' If you have misunderstood, the speaker can then correct you. Maybe you don't disagree after all. But even if you do, *hear the person out.* Don't cut the speaker off. If you won't listen, others won't speak, and then your team will lose their good ideas as well as their not-so-good ones.

You can also provide *physical* signals that show you are interested in what a speaker has to say. In one study, researchers arranged to have one group of individuals speak to people who gave them little eye contact and another group speak to people who gave them more eye contact.[1] The researchers discovered that when speakers think they are receiving less eye contact, they also think that their listeners are less attentive. In another study, researchers showed films of actors who maintained eye contact while listening 15 percent or 80 percent of the time.[2] People who watched the films judged the actors who maintained more eye contact to be friendlier. Researchers have also studied the effect of gestures on speakers.[3] They found that speakers feel uncomfortable if the listeners engage in nervous gestures, such as cleaning their fingernails, drumming their fingers, or holding their hands over their mouths. Thus, even where you look and what you do while others speak can encourage—or discourage—the kind of open discussion that is important in team writing. Be sure that you *look* interested in what the other people are saying.

The listening strategies just described are sometimes called *active listening skills.* That's because they emphasize the positive actions you can take to be sure that you understand other people and to show the other people that you are striving to hear their message accurately. The name *active listening* is particularly appropriate because it emphasizes that in group meetings you should put as much positive effort into listening as you put into speaking.

Be Considerate When Discussing Drafts

Team meetings can become particularly awkward when drafts produced by members working independently are reviewed by the group as a whole. Many people almost automatically resist any suggestion for change in their drafts. This resistance is entirely understandable. The writer has invested considerable personal creativity and effort in the project, and so is reluctant to see that work undone. Furthermore, writers sometimes feel that the team is criticizing their overall writing ability when it requests changes. The resulting defensive responses by the writer can prevent needed improvements and undermine the mutual goodwill that is essential to effective teamwork.

On the other side, the people reviewing another team member's work often avoid suggesting changes because they sympathize with the drafter's anxiety and fear a conflict. "We just can't think of any way to make this better," they say—no matter how badly improvements are needed.

Such resistance by the writer and other team members can be very counterproductive. Most drafts can be improved, and open discussion of them can discover how that can be done.

The most effective way to promote an open discussion is to be considerate of one another's feelings. Present your ideas for changing a draft as suggestions or options: "Here's another way you could say that." When discussing options, focus on the *positive* reasons for choosing one option over another. Avoid making statements that sound like criticism of the writer's work. Also, accompany suggestions with praise for what is strong in the writer's draft. Indicate that you are offering ideas for improving a draft that is basically good, not corrections for a draft that is fundamentally bad.

You can also be considerate of others' feelings when your draft is being reviewed by the team. Help them feel comfortable by being open, not defensive. Make them feel that their ideas are welcomed rather than resented.

Treat Drafts As Team Property, Not Individual Property

Another way to promote open discussion is to encourage team members to give up their sense of personal "ownership" of the material they draft. While still taking pride in your contributions, you and your teammates can try to see your drafts as something you've created for the team so that the drafts are the group's property, not your personal property. This doesn't mean that you and your teammates should give up supporting the strong points of your own drafts if you think they have been underestimated. It means that when the team is reviewing your draft, you should join in a *reasoned* discussion of the best way to write "our" communication rather than struggle to protect "my" writing from assaults by others. Your openness about your draft will encourage others to take the same attitude concerning theirs.

Your team can also reduce the troublesome sense of personal ownership by agreeing to swap responsibilities at a certain point. For example, after second drafts are written, the first section might become the responsibility of the person who drafted the second section, the second section might become the responsibility of the person who drafted the third section, and so on. Also, the team can agree that at some point everyone's drafts will be combined for editing and polishing by the whole group or by an individual the group designates. At that point, the draft clearly becomes *group* property, not *individual* property.

GUIDELINE 2	**IN MEETINGS, EMPHASIZE EFFICIENCY**

While it's important to encourage a free and open discussion of ideas, it is also important to conduct team meetings efficiently.

One tool for efficiency is an agenda. An agenda helps to focus discussion by identifying the specific issues that the team needs to consider and by reminding the team of the total amount of work it must accomplish during the meeting. Such a reminder is helpful for teams that tend to spend so much time on one issue that they don't have time to get to the others.

Another way to promote efficiency is to stick to the topic. Especially during the brainstorming sessions and freewheeling debates that are so important to successful group work, it is possible to digress far from the topic without realizing it. It's important for all team members to try to keep their own comments focused on the topic, and to help others do the same. Any member of the team can help by saying, ''I think we're drifting away from our subject a little,'' or ''Let's try to settle our main question before going on to other matters.''

On the other hand, groups sometimes continue talking about a topic even after a consensus has been reached. One way to help the group close such a discussion is to formalize the agreement by saying, ''I think I hear everyone agreeing that . . .'' If the others don't agree with your interpretation, then you will have helped the team focus on the points that still need to be resolved. If they do agree, you have helped move them on to the next item of business.

Agenda setting, reestablishing the focus of a discussion, and bringing a discussion to a close are all functions we often imagine a project manager or team leader will perform. That doesn't mean, however, that they can be performed by only one person. Meetings work best when all participants take responsibility for efficiency. When you notice that the team needs a nudge, speak up. Help your team get back on track.

One caution is in order. Even at work, a team is a social group. Friendliness and personal interest in one another are natural and desirable. In your attempts to be efficient, don't take the fun out of meeting together. On the other hand, don't let the fun take up so much of the meeting that everyone leaves feeling frustrated because so little was accomplished.

PROVIDE DETAILED GUIDANCE FOR INDIVIDUAL WORK

The two guidelines you have just read concern things you and your teammates can do to work productively together *during* meetings. As explained at the beginning of this chapter, on many team projects much of the work is done *between* meetings by individuals working independently. On those projects, the writing team must coordinate group and independent work in a way that enables all team members to use their time as effectively as possible.

On the job, the following three-phase pattern for coordinating group and individual work is widely used:

1. The group meets to generate ideas, make basic decisions, and plan strategy.
2. One or more team members work independently, each guided by the strategy developed in the meeting.
3. The group meets again to review the work done independently, to generate new ideas, and to plan strategy for the next round of individual work.

If necessary, this pattern can be repeated for many meetings. In this pattern, the whole team works together on basic decision-making, an area where a diversity of opinions can sharpen the final result. The pattern promotes efficiency in such follow-up activities as gathering data the group needs and drafting the ideas the group has settled upon by assigning them to individuals. Figure 26–1 illustrates two versions of this pattern, and Figure 26–2 tells how Jack's group followed it to prepare the site selection report described at the beginning of the chapter.

One key to coordinating group and individual work is to be sure that the group gives detailed guidance to the individual team member or members about what it wants them to do when they are working independently. This means, first, that the group must let every team member know exactly what he or she is to bring to the next meeting—a folder of facts, an outline, a draft of section three, whatever. Otherwise, when the team meets again, critical jobs may be left undone because everybody thought someone else would do it.

Additionally, as the team makes assignments, it should provide each individual with clear directions about *how* it wants the person to carry out his or her work. For instance, during its first meeting, Jack's team not only assigned three people to investigate each of the three possible sites for the new manufacturing plant, but it also made a list of the specific topics (such as construction costs and availability of railroad transportation) to be investigated, and it even discussed the resources that each individual researcher might use. Without such guidance, an individual may find out at the next team meeting that all of his or her work was for naught because the other team members wanted the work done another way. Such misunderstandings bruise feelings and waste everybody's time.

Clear guidance for individuals is particularly important before drafting

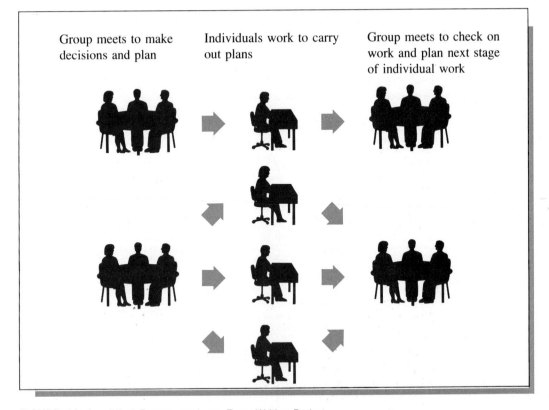

Group meets to make decisions and plan

Individuals work to carry out plans

Group meets to check on work and plan next stage of individual work

FIGURE 26–1 Work Patterns on Large Team-Writing Projects

begins. Outlining can be very helpful in providing that guidance, so you should consider using outlines when writing on teams even in situations where you wouldn't find it helpful to outline if you were writing the communication by yourself. When working alone, you can often create and keep your writing plan in your head. Teams, however, have many heads. Even when they make explicit oral agreements, team members may have in mind a variety of understandings of what the team agreed to. Because outlining forces the team to agree on the specific words that it will set down on paper, it helps to uncover these misunderstandings early—before anyone wastes time drafting in accordance with a misunderstood plan. Outlining also enables everyone to leave a meeting with a written record of what was decided, so that no one forgets while doing independent work.

Another way to provide guidance to drafters is to use a style guide. Style guides tell how to design tables, handle headings, use abbreviations, and deal with other stylistic features of a communication. If everyone follows this style guide from the outset, the team will greatly reduce the amount of editing that must be done later to achieve consistency among the drafts. Some employers

Schedule Used by Jack's Team for Preparing Its Site Evaluation Report

1. The six members of Jack's team met to discuss the project and its objectives. The team made a broad outline that identified the general kinds of information it wanted to include about each site in the final report. It assigned each of three team members responsibility for researching one of the sites and for bringing results to the next meeting.

2. The three team members conducted their research by gathering the information designated by the team and by noting other important facts they encountered.

3. The team met to review the results of the initial research. It developed a much more detailed outline based on its impressions of the facts gathered by the three researchers. In the process, it identified some additional kinds of information that needed to be gathered about certain sites. The team discussed the writing style it would use, the format it would employ for tables, and similar issues that would help the writers create consistent drafts. Finally, the team asked each of the people who were investigating a site to draft a chapter about it, and it asked a fourth member to write up an introduction.

4. The team members who were assigned drafting responsibilities performed their work individually.

5. The team met to review the drafts. It decided to alter the structure of the chapters about the individual sites to highlight certain information. It also decided the points it wanted to make in the chapter that would present its conclusions and recommendations. It then asked the drafters of the site descriptions to revise them, and it assigned a fourth member to write the chapter with the conclusions and recommendations agreed on by the team.

6. The team members responsible for writing and revising carried out their assignments independently.

7. The team met to review the entire draft. It settled on some additional revisions that were needed to make the report clear and persuasive. The team also discussed some inconsistencies in style that had crept in despite their earlier discussions. Two individuals who hadn't participated in drafting or revising up to this point were asked to make the revisions. Another team member was assigned to write the executive summary.

8. The individuals responsible for revising carried out their assignments and circulated their drafts through the company mail. All team members read the drafts in preparation for the next meeting.

9. The team met one last time to discuss the final draft. Some minor editorial changes were determined. One team member was assigned responsibility for preparing the final draft.

FIGURE 26–2   Schedule Used by Jack's Team for Preparing Its Site Evaluation Report

publish style guides that writing teams can use. If your employer doesn't, your team can create one quickly, perhaps by delegating that responsibility to one member.

<table>
<tr><td>GUIDELINE
4</td><td></td></tr>
</table>

MAKE A PROJECT SCHEDULE

Schedules are helpful for almost any team-writing project that requires more than one meeting, but schedules are especially helpful for projects that involve both group and individual efforts that must be coordinated and completed on time. The schedule can help motivate the team to finish each task on time because it will help them see the consequences of slipping at any point along the way. The schedule can also enable team members to set aside specific blocks of time on their calendars for the group and individual work that will be required.

Three important elements to include in your schedule are the following:

- **Time to define the project's objectives, probably at the first meeting.** If the project leader has already defined the objectives, he or she should take time to discuss them with the team. Alternatively, the entire team can define the communication's objectives by discussing its purpose and readers. In either case, this definition of objectives can serve as a beacon to guide both group discussions and individual work throughout the entire project.
- **Frequent checkpoints.** The team needs to meet often enough to see that the work that individual members are doing independently is proceeding in the way the team wants. The early discovery of problems helps the individual as much as it helps the team because the individual is saved from investing additional time doing work that will have to be redone. Frequent meetings also give the team ample opportunity to refine or alter its plans *before* team members have devoted overly large amounts of energy and creativity to work that later would be changed substantially. As Figure 26–2 shows, Jack's team had four meetings that served as checkpoints. At one, it checked on research and at the other three it checked on drafts.
- **Time to edit the drafts for consistency and coherence.** When you write alone, you usually don't need to edit for consistency because you will probably have written consistently in the first place. But on a group project, different sections are written by different people, each with a distinctive, individual style. Even if the team uses a style guide, sharp differences may appear between sections. The editing required to smooth out these differences might be assigned to one team member or it might be undertaken by the whole team acting as an editorial committee. Whichever approach your team takes, it will need to provide time in its schedule for this work.

The best time to make a schedule is at the very beginning of the specialized, technical work that gives rise to the communication. For instance, Jack's team made its schedule when it first received its assignment to investigate the sites. That enabled the team to create a schedule that integrated writing activities with the research and other work the team was to conduct. When teams instead plan to do all of the writing work at the end of the project, they frequently leave too little time for writing. Consequently, their communication is either done poorly or late—outcomes that you and your teammates will surely want to avoid. Figure 26–3 shows how Jack's team coordinated its specialized work and its writing work in its schedule, and it also shows the schedule for another project in which the two types of work were not coordinated effectively.

CONCLUSION

Team writing is very common in the workplace. While all the guidelines in this book apply to team writing as well as to individual writing, team projects require some additional skills. The four guidelines presented in this chapter will help you create team efforts in which the team members work together productively and enjoy their mutual effort.

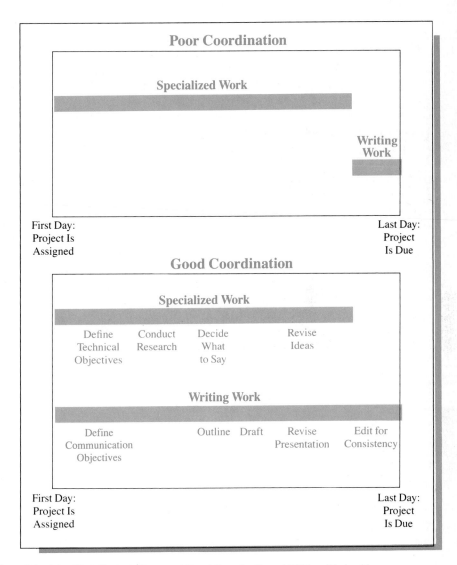

FIGURE 26–3 Schedules That Illustrate Poor and Good Coordination of Writing Work with Specialized Work

Based on Thomas N. Trzyna and Margaret M. Batschelet, *Writing for the Technical Professions* (Belmont, Calif.: Wadsworth, 1987), 373.

CHAPTER 27

PREPARING ORAL PRESENTATIONS

Tony is thrilled—and terrified. Today, he learned that a group of upper-level managers wants to hear him talk about a project that he and three other people recently completed. The vice presidents will give him ten minutes to explain the aims, methods, results, and significance of the project. And then he should be ready to field questions.

"How nervous I'll be," he says, as he imagines himself standing before that audience, alone, with all eyes on him.

Then he thinks of the pride he will feel as he presents his ideas to such an audience. "What an opportunity to shine," he thinks.

Besides the opportunity to shine, Tony has some other pleasures to look forward to—pleasures that are not available when he communicates in writing. These include the enjoyment of talking *personally* to people who have come to hear what he has to say and the satisfaction of feeling *immediately* his audience's appreciation of his efforts.

If you are like most people, when you look forward to giving a talk at work you will experience much the same mixture of emotions that Tony is experiencing: great eagerness and anticipation, with some nervousness.

You can calm your nerves and increase the effectiveness of your talks by following the guidelines given below. They tell you how to prepare successful talks. In the next chapter, you will find nine guidelines that will help you present those talks effectively.

ORGANIZATION OF THIS CHAPTER

The first seven guidelines in this chapter follow the sequence of activities involved with preparing a talk, activities comparable to those involved with preparing a written communication:

- Defining your objectives (Guideline 1)
- Planning (Guidelines 2 through 4)
- Drafting (Guidelines 5 through 7)

Guidelines 8 and 9 concern preparations that apply specifically to speaking, not to writing.

GUIDELINE 1	**DEFINE YOUR OBJECTIVES**

You should begin working on an oral presentation in the same way you begin working on written communications: by figuring out exactly what you want to accomplish. In particular, you should consider the following three things.

- **Purpose and audience.** First, think about who your listeners are and how you want to affect them. You can do this in exactly the same ways you think about the audience and purpose of a written communication. If you wish to review these procedures, turn to Chapter 3.

● **Expectations.** Next, consider people's expectations about what you will say and how you will say it. To do this you should consider the same factors that you consider when preparing a written communication (see Chapter 5, pp. 168–69), plus one other.

That additional factor is the length of time you are expected to speak. Often, you will be given a time limit. Even when you are not, your listeners are likely to have some firm expectations about the time you will take. Keep in mind that your talk will be only one part of your listeners' schedules. Do not encroach upon the time they have committed to other speakers or other business. You can ruin a good presentation by talking past your time limit.

● **Scene.** Finally, as you are defining objectives, consider the circumstances, or *scene,* of your talk. Three aspects of the scene are especially important:

> *Size of audience.* The smaller your audience, the smaller your visual aids can be, the more likely your audience is to expect informality from you, and the more likely your listeners will be to interrupt you to ask questions.
>
> *Location of your talk.* If the room has fixed seats, you will have to plan to use visual aids in spots where they are visible to everyone. If you can move the seats, you will be able to choose the seating arrangement best suited to your particular presentation.
>
> *Equipment available.* The kinds of equipment that are available determine the types of visual aids you can use. You can't use a blackboard or overhead projector if you can't obtain one for the room.

Once you have defined your objectives you are ready to plan the talk through which you will achieve your objectives. The next three guidelines will help you plan effectively by providing advice about the form of oral delivery you will use (Guideline 2) and about visual aids (Guidelines 3 and 4).

SELECT THE FORM OF ORAL DELIVERY BEST SUITED TO YOUR PURPOSE AND AUDIENCE

When people make presentations at work, they generally select from three major forms of oral delivery: the scripted talk, the outlined talk, and the impromptu talk. To choose among them, you should consider first your audience's expectations. In most of the situations in which you will speak at work, your listeners will have some fairly firm expectations concerning the form of talk you will give. You may reduce your chance of achieving your purpose if you disappoint these expectations.

In some situations, however, you will have complete freedom to select the form of delivery you desire. The following paragraphs provide information that will help you choose among the types.

Scripted Talk

With a scripted talk, you write out your entire talk, word for word, in advance. Then you deliver your talk by reading the script or by reciting it from memory.

The scripted talk is well suited for situations where you want to be very precise, because it lets you work out ahead of time exactly the phrasing you will use. This is a great advantage when you must present complex or detailed information clearly and when small slips in phrasing could be embarrassing or damaging. Government and corporate officials often use scripted talks when making formal statements, and speakers at professional conferences use scripted talks to ensure that they communicate clearly and concisely.

Besides precision, a script offers a great deal of security. It ensures that you will be able to say exactly what you had planned to say. You won't leave something out, digress, speak in confusing sentences, or forget to show your visual aids. And you won't exceed your time limit because you will know from your rehearsals precisely how long your talk will take.

Because of the security it offers, the scripted talk is a good choice when you expect to be nervous during your delivery. That might happen, for instance, when you give your first few talks on the job, before you are comfortable speaking in front of your co-workers and bosses. Even if you become too nervous to think straight, you will have your words spelled out in front of you.

A major disadvantage of a scripted talk is that you need to spend a long time preparing it. Another disadvantage is that when you are delivering your talk, you cannot easily adjust it in light of the reactions you see from your audience. Being thoroughly planned, the scripted talk is also rigid.

When delivering a scripted talk, you must be careful to avoid keeping your eyes riveted to your page. That will prevent you from establishing a good rapport with your audience. You should rehearse your script enough times so that you can deliver your talk smoothly by looking down at your notes only occasionally.

One way to avoid the temptation to keep your eyes on your script rather than your audience is to deliver your talk from memory. By doing so, you can create an impression of spontaneity while still retaining great precision. When you speak from memory, however, you risk forgetting your script. Therefore, you may want to keep a copy of your script (or an outline of your talk) at your side for safety's sake.

For many people, the greatest challenge of writing a scripted talk is that of achieving a style that sounds natural when read aloud. Most of us have a casual, informal speaking voice, and another voice—more formal, less lively— that we use when writing. Even when we have written a script that uses a natural style, some of us have trouble reading or reciting the script aloud in a manner that sounds natural. When rehearsing and delivering a scripted talk, you should concentrate on using your voice in the ways you usually do in conversation.

Outlined Talk

To prepare an outlined talk, you do what the name implies: prepare an outline, perhaps very detailed, of the things you plan to say. You should, in addition, practice your talk, unless it is on a subject you have treated in oral presentations several times before. When practicing, you can work out your general treatment for each portion of your talk, decide how to emphasize your main points, and develop transitions that are clear and concise. You can also use this practice to judge how much you can say on each point and still keep within your time limit.

At work, outlined talks are much more common than scripted talks. They can be created much more quickly, and the presenter speaks with a "speaking voice," which helps increase interest and appeal. Furthermore, outlined talks are very flexible. Based upon your listeners' reactions, you can speed up, slow down, eliminate unnecessary material, or add something you discover is needed. Outlined talks are ideal for situations in which you speak on familiar topics to small groups, such as on occasions when you are meeting with other people in your department.

The chief weakness of the outlined talk is that it is so flexible that unskilled speakers can easily run over their time limit, leave out crucial information, or encounter difficulty finding the phrasing that will explain their meaning clearly. For all these reasons, you may want to avoid giving outlined talks on unfamiliar material or to audiences you don't know well. Also, you may want to avoid them in situations where you might be unusually nervous and therefore likely to become tongue-tied or to forget your message.

Impromptu Talk

An impromptu talk is one you give on the spur of the moment with little or no preparation. At most, you might jot down a few notes about the points you want to cover.

The chief advantage of the impromptu talk is that it takes little preparation time. It is well suited for situations in which you are speaking on subjects so familiar that you can express yourself clearly and forcefully with little or no forethought.

The chief disadvantage of the impromptu talk is that you prepare so little that you risk treating your subject in a disorganized, unclear, or incomplete manner. You may even miss the mark entirely by simply failing to address your audience's concerns at all.

For these reasons, the impromptu talks given at work are usually short, and they are usually used in informal meetings where the listeners can interrupt to ask for additional information and clarification. The impromptu talk is most common in meetings among people who work together regularly or who are at the same level within an organization. It is used much less often for formal presentations to people at higher levels or to large groups.

Of the three types of talks, the scripted and outlined talks are the most useful to you when you are trying to learn to give effective oral presentations. They require you to pay conscious attention to all the elements of a good presentation. In addition, they enable others to give you fairly precise advice about your plans before you give your talk, because your plans are evident in your written script or outline. If your instructor asks you to make an oral presentation in your class, it is likely to be a scripted or outlined talk.

GUIDELINE 3	USE VISUAL AIDS

Your talks at work will be much more effective if you present your message to your audience's eyes as well as to their ears. You can do this by incorporating visual aids and by making them an integral part of your presentation.

Benefits of Using Visual Aids

Visual aids can increase the effectiveness of your presentations in several ways:

- **Visual aids help you attract and hold your audience's attention.** Listeners are more likely than readers to let their minds wander. When they look away from a speaker, they may also be letting their thoughts drift off. By using visual aids, you provide listeners with a second place to gaze that is directly related to your message. Also, when you switch from one poster or slide to the next, your activity can draw wandering eyes back to you.

- **Visual aids help your listeners follow the organization of your talk.** You can use visual aids to provide headings that map your talk. For instance, you can display a poster that names the four topics you will cover, or the three recommendations you will make. If you leave such a poster up throughout the appropriate portion of your talk and point to it when you shift from one part of your talk to the next, you can help your listeners know at all times where you are in your presentation.

- **Visual aids help you emphasize key points.** By displaying your main points on a visual aid, you make them more prominent than the many points that you do not place on one.

- **Visual aids help you explain your material.** Just as in a written communication, you can use drawings, graphs, charts, and other visual aids to communicate your material with an economy and effect that words alone cannot achieve.

- **Visual aids help your listeners remember what you have said.** Because visual aids communicate so succinctly and forcefully, the information they present is easy to recollect.

- **Visual aids help you remember what you want to say.** Visual aids that are carefully integrated into your talk can serve as your notes while you speak. If you need to be reminded about what you intended to say next, you need only glance at the visual aid you are displaying or the one that will be displayed next.

Using a Storyboard to Integrate Visual Aids into Your Presentation

For your visual aids to contribute as effectively as possible to your presentation, you must coordinate them with the words you will speak. You can do this by using a *storyboard,* a planning tool developed by people who write movie and television scripts.

Storyboards have two columns. One shows the words and other sounds that an audience will hear. The other describes (in words or sketches) the things the audience will see as those words are spoken. By reading a storyboard, someone can tell what an audience will hear and see at each moment.

Figure 27–1 shows one kind of storyboard you can create for planning your talks. However, a storyboard doesn't have to be this elaborate to be effective. If you are going to use an outline rather than a script, you can make a few notes in the margins to help you combine the audio and visual aspects of your presentation into a single, effective unit.

GUIDELINE 4	SELECT THE VISUAL MEDIUM BEST SUITED TO YOUR PURPOSE, AUDIENCE, AND SITUATION

The preceding guideline suggests that you integrate visual aids into your talks. The guideline you are now reading concerns your decision about what kind of visual aid to use.

A *visual aid* is anything you give your audience to look at during your presentation. It might be something they touch and hold, such as a sample of your company's product. It might be a table, graph, drawing, or photograph. Whenever you use visual aids that can be printed or projected—such as tables, graphs, and drawings—you can choose from among a variety of media for presenting them to your listeners. At work, the media used most often are handouts, blackboards, overhead transparencies, posters, and 35-mm slides.

As you choose among these alternatives, you should think first about your purpose and audience. For some purposes, such as presentations at sales meetings, you may need the high polish that 35-mm slides can give. For other purposes, such as teaching, you may need to be able to make spontaneous drawings—so that 35-mm slides would not be appropriate, although a blackboard, overhead transparencies, or posters might work well.

You should also consider your audience's expectations. If your audience expects you to draw lines on a blackboard but you present a slickly packaged slide show, your choice of visual medium may distract your audience and seriously reduce the effectiveness of your presentation. Conversely, if the members of your audience expect you to arrive with your visual aids all prepared, they may become impatient if you pause periodically in your talk to draw diagrams on the blackboard.

Finally, you should consider the situation in which you are to prepare and deliver your talk. How many days or weeks are there until your talk? How much of that time can you devote to preparing your visual aids? How big is the room in which you will speak? Will special equipment be available to you? How big is your budget?

TALK TO COMPANY STEERING COMMITTEE ON
INTEGRATED WORK STATION

Good morning. I want to thank Mr. Chin for
inviting me to speak to you this morning
about a project that my staff in the
Computing Services Department thinks will
increase our company's productivity
considerably.

As you know, people in our company use   Show poster entitled
computers for many different purposes: our   "Uses of Computers."
managers use computers to create budgets and   List: Management
schedules; our researchers use them to   Functions, Research
analyze their data; our marketing staff uses   Analysis, Marketing
them to perform their duties; our   Analysis, Design,
development teams use them to design   Production, Word
products; our manufacturing people use them   Processing.
to control production lines; and everybody
uses them for word processing. In addition,
people throughout the organization have
expressed a desire for electronic mail.

We estimate, however, that at present the
company is realizing only about two-thirds
of the labor-saving capability of present-
day computer technology. To a large extent,   Show poster entitled
that's because of barriers that prevent the   "Computer Barriers."
easy flow of data from one computer   List: Program to
application to another and from one location   Program, Location to
to another. For example, the data prepared   Location.
by a researcher must be retyped by a
secretary to be included in the research
report. When our corporate planners want to
use information developed by marketing
analysts, they must obtain printouts of that
data and then retype it into their own
computers. And memos prepared through word
processing that we could send almost
instantly over telephone lines are instead
printed out and sent via the Post Office or
similar services.

[NOTE: This storyboard continues for six
more pages.]

FIGURE 27–1   Portion of a Storyboard for a Talk to a Group of Decision-Makers

When planning the visual aids for your talks, keep in mind that you can
use more than one type in the same talk. For instance, you may find it helpful
to distribute handouts that outline your talk and to use one or more other types
of visual aid to show your other materials.

<table>
<tr><td>GUIDELINE
5</td><td>

TALK *WITH* YOUR LISTENERS
</td></tr>
</table>

Guidelines 5 through 7 concern your work at drafting your talk. This advice applies whether your talk will be scripted, outlined, or impromptu.

Guideline 5, ''Talk *with* your listeners,'' should sound familiar to you. It is a variation of some advice you read in the first chapter of this book: ''When writing, talk with your readers.'' Both versions emphasize the importance of responding to your audience's moment-by-moment reactions to what you are saying. You must respond effectively to those reactions because your audience's final reaction to your communication—whether written or oral—will be determined by these smaller reactions to one statement, then the next, and so on.

Of course, when applied to *writing,* this advice to talk with your readers isn't meant to be taken literally. To follow it, you create an imaginary portrait of your readers, then use that portrait to picture your readers' responses. When you *speak,* your listeners are present and it therefore might seem to be easy to talk with them. However, such is not necessarily so. The following paragraphs explain why.

Giving a Talk Differs from Conversing with People

Although your audience is present when you *deliver* a talk, they are not present when you *prepare* it. You prepare your talk in isolation, away from your listeners, who are then no closer to you than your readers are when you write.

The more detailed your preparations, the more difficult it will be for you to respond to your listeners' reactions. Thus, you will have the most difficulty talking with your listeners in scripted talks, where you work out beforehand even the phrasing of each individual sentence. Consequently, when you are preparing a scripted talk, you can benefit greatly from creating imaginary portraits of your listeners in the same way that you create such portraits of your readers (see Chapter 3).

You will encounter less difficulty responding to your listeners' reactions when you make outlined talks; you will have the least difficulty with impromptu talks. Even in an impromptu talk, however, speakers sometimes fail to respond to their listeners' reactions, either because the speakers are too nervous to observe those reactions or because they focus their attention on themselves or their subject matter, to the exclusion of their audience.

Thus, regardless of the form of oral delivery you choose, you should talk with your listeners by anticipating (or observing) their reactions and then responding accordingly. In this respect, preparing a talk is very much like drafting a written communication.

The Importance of Building Rapport

When applied to oral presentations, the advice to talk with your listeners means something more than it does when applied to writing. When you give a talk, you and your listeners are right there in the same room, looking at each other.

They expect you not merely to present information and ideas, but also to talk *to them,* to show an interest *in them.* Accordingly, to talk effectively with your listeners you must do more than respond to their reactions in a formal way. You must build some personal connection, a rapport, with your audience.

The following paragraphs list some specific suggestions for building rapport with your listeners:

- **While preparing your talk, imagine your listeners in the act of listening.** Your imaginary portrait of a typical listener can help you build rapport. By imagining that you are *speaking* to this person, you will increase your ability to express your message in a way that sounds like one person speaking (rather than writing) to another.

- **Use the word *you* or *your* in the first sentence.** With those words, you signal that you intend to speak directly to your audience. You will also set a pattern of phrasing that you can follow in the rest of your preparations.

   There are many ways you can include *you* or *your* in the first sentence: thanking your listeners for coming to hear you, praising something you know about them (preferably something related to the subject of your talk), stating the reason they asked you to address them, or talking about the particular goals of theirs that you want to help them achieve.

- **Use a conversational style throughout.** Use shorter, less complex sentences than you might use in writing. Use personal pronouns *(I, we, you)* and the active voice.

- **Stick to points that are directly and clearly relevant to your listeners.** When you give a talk, you are making an implicit promise to provide information that your listeners will find useful or interesting. If you break that promise, you have broken the social connection between you and them. This point is especially important because it is tempting to digress in oral presentations, especially in impromptu and outlined talks. To maintain a good relationship with your listeners during your talk, give them what they want and can use.

- **Emphasize the implications of your points for your listeners.** In this way, you can assure your listeners that you have prepared remarks that are, in fact, for them.

- **Be very clear.** Nothing is more likely to make your listeners feel ill-treated than to speak in a way they cannot understand.

GUIDELINE
6

HIGHLIGHT YOUR MAIN POINTS

By emphasizing your main points, you improve your talk in two ways. First, your main points—not your incidental thoughts, not good stories—are what your *listeners* most want to hear. By emphasizing your main points, you draw your listeners' attention to the material they most desire.

Second, your main points are what *you* most want your listeners to take away from your talk. By emphasizing those points, you increase the chance that your listeners will notice and remember them.

In many situations at work, you will prepare talks that complement written reports. For instance, if you are proposing a major project, you might prepare a sixty-page written communication that you would introduce to your readers by making a fifteen-minute oral presentation. In your presentation, you would focus on the key points about your project, relying upon the written communication to fill in the details.

Here are some ways you can emphasize your main points:

●   **Tell the listeners in advance that the main points are coming.** "This process has three major parts." "I will make three recommendations."

●   **Announce each main point as you come to it.** "The second major step is as follows. . . ." "My second recommendation is. . . ."

●   **State your main points in your visuals.** In this way, your listeners will see as well as hear them.

When presenting your main points, remember to make clear their significance to your listeners.

GUIDELINE 7	MAKE THE STRUCTURE OF YOUR TALK EVIDENT

To understand and remember your presentation fully, your listeners must be able to organize your points in their own minds. They stand the best chance of doing this if they can follow the structure of your talk.

Listeners must figure out your talk's structure without several of the aids that *readers* have. Listeners have no paragraph breaks, headings, or indentations to help them. Further, if they forget the overall structure or where they are in it, they cannot flip back through the pages to reorient themselves.

Here are two strategies you can use to help your listeners follow your organization.

●   **Forecast the organization.** Tell your listeners in advance what your organization is. Your forecast will be especially effective if you both *tell* and *show* your organization. You can show it, for instance, by using a poster that contains a key word for each of the parts of your talk, or a handout that contains an outline of your talk. For detailed advice about forecasting the organization of your talk, see the discussion of forecasting statements in Chapter 7.

●   **Clearly indicate the transitions from one topic to another.** Listeners can have real difficulty determining when you shift from one topic to another. The problem is especially critical because if they miss a transition, they will have trouble figuring out what your current topic is. Some ways you can make your shifts obvious are as follows:

1. **Announce the shift.** "Now I would like to turn to my second topic."

2. **Pause before beginning the next topic.** The pause in the flow of your talk will signal a shift.

3. **Change visuals.** Take down one poster and put up another. Move to the next slide. Refer to a visual that displays the overall organization of your talk.

4. **Move.** If you are giving a talk in a setting where you can move about, signal a shift from one topic to another by moving from one spot to another.

GUIDELINE 8

PREPARE FOR INTERRUPTIONS AND QUESTIONS

The seven guidelines you have read so far all concern composing activities you must perform whether you are preparing a written or an oral communication. In contrast, the final two guidelines concern some special kinds of preparations that you should make for a talk but that are irrelevant when you are creating a written communication.

At work, your talks will often be followed by a period for questions and discussion. This period provides your listeners with an opportunity to ask you for further information, to think through the significance of what you have said, and even to argue with you when they disagree. Such interchanges are a normal part of talks at work, and you should prepare for them.

To do this, think of the kinds of questions that might arise from your audience. In a sense, you do this when you plan the presentation itself, at least if you follow Chapter 4's advice that you begin planning your communication by thinking about the various questions your reader will want it to answer. Usually, however, you will not have time in your talk to answer all the questions you expect that your listeners might ask. The questions you can't answer in your talk are ones your listeners may raise in a question period. Prepare for them by planning your responses.

GUIDELINE 9

REHEARSE

All of your other good preparations can go for naught if you are unable to deliver your message in a clear and convincing manner. The next chapter provides you with guidelines for delivering your talk, but you should be *practicing* those guidelines well before you stand in front of your intended audience to give your talk.

- **Rehearse in front of other people.**
- **Pay special attention to your delivery of the key points.** These are the points where stumbling can cause the greatest problems.
- **Rehearse with your visuals.** You need to practice coordinating your visual aids with your talk.
- **Time your rehearsal.** In this way, you can ensure that you keep within your time limit. Be sure to speak at a normal pace in rehearsal. If you race through your talk when you actually give it, you will make it difficult for

your audience to follow. If you slow down when you deliver it, then you will take longer than you planned, perhaps exceeding your time limit.

CONCLUSION

This chapter has described a listener-centered approach to preparing oral presentations. First, it has suggested ways to adapt this book's advice for creating written messages to situations where you address your audience orally. Also, it has discussed several additional considerations—such as time limits, the scene of your presentation, and the need for rehearsal—that you must take into account when preparing spoken messages. In the next chapter, you will learn what to do as you stand before your audience to deliver your oral presentations.

EXERCISES

1. Imagine three situations in which you might have to prepare talks at work: one in which the talk would be scripted, one outlined, and one impromptu. Write a paragraph or two describing each of these situations. Tell the following:
    - What your topic will be
    - Who will have asked you to talk
    - Why you have been asked to talk
    - What the purpose of your talk will be
    - Who your listeners will be
    - How you want to modify your audience's attitudes
    - What your listeners will do with the information you provide
    - Where you will speak
    - What your time limit will be
    - What visual aids you will use

2. Make an outline for a talk to accompany a written communication you have prepared or are preparing in one of your classes. The audience for your talk will be the same as for your written communication. The time limit for the talk will be ten minutes. Be sure that your outline indicates the following:
    - The way you will open your talk
    - The overall structure of your talk
    - The main points from your written communication that you will emphasize
    - The visual aids you will use

    Be ready to explain your outline in class.

3. Imagine that you must prepare a five- to ten-minute talk on some equipment, process, or procedure. Identify your purpose and readers. Then write a script or outline for your talk (whichever your instructor assigns). Be sure to plan what visual aids you will use and when you will display each of them to your listeners.

CHAPTER 28

DELIVERING ORAL PRESENTATIONS

Angela sits in one of the soft, leather chairs placed against the wall. In the center of the wood-paneled room is a long table, at which ten men and women sit, all managers or staff advisers in her employer's organization. As soon as they have completed one last piece of business, Angela will speak to them, at their request, concerning a new marketing plan she has developed.

Angela is nervous. For the past week she's concentrated most of her attention on preparing the ten-minute talk during which she will explain her plan. She's carefully written a script, designed a handout, created a set of twelve posters, sought advice and approval from her boss, and practiced over and over again. In a moment, all that work—and the months of labor that preceded it—will be on the line.

As Angela rises from her chair and begins her presentation, what can she do to make her presentation effective? What kinds of things should you try to do when you find yourself in a similar situation? This chapter presents eight guidelines that will help you deliver your oral presentations effectively.

IMPORTANCE OF COMMUNICATING WITH EACH INDIVIDUAL IN YOUR AUDIENCE

The eight guidelines in this chapter touch on many activities involved with delivering talks, including such things as deciding where to stand, using your voice, and responding to interruptions, questions, and comments. The guidelines concerning all these activities deserve your careful attention. There is, however, one piece of advice that merits special emphasis: when you are presenting a talk, above all else concentrate on communicating with each individual in your audience. Each has come to hear you. Each is looking for something in particular from you. Each wants to feel that you are striving to address his or her individual interests and concerns.

"But," you may be thinking, "why tell me to concentrate on communicating with my audience? What else would I be trying to do?"

Actually, as they stand to speak, many people seem to forget that their purpose is to communicate. They seem to focus mainly on following their notes flawlessly and handling their visual aids without mishap. They speak in a flat, lifeless manner, wear wooden expressions, and make foolish mistakes, such as standing in front of their visual aids. Instead of acting like people with something helpful or significant to tell someone else, they act as if they were indifferent toward their audience and subject matter.

You know from your own experience how such speakers affect people. As a listener, you become irritated or bored. You look out the window or at the floor. You begin thinking about something else.

To avoid such reactions from the listeners to your talks, you must strive to make each person feel that you have come for the express purpose of saying something important to him or her. Concentrate on those listeners, not on anything else. Don't simply read your notes; speak to your listeners. Don't simply display your visual aids; use them to help you say something to your listeners.

The various guidelines in this chapter will assist you in many other ways, but all will also help you achieve this very important goal of communicating directly and effectively with each individual in your audience.

GUIDELINE 1   ARRANGE YOUR STAGE SO YOU AND YOUR VISUAL AIDS ARE THE FOCUS OF ATTENTION

When you deliver a talk at work, you are a little like a stage performer, with some words to deliver and some props (your visual aids) to manage. For your performance to go smoothly, you must be sure that your "stage" is set carefully. Any difficulties with your stage can interfere with your listeners' ability to concentrate on and understand your message.

To arrange your stage in a way that will help your listeners focus their attention on your message, you can do three things. First, arrange your stage so that you are the focus of attention. Do not allow yourself to be dwarfed or obscured by your posters or other visual aids. Second, place your visual aids where they are visible to all members of your audience. Third, place your visual aids where you can manipulate them easily.

For most of your talks, these arrangements will require little thought or effort. Usually you will be speaking in familiar rooms that are often used for such presentations. You will be able to set your stage effectively if you imitate the arrangements you have seen others use or those you yourself have used successfully in the past. However, you will benefit from thinking more deliberately about the arrangement of your stage when you find yourself in any of the following situations:

- **You are going to speak in an unfamiliar location.** Visit the room beforehand to arrange your stage.
- **You are speaking in a room that is larger than you are used to.** Be sure that your visual aids will be visible to all and that everyone in your audience will be able to hear you.
- **You are going to use a 35-mm slide projector.** Be sure that you will have a source of light for reading your notes even when the room is darkened to show the slides.

Of course, an essential part of arranging your stage is being sure that your props will be available. Make arrangements ahead of time if you are going to need a projector, blackboard, easel (for holding posters), or any other piece of equipment for your talk.

GUIDELINE
2

LOOK AT YOUR AUDIENCE

One of the most important ways to let your listeners know that you want to communicate with each of them is to look at them while you talk. By looking at them you create a personal bond. You show them that you are interested in them as individuals, both personally and professionally.

Also, by looking at your listeners you can get them to pay attention to you. Social convention requires people to pay attention to people who are talking to them. If you look at your listeners, they are much more likely to mind their manners by paying close attention to you.

In addition, by looking at your listeners you can see how your talk is going. You can see the eyes fastened on you with interest, the nods of approval, the smiles of appreciation. And you can see the puzzled looks, the wandering attention. As you observe these reactions, you can adjust your talk appropriately. If you are presenting an outlined talk, you can add needed explanations or drop unnecessary elaborations. Even with a scripted talk, you can make some adjustments, changing your rate of speech, speaking more loudly, giving your audience more time to study a visual aid.

Of course, you don't need to look at your audience constantly. If you are using notes (either an outline or a script), you will want to refer occasionally to them. You should not, however, rivet your gaze on your notes so that you never look at your audience.

If you have difficulty looking at your listeners when you speak, here are some strategies you can use:

- **Look at your audience before you start to speak.** Many people have difficulty looking at others when they are nervous or insecure. Perhaps you are one of them. Often these people are afraid that if they look at their listeners they will see signs of disagreement or disinterest. If such is the case, this first look before you begin to speak can help you greatly because it will come at a time when you cannot possibly fear seeing an adverse reaction to your talk. Having built up your confidence at the beginning of your talk, you may be more willing to look at your audience in the midst of it.

- **Follow a plan for looking.** For instance, at the beginning of each paragraph of your talk look at a particular part of your audience—to the right for the first paragraph, to the left for the second, and so on.

- **Target a particular part of your listeners' faces for your glances.** If their eyes are too threatening for you, plan to look at their noses or foreheads.

● **When rehearsing, practice looking at your audience.** For instance, develop a rhythm of looking down at your notes then up at your audience, down then up. Establishing this rhythm in rehearsal will help you avoid holding your head down throughout your talk.

When looking at your audience, try to avoid skimming over their faces without focusing your eyes on any individual. You must focus on an individual to make that person feel that you are paying attention to him or her. Try setting the goal of looking at a person for four or five seconds—long enough for you to state one sentence or idea. Some speakers find that such a pattern of looking helps them establish an even pace of speaking and also helps them avoid using distracting ''filler'' words (''umm,'' ''you know,'' and so on).

In sum, by looking at your listeners you can command their attention and interest in your message.

| GUIDELINE 3 | SPEAK IN A NATURAL MANNER, USING YOUR VOICE TO CLARIFY YOUR MESSAGE |

Another way to persuade your individual listeners that you are speaking directly to each of them is to speak like a person having a conversation. Listen to yourself and your friends converse. Your voices are lively and animated. To emphasize points, you change the pace, rhythm, and volume of your speech. You draw out words. You pause at key points. Your voice rises and falls in the varied cadence of natural speech.

When they make oral presentations, however, many people drain all of the natural life from their voices. They speak in a monotone, at a deadeningly even pace, without pause or variation of any sort. You know from your own experience how difficult it is to listen to such speakers. You rely on variations in the speaker's voice to maintain your interest, to help you distinguish major points from minor ones, to clarify the connections between ideas, and to identify the transitions and shifts in topic that indicate the structure of the speaker's talk.

Here are a few strategies that will help you use your voice to keep your listeners' interest and clarify your message:

● **In rehearsal and when delivering your talk, focus your attention on your listeners.** Keep in mind that you are talking to people, not reciting a well-practiced string of words. When you focus your attention on your listeners, you are much more likely to fall into natural speech patterns.

● **In rehearsal, concentrate on your manner of speaking.** This will be easiest to do if you practice in front of people, but you can also do it if you *imagine* that your audience is in front of you. If you find a passage that is difficult for you to say in a natural voice, rephrase it.

● **In rehearsal, decide what points you want to emphasize, then practice presenting them emphatically.** In your notes (whether an outline or a script), mark these points so that you remember to emphasize them when you stand in front of your audience.

| GUIDELINE 4 | **EXHIBIT ENTHUSIASM AND INTEREST** |

When rehearsing and presenting your talk, keep in mind that even at work feelings are contagious. People who are enthused about projects often engender enthusiasm in others. At the same time, people who are indifferent often infect others with their indifference. For that reason, be sure to show your interest and enthusiasm when you speak.

What can you do to show your commitment to your subject? Certainly, speaking in a natural, forceful fashion will help. So will your facial expressions. An occasional smile can be your best visual aid. Remember that you must communicate more than information and ideas: you must communicate your *attitude* toward your subject.

| GUIDELINE 5 | **DISPLAY VISUAL AIDS EFFECTIVELY** |

When you give a talk, you offer your audience information through two senses: hearing and seeing. This is a real advantage if you provide essentially the same message to both senses. However, if you talk about one thing but display a visual aid about something else, each person in your audience must choose either to listen to your words or to read your visual aid. Either way, your audience may miss part of your message.

Therefore, you should be very careful to show each of your visual aids only when it supports and reinforces your words. Don't show your visual aids too soon. Don't leave them up too long.

One way you can coordinate your visual aids with your words is to prepare visual aids that you can reveal *gradually*. For instance, you may have a poster or overhead transparency that lists each of the three anticipated outcomes of a certain course of action. You can cover the last two outcomes with a sheet of paper while you state and discuss the first, then uncover the second as you state and discuss it, and so on.

| GUIDELINE 6 | **MAKE PURPOSEFUL MOVEMENTS** |

Another point to remember as you give your talk is that your movements can support or detract from your presentation. In conversation, you naturally make many movements—nodding or shaking your head, holding out your arms to show the size of something, and so on. When making oral presentations, you can also use such movements to make your meaning and feelings clear.

Also helpful are the movements you make to control your visual aids. In fact, one of the advantages of using visual aids is that they give you a good reason for moving, so that you don't stand stiffly and unnaturally throughout your talk.

Unhelpful are movements that express nervousness: pacing, rocking, fidgeting of any sort. These movements distract your audience's attention from your message. You should strive to eliminate them.

GUIDELINE
7

RESPOND COURTEOUSLY TO INTERRUPTIONS, QUESTIONS, AND COMMENTS

Audiences at work often ask questions and make comments to a speaker. In fact, most of the talks you give at work will be followed by discussion periods during which members of your audience will ask you for more information, discuss the implications of your talk, and even argue with you about points you have made. This give and take helps explain the widespread popularity of oral presentations at work: they permit speaker and audience to engage in a mutual discussion of what the speaker has to say.

When you respond to your audience, try to sustain your good relationship with all of your listeners. Be courteous even if the questions are annoying or hostile. Remember that the questions and comments are important to the people who ask them, even if you don't see why. Give the requested information if you can. If you don't know how much detail the questioner wants, offer some and ask the questioner if he or she wants more. If you don't know the answer to a question, say so.

Although questions and comments are often held until after the talk, in some situations people will interrupt you to say or ask something. Interruptions require special care. First, speak to the person immediately. If you are already planning to address later in your talk the very matter that your questioner has raised, you may want to ask the person to wait for your response. If not, you may want to respond right away. Once you are done with your response and again pick up the thread of your talk, be sure to remind your listeners where you are in your talk: "Well, now I'll return to my discussion of the second of my three recommendations."

GUIDELINE
8

LEARN TO ACCEPT AND WORK WITH YOUR NERVOUSNESS

This final guideline is the most difficult for many novice speakers to follow. But it is very important. Not only is nervousness unpleasant for you to experience, but it can lead to distracting and unproductive behaviors that greatly diminish the effectiveness of your talk. Furthermore, some of these behaviors can cause your listeners to believe mistakenly that you are not interested in them or in your subject. The following list identifies some of the most common nervous behaviors that you should strive to avoid:

- Looking away from your audience instead of looking into their eyes.
- Speaking in an unnatural or forced manner.
- Exhibiting a tense or uninterested facial expression.
- Fidgeting and pacing.

What can you do to avoid these problems? It would be nice if you could simply banish the nervousness that causes them, making it go away forever. Unfortunately, nervousness cannot be eliminated through a simple act of will-power. Even highly polished speakers with decades of experience may still feel nervous as they rise to make each of their presentations.

If you can't stop your nervousness, what can you do about it? First, accept it. It's natural. It's going to be there whether you want it or not. If you start to worry about being nervous, you merely compound the emotional tension you must deal with when you speak. Keep in mind that your nervousness is not nearly so obvious to your listeners as it is to you. Even if your listeners do notice that you are nervous, they are more likely to be sympathetic than displeased. Furthermore, a certain amount of nervousness can help you. The adrenaline it pumps into your system will make you more alert and more energetic as you speak.

Besides learning to accept your nervousness, you can learn to control the undesirable mannerisms it can foster—the pacing, the fidgeting, and so on. They can be eliminated through conscious effort. Perhaps by asking for help from your friends, classmates, or co-workers, you can identify any distracting mannerisms you have and concentrate on controlling them.

In sum, you should remember that nervousness is natural. Instead of being embarrassed by it you should accept it and work to overcome the unproductive behaviors it fosters.

CONCLUSION

When making an oral presentation, you should strive to make each member of your audience believe that you have presented your message clearly and effectively. To achieve this goal, you must consider many things, including the arrangement of your stage, the places you look, the ways you employ your voice and control your movements, and the manner in which you display your visual aids.

This chapter has presented advice on all these matters. It has also provided suggestions about how to respond to interruptions, questions, and comments and about how to work with any nervousness you might feel. By following the chapter's guidelines and by gaining practice at making oral presentations, you will quickly polish your speaking skills.

EXERCISES

1. At work you will sometimes be asked to contribute to discussions about ways to make improvements. For this exercise, you are to deliver a five-minute impromptu talk describing some improvement that might be made to some organization you are familiar with. Topics you might choose include ways of improving efficiency at a company that employed you for a summer job, ways of improving the operation of some club you belong to, and ways that an office on campus can provide better service to students.

   In your talk, clearly explain the problem and your solution to it. Your instructor will tell you which of the following audiences you should address in your talk.

   ● Your classmates in their role as students. You must try to persuade them of the need for, and the reasonableness of your suggested action.

   ● Your classmates, playing the role of the people who actually have the authority to take the action you are suggesting. You should take one additional minute at the beginning of your talk to describe these people to your classmates.

2. At work you will sometimes have to tell your boss and co-workers about your plans for a communication you are writing. For this exercise, you are to describe in a five-minute outlined talk one of the written projects you are preparing for class. Let your talk fall into two parts. In the first, describe the background and objectives of your project by talking about such things as the following:

   ● What your topic will be

   ● Who asked you to write the communication

   ● Who your readers will be

   ● What task your readers will perform while reading your communication

   ● How you want your communication to affect your readers' attitudes

   ● What limitations and expectations apply to your communication (including such things as limits on length and expectations about appearance)

In the second part of your talk, describe your plans for your communication in a way that shows how your plans are designed to achieve your objectives.

Be sure to use at least two visual aids. These might include, for instance, a poster outlining the main points of your talk, a handout showing your proposed table of contents, or a sample page showing some of the important features of your design.

Be prepared to answer questions from your audience.

3. Standing in front of your class, deliver the talk that you prepared in Exercise 2 or 3 of Chapter 27. Let your classmates and instructor play the role of your intended audience. When preparing your talk, be sure to rehearse it more than once, timing it to see that you can stay within your time limit. After your talk, take questions from your audience, who will be asking the types of questions your intended listeners would ask.

APPENDIXES

A

FORMATS FOR LETTERS, MEMOS, AND BOOKS

Formats are conventional packages for presenting messages. In this appendix, you will learn how to use the three formats most often employed at work: letter format, memo format, and book format.

Two of these three formats are familiar to you already, although you may never have written them yourself. You have surely read business letters sent to you by your college and by advertisers. And even if you've never read a memo addressed specifically to you, you have seen many sample memos in other chapters of this book. Perhaps much less familiar to you is the book format used at work. In many ways it resembles the format used for the novels and textbooks that you are accustomed to calling "books." For instance, it includes front and back covers, title page, table of contents, and separate chapters or sections. However, the book format includes some features not usually found in novels and textbooks, such as a summary placed at the beginning and appendixes placed at the end. And it is used not only with communications hundreds of pages long but also with ones as short as ten or twenty pages.

The letter, memo, and book formats are worth studying for several reasons. First, by using these conventional patterns successfully, you demonstrate to your audience that you know some of the basic customs of the workplace, a demonstration that helps establish your credibility. Also, the formats aid communication by reminding you to include key pieces of information (such as your return address in a letter) that are important to your readers. They also aid communication by directing you to put these pieces of information where any reader familiar with the format can readily find them.

FORMATS AND EMPLOYERS' STYLE GUIDES

Standard formats contribute so significantly to the success of writing that some employers distribute instructions that tell employees how to use them. These instructions, sometimes called *style guides,* describe everything from the size of margins to the placement of headings. Of course, from company to company these instructions vary somewhat. For instance, on the cover of a report, one employer may want the title placed in a certain location and another employer may want it placed somewhere else. In a letter, one employer indents the first line of a paragraph and another does not.

Despite these variations, however, most business letters, memos, and formal reports look very much alike. The versions of the formats described in this chapter reflect common practices in the business world. You can use them with confidence in situations where no one has specified that you use some other version. Also, if you learn these versions, you will have little trouble adapting to other ones, because the differences involved will be minor.

CHOOSING THE BEST FORMAT FOR YOUR COMMUNICATIONS

Conventional formats are *rigid* enough that, for example, most letters written at work look similar. From another point of view, however, these formats are very *flexible.* Each can convey any type of message. Imagine, for instance, that you want to report on a project that you have completed. You can do so in the letter format, the memo format, or the book format—just as you could use any of these formats to propose a project or provide instructions. This is because formats operate independently of the messages they contain. They are like packages in which you can present any type of message.

"If formats and their messages are so independent of each other," you may wonder, "how can I decide which format to use for some particular communication?" Often, you will not need to choose at all. Your boss will have said, "Send a letter to Kettering," or "Write a memo to Horton." But in situations where you must choose, the following notes will be helpful:

- **Letters and memos are usually used for shorter communications, and books for longer communications.** The book format provides the protective cover, table of contents, summary, and other features that help readers use long communications (about ten or more pages). For shorter communications, those aids would not be needed. They would merely make the communications unnecessarily complicated and time-consuming to read.

- **Letters are usually sent to people outside the writer's organization, and memos to people inside.** The format for letters includes your address so that your readers can find it readily. Each letter also contains its reader's address, so that you can find the address readily for follow-up communications. In addition, because the style of letters is customarily more gracious and formal than that of memos, the letter is a more appropriate format for addressing customers, clients, and people you don't know well. On the other hand, the memo is designed to be written quickly, so it is well suited for internal communications, in which addresses are not necessary and formality might seem odd.

- **Format preferences vary from organization to organization.** Some organizations use letters for communications that other organizations would send as memos. Some use memos to send messages that others would communicate in the book format. Let your choices concerning format be guided by the usual practices in the organization that employs you.

Notice that decisions about which format to use are based on the lengths of your messages, the location of your readers, and the preferences of your employer, not on the subjects or purposes of your messages.

PREPARING NEAT, LEGIBLE COMMUNICATIONS

To use any of these formats well, you must prepare your communications neatly and legibly. If you fail to do so, you make communications that are hard work for your readers to read, and you risk offending your readers, who may resent that you have been careless when writing to them. In addition, if your readers think that your written communications are careless, they may suspect that you are sloppy in other aspects of your work as well.

Here are a few brief pointers about the appearance of the communications you prepare at work. They apply equally to communications you prepare on typewriters and those you prepare on word processors.

- **Use clean, straight keys.** The characters printed by your typewriter or printer should be sharp and evenly aligned.
- **Use a dark ribbon.** If your ribbon prints gray letters rather than black ones, you should replace it.
- **Make all corrections neatly.** Before making corrections, carefully and completely erase the mistakes, or cover them thoroughly with white correction fluid. If you can't make the corrections neatly, start the page again.
- **Use ample, consistent margins.** Margins at work usually are about one inch on the sides. Top and bottom margins vary among the formats.

If you use a word processor, be sure the printer produces copies good enough to meet your readers' expectations and preferences. Some dot-matrix printers create communications that are difficult to read. (Dot-matrix printers create each letter not as a solid image but as a pattern of dots.) Also, keep in mind that while some people feel that letters prepared on dot-matrix are perfectly acceptable, others think they look unprofessional. Find out the preferences of your particular readers.

ORGANIZATION OF THE REST OF THE CHAPTER

The rest of this chapter is designed for you to use as a reference manual. When writing a communication in one of the three formats described here, you can turn directly to the appropriate discussion and read the advice given there without worrying about what is said about the other formats.

Because the discussion of each format provides so many detailed pieces of advice, the discussions are not organized around guidelines like those provided in much of the rest of this book. Instead, each format is discussed one part at a time. For instance, the discussion of the book format tells you how to write the cover, then the title page, and so on. This arrangement enables you to consult the discussion of each part as you are writing it. You should note, however, that although the parts are presented in the order in which they appear in a finished communication, you may have good reasons for writing them in another order. For instance, although a report in the book format usually begins with a summary, you will usually be wisest to write your summary last, so that you will have decided fully what you want to say before you try to summarize your message.

USING THE LETTER FORMAT

Letters are widely used for communicating relatively short messages to customers, clients, government agencies, and other readers outside the writer's organization. Three variations of the letter format are shown in Figures A–1 through A–3. All three variations have the same parts, which are described below. Letters written at work are almost always typed.

Letterhead

Superior Fabrication Company

176 Lafayette Court
Baton Rouge, Louisiana 70816
517/235-9008

Date

October 17, 19--

Inside address

Mr. Anthony Fazio
Capra Products, Incorporated
9223 Taft Street
Grand Rapids, Michigan 49507

Subject line

Subject: <u>Anticipated Orders for Next Quarter</u>

Salutation

Dear Mr. Fazio:

XXXXXXXXXXXXXXXXXXXXXXXXXXXXXXXXXXXXXXXXXXXXXXXXXXXXXXXXXXXXXXXXXXX
XXXXXXXXXXXXXXXXXXXXXXXXXXXXXXXXXXXXXXXXXXXXXXXXXXXXXXXXXXXXXXXXXXX
XXXXXXXXXXXXXXXXXXXXXXXXXXXXXXXXXXXXXXXXXXXXXXXXXXXXXX

XXXXXXXXXXXXXXXXXXXXXXXXXXXXXXXXXXXXXXXXXXXXXXXXXXXXXXXXXXXXXX
XXXXXXXXXXXXXXXXXXXXXXXXXXXXXXXXXXXXXXXXXXXXXXXXXXXXXXXXXXXXXXXXXXX
XXXXXXXXXXXXXXXXXXXXXXXXXXXXXXXXXXXXXXXXXXXXXXXXXXXXXXXXXXXX
XXXXXXXXXXXXXXXXXXXXXXXXXXXXXXXXXXXXXXXXXXXXXXXXXXXXXXXXXX
XXXXXXXXXXXXXXXXXXXXXXXXXXXXXXXXXXXXXXXXXXXXXXXXXXXXXXXXXX
XXXXXXXXXXX

XXXXXXXXXXXXXXXXXXXXXXXXXXXXXXXXXXXXXXXXXXXXXXXXXXXXXXXXXXXXX
XXXXXXXXXXXXXXXXXXXXXXXXXXXXXXXXXX

Sincerely,

Francis V. Sullivan

Signature block

Francis V. Sullivan
Purchasing Agent

Enclosure
notation

Enclosures (2)

Distribution list

cc: T.L. Klein

Typist
identificaton

FVS/tm

Computer file
number

0654S

FIGURE A–1   Block Format for Letters

Heading

The heading gives your address (but not your name) and the date. In business, people usually do not abbreviate words in the heading; they spell out *Street, Avenue,* and so on. There is one exception: people often use the Post Office's two-letter abbreviations for the states (for example, *NY* for *New York* and *TX* for *Texas*).

Super M Printing   *2783 High Street   Miami, Florida 33143*

January 4, 19--

Joshua Kendall, Director
Public Relations Department
Harper Industries
13200 Constitution Avenue, NW
Seattle, Washington 98038

RE: Modifications to Contract

Dear Josh,

XXXXXXXXXXXXXXXXXXXXXXXXXXXXXXXXXXXXXXXXXXXXXXXXXXXXXXXXXXXXXXXX
XXXXXXXXXXXXXXXXXXXXXXXXXXXXXXXXXXXXXXXXXXXXXXXXXXXXXXXXXXXXXXXXX
XXXXXXXXXXXXX

XXXXXXXXXXXXXXXXXXXXXXXXXXXXXXXXXXXXXXXXXXXXXXXXXXXXXXXXXXXXXXXX
XXXXXXXXXXXXXXXXXXXXXXXXXXXXXXXXXXXXXXXXXXXXXXXXXXXXXXXXX
XXXXXXXXXXXXXXXXXXXXXXXXXXXXXXXXXXXXXXXXXXXXXXXXXXXXXXXXXXXXXXXXX
XXXXXXXXXXXXXXXXXXXXXXXXXXXXXXXXXXXXXXXXXXXXXXXXXXXXXXXXXXXXXX
XXXXXXXXXXXXXXXXXXXXXXXXXXXXXXXXXXXXXXX

XXXXXXXXXXXXXXXXXXXXXXXXXXXXXXXXXXXXXXXXXXXXXXXXXXXXXXXXXXXXXXXXX
XXXXXXXXXXXXXXXXXXXXXXXXXXXXXXXXXXXXXXXXXXXXXXXXXX

With best wishes,

Gladys

Gladys K. Lister
President

cc: Ernest Stevenson

GKL:lc

305/345-9292

FIGURE A–2   Modified Block Format for Letters

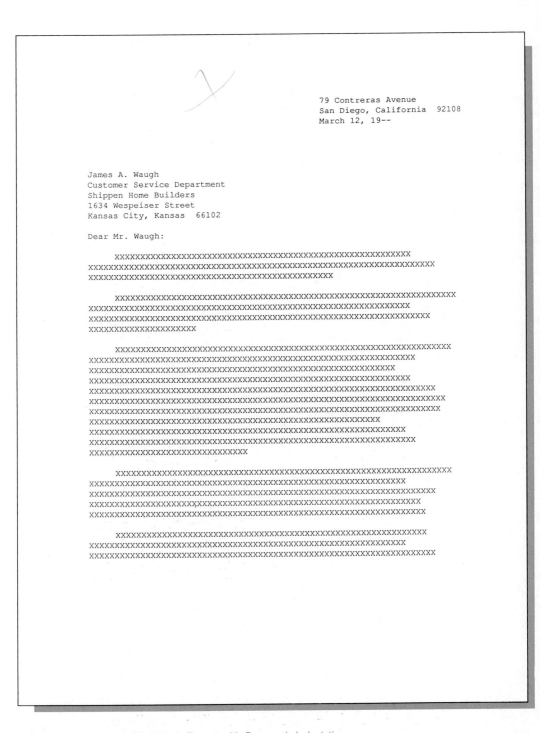

James A. Waugh   - 2 -   March 12, 19--

XXXXXXXXXXXXXXXXXXXXXXXXXXXXXXXXXXXXXXXXXXXXXXXXXXXXXXXXXXXXXXXXXXXXXX
XXXXXXXXXXXXXXXXXXXXXXXXXXXXXXXXXXXXXXXXXXXXXXXXXXXXXXXXXXXX
XXXXXXXXXXXXXXXXXXXXXXXXXXXXXXXXXXXXXXXXXXXXXXXXXXXXXXXXXXXXXXXXXX

    XXXXXXXXXXXXXXXXXXXXXXXXXXXXXXXXXXXXXXXXXXXXXXXXXXXXXXXXXXXXXXXX
XXXXXXXXXXXXXXXXXXXXXXXXXXXXXXXXXXXXXXXXXXXXXXXXXXXXXXXXXXXXXXXXXX
XXXXXXXXXXXXXXXXXXXXXXXXXXXXXXXXXXXXXXXXXXXXXXXXXXXXXXXXXXXXXXXXXXXXXX
XXXXXXXXXXXXXXXXXXXXXXXXXXXXXXXXXXXXXXXXXXXXXXXXXXXXXXXXXXXXXXXXXX
XXXXXXXX

                                 Sincerely,

                                 Justin L. Kaputska

                                 Justin L. Kaputska

Enclosures: 2 Repair Bills

FIGURE A–3   *(continued)*

Most likely, the letters you send at work will be typed on "letterhead" stationery that has your organization's name, address, and phone number already printed on it. The date will be the only part of the heading that you will have to provide. Figure A–4 shows an example.

Inside Address

The inside address gives the name and address of your reader. When typing your reader's name, include the person's title and position. By custom, the titles *Mr., Mrs., Ms.,* and *Dr.* are abbreviated, but other titles are usually written out in full: *Professor, Senator.* The reader's title is usually typed before his or her name, and the reader's position is typed after it. If your reader uses a middle initial, include it:

Reader's title————— Dr. Helen R. Reed, Manager ——————————Reader's position
                    Consumer Products Division

Reader's title————— Professor John O. Jaworski, Chair ————Reader's position
                    Accounting Department

In the two examples above, notice that the reader's department or unit within the overall organization is placed on a line below the person's title, name, and position. Sometimes the reader's position is also placed on a separate line:

Reader's title————— Mr. Harish Batterjhargee
                    Editor ———————————————————— Reader's position
                    Journal of Business Investment

If you are writing to an organization but do not know the name of the appropriate person to contact, you may address your letter to the organization or a specific department in it:

Customer Relations Department

Spindex Corporation

Salutation

The salutation is the letter's equivalent of "hello." By custom, a salutation includes the word *Dear,* usually followed by the reader's title, last name, and a colon:

Dear Mr. Dobcheck:

If you know your reader well enough to use his or her first name in conversation, then in your salutation you may use the reader's first name rather than his or her title and last name. When writers use the reader's first name, they sometimes follow the salutation with a comma rather than a colon. A semicolon is never used.

Dear Leon,

MIAMI UNIVERSITY

Department of English
Bachelor Hall
Oxford, Ohio 45056
513 529-5221

October 12, 19--

Mr. Paul Ring
P. O. Box 143
Watson, Illinois 62473

Dear Mr. Ring:

I am delighted that you wish to learn about Miami University's master's degree program in technical and scientific communication. This professional, practice-oriented program prepares people for careers in which they will help specialists in scientific, technical, and other fields communicate their knowledge in an understandable and useful way. The job market for our graduates is excellent.

People studying with us complete three semesters of course work and an internship. Within this framework, we strive to tailor each student's course of study to his or her particular interests and career objectives. Consequently, our graduates work in a wide variety of jobs and deal with many types of communication. These types include instruction manuals for computers and other high-tech equipment, informational booklets given to cancer patients, technical advertising, corporate procedure books, and technical reports and proposals in many fields, such as chemistry, aerospace engineering, pharmaceuticals, environmental protection, and health care. A special feature of our program is that it prepares people to advance rapidly in the profession--to management, policy-making roles, or ownership of their own communication companies.

We welcome applications from people with undergraduate degrees in many different subjects, including English and the other humanities, communication, natural and social sciences, engineering and other technical fields, art, business, and education. We strive to obtain graduate assistantships or other financial aid for every student accepted into the program who requests it.

I am enclosing a booklet describing our MTSC program in detail. If you wish to learn about similar programs at other schools, you may want to purchase a copy of Academic Programs in Technical Communication, which is sold by the Society for Technical Communication, 815 Fifteenth Street, Northwest, Washington, D. C. 20005.

If you have any questions about the MTSC program, please feel welcome to write, call, or visit.

Sincerely,

C. Gilbert Storms, Director
Master's Degree Program in Technical
and Scientific Communication

Enclosure: Booklet

Excellence is Our Tradition

FIGURE A–4   Letter Prepared on Preprinted Stationery

If you do not know the name of the appropriate person to address, you may use the name of the department or organization:

```
Dear Customer Relations Department:
```

When you do not know the name of the person, you should avoid writing *Dear Sir* or *Gentlemen*. These salutations are now considered old-fashioned and even objectionable because of the assumption they make about the sex of the person who will read and act on the letter.

Subject Line

A subject line is like a title, typically using eight words or fewer. Usually the subject line begins with the word *Subject* or *Re*, followed by a colon (Figure A–1). You may place the subject line above or below the salutation:

```
Subject: Response to Your Letter of March 8, 19--
Re: Continuing Problems with the SXD
```

A subject line can be very helpful to your reader. It not only focuses the reader's attention upon the topic of your message but also helps your reader relocate your letter quickly in his or her files. For that reason, you should write your subject line in the same reader-centered way you compose the rest of your communication. General phrases, such as "Responses to Your Questions," are not nearly as helpful as more precise ones, such as "Near-Term Risks of Investing in Southeast Asia." Be brief but specific.

Although subject lines are very helpful to readers, many letters are sent without them. If you are unsure about whether to include one, consider the custom in your organization and the extent to which a subject line will help the reader of the particular letter you are writing.

Body

The body of a letter contains your message. Except in rare instances, the body is single-spaced with a double-space used between paragraphs. The paragraphs are usually short—ten lines or fewer. However, it is not incorrect to use longer paragraphs if they provide the most effective way to convey your message.

Customarily the body of a letter has a three-part structure consisting of a beginning, middle, and end. In most letters, the beginning is one paragraph long. It announces the writer's reason for writing. In letters between people who have communicated frequently, it may include some personal news. The entire discussion of the writer's topic is usually contained in the middle section. The final paragraph usually includes a social gesture—thanking the other person for writing, expressing a willingness to be of further assistance, or the like.

Complimentary Close

By custom, the complimentary close is the letter-writer's way of saying "goodbye." It begins with one of several familiar phrases, such as *Yours truly, Sincerely,* or *Cordially.* The first letter of the first word is always capitalized, but not the first letters of any other words. The complimentary close ends with a comma.

When choosing the phrase you will use in your complimentary close, consider your personal relationship with your reader. When writing to people you do not know, select one of the more formal phrases, such as *Sincerely* or *Sincerely yours.* When writing to acquaintances, use the more informal phrases, such as *Cordially* or *With best wishes.*

Signature Block

You give your name twice in the signature block, once in handwriting (with a pen) and once typed. When writing as a student, however, you should not give yourself a title.

Together, the complimentary close and signature block look like this:

Complimentary close	Sincerely yours,	Cordially,
Signature	*Raphael Goodman*	*Constance Idanopolis*
Typed name Title	Raphael Goodman Senior Auditor	Constance Idanopolis Head, Sales Division

Special Notations

Following the signature block, you may include five kinds of notations:

● **Identification of typist.** If the letter is typed by someone other than you, the typist may include your initials and his or hers, usually against the left-hand margin. Your initials appear first, in uppercase letters, followed by the typist's initials in lowercase letters. A colon or slash usually appears between the two:

    TLK:smc   TLK/smc

Sometimes, only the typist's initials appear, always in lowercase letters.

● **Identification of word-processing or computer file.** Letters typed on a word processor or computer may include on a separate line the name or number of the file in which the letter is stored (Figure A–1). Such a notation is especially common where printed copies of the letter will be distributed for review before the letter is sent to its intended reader and where the letter might, with some slight modification, also be sent to other readers. The file identification helps the writer or typist find the file again to make the revisions.

● **Enclosure.**   If you are enclosing materials with your letter, you may want to note that. You may also specify just how many items you are enclosing or what the enclosures are:

```
Enclosure

Enclosures (2)

Enclosure: Brochure
```

● **Distribution.**   If you are going to distribute copies of the letter to other people, you may list those people in alphabetical order. The abbreviation *cc,* followed by a colon, appears before the names. The abbreviation *cc* stands for "carbon copy" but is used also when the copies are made on a photocopy machine:

```
cc: T.K. Brandon
    F. Lassiter
    P.B. Waverly
```

Sometimes it may be appropriate to include the titles, positions, and locations of the individuals in the distribution list. If you work in a large organization, some of your readers may not know who the other readers are unless you identify them in this way. Also, because people often change positions and employers, this information will help future readers know what departments were initially given copies of the communication.

```
cc: T. K. Brandon, Vice President for Research
    Dr. F. Lassiter, Manager, Cryogenics Laboratory
    P. B. Waverly, Purchasing Department
```

● **Postscript.**   A postscript is a note added to the end of a letter. Postscripts are often useful for emphasis, as in the following example from a sales letter:

```
P.S. Our service contracts cost 15% less than those
     offered by our competitors.
```

Postscripts are also useful for conveying personal notes in business correspondence addressed to friends:

```
P.S. Perhaps after Thursday's meeting we could spend an
     hour in the Metropolitan Museum.
```

If you find yourself writing a postscript because you forgot to make a key point in the body of your letter, you should probably redraft the letter.

Placement of Text on the Page

The customary margins for page-length letters are 1 to 1½ inches on the top and sides. Typically, the bottom margin is ½ inch longer than the top margin. Many letters are shorter than a full page, however, and you will want to avoid making them appear as if they are unusually high or low on the page. You can usually avoid these difficulties by placing the middle line of your letter a few lines above the middle of your page. Unless you are an experienced typist, you may find yourself typing a letter twice, once to see how many lines long it will be and once to place it pleasingly on the page.

When you write a letter that is longer than one page, you should prepare the second and subsequent pages on plain paper, not on letterhead stationery. At the top of these additional pages, type a heading that includes the name of your reader, the page number, and the date of your letter. Here are two commonly used ways of arranging this heading:

```
Hasim K. Lederer              2              November 16, 19--
```

```
Katherine W. Hodges                                    Page 2
November 16, 19--
```

Continue the text of your letter with a double-space or triple-space below the heading. On additional pages, the heading is always placed ½ to 1 inch from the top of the sheet, no matter how much blank space might appear at the bottom of the page. If little except your complimentary close and signature block will appear on the final page, you might retype the preceding page with large top and bottom margins so that you can carry over more material to the final page.

Format for Envelope

The letters you write at work should be sent in business-sized envelopes (9½ by 4⅛ inches). Figure A–5 shows where to place the reader's address and the return address on the envelope.

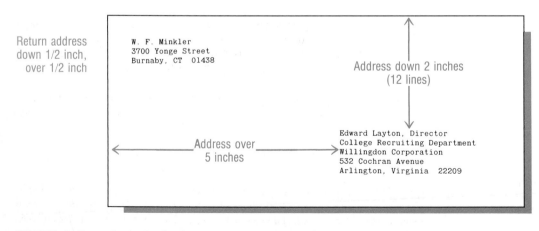

Return address down 1/2 inch, over 1/2 inch

W. F. Minkler
3700 Yonge Street
Burnaby, CT 01438

Address down 2 inches (12 lines)

Address over 5 inches

Edward Layton, Director
College Recruiting Department
Willingdon Corporation
532 Cochran Avenue
Arlington, Virginia 22209

FIGURE A–5   Format for Envelope

Conventions about Style

You can use a letter to communicate any type of message. This means, also, that you can use a letter to communicate in any writing style that is appropriate at work. Nevertheless, there are some conventions about the style for letters that are widely—though not universally—observed.

Of the three formats discussed in this chapter, the letter usually uses the most gracious style. The polite phrasing of the salutation *(Dear . . .)* and the complimentary close *(Sincerely yours)* are among the signs that people associate the letter with a polite conversation. In a similar fashion, the opening sentence of a letter often contains the words *you* and *I*.

This style of direct, well-mannered exchange is often carried into the body of a letter, where writers often use a conversational style. They continue to use *you* and *I*, employ relatively short sentences, and generally write as if they were talking in person with the reader. In a similar fashion, the final paragraph usually refers to the relationship between the writer and reader, for instance by expressing the writer's gratitude or eagerness to be of further assistance to the reader.

Of course, some conversations are much more formal than others—and so are some letters. Among the most formal are official letters that communicate policy or that serve as legal rulings or contracts. In the end, you have to decide upon the style of your letter in light of your purpose and reader.

USING THE MEMO FORMAT

Memos are communications written on preprinted forms like the one shown in Figure A–6. They are most often addressed to the writer's co-workers, managers, and other readers inside the writer's own organization.

By custom, memos may be typed or handwritten, with handwriting usually reserved for brief messages that will not be preserved in files for future reference.

Heading

The memo's distinguishing characteristic is its heading, which has preprinted spots for you to enter your name, your reader's name, and the date. In some organizations it is customary to include the writer's and reader's positions and departments. If you do this, place them on the same lines as the names, or on the next lines. Do not include addresses.

With the increasing use of word-processing equipment that can store format material, more and more writers are using word processors to create memo forms as they write. Figure A–7 shows such a form.

Subject Line

As in a letter, you can use a subject line to indicate briefly the topic of your communication. Subject lines are much more common in memos than in letters. In fact, many organizations include a preprinted spot for a subject line as part of the headings on their memo forms.

Rentscheller Company
Interoffice Memorandum

FROM: Natalie Sebastian

TO: Dwight Levy

DATE: June 16, 19--

SUBJECT: Questions about Peptides

   XXXXXXXXXXXXXXXXXXXXXXXXXXXXXXXXXXXXXXXXXXXXXXXXXXXXXXXXXXXXXXXXXXXXXXX
XXXXXXXXXXXXXXXXXXXXXXXXXXXXXXXXXXXXXXXXXXXXXXXXXXXXXXXXXXXXXXXXXXXXXXXXX
XXXXXXXXXXXXXXXXXXXXXXXXXXXXXXXXXXXXXXXXXXXXXXXX

  1. XXXXXXXXXXXXXXXXXXXXXXXXXXXXXXXXXXXXXXXXXXXXXXXXXXXXXXXXXXXXXX
     XXXXXXXXXXXXXXXXXXXXXXXXXXXXXXXXXXXXXXXXXXXXXXXXXXXXXXXXXXXXXX
     XXXXXXXXXXXXXXXXXXXXXXXXXXXXXXXXXXXXXXXXXXXXXXXXXXXXXXXXXXXXXX
     XXXXXXXXXXXX

  2. XXXXXXXXXXXXXXXXXXXXXXXXXXXXXXXXXXXXXXXXXXXXXXXXXXXX

  3. XXXXXXXXXXXXXXXXXXXXXXXXXXXXXXXXXXXXXXXXXXXXXXXXXXXXXXXX
     XXXXXXXXXXXXXXXXXXXXXXXXXXXXXXXXXXXXXXXXXXXXXXXXXXXXXXXX
     XXXXXXXXXXXXXXXXXXXXXXXXXXXXXXXXXXXXXXXXXXXXXXXXXXXXXXXX
     XXXXXXXXXXXXXXXXXXXXXXXXXXXXXXXXXXXXXXXXXXXXXXXXXXXX

   XXXXXXXXXXXXXXXXXXXXXXXXXXXXXXXXXXXXXXXXXXXXXXXXXXXXXXXXXXXXXXXXXXXXXXX
XXXXXXXXXXXXXXXXXXXXXXXXXXXXXXXXXXXXXXXXXXXXXXXXXXXX

cc: Marty Gonzales

MEMO397

FIGURE A–6   Memo Prepared on Preprinted Stationery

INTEROFFICE MEMORANDUM
Mathematics Department

                                                  March 6, 19--

FROM: Trina May, Chair, Committee on Undergraduate Excellence

TO: Aldon Butler, English Department

RE: MEETING ON MARCH 18

     Thank you for agreeing to talk with the Committee on
Undergraduate Excellence on March 18. I have put you third on
the agenda for that day's meeting.

     After reviewing the various topics you mentioned in your
recent memo, the committee has decided that it would like you
to address the following issues:

     1. Ways of giving recognition to excellent student
        accomplishments in all aspects of university life.

     2. Strategies for encouraging students to develop their
        academic strengths through activities not tied to their
        classes.

     The committee will meet at 3 p.m. in Room 112 of Upham Hall.
We all look forward to hearing your remarks.

     192A

FIGURE A–7   Memo Format Typed along with the Message

Body

The body of a memo closely resembles the body of a letter. The beginning, usually brief, states the purpose of the communication, and the middle presents the gist of the writer's message.

With respect to the ending of memos, practice varies from organization to organization. In many, writers include final paragraphs just like those found at the end of most letters—ones that express thanks, indicate an eagerness to work further with the reader, or the like. In many other organizations, those paragraphs are omitted on the grounds that they are redundant: everyone knows that fellow employees are willing to help one another and that they are grateful for the assistance they receive.

Signature

There are several conventional ways of signing memos. In some organizations, writers sign only their initials, while in others they sign their full names. In some organizations they sign at the bottom of the memo, and in others they sign by the line in the heading that gives their name. In memos, writers rarely use the kind of complimentary close and signature block that are used in letters.

Special Notations

Memos can include any of the five kinds of notations that letters sometimes have: identification of the typist, identification of the word-processing or computer file, enclosure list, distribution list, and postscript. See the discussion of these notations in the preceding section on letters.

Placement of Text on the Page

The body of a memo begins two or three lines below the heading, regardless of the amount of blank space that this might leave at the bottom of the page.

When a memo extends beyond the first page, the additional pages use the same sort of heading that is used for the second and subsequent pages of letters. See the description of these headings in the preceding section on letters.

Conventions about Style

The memo format is designed to be efficient. The writer can write quickly, without needing to take time to spell out full addresses or even the words *To* and *From*. The writer usually does not write out a complimentary close and may sign with initials instead of writing out his or her whole name.

This same concern with efficiency sometimes carries over into the writing style of the memo. Whereas letters are often (not always) gracious, memos are often (not always) brisk. They often begin and end without the social amenities found in the beginnings and endings of many letters. Throughout, they are often crisp, though not to the point of intentional rudeness or lack of clarity.

You must determine the best style for each memo you write in light of your purpose and your readers. Although memos are well suited to situations in which you want to communicate with your co-workers with as little fuss as possible, you can also use memos in situations where more sociability is desirable.

USING THE BOOK FORMAT

The last of the three formats most often used at work is the book format. Communications in this format resemble textbooks, novels, and similar publications. Generally, they have a strong binding, a protective and informative cover, a table of contents, and separate chapters or sections for each of the major blocks of information.

However, the book format differs from novels and textbooks in several significant ways. Often, communications prepared in this format at work are very brief—ten pages or fewer—although some are hundreds of pages long. At work, communications prepared in the book format are often typewritten on 8½-by-11-inch paper. Sometimes, the text is printed on only one side of the sheets. The binding might be a set of staples or a plastic spine that grips perforations along one edge of the pages. Furthermore, the book format has several special elements such as summaries and appendixes that are not usually found in novels and textbooks. So, although this format used at work resembles the format we usually associate with books, it differs from that format in significant ways also.

In fact, the term *book format* is a special one devised for the textbook you are now reading. In the workplace, there is no single term for this format. When reports and proposals are prepared in this format, they are often called *formal* reports and *formal* proposals to distinguish them from reports and proposals presented in letters and memos. When instructions are presented in this format, they are usually called *instruction manuals,* to distinguish them from instructions presented in letters, memos, or single-page instruction sheets.

Why, then, should the book format be given a special name in the textbook you are now reading? Giving the format its own name enables you to focus your attention on it, to study it separately from the superstructures for reports, proposals, and instructions that are discussed in Chapters 20 through 25. Separate study has two advantages.

First, it helps you learn efficiently. By studying the book format as a separate subject, you will learn the basic principles that apply to its use with all kinds of messages. You won't have to study the format once to learn about formal reports, again to learn about formal proposals, and a third time to learn about instruction manuals.

Second, by studying formats separately from types of messages, you will learn to distinguish your decisions about the format you are using (for example, letter, memo, book) from your decisions about the conventional patterns for organizing the type of messages you are writing (for example, report, proposal, instructions). As a result, you will become a more creative and resourceful writer, able to see the full range of ways you can combine and adapt conventional formats with conventional superstructures to achieve your purposes.

Overall, the various elements of the book format fall into three groups: front matter, body, and supplementary material. In the following discussion, you will find detailed information about each of those parts, followed by advice about typing, page numbering, and style. Finally, you will learn how to write a letter of transmittal, which often accompanies documents using the book format.

Front Matter

Front matter consists of all the materials that precede the body of the communication: cover, title page, summary, table of contents, and list of figures and tables.

Cover

Figures A–8 and A–9 show typical covers, one for a report and one for an instruction manual. Covers are usually printed on heavy but flexible paper, usually called *cover stock*.

On a cover you should give your communication's title, along with other information that will help people who file and later want to find your communication. For instance, if you have written a report for readers in some other organization, you might give both your organization's name (the name under which your readers might file it) and the name of your readers' organization (the name under which your organization might file it.)

You can prepare your cover in a variety of ways. One simple method is to make a master copy on regular typing paper, then print the material onto the heavier cover paper using a photocopy machine. To make words in large type, you can use press-on letters or a Kroy lettering machine (both available in many copy centers). Some companies have covers typeset by a printer, in which case you will simply have to tell the printer what you want the cover to say. If you are writing on a word processor, you may be able to use it to make larger letters.

Title Page

The title page of a formal report contains all of the information on the cover. You may also add the following items:

- Names of the authors and other contributors.
- Addresses of the authors and of the organization for which they work.
- Cross-referencing material. In a report, for example, you might include the name of the contract under which the work was done (for example, Contract Number WEI-377-B), or the titles of related reports. In proposals, you might tell the title or number of the document issued by your readers when they asked you to submit your ideas to them. In an instruction manual, you might give the model numbers of the equipment described.
- File number. Some organizations assign numbers to the reports, manuals, and other communications they produce.
- Copyright notice. If your employer claims a copyright on the material in your communication, you should include a copyright notice on the title page. Such notices include the word *Copyright* or the symbol ©, followed by the date of the copyright and the name of the organization.
- Notice of restrictions on distribution or photocopying. You would include such a notice, for instance, when writing a communication that contains information your employer does not want its competitors to obtain.

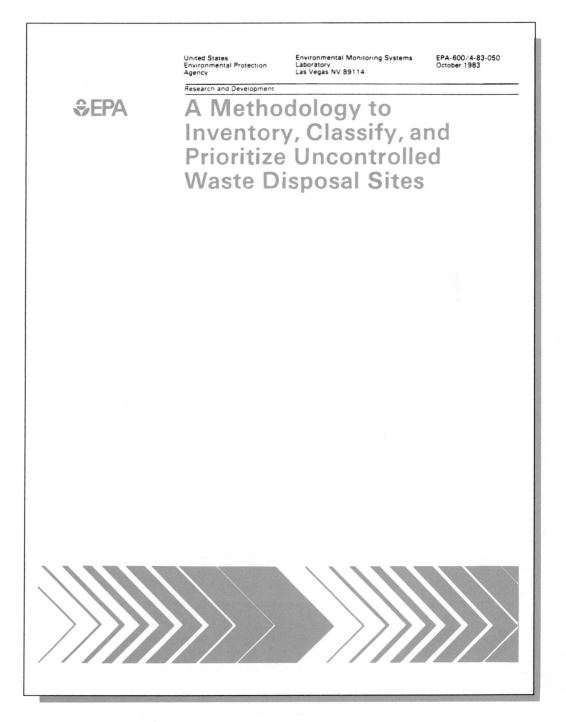

FIGURE A–8   Cover of a Report

DETROIT DIESEL
Series 53 Service Manual
Sections 1-3

FIGURE A–9   Cover of an Instruction Manual
Courtesy of Detroit Diesel Allison Division, General Motors Corporation.

Figure A–10 shows the title page from the report whose cover is shown in Figure A–8. This report was prepared by two researchers working under a contract with the U.S. Environmental Protection Agency (EPA). As you can see, in addition to the information shown on the cover, the title page gives the names and addresses of the authors, the name and address of the person in the EPA who was responsible for overseeing the report, and the contract number under which the authors prepared the report.

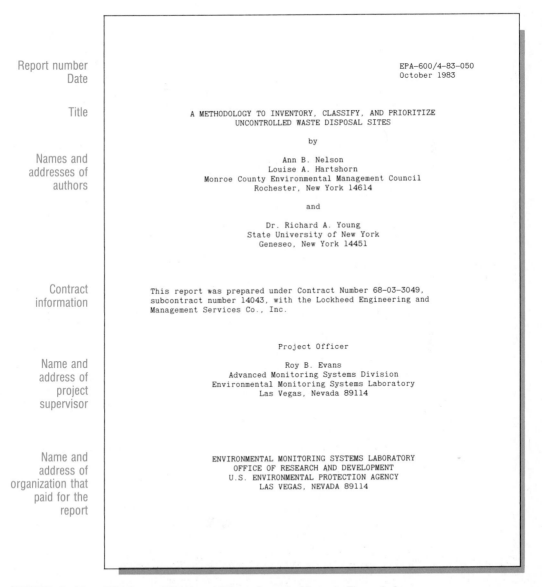

Report number
Date

Title

Names and
addresses of
authors

Contract
information

Name and
address of
project
supervisor

Name and
address of
organization that
paid for the
report

EPA–600/4–83–050
October 1983

A METHODOLOGY TO INVENTORY, CLASSIFY, AND PRIORITIZE
UNCONTROLLED WASTE DISPOSAL SITES

by

Ann B. Nelson
Louise A. Hartshorn
Monroe County Environmental Management Council
Rochester, New York 14614

and

Dr. Richard A. Young
State University of New York
Geneseo, New York 14451

This report was prepared under Contract Number 68–03–3049,
subcontract number 14043, with the Lockheed Engineering and
Management Services Co., Inc.

Project Officer

Roy B. Evans
Advanced Monitoring Systems Division
Environmental Monitoring Systems Laboratory
Las Vegas, Nevada 89114

ENVIRONMENTAL MONITORING SYSTEMS LABORATORY
OFFICE OF RESEARCH AND DEVELOPMENT
U.S. ENVIRONMENTAL PROTECTION AGENCY
LAS VEGAS, NEVADA 89114

FIGURE A–10   Title Page of the Report Whose Cover Is Shown in Figure A–8

Figure A–11 shows another title page, this one for the instruction manual whose cover is shown in Figure A–9. In this case, the title page adds the following information: the file number of the report, the date of the most recent revision, the address of the manufacturer, a note telling where readers can obtain additional copies, and a copyright notice.

Executive Summary

Summaries convey the essence of a communication in a very brief space, usually one page or less. Almost all reports and proposals written in the book format begin with them. Those that don't are usually less effective as a result.

How Summaries Help Readers   By providing a summary, you help your readers in one or more of the following ways. First, your summary enables busy people (such as decision-makers) to learn the *essential contents* of your communication without reading more than a page or so. Having learned the essential message of the report, readers can then read selectively through the report for more information on the specific topics that are important to them.

Second, your summary provides your readers with a *preview* of the main points of your communication, enabling them to build a mental framework for organizing and understanding the detailed information you provide in the body of your communication. Such a preview is especially useful to those readers who will read sequentially through the report from beginning to end.

Third, your summary can help people determine the *key results* reported in your communication. This particular use of summaries occurs most often among scientists, engineers, and other specialists who want to determine what others have found when addressing problems like the ones that they now face. Accordingly, they go to corporate, university, and public libraries where research reports relevant to their field are stored. If you write reports that will be placed in a library, your summary will probably be collected with other summaries in card files, bound volumes, computerized data bases, or other research aids. By looking through these resources, researchers can determine quickly which reports have information they can use.

Features of a Good Summary   What makes a good summary? No single pattern or formula is always appropriate. You must shape each one according to your purpose and readers, asking yourself, "If I were allowed only 200 words (for example) to affect my readers in the way I desire, what would I say?"

Your challenge is particularly great because, by convention, summaries must be entirely self-contained. This means, for instance, that you must not only list your recommendations, but also introduce the problem, tell how you conducted your investigation, describe the significant results, and answer such questions about your recommendations as, How much will they cost? What will be the benefit? How will the work be managed? and, What will the schedule be? These sample questions, as you can tell, concern the summary for a report addressed primarily to a decision-maker. In a report addressed to other researchers or designers, the emphasis would be very different. The summary would still have to introduce the problem, but would emphasize the methods used, the key results obtained, and the implications for future work.

Because summaries are self-contained, you must also present your material very

Title

DETROIT DIESEL
Series 53 Service Manual
Sections 1-3

Name and
address of
the company

DETROIT DIESEL
CORPORATION

13400 Outer Drive, West / Detroit, Michigan 48239-4001
Telephone: 313-592-5000
Telex: 4320091 / TWX: 810-221-1649
FAX: 313-592-7288

Copyright
notices
File number
of manual

Detroit Diesel® and the spinning arrows are registered trademarks of Detroit Diesel Corporation
6SE202 ©Copyright 1990   Detroit Diesel Corporation   Litho in USA

FIGURE A–11   Title Page of the Instruction Manual Whose Cover Is Shown in Figure A–9.
Courtesy of Detroit Diesel Allison Division, General Motors Corporation.

clearly and precisely, so that your reader will not have to look at the body of your report to see what you mean.

Sample Summaries   Figure A–12 shows the summary (called an *abstract* in this case) from the EPA report mentioned above. This summary identifies the contents of the report for readers who might be looking for the kind of tools that the report describes for identifying and evaluating waste disposal sites. It stresses the capabilities of the methodology described in the report.

Figure A–13 shows the type of summary often written for reports addressed to decision-makers. It focuses on the practical results of the writer's study and on the action recommended by the writer.

Figure A–14 shows the kind of summary often written for other researchers. It focuses on the method of the research and on the new knowledge generated by it.

The variety of these summaries should help you appreciate the extent to which you must adapt each of the summaries you write to your particular purpose and readers. At the same time, all the summaries shown share some features. All provide background information about the purpose of the work undertaken, all describe in specific terms the nature of that work, and all tell the results obtained.

When to Write Your Summary   Generally, you will be able to write your summaries most easily if you prepare them after you have completed a full draft of your communication—perhaps even after you have completed the *final* draft of it. This is because the summary should only make points made in the report itself. Even though it is placed at the beginning of a communication, it is not an introduction but a condensation of the body of the communication. Only after you have worked out fully the contents of the body are you ready to write the summary.

How Long Should Your Summary Be?   As you sit down to write a summary, you may find yourself wondering how long you should make it. In many situations, someone (like your boss) will set your limit by specifying the maximum number of words or the maximum amount of space it can take. If you have no such guidance, you might follow this rule of thumb: make your summary roughly 5 to 10 percent of the length of your overall communication. Thus, if you are writing a 20-page communication, make the summary between 1 and 2 pages long. However, summaries are rarely longer than 5 pages, regardless of the length of the report.

Table of Contents

A table of contents serves two purposes. First, it helps readers who do not want to read the whole communication in sequence but rather want to find particular parts of it—the description of the research method, the proposed schedule, the budget, and so on. To these readers, the table of contents is like an expressway through your document that leads them directly to their destination.

A table of contents also helps readers who will read sequentially through your text. This is because, like a summary, a table of contents enables readers to see very quickly the general scope and arrangement of the material you cover, so they can begin to build the mental framework in which to organize the various pieces of information they will gain from the communication. Because such frameworks are so helpful to readers, many people read the table of contents before they look at anything else in a communication.

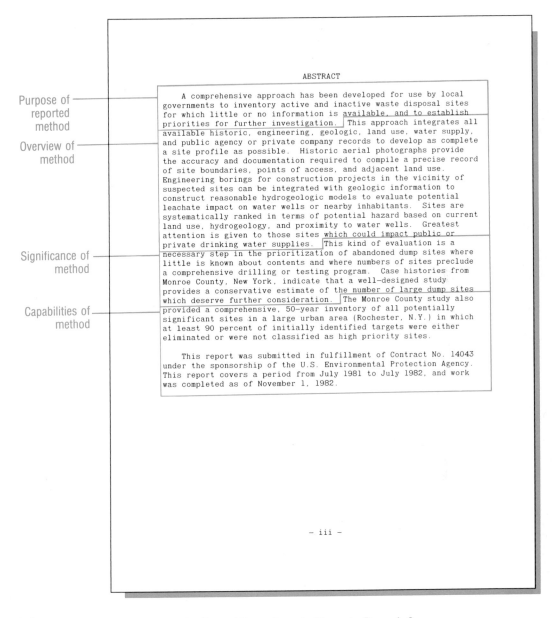

Purpose of reported method

Overview of method

Significance of method

Capabilities of method

ABSTRACT

A comprehensive approach has been developed for use by local governments to inventory active and inactive waste disposal sites for which little or no information is available, and to establish priorities for further investigation. This approach integrates all available historic, engineering, geologic, land use, water supply, and public agency or private company records to develop as complete a site profile as possible. Historic aerial photographs provide the accuracy and documentation required to compile a precise record of site boundaries, points of access, and adjacent land use. Engineering borings for construction projects in the vicinity of suspected sites can be integrated with geologic information to construct reasonable hydrogeologic models to evaluate potential leachate impact on water wells or nearby inhabitants. Sites are systematically ranked in terms of potential hazard based on current land use, hydrogeology, and proximity to water wells. Greatest attention is given to those sites which could impact public or private drinking water supplies. This kind of evaluation is a necessary step in the prioritization of abandoned dump sites where little is known about contents and where numbers of sites preclude a comprehensive drilling or testing program. Case histories from Monroe County, New York, indicate that a well-designed study provides a conservative estimate of the number of large dump sites which deserve further consideration. The Monroe County study also provided a comprehensive, 50-year inventory of all potentially significant sites in a large urban area (Rochester, N.Y.) in which at least 90 percent of initially identified targets were either eliminated or were not classified as high priority sites.

This report was submitted in fulfillment of Contract No. 14043 under the sponsorship of the U.S. Environmental Protection Agency. This report covers a period from July 1981 to July 1982, and work was completed as of November 1, 1982.

– iii –

FIGURE A–12   Summary from the Report Whose Cover Is Shown in Figure A–8

When writing your table of contents, think of it as an outline of your communication. Construct it from headings you have included in the text. However, you don't need to include *all* levels of headings. In deciding which headings to include, think of your readers. If your table of contents becomes too long (over a page or two), it may be difficult for your readers to find the organizational pattern in it. On the other hand,

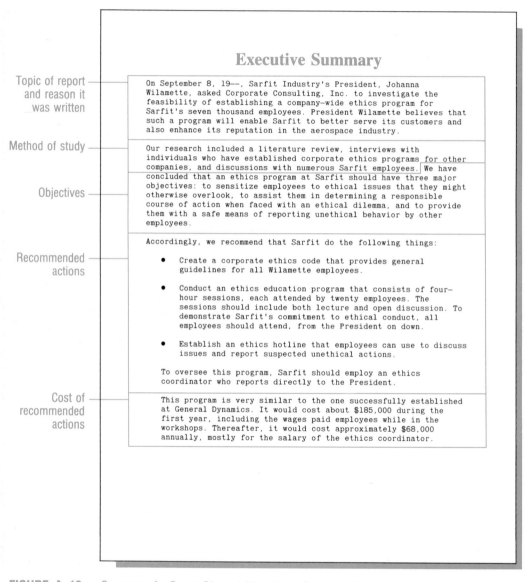

Executive Summary

Topic of report and reason it was written

On September 8, 19--, Sarfit Industry's President, Johanna Wilamette, asked Corporate Consulting, Inc. to investigate the feasibility of establishing a company-wide ethics program for Sarfit's seven thousand employees. President Wilamette believes that such a program will enable Sarfit to better serve its customers and also enhance its reputation in the aerospace industry.

Method of study

Our research included a literature review, interviews with individuals who have established corporate ethics programs for other companies, and discussions with numerous Sarfit employees. We have concluded that an ethics program at Sarfit should have three major

Objectives

objectives: to sensitize employees to ethical issues that they might otherwise overlook, to assist them in determining a responsible course of action when faced with an ethical dilemma, and to provide them with a safe means of reporting unethical behavior by other employees.

Accordingly, we recommend that Sarfit do the following things:

Recommended actions

- Create a corporate ethics code that provides general guidelines for all Wilamette employees.

- Conduct an ethics education program that consists of four-hour sessions, each attended by twenty employees. The sessions should include both lecture and open discussion. To demonstrate Sarfit's commitment to ethical conduct, all employees should attend, from the President on down.

- Establish an ethics hotline that employees can use to discuss issues and report suspected unethical actions.

To oversee this program, Sarfit should employ an ethics coordinator who reports directly to the President.

Cost of recommended actions

This program is very similar to the one successfully established at General Dynamics. It would cost about $185,000 during the first year, including the wages paid employees while in the workshops. Thereafter, it would cost approximately $68,000 annually, mostly for the salary of the ethics coordinator.

FIGURE A–13   Summary of a Report Directed Primarily to Decision-Makers

you need to include enough information to allow your readers to create a general picture of the organization and also to find the specific pieces of information they are likely to seek.

One way to determine which levels of headings to include in a table of contents is to ask which sections might be sought by readers who want to find only *one* part of the communication. The sections you identify in this way should be included in the table

Description of
study: focus on
method

Summary

Canadian respondents from the 1981 Edmonton Area
Study (mean age 35 years) and the 1981 Winnipeg Area
Study (mean age 43 years) completed indexes of loneliness
(e.g., UCLA Loneliness Scale), satisfaction, and fear of crime.
The LISREL program was used to construct a model
explaining the relationship between these endogenous
variables and age, gender, and the number of people living
in the respondent's household.

New knowledge
produced by
study

Findings indicate that males and females were so
different in their responses that separate models had to be
built. In the male model, number of people in the household
contributed to loneliness, and age had a direct effect on fear.
In the female model, age had a significant effect on
loneliness, and loneliness determined satisfaction. Younger
females expressed greater feelings of loneliness than older
females. For females, it was living alone rather than being
lonely that created fear. For males, living alone was related
to loneliness, but loneliness did not necessarily lead to fear.

FIGURE A–14   Summary from a Report Directed Primarily to Other Researchers
From Robert A. Silverman and Leslie W. Kennedy. "Loneliness, Satisfaction and Fear of Crime: A
Test for Non-recursive Effects," *Canadian Journal of Criminology* 27 (1985): 1–13.

of contents, along with all the other elements at the same and higher levels within the
organizational hierarchy. Also, you should keep in mind that readers almost always find
it useful to have more than one level of organization included in the table of contents.

Tables of contents can be just as useful to the readers of short communications as
they are to the readers of long ones. Although some writers don't consider including

tables of contents in communications shorter than 20 or 30 pages, readers often welcome them in communications as brief as 10 pages.

Figure A–15 shows a sample table of contents, taken from the report mentioned earlier. Notice how the writers have used blank lines to help indicate the organization of the communication: two blank lines set off the front matter from the body of the report, and other blank lines set off other major parts. If the writers had included information about the major sections of each chapter, they might also have placed blank lines between the chapters. You will find another table of contents in Figure 7–13, this one set by a printer.

List of Figures and Tables

When your readers are looking for some part of your communication, they may be looking not for a certain paragraph, but for a particular table, drawing, or other graphic aid. You can help them by including in the front matter a list of figures and tables. In some communications, it might even be helpful to list tables separately from figures. By custom, the list of figures and tables (or the separate lists) follows the table of contents. Figure A–16 shows the list of figures from the EPA report mentioned above. That report also contains a list of tables, which has the same format.

Body

The body of your communication is the heart of your treatment of your subject. In communications written in the book format, the body consists of the communication's chapters (often called sections).

What should these chapters be about and how should they be developed? The answer depends entirely on your purpose and readers. For advice about how to organize the body, see Chapter 4, "Planning to Meet Your Readers' Informational Needs," and Chapter 5, "Planning Your Persuasive Strategies." Based on the advice given there and on your study of your particular purpose and readers, you might organize the body using one of the conventional superstructures described in Chapters 20 through 25. Or, you might find that you are best able to achieve your objectives by devising your own, original structure.

Writing the Introduction

Despite the many differences among communications written in the book format, all share one structural feature: an introduction. This introduction may appear under many titles: "Introduction," "Problem," "Need," "Background," and so on. The contents of introductions vary widely, depending on the writer's purpose and readers. For general guidelines about writing introductions, see Chapter 9 on "Beginning a Communication." For suggestions about the contents of introductions for the specific kind of communication you are writing, see Chapters 20 through 25 on conventional superstructures.

As you write an introduction, you may wonder whether to repeat the introductory material that will go into your summary. The answer is *yes*. It's true that most readers will read your summary immediately before reading the introduction. Nevertheless, the universal custom is to write the introduction as if the readers were beginning there—

TABLE OF CONTENTS

–v–

FIGURE A–15   Table of Contents of the Report Whose Cover Is Shown in Figure A–8

FIGURE A–16 List of Figures from the Report Whose Cover Is Shown in Figure A–8

even if it means that the first sentence of the summary and the first sentence of the introduction are exactly the same. To put it another way (and to repeat an observation made above in the discussion of summaries): a summary should not provide any information—including introductory information—that is not also provided in the body of the communication.

Writing the Conclusion

Because you have just read that all communications written in the book format begin with an introduction, you may wonder whether they all also end with a conclusion. Many do, but not all. Instruction manuals, for instance, often end after conveying the last bit of information the reader needs to operate the equipment or perform the task being described. For general advice about how to conclude the communications you write in the book format, see Chapter 10 on "Ending a Communication."

Writing the Chapters

No matter how you structure the body of a communication you are writing in the book format, you should begin each of its chapters on a new page and give each its own chapter number. Usually, arabic numbers (1, 2, . . .) are used for chapters, although roman numbers are sometimes employed. Figure A–17 shows the first page of a chapter in the sample EPA report, and Figure A–18 shows the first page from a chapter in the instruction manual whose cover is shown in Figure A–9. The chapters in this manual (which are called sections) are so long that each has its own table of contents.

Most likely your chapters will vary considerably in length. For instance, if you look again at the sample table of contents shown in Figure A–15, you will see that the "Introduction" to that report is four pages long, the chapter on "Application of Methodology to Rank Sites" is eight pages long, and the chapter on "Hydrogeologic Hazard Analysis" is seventeen pages long.

If you find that chapters in your communications are similarly varied in length, do not worry—even if one or more of your chapters is less than a page long. Remember that the chapters are supposed to help the reader find information and understand the structure of your communication. The chapters should reflect the communication's logic, even if this means that they do not divide it into approximately equal parts.

Supplementary Elements

Almost every communication written in the book format contains all of the elements you have read about so far—cover, title page, summary, table of contents, list of figures and tables, and body. The major exceptions are that shorter communications often omit the list of figures and tables, and that instruction manuals do not include summaries.

Many communications in book format also contain one or more of the following supplementary elements: appendixes; list of references, endnotes, or bibliography; glossary or list of symbols; index; and letter of transmittal.

Appendixes

Appendixes can help you overcome one of the more common problems that vex people who are writing long communications. Sometimes you may find that you want to make

Chapter 1

INTRODUCTION

In the late 1970's the public became concerned about uncontrolled waste disposal sites that could pose a hazard to human health. In order to determine the location and impact of these sites, accurate information is needed on site locations, boundaries, contents, subsurface hydrogeologic conditions, and proximal land uses. Documentation of past waste disposal activities is, at best, incomplete and, in many instances, nonexistent. An accurate and inexpensive method is needed to develop site information based on existing data so that expensive drilling and testing programs can be focused on those sites of greatest potential hazard to human health.

This report describes a comprehensive approach that can be used by local governments, particularly counties and large municipalities, to inventory active and inactive sites for which little or no information is available, and to establish priorities for further investigation. The methods were designed by agencies in Monroe County, New York, in response to a 1978 county legislature request to locate hazardous sites and a 1979 New York State law requiring counties to identify suspected inactive hazardous waste sites and to report their locations to the New York State Department of Environmental Conservation (DEC). The study has at various times been financially supported by the County of Monroe, the State of New York, and the United States Environmental Protection Agency. This broad base of support illustrates the concern felt at all levels of government that uncontrolled hazardous waste sites be identified and their impacts accurately assessed so that potential health hazards can be identified and corrected. It also reflects the fact that limited available resources must be committed to cleaning up the worst sites.

The study was conducted under the direction of the Monroe County Landfill Review Committee (LRC). This committee, chaired by the County Director of Health, includes representatives from the county departments of Health, Planning, and the Environmental Management Council (EMC); The New York State Departments of Health and Environmental Conservation; The City of Rochester; and the local Industrial Management Council. The participation of individuals from this broad range of interests facilitated access to information and

— 1 —

FIGURE A–17   First Page of a Chapter in the Report Whose Cover Is Shown in Figure A–8

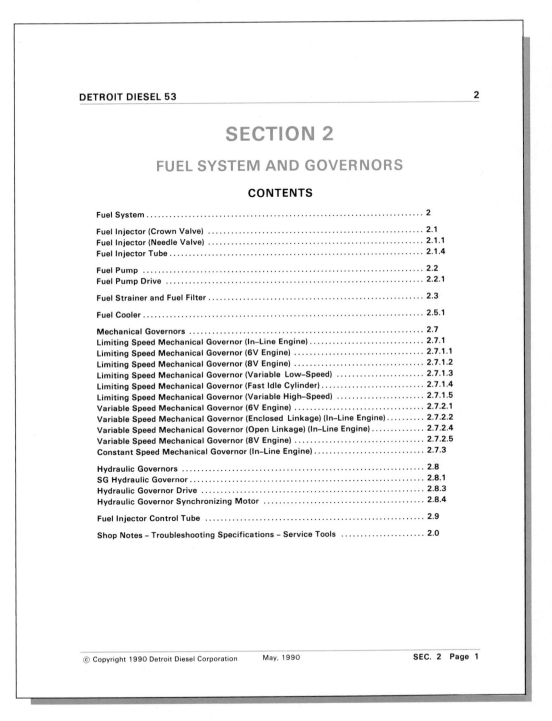

DETROIT DIESEL 53   2

SECTION 2

FUEL SYSTEM AND GOVERNORS

CONTENTS

© Copyright 1990 Detroit Diesel Corporation   May, 1990   **SEC. 2   Page 1**

FIGURE A–18   First Page of a Chapter in the Instruction Manual Whose Cover Is Shown in Figure A–9
Courtesy of Detroit Diesel Allison Division, General Motors Corporation.

certain information available to your reader in order to achieve your purpose, but that if you include the information in the body of your communication you will *reduce* your communication's effectiveness.

How can this happen? The following paragraphs describe three typical situations in which it might.

1. **When you must include detailed information that would otherwise interfere with your general message.** Imagine, for instance, that you are preparing a report on a research project. You realize that you need to include a two-page account of some calculations that you used to analyze your data. Some of your readers might want to check those calculations, and others might want to use them in another experiment. At the same time, you realize that your readers will have trouble following the general thread of your overall research strategy if they are diverted into studying the details of your calculations.

    You can solve this problem by placing the calculations in an appendix, where they are available but won't interfere with your presentation of the body of the report.

    In addition to lengthy calculations, other kinds of detailed information that are often placed in appendixes include the following:

    ● Detailed data obtained in a research study.
    ● Detailed drawings and illustrations.
    ● Lengthy tables of values that a reader would refer to while operating some equipment or performing some procedure.
    ● Detailed descriptions of the professional qualifications of the people who will work on a proposed project.

2. **When your readers are unlikely to read the body of your communication if it exceeds a certain length.** Sometimes, you may find yourself addressing a group of readers that includes some key readers who will not read the body of a communication if it appears to be too long. In fact, you may even find situations in which your readers state a limit to the length of the body of your communication.

    In situations such as this, you can create a communication of the appropriate length by saying as much as you can in the number of pages you think your primary readers will read—and then including everything else of importance in appendixes. The result might be, for example, a report that has a one-page summary, a fifteen-page body, and sixty (or more) pages of appendixes. To signal your readers that the body of your communication is relatively short, you can print the appendixes on paper of a different color, so that your readers will know at a glance how much of the thick communication is devoted to its body.

3. **When you must write one communication that will meet the diverse needs of different readers.** For instance, when writing a proposal that will be read by both a decision-maker and a technical adviser, you may find that some information that will be largely irrelevant to the decision-maker may be essential to the technical adviser. To use appendixes effectively in this situation, you first decide who your primary reader will be. Then, place material directed mainly at your primary reader in the body, and place material mainly for other readers in appendixes.

When planning appendixes, there is one other important consideration. Some writers find it tempting to use appendixes to present all kinds of material that *no* reader is likely to need. Avoid using appendixes to collect material that doesn't belong *anywhere* in your communication.

If you decide to use one or more appendixes, be sure to tell your readers that they exist. List them in your table of contents (see Figure A–15). Give each appendix an informative title that, when listed in the table of contents, indicates clearly what the appendix contains. Also, mention each appendix in the body of your report at the point where your readers might want to refer to it:

```
Printouts from the electrocardiogram appear in
Appendix II.

Appendix A contains the names and addresses of
companies in our area that provide the type of
service contract that we recommend.
```

In the book format, each appendix begins on its own page, just as each chapter does. It is customary to arrange and label the appendixes in the same order in which they are mentioned in the body of the communication. If you have only one appendix, it is sufficient to label it "Appendix." If you have more than one appendix, you may use roman numerals, arabic numerals, or capital letters to label them:

```
Appendix I, Appendix II . . .
Appendix 1, Appendix 2 . . .
Appendix A, Appendix B . . .
```

List of References, Endnotes, or Bibliography

Whenever you are writing at work, you may want to direct your readers' attention to other sources of information on your subject. Perhaps you want to do so to acknowledge the sources of your information or to help your readers learn more about your topic. In the shorter formats (letters and memos), these references are often worked right into the body of the communication. In the book format, however, they are often gathered in one place. Depending on the format being used, this place might be a list of references, an endnotes section, or a bibliography.

In the book format, reference lists and bibliographies usually follow the body of the communication, or they go immediately after the appendixes. Sometimes, however, a separate list or bibliography goes at the end of each chapter.

To learn what sources to mention and how to construct reference lists, endnotes, and bibliographies, see Appendix B.

Glossary or List of Symbols

When you are writing at work you may sometimes use specialized terms or symbols that are not familiar to some of your readers. Sometimes, you will want to explain these terms and symbols in the body of your communication. At other times you may want to present your explanations in a separate glossary or list of symbols.

How can you decide which strategy to use? Think about your readers in the act of reading. If you are using a special term or symbol only once or in only one small part of your communication, your readers probably have no use for a glossary or list of symbols. If you include your explanation in the text, they will be able to remember it or relocate it quickly when the same term or symbol appears a few sentences later. For that reason, glossaries and lists of symbols are rarely used with short communications such as letters and memos.

On the other hand, if you are going to use the same term or symbol throughout a long communication, you may be able to help your readers greatly by including a glossary or list of symbols. Imagine, for instance, that you are writing an instruction manual in which special terms appear on pages 3, 39, and 72, and twelve other places. You might define the term the first time that it appears, in the hope that your readers will remember the definition when they encounter it again. However the readers may not be able to remember. Or they may go directly to one of the other pages without reading page 3 because the particular information they want is located elsewhere in your manual.

In either case, your readers will have to scan through pages of your text to find the definition you provided earlier. You could save your readers this search by defining the term in every place it appears, but such repetition could annoy readers once they have learned the term. By defining the term in a glossary, you provide the definition in a place that is easy to find but that keeps the definition out of the readers' way if they do not need it.

Because of the way they make explanations available without requiring readers to read them, glossaries and lists of symbols are also useful in communications where some of your readers need explanations and some don't.

Figure A–19 shows the first page of the glossary from the EPA report previously mentioned. Notice how the author has underlined the terms being defined, so that readers can find them easily. You can use many other strategies for achieving this same goal, such as using boldface type, or placing the terms in a column of their own on the left-hand side of the page while putting the explanations of the terms on the right-hand side.

By custom, glossaries and lists of symbols can be placed at the beginning of a communication (for instance, directly before the introduction) or at the end. In either case, be sure to include them in your table of contents.

Index

Like a table of contents, an index can provide readers with a quick route to specific pieces of information. In fact, in long documents an index can be much more useful than a table of contents. In an index, readers can locate items alphabetically but in a table of contents they must scan through many entries, which can be a time-consuming chore if the table of contents is very long.

The procedure for determining which items to include in your index is as follows:

1. Identify all the specific kinds of information your readers might seek when using your communication as a reference document.

2. List the words your readers might use while searching for those pieces of information. Where readers might use any of a variety of words to search for particular information, include *all* the most likely terms with cross-references to the main

GLOSSARY

ablation till: Loosely consolidated rock debris that accumulated in place as the glacial ice melted. (see also glacial till)

acid magenta: A dye that is produced by oxidation of a mixture of aniline and toluidines and yields a brilliant bluish red.

aeration booms: Mechanical aeration or mixing booms visible above settling tanks at sewage treatment facilities. Allows discrimination of sewage treatment facility from similar-sized circular storage tanks on aerial photographs.

alluvial deposits: A general term for unconsolidated detrital material deposited by a stream in its channel or on a floodplain.

aquifer: A body of rock or soil sufficiently permeable to conduct groundwater and to yield significant quantities of water to wells or springs.

attenuation: The reduction of ionic concentrations in solutions by natural earth materials by the processes of cation exchange with clays or similar mechanisms.

backshore storm beach ridges: The shore zone between mean high water and the upper limit of shore-zone processes that is acted upon only by severe waves or during unusually high tides.

bathtub effect: An overflow effect commonly seen in landfills located in impermeable clay soils where infiltration of precipitation through waste and cover materials exceeds the capacity of the soil to absorb the normal rainfall. Springs of leachate may appear around the site perimeter.

FIGURE A–19   First Page of the Glossary of the Report Whose Cover Is Shown in Figure A–8

entry. Be sure to include words they might use even if you have not used those words in your text. Imagine for example, that you are writing a marketing brochure for a chain of garden shops. If you have a section on trees and expect that some readers may look under the word *Evergreen* in the index for information you've indexed under *Conifers,* then include an entry for *Evergreen* that directs readers to the *Conifer* entry.

Some word-processing programs help you create an index by generating a list of the words used in your communication. From this list, you can select the words that will help your readers find the information they desire.

3. Where possible, use headwords to gather related terms (for example, use *Conifers* as a headword for *Fir, Cedar, Redwood,* and so on). If an entry is under a headword where some readers may not think to look for it, be sure to use cross-references. One cookbook has made it difficult to find the recipe for pancakes because in the index the writer listed *Pancakes* not under *P* but as the subgroup of *Breads.*

Figure A–20 shows the first page of the index of the instruction manual whose cover is shown in Figure A–9.

Typing

Communications written in the book format are always typed, word processed, or typeset. If you single-space, double-space between paragraphs. If you double-space, do not leave any additional blank lines between paragraphs, but be sure to indent the first line of each paragraph.

Page Numbering

You can handle the page numbering in the body of your communication in either of two ways. First, you can give the first page of your introduction the number *1* and then number all your following pages in one long sequence.

Alternatively, you can begin a new sequence at the first page of each chapter. To do this, give each page a number that has two parts. The first part tells the chapter number and the second part tells the number of the page within the chapter. Thus, page 1–2 is the second page of Chapter 1 and page 3–4 is the fourth page of the third chapter.

The advantage of this system is that it lets writers and typists prepare each chapter separately. Page numbers can be placed on the pages in Chapter 4, for instance, before the number of pages in Chapter 3 has been determined. In addition, if some material needs to be added or deleted from a chapter at the last minute, the only pages that would need to be renumbered are those in that chapter.

In the book format, pages that appear before the first page of the body either are given no numbers or are numbered in lowercase roman numerals (i, ii, iii, and so on). Page numbering in the appendixes and other supplementary materials that follow the body depends partly upon the style of numbering used in the body of the report. If the chapters in the body are numbered separately, the supplementary materials usually are also. Thus, the second page of the first appendix might be A–2 or I–2. The same method may be used even if the pages in the body are numbered in one long sequence. Or the paging sequence of the body might be continued through all of the supplementary parts.

DETROIT DIESEL 53

ALPHABETICAL INDEX

*General Information and Cautions Section

Ⓒ Copyright 1990 Detroit Diesel Corporation   May, 1990   **Page 1**

FIGURE A–20   Index of the Instruction Manual Whose Cover Is Shown in Figure A–9
Courtesy of Detroit Diesel Allison Division, General Motors Corporation.

Letter of Transmittal

When you prepare communications in the book format, you will often send them (rather than hand them) to your readers. In these cases, you will want to accompany them with letters (or memos) of transmittal.

The exact contents of a letter of transmittal will depend greatly on your purpose, but will usually conform to these general observations:

● **A letter of transmittal will usually begin by mentioning the enclosed communication.** If appropriate, it will also explain or remind the reader why the communication was written:

To the director of marketing:

> Today the printer delivered the instruction manuals for our new Model 100 voltage regulator. I attach a copy.

To a prospective client:

> In response to your recent request for proposals, BioLabs is pleased to submit the enclosed plan for continuously monitoring effluent from your Eaton plant for various pollutants, including heavy metals.

To the director of manufacturing operations:

> As you know, during the past six months our Number 1 machine has been shut down three times because of damaged valves. The enclosed report details my study of this problem.

● **The letter may then say something about the purpose, contents, or special features of the communication.** For instance, the writers of a research report who want their readers to note the most important consequences of their reported findings might briefly explain their main findings and list their most important recommendations. (In this case, the letter would repeat some of the information contained in the summary; such repetition is common in letters of transmittal.)

● **The letter may acknowledge the assistance of key contributors to the communication.**

● **Like most letters of any kind, a letter of transmittal usually ends with a short paragraph (often one sentence) that does something like one of the following:** expresses the hope that the reader will find the communication helpful or satisfactory, states his or her willingness to work further with the reader, or promises to try to answer any questions the reader may have.

Figure A–21 shows a letter of transmittal prepared in the workplace; Figure A–22 shows one prepared by a student.

ELECTRONICS CORPORATION OF AMERICA
MEMORANDUM

TO: Myron Bronski, Vice-President, Research

FROM: MCB
               Margaret C. Barnett, Satellite Products Laboratory

DATE: September 30, 19--

RE: REPORT ON TRUCK-TO-SATELLITE TEST

On behalf of the entire research team, I am pleased to submit the attached copy of the operational test of our truck-to-satellite communication system.

The test shows that our system works fine. More than 91% of our data transmissions were successful, and more than 91% of our voice transmissions were of commercial quality. The test helped us identify some sources of bad transmissions, including primarily movement of a truck outside the "footprint" of the satellite's strongest broadcast and the presence of objects (such as trees) in the direct line between a truck and the satellite.

The research team believes that our next steps should be to develop a new antenna for use on the trucks and to develop a configuration of satellites that will place them at least 25° above the horizon for trucks anywhere in our coverage area.

We're ready to begin work on these tasks as soon as we get the okay to do so. Let me know if you have any questions.

Encl: Report (2 copies)

FIGURE A–21   Letter of Transmittal Written at Work

Box 114, Bishop Hall
Miami University
Oxford, Ohio 45056
December 10, 19——

Professor Thomas P. Weissman
Department of English
Miami University
Oxford, Ohio 45056

Dear Professor Weissman:

I am enclosing my final project for your technical writing course, a proposal for a crime prevention program directed to elderly citizens of Oxford. As you recall, I have developed this proposal at the request of the City Manager, Tom Dority.

While working on this project, I learned that persons 65 and older comprise 18% of the nonstudent population of Oxford. Experience nationwide suggests that these individuals are particularly vulnerable to crime, but that much of that crime can be prevented through simple precautions. I propose that the City of Oxford help protect its elderly by participating in two nationwide programs. Whistle Alert and Operation Safe Return, and that the city offer a series of nine presentations on crime prevention for the elderly. The entire effort could be supported by donated supplies and services, with no cost to the City of Oxford.

Throughout my work on this project, I received much help from the Oxford Police Department's Crime Prevention Officer, Dwight Johnson. I have also been assisted by the staff at the Oxford Senior Citizens Center.

I believe there is a reasonable chance the city will accept my plan. Officer Johnson has already said he likes it.

Thank you very much for your help and encouragement.

Sincerely,

Tricia Daniels

Tricia Daniels

Enclosure: Final Project

FIGURE A–22   Letter of Transmittal Written by a Student

Conventions About Style

There are no conventions about writing style—tone of voice, level of formality, and so on—that apply generally to *all* communications prepared in the book format. Some are quite formal, almost stiff, while others are very informal. As with every other feature of your communication, you will have to select a style that is suited to your audience and purpose.

B

FORMATS FOR FOOTNOTES, REFERENCE LISTS, AND BIBLIOGRAPHIES

I n many of the communications you write at work, you will want to tell your readers about other sources of information concerning your subject. This chapter will help you do this. First, you will learn when to provide documentation, and when not to. Then, you will learn how to use the forms of documentation most widely employed in the workplace. Throughout, you will find easy-to-imitate models designed to make your work at documenting sources as simple as possible.

PURPOSES OF DOCUMENTATION

At work, you may provide documentation for any of four reasons:

● **To acknowledge the people and sources that have provided you with ideas and information.** At work, as at school, such acknowledgment is more than a courtesy: it is an ethical obligation.

● **To help your readers find additional information about something that you have discussed.** In many situations, one of your readers' major questions will be: "Where can we learn more about that?" You can answer this question through documentation.

● **To persuade your readers to consider seriously a particular idea.** By showing that an idea was expressed by some respected person or in some respected publication, you are arguing that the idea merits acceptance.

● **To explain how your research relates to the development of new knowledge in your field.** In research proposals and in research reports published in professional journals, writers often include literature survey sections to help demonstrate how their research projects contribute to the knowledge and capabilities in their fields. For more information about how to use references in this way, see the discussion of literature review sections in Chapter 21.

DECIDING WHAT TO ACKNOWLEDGE

The first of the three reasons for providing documentation deserves your special attention. At work you will have the same sort of ethical obligation that you have at school to acknowledge the sources of your information. However, the standard used for determining which particular sources you need to acknowledge is somewhat different at work than at school.

In both places, you must document material (1) that you have derived from someone else and (2) that is not "common knowledge." However, what's considered to be common knowledge at school is different from what is considered common knowledge at work. At school, common knowledge is knowledge that every person possesses without doing any special reading. Thus, you must document any material you find in print. In contrast, at work common knowledge is the knowledge usually possessed by or readily available to the people in your field. Thus, you do not need to acknowledge material you obtained through your college classes, your textbooks, the standard reference works in your field, or similar sources.

CHOOSING A FORMAT FOR DOCUMENTATION

Once you have decided which sources to acknowledge, you must determine where to place information about them and how to present that information to your readers. Your decisions about these matters are greatly simplified because there are standard formats for documentation. Unfortunately, however, there is no single format that is correct for all situations. Instead, there are many different formats, some very distinct from one another, some differing only in small details. Consequently, to document correctly you must do two things:

1. Find the particular format that is required or most appropriate for the communication you are writing.
2. Follow that format to the last detail.

How can you find the appropriate format? First, ask someone: "Is there a specific format I must use for documentation?" Your employer may specify a particular format. If you are writing to people in another organization, that organization may have its own preferences or requirements about documentation.

To describe the documentation format they want their employees to use, many employers have created and issued style guides that include rules for documentation together with sample citations. Many other employers—and almost all professional journals—ask writers to follow one of the popular published style guides. These include the *APA Style Guide* (published by the American Psychological Association), the *CBE Style Guide* (published by the Council of Biological Editors), and the *MLA Style Guide* (published by the Modern Language Association). Each of these style guides is used widely in particular fields: the APA in the social sciences, the CBE in the life sciences, and the MLA in the humanities.

If it turns out that you are not required to use some specific style guide, you may employ one of the formats described below, which are based upon *The Chicago Manual of Style*. To decide which of these styles to choose, determine which seems to be closest to what your audience is accustomed to seeing. In situations where no style is specified, consistency (even in detail) is usually more important than the particular style itself.

USING AUTHOR–DATE CITATIONS COMBINED WITH A REFERENCE LIST

In the author–date format, you cite a source by putting the name of the author and the year of publication at the appropriate place in the body of your communication. This citation refers your readers to the full bibliographic information that you provide in an alphabetical list of sources at the end of your communication. The following sections tell you how to write the citation, where to put it in the body of your communication, and how to write the entry in your list of references.

The information provided below about the author–date format follows *The Chicago Manual of Style* almost exactly, varying only in a few places where simplification seems both possible and desirable for the writing you will do on the job.

Writing Author–Date Citations

To write an author–date citation, enclose the author's last name and the year of publication in parentheses *inside* your normal sentence punctuation. Do not put any punctuation between the author's name and the date:

```
The first crab caught in the trap attracts others to it
(Tanner 1985).
```

If you are referring to one particular part of your source, you can help your reader find that part by including its page numbers in your citation. Place a comma between the year and the page numbers:

```
(Angstrom 1982, 34–49)
```

If you use the author's name in part of your sentence, place the year of publication and page numbers immediately afterwards in parentheses. Do not repeat the author's name:

```
Angstrom (1982, 34–49) showed that the strength of these
desires is inversely related to the person's level of
self-confidence.
```

In some of your communications, you may cite two or more sources by the same author. If they were published in *different* years, your reader will have no trouble telling which work you are referring to. If the works were published in the *same* year, you can distinguish between them by placing lowercase letters after the publication dates in your citations and in your reference list:

```
(Burkehardt 1981a)
(Burkehardt 1981b)
```

If you are citing a work with two or three authors, give the names of all authors. Notice that no comma precedes the word *and:*

```
(Hoeflin and Bolsen 1986)
(Wilton, Nelson and Dutta 1978)
```

If you are citing a work with more than three authors, give the first author's name, followed by *et al.,* which is an abbreviation for the Latin phrase *et alii* (''and others''):

```
(Dutta et al. 1984)
```

If you are citing a source that does not name an individual or set of individuals as authors, give the name of the organization that publishes or sponsors the source:

```
(National Cancer Institute 1983)
```

If you need to name two or more sources in one place, enclose them all within a single pair of parentheses, separating them from one another with semicolons. Do not use the word *and* between sources:

(Justin 1984; Skol 1972; Weiss 1986)

In this example, the three sources are arranged alphabetically. They could also be arranged chronologically.

Deciding Where to Place Citations

Your primary objective when deciding where to place your citations should be to make clear to your readers what part of your text is being referred to by each citation. This will be easy with citations that pertain to a single fact, sentence, or quotation. You simply place your citation immediately after the appropriate material:

According to D. W. Orley (1978, 37), "We cannot tell how to interpret these data without conducting further tests."

Researchers have shown that a person's self-esteem is based upon performance (Dore 1964), age (Latice 1981), and weight (Swallen and Ditka 1970).

If your note refers to material in several sentences, you can place your citation in what your readers will clearly see as a topic sentence for the affected material. Your readers will then understand that the citation covers all the material that relates to that topic sentence. To assist your readers even more, you may want to use the author's name (or a pronoun) in more than one sentence:

A much different account of the origin of oil in the earth's crust has been advanced by Thomas Gold (1983). He argues that. . . . To critics of his views, Gold responds. . . .

Making the List of References

In the list of references you should describe each source in enough detail to help your readers find the source quickly in a library. If you are using author–date citations, arrange the list alphabetically by author. No matter how many times you cite the same work, place it in your list only once.

For your convenience, the following discussion groups three types of listings: those for books, those for journal articles, and those for other kinds of sources you are likely to cite on the job.

References to Books

When describing a book, you should provide the following information:

● **Author.** Give the author's last name first, followed by the author's initials. Alternatively, you may use the author's given name and middle initials, copying exactly the way they appear in the book you are citing. Follow the author's name with a period.

   If there is more than one author, give the names of the second and additional authors in their natural order (first initial, middle initial, last name). Separate the names of the various authors with commas, and place *and* before the last author. Follow the last author's name with a period. (If an author has two or more initials, leave a space between them.)

● **Year of publication.** Follow the year of publication with a period.

● **Title.** Capitalize only the first word of the title, the first word of the subtitle, if there is one, and proper nouns. Underline and follow with a period.

● **Edition.** If you are using the second or subsequent edition, tell which one you used. Abbreviate *edition* to *ed.* Use *2nd* and *3rd* rather than *second* and *third*.

● **Publisher.** Write the city of publication, a colon, and the name of the publisher. Follow with a period. (Be sure to leave a space after the colon.)

Below you will find examples that show how this information is presented for various commonly cited types of books. In these samples, the authors' first and middle names are represented only by their initials even if their names are given in full on the title page. As you study the examples, note carefully their capitalization and punctuation.

One author
: Ayers, R. U. 1969. <u>Technological forecasting and long-range planning</u>. New York: McGraw-Hill.

Two or three authors
: Harris, J., and R. Kellermayer. 1970. <u>The red cell: Production, metabolism, destruction, normal and abnormal</u>. Cambridge, Mass.: Harvard University Press.

: Middleditch, B. S., S. R. Missler, and H. B. Hines. 1981. <u>Mass spectrometry of priority pollutants</u>. New York: Plenum Press.

(Notice that a comma follows as well as precedes the first author's initials or given name).

More than three authors
: Cooke, R. U., D. Brunsden, J. C. Doornkamp, and D. K. C. Jones. 1982. <u>Urban geomorphology in drylands</u>. New York: Oxford University Press.

In your reference list, be sure to name *all* the authors of a source, even though in the body of your communication you give only the first author's name followed by *et al.*

Corporate author or sponsor
: National Institutes of Health. 1976. <u>Recombinant DNA research</u>. Washington, D. C.: U. S. Government Printing Office.

Editor   Sabin, M. A., ed.   1974.   <u>Programming techniques in computer aided design</u>.   London: NCC Publications.

Second or subsequent edition   Hay, J. G. 1978.   <u>The biomechanics of sports techniques</u>.   2nd ed. Englewood Cliffs, N.J. : Prentice—Hall.

References to Articles and Essays

When describing a journal article, you should provide the following information:

- **Author.** Give the names of authors of articles and essays in the same way you give the names of authors of books (see above).
- **Year of publication.** Follow with a period.
- **Title of article or essay.** Capitalize only the first word of the title, the first word of the subtitle, if any, and proper nouns. Do *not* underline or enclose in quotation marks. Follow with a period.
- **Publication information.** When citing an article, capitalize all major words in the periodical's name. Underline. Next, give the volume number in arabic numbers, even if the periodical uses roman numerals. Do not place any punctuation between the periodical's name and the volume number. Follow the volume number with a colon. If that periodical numbers its pages in a single sequence throughout the volume, omit the issue number. However, if the periodical begins numbering pages anew with each issue, identify the issue by placing its date in parentheses directly after the volume number (no space) and before the colon. Finally, give the first and last pages of the article or essay, leaving no space between the colon that follows the volume or issue number and the first page number. End with a period.

  When citing an essay in a book, begin with the word *In,* followed by the book's title, capitalizing only the first word of the title, the first word of the subtitle, if there is one, and proper nouns. Underline and follow with a comma. Next, write the abbreviation ''ed.'' (for ''edited by'') and give the editor's name, followed by a comma. Then write sthe first and last pages of the essay, followed by a period. Finally, write the city of publication, a colon, a space, and the name of the publisher. End with a period.

The following samples show how this information is presented.

Magazine or journal article   McNerney, W. J.   1980.   Control of health care costs in the 1980's.   <u>New England Journal of Medicine</u> 303:1088—95.

Hoeflin, R., and N. Bolsen.   1986.   Life goals and decision making: Educated women's patterns.   <u>Journal of Home Economics</u> 78(Summer):33—35.

(Note that the *Journal of Home Economics* begins numbering its pages anew with each issue, unlike the *New England Journal of Medicine*. This explains why the entry for the *Journal of Home Economics* article gives the year and date, but the entry for the other article gives only the year.)

Newspaper article   Lewis, P. H. 1986. UNIX and MS–DOS: dueling for dominance in computers. <u>New York Times</u>, 13 May, sec. C, 9.

Essay in a book   Tuchman, G. 1979. The impact of mass–media stereotypes upon the full employment of women. In <u>Women in the U.S. labor force</u>, ed. A. F. Chan, 249–68. New York: Praeger.

(Notice that the pages of the essay follow a comma after the editor's name.)

Paper in a proceedings   Stover, E. L. 1982. Removal of volatile organics from contaminated ground water. In <u>Proceedings of the second national symposium on aquifer restoration and ground water monitoring</u>, ed. D. M. Nielsen, 77–84. Worthington, Ohio: National Well Water Association.

Encyclopedia article   Aller, L. H. 1982. Astrophysics. <u>McGraw–Hill encyclopedia of science and technology</u>, 5th ed.

References to Unpublished Sources

Here are some sample bibliography entries for unpublished sources:

Unpublished corporate reports   Belkaoui, M. and E. Cohen. 1985. Three–Year Market Projections for the Home Appliance Industry. New York: General Appliance Company.

Thesis or dissertation   Charron, D. C. 1985. Individual Moral Responsibility and the Business Firm. Ph.D. diss., Washington University, 1985.

Letter   McConnell, D., Director of Corporate Recruiting, Sterling Corporation. Letter to the author, 10 August 1990.

Interview   Cawthorne, L., Attorney at Law. Interview with the author. New York City, 18 March 1990.

Sample List of References

Figure B–1 shows a sample list of references. Notice that if your list includes two entries by the same author, the author's name is given only for the first. The second and subsequent entries begin with a short, dashed line. (See the entries for the two books by R. Ayers.)

USING FOOTNOTES AND ENDNOTES

You are undoubtedly familiar with the general strategy of footnoting. To signal readers that you are citing a source, you place a superscript number at the appropriate place in your prose. Readers then know to look for information about the source in a note with the corresponding number.

REFERENCES

Aller, L. H. 1982. Astrophysics. <u>McGraw-Hill encyclopedia of</u> <u>science and technology</u>, 5th ed.

Asparagus. 1986. <u>Encyclopaedia Britannica</u>. 15th ed.

Ayers, R. U. 1969. <u>Technological forecasting and long-range</u> <u>planning</u>. New York: McGraw-Hill.

——. 1984. <u>The next industrial revolution: Reviving industry through</u> <u>innovation</u>. Cambridge, Mass.: Ballinger.

Cawthorne, L., Attorney at Law. Interview with the author in New York City, 18 March 1990.

Cooke, R. U., D. Brunsden, J. C. Doornkamp, and D. K. C. Jones. 1982. <u>Urban geomorphology in drylands</u>. New York: Oxford University Press.

Davis, D. 1985. Furniture made fun. <u>Newsweek</u> 106(4 November):82–83

Feldhamer, G. A., J. E. Gates, D. M. Harman, A. J. Loranger, and K. R. Dixon. 1986. Effects of interstate highway fencing on white-tailed deer activity. <u>Journal of Wildlife Management</u> 50:497–503.

Harris, J., and R. Kellermayer. 1970. <u>The red cell: Production,</u> <u>metabolism, destruction, normal and abnormal</u>. Cambridge, Mass.: Harvard University Press.

Hay, J. G. 1978. <u>The biomechanics of sports techniques</u>. 2nd ed. Englewood Cliffs, N.J.: Prentice-Hall.

Hoeflin, R., and N. Bolsen. 1986. Life goals and decision making: Educated women's patterns. <u>Journal of Home Economics</u> 78(Summer): 33–35.

Lewis, P. H. 1986. UNIX and MS-DOS: Dueling for dominance in computers. <u>New York Times</u>, 13 May, sec. C, 9.

Lynn, S. J., and J. W. Rhue. 1986. The fantasy-prone person: Hypnosis, imagination, and creativity. <u>Journal of Personality and</u> <u>Social Psychology</u> 50:404–408.

McNerney, W. J. 1980. Control of health care costs in the 1980's. <u>New England Journal of Medicine</u> 303:1088–95.

Middleditch, B. S., S. R. Missler, and H. B. Hines. 1981. <u>Mass</u> <u>spectrometry of priority pollutants</u>. New York: Plenum Press.

FIGURE B–1   List of References

Morris, M. G. 1976. Conservation and the collector. In <u>Moths and butterflies of Great Britain and Ireland</u>, ed. J. Heath, vol. 1, 107–16. London: Curwen.

National Institutes of Health. 1976. <u>Recombinant DNA research</u>. Washington, D.C.: U.S. Government Printing Office.

Sabin, M. A., ed. 1974. <u>Programming techniques in computer aided design</u>. London: NCC Publications.

Stover, E. L. 1982. Removal of volatile organics from contaminated ground water. In <u>Proceedings of the second national symposium on aquifer restoration and ground water monitoring</u>, ed. D. M. Nielsen, 77–84. Worthington, Ohio: National Well Water Association.

Torgul, D., Director of Research and Development, Sterling Corporation. Letter to the author, 10 August 1990.

U.S. Environmental Protection Agency. 1977. <u>Is your drinking water safe?</u> Washington, D.C.: U.S. Government Printing Office.

FIGURE B–1 *(continued)*

Handling the Superscripts

Whenever possible, place superscripts at the end of a sentence:

> Thomas Gold has a new theory about the origins of oil.[1]

However, to avoid confusion you may place superscripts within a sentence—but still after any punctuation:

> According to Haljmer Sunderstan,[1] whose truthfulness has been questioned in official proceedings,[2] the events were as follows.

If your note refers to information conveyed in several sentences, you can place your citation in what your readers will clearly see as a topic sentence for the affected material. Your readers will then understand that the citation covers all the material that relates to that topic sentence. To assist your readers even more, you may want to use the author's name (or a pronoun) in more than one sentence.

> George A. Steiner claims that in the early 1960s the U.S. developed a new type of government regulation, which has evolved and expanded rapidly.[1] According to Steiner, the old regulation focused upon specific industries and their practices. Under the new regulation, government. . . .
>
> To illustrate this new regulation, Steiner offers three examples. First, ...

You should number the superscripts sequentially. If you are writing a communication with several chapters (often called ''sections'' in the writing done at work), you may use one continuous numbering sequence throughout your communication, or you may begin anew with the number *1* in each chapter.

Notes about sources for tables, charts, and other visual aids are not included in this sequence. They are provided at the bottom of the visual aid. Usually, they are preceded by the word *Source*, which may be underlined, and followed by a colon.

Placing the Notes

The notes themselves may appear at the bottom of the page (in which case they are called *footnotes*), or they may be gathered in a separate section at the end of each chapter or at the end of the entire communication (in which case they are called *endnotes*). Footnotes save the reader the trouble of turning to a separate page for source information, but they are much more troublesome to type. Consequently, endnotes are preferred in many companies. Nowadays, many word-processing programs automatically make the appropriate amount of space for footnotes, so they may become more popular.

When using footnotes, use the same bottom margin as on pages without footnotes. This will require some planning, so that you can shorten the text enough to make room for the footnote(s). Footnotes are separated from the text by one or two blank lines plus a short underscore placed flush against the left-hand margin. Here is an example:

```
    Imagine that this is the last sentence on a page of
text.¹
```

————————————

```
    1. This is the footnote.
```

Writing the Notes

Whether you place your notes at the bottom of the page or gather them in a separate section, you should type them single-spaced even if the text is typed double-spaced. Each should begin with the note's number. In typed footnotes, the number is usually aligned with the rest of the note, not raised into a superscript.

Notes usually have some indentation. Sometimes only the first line is indented:

```
    1. Peter S. Rose, The Changing Structure of American
Banking (New York: Columbia University Press, 1987), 123.
```

Sometimes the later lines are indented:

```
    2. Peter S. Rose, The Changing Structure of American
       Banking (New York: Columbia University Press, 1987),
       123.
```

Indenting the later lines makes it easier for readers to pick out the number of a particular note when they are searching through a series of notes.

To reduce redundancy, the second and subsequent notes referring to a particular work differ from the first note. The following sections tell you how to write first notes for books, articles and essays, and the unpublished sources you are most likely to use on the job. Another section then tells you how to write additional citations of the same source. The formats presented here are based upon *The Chicago Manual of Style* (with some simplification) and reflect common practice in the workplace.

First Notes for Books

When citing a book for the first time, you should provide the following information:

● **Author.** Give the author's names in natural order: first initial (or name), middle initial, and last name. If your source has two authors, place *and* between them. If the source has three or more authors, separate the names with commas and place *and* before the last one. Place a comma after the names of the author or authors and before the title.

Give the author's names exactly as they appear on the title page: spell out first and middle names where the title page spells them out, but use initials if the title page provides only initials.

If your source gives no author or editor, cite the sponsoring agency (such as the company or government agency that sponsored the book).

● **Title.** Capitalize all major words. Underline the title or type it in italics. Include the subtitle, separating it from the main title with a colon. Use no punctuation after the title. If you used the second or subsequent edition, note that fact with the appropriate abbreviations: 2nd ed. Place a comma between the title and the edition, but none after the edition.

● **Publication information.** In parentheses, write the city of publication, colon, publisher's name, comma, and year of publication. Follow the parentheses with a comma. (Leave a space after the colon that precedes the publisher's name.)

When naming well-known cities (such as New York, Boston, or San Diego), omit the state, but for less well-known cities, identify the state (for example, Geneva, Illinois).

● **Pages.** Give the page or pages you are citing, followed by a period. If you are citing an entire book, omit page numbers and place a period after the publication information.

Here are some sample footnotes that show how this information is presented for various commonly cited types of books:

One author

3. Robert U. Ayers, <u>The Next Industrial Revolution: Reviving Industry through Innovation</u> (Cambridge, Mass.: Ballinger, 1984), 172.

Two authors

4. J. Harris and R. Kellermayer, <u>The Red Cell: Production, Metabolism, Destruction, Normal and Abnormal</u> (Cambridge, Mass.: Harvard University Press, 1970), 552–73.

Three authors

5. Brian S. Middleditch, Stephen R. Missler, and Harry B. Hines, <u>Mass Spectrometry of Priority Pollutants</u> (New York: Plenum Press, 1981), 237.

Four or more authors

6. R. U. Cooke et al., <u>Urban Geomorphology in Drylands</u> (New York: Oxford University Press, 1982), 86.

(Note that *et al.* is an abbreviation for the Latin phrase *et alii*, which means "and others.")

Corporate author or sponsor

7. American Broadcasting Companies, <u>Annual Report to Stockholders</u> (New York: American Broadcasting Companies, 1984), 17.

8. National Institutes of Health, <u>Recombinant DNA Research</u> (Washington, D.C.: U.S. Government Printing Office, 1976), 42–47.

Editor

9. M. A. Sabin, ed., <u>Programming Techniques in Computer Aided Design</u> (London: NCC Publications, 1974), 110–42.

<table>
<tr><td>Second or
subsequent edition</td><td>10. James G. Hay, <u>The Biomechanics of Sports Techniques</u>,
2nd ed. (Englewood Cliffs, N.J.: Prentice-Hall,
1978), 382–405.</td></tr>
</table>

First Notes for Articles and Essays

When citing articles and essays for the first time, you should supply the following information:

- **Author.** Present the names of authors of journal articles in the same way you present the names of authors of books (see the preceding discussion).

- **Title of article or essay.** Capitalize all major words. Follow the article title with a comma. Place the title and the comma in quotation marks.

- **Publication information.** When citing an article in a journal or magazine, capitalize the major words of the title of the periodical, and underline or italicize it. Follow the title with the number of the volume in which the article appears. Skip a space, then give the year of the volume in parentheses, followed by a colon and a space. If the periodical begins numbering its pages anew in each issue (rather than using a single sequence throughout the entire volume), give the month (or month and day) of the issue in which the article appears (as in Footnote 12 below).

  When citing an essay, follow the essay's title with the word ''in,'' the title of the book (underlined or italicized and with major words capitalized), and a comma. Next, give the abbreviation ''ed.'' (for ''edited by'') and the name of the editor or editors. Then in parentheses write the city of publication, colon, publisher's name, comma, and year of publication. Follow the parentheses with a colon and a space.

- **Pages.** List the page or pages you are citing, followed by a period. Unless you want to acknowledge the entire article or essay, give the numbers of only those pages that contain the information you are using.

Here are some sample footnotes:

<table>
<tr><td>Magazine or journal
article</td><td>11. Walter J. McNerney, "Control of Health Care Costs in
the 1980's," <u>New England Journal of Medicine</u> 303
(1980): 1093.</td></tr>
<tr><td></td><td>12. Ruth Hoeflin and Nancy Bolsen, "Life Goals and
Decision Making: Educated Women's Patterns," <u>Journal
of Home Economics</u> 78 (Summer 1986): 33–35.</td></tr>
</table>

(Note that the *Journal of Home Economics* begins numbering its pages anew with each issue, unlike the *New England Journal of Medicine*. That explains why Footnote 12 gives the full date but Footnote 11 gives only the year.)

<table>
<tr><td>Newspaper article</td><td>13. Peter H. Lewis, "UNIX and MS–DOS: Dueling for
Dominance in Computers," <u>The New York Times</u>, 13 May
1986, sec. C, 9.</td></tr>
<tr><td>Essay in a book</td><td>14. Gaye Tuchman, "The Impact of Mass–Media Stereotypes on
the Full Employment of Women," in <u>Women in the U.S.
Labor Force</u>, ed. Ann Foote Chan (New York: Praeger,
1979), 257.</td></tr>
</table>

Pamphlet or booklet

15. U.S. Environmental Protection Agency, <u>Is Your Drinking Water Safe?</u> (Washington, D.C.: U.S. Government Printing Office, 1977), 15.

Paper in a proceedings

16. Enos L. Stover, "Removal of Volatile Organics from Contaminated Ground Water," in <u>Proceedings of the Second National Symposium on Aquifer Restoration and Ground Water Monitoring</u>, ed. David M. Nielsen (Worthington, Ohio: National Well Water Association, 1982), 77–84.

Encyclopedia article

17. Lawrence H. Aller, "Astrophysics," <u>McGraw-Hill Encyclopedia of Science and Technology</u>, 5th ed.

First Notes for Unpublished Sources

The following examples show how to describe some other sources you might use when writing on the job:

Unpublished corporate reports

18. Michael Belkaoui and Elihu Cohen. "Three-Year Market Projections for the Home Appliance Industry" (New York: General Appliance Company, 1985), 21.

Theses and dissertations

19. Donna C. Charron, "Individual Moral Responsibility and the Business Firm," (Ph.D. diss., Washington University, 1985), 77–78.

Letter

20. Dominic McConnell, Director of Corporate Recruiting, Sterling Corporation, letter to the author, 10 August 1990.

Interview

21. Linda Cawthorne, Attorney at Law, interview with the author, New York City, 18 March 1990.

Subsequent Notes

In the second and subsequent citations of a given source, you should use an abbreviated note that signals the readers to look to the first citation for a full description of the source. These subsequent notes usually include only the author's last name and the page number:

22. Wagner, 179.

If you cite two or more works by the same author, however, you must let your readers know which work you refer to in each subsequent note. Do this by adding a short version of the title. The following example is a second citation for Robert V. Ayers' book *The Next Industrial Revolution: Reviving Industry through Innovation*.

23. Ayers, <u>Next Industrial Revolution</u>, 207.

If you refer to the same source in two or more adjacent notes, you have the option of using the abbreviation *Ibid*. to replace the author's name and the publication's short title (if needed) in the second and subsequent notes. For instance, Footnote 24 might look like this if it refers to the Ayers book already cited in Footnote 23:

```
24. Ibid., 92.
```

Sample Page of Endnotes

Figure B–2 shows a sample page of endnotes.

USING BIBLIOGRAPHIES

A bibliography is an alphabetized list of sources. In the communications you write at work, you can use a bibliography for two purposes. The first is to provide your readers with a *longer* list of sources on your subject than you cited in the body of your communication. You might do this, for instance, to indicate general sources of information that you used but did not refer to specifically in the body.

The second purpose of a bibliography pertains to communications in which you use the footnote format. With footnotes, you present your sources out of alphabetical order and, if you place your notes on the bottom of the page, you scatter your notes throughout your communication. These circumstances make it difficult for your readers to determine quickly what sources you used. Readers might want to check your sources in order to guide their own further reading or else to see whether or not you overlooked some particular source they know to be important. You can help your readers quickly find all your sources if you list all your sources alphabetically in a bibliography.

The Form for Bibliography Entries Varies

The form of the entries in a bibliography will depend upon the format you use to cite sources in the body of your communication. If you are creating a bibliography for a communication in which you use one of the reference list formats (with either author–date citations or numbered citations), write the entries in the bibliography in the same way you write the entries in the reference list. If you use the footnote format, you should write the entries in your bibliography in the same way that you write the footnotes—with the following three exceptions:

- Write the first author's name with the last name first.
- Use periods (not commas or parentheses) to separate the major parts of the citation. For a book, place these periods after the author's name, the book's title, and the publication information (city, publisher, and date). For an article, place the periods after the author's name, the article's title, and the publication information (journal name, volume, and page numbers).
- Include page numbers *only* when you are indicating the first and last pages of an article in a journal, book, proceedings or similar collection. Precede any page numbers with a colon and space.

ENDNOTES

1. National Institutes of Health, <u>Recombinant DNA Research</u> (Washington, D.C.: U.S. Government Printing Office, 1976), 342–75.

2. J. Harris and R. Kellermayer, <u>The Red Cell: Production, Metabolism, Destruction, Normal and Abnormal</u> (Cambridge, Mass.: Harvard University Press, 1970), 552–73.

3. Brian S. Middleditch, Stephen R. Missler, and Harry B. Hines, <u>Mass Spectrometry of Priority Pollutants</u> (New York: Plenum Press, 1981), 86–101.

4. Robert U. Ayers, <u>Technological Forecasting and Long–Range Planning</u> (New York: McGraw–Hill, 1969), 216–34.

5. Robert U. Ayers, <u>The Next Industrial Revolution: Reviving Industry through Innovation</u> (Cambridge, Mass.: Ballinger, 1984), 14.

6. Ayers, <u>Technological Forecasting</u>, 237.

7. Walter J. McNerney, "Control of Health Care Costs in the 1980's," <u>New England Journal of Medicine</u> 303 (1980): 1093.

8. M. A. Sabin, ed., <u>Programming Techniques in Computer Aided Design</u> (London: NCC Publications, 1974), 110–42.

9. Dwight Torgul, Director of Research and Development, Sterling Corporation, letter to the author, 10 August 1990.

10. James G. Hay, <u>The Biomechanics of Sports Techniques</u>, 2nd ed. (Englewood Cliffs, N.J.: Prentice–Hall, 1978), 382–405.

11. Harris and Kellermayer, 313.

12. Ibid., 310.

FIGURE B–2   Endnotes

Because bibliography entries are so similar to footnotes, you may find it helpful to review discussions given above for the types of communications you are citing.

Writing Bibliography Entries

The rest of this section presents sample entries for a bibliography appearing in a communication that uses footnotes.

Bibliography Entries for Books

One author
: Ayers, Robert U. <u>Technological Forecasting and Long-range Planning</u>. New York: McGraw-Hill, 1969.

Two authors
: Harris, J., and R. Kellermayer. <u>The Red Cell: Production, Metabolism, Destruction, Normal and Abnormal</u>. Cambridge, Mass.: Harvard University Press, 1970.

Three authors
: Middleditch, Brian S., Stephen R. Missler, and Harry B. Hines. <u>Mass Spectrometry of Priority Pollutants</u>. New York: Plenum Press, 1981.

Four or more authors
: Cooke, R. U., D. Brunsden, J. C. Doornkamp, and D. K. C. Jones. <u>Urban Geomorphology in Drylands</u>. New York: Oxford University Press, 1982.

Corporate author or sponsor
: National Institutes of Health. <u>Recombinant DNA Research</u>. Washington, D. C. : U. S. Government Printing Office, 1976.

Editor
: Sabin, M. A., ed. <u>Programming Techniques in Computer Aided Design</u>. London: NCC Publications, 1974.

Second or subsequent edition
: Hay, James G. <u>The Biomechanics of Sports Techniques</u>, 2nd ed. Englewood Cliffs, N. J.: Prentice-Hall, 1978.

Bibliography Entries for Articles and Essays

Magazine or journal article
: McNerney, Walter J. "Control of Health Care Costs in the 1980's." <u>New England Journal of Medicine</u> 303 (1980): 1088-95.

Hoeflin, Ruth, and Nancy Bolsen. "Life Goals and Decision Making: Educated Women's Patterns." <u>Journal of Home Economics</u> 78 (Summer 1986): 33-35.

Newspaper article
: Lewis, Peter H. "UNIX and MS-DOS: Dueling for Dominance in Computers." <u>New York Times</u>, 13 May 1986, sec. C, 9.

Essay in a book
: Morris, M. G. "Conservation and the Collector." In <u>Moths and Butterflies of Great Britain and Ireland</u>, ed. John Heath, vol. 1, 107-16. London: Curwen, 1976.

Pamphlet or
booklet
U.S. Environmental Protection Agency. <u>Is Your Drinking Water Safe?</u> Washington, D.C. : U. S. Government Printing Office, 1977.

Paper in a
proceeding
Stover, Enos L. "Removal of Volatile Organics from Contaminated Ground Water." In <u>Proceedings of the Second National Symposium on Aquifer Restoration and Ground Water Monitoring</u>, ed. David M. Nielsen, 77–84. Worthington, Ohio: National Well Water Association, 1982.

Encyclopedia article
Aller, Lawrence H. "Astrophysics." <u>McGraw–Hill Encyclopedia of Science and Technology</u>, 5th ed., 1982.

Bibliography Entries for Unpublished Sources

Letter
Torgul, Dwight, Director of Research and Development, Sterling Corporation. Letter to the author, 10 August 1990.

Interview
Cawthorne, Linda, Attorney at Law. Interview with the author in New York City, 18 March 1990.

Sample Bibliography

Figure B–3 shows a sample bibliography for a communication that uses footnotes.

EXERCISES

1. Pick a topic in your field and find six articles, books, and other printed sources of information about it. For each, do the following:
   - Write an entry for use in a list of references.
   - Write a footnote (cite some particular page or pages in each source).
   - Write a bibliography entry.

   Alternatively, you may find six sources concerning the subject of an assignment you are preparing in your writing course.

2. Find a journal and a book concerning topics in your field. Compare the journal's documentation format with the format closest to it described in this chapter. Do the same for the book. Note similarities and differences in detail.

3. For each of the following sources do these three things:
   - Write an entry for use in a list of references.
   - Write a footnote (cite some particular page or pages in each source).
   - Write a bibliography entry.
   a. The third edition of a book entitled *Occupational Safety Management and Engineering*, published by Prentice-Hall, which has its headquarters in Englewood Cliffs, New Jersey. The author is Willie Hammer. This edition was issued in 1976.

BIBLIOGRAPHY

Aller, Lawrence H. "Astrophysics." <u>McGraw-Hill Encyclopedia of Science and Technology</u>, 5th ed., 1982.

"Asparagus." <u>Encyclopaedia Britannica</u>, 15th ed., 1986.

Ayers, Robert U. <u>Technological Forecasting and Long-range Planning</u>. New York: McGraw-Hill, 1969.

----. <u>The Next Industrial Revolution: Reviving Industry through Innovation</u>. Cambridge, Mass.: Ballinger, 1984.

Cawthorne, Linda, Attorney at Law. Interview with the author in New York City, 18 March 1990.

Cooke, R. U., D. Brunsden, J. C. Doornkamp, and D. K. C. Jones. <u>Urban Geomorphology in Drylands</u>. New York: Oxford University Press, 1982.

Davis, Douglas. "Furniture Made Fun." <u>Newsweek</u> 106 (4 November 1985): 82–83.

Feldhamer, George A., J. Edward Gates, Dan M. Harman, Andre J. Loranger, and Kenneth R. Dixon. "Effects of Interstate Highway Fencing on White-Tailed Deer Activity." <u>Journal of Wildlife Management</u> 50 (1986): 497–503.

Harris, J., and R. Kellermayer. <u>The Red Cell: Production, Metabolism, Destruction, Normal and Abnormal</u>. Cambridge, Mass.: Harvard University Press, 1970.

Hay, James G. <u>The Biomechanics of Sports Techniques</u>, 2nd ed. Englewood Cliffs, N.J.: Prentice-Hall, 1978.

Hoeflin, Ruth, and Nancy Bolsen. "Life Goals and Decision Making: Educated Women's Patterns." <u>Journal of Home Economics</u> 78 (Summer 1986): 33–35.

Lewis, Peter H. "UNIX and MS-DOS: Dueling for Dominance in Computers." <u>New York Times</u>, 13 May 1986, sec. C, 9.

Lynn, Steven J., and Judith W. Rhue. "The Fantasy-Prone Person: Hypnosis, Imagination, and Creativity." <u>Journal of Personality and Social Psychology</u> 50 (1986): 404–408.

McNerney, Walter J. "Control of Health Care Costs in the 1980's." <u>New England Journal of Medicine</u> 303 (1980): 1088–95.

Middleditch, Brian S., Stephen R. Missler, and Harry B. Hines. <u>Mass Spectrometry of Priority Pollutants</u>. New York: Plenum Press, 1981.

FIGURE B–3   Bibliography for Use with Footnotes

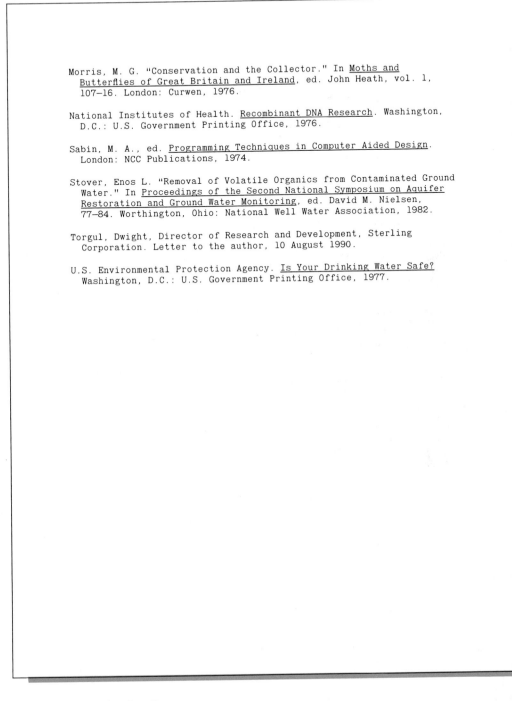

Morris, M. G. "Conservation and the Collector." In <u>Moths and Butterflies of Great Britain and Ireland</u>, ed. John Heath, vol. 1, 107–16. London: Curwen, 1976.

National Institutes of Health. <u>Recombinant DNA Research</u>. Washington, D.C.: U.S. Government Printing Office, 1976.

Sabin, M. A., ed. <u>Programming Techniques in Computer Aided Design</u>. London: NCC Publications, 1974.

Stover, Enos L. "Removal of Volatile Organics from Contaminated Ground Water." In <u>Proceedings of the Second National Symposium on Aquifer Restoration and Ground Water Monitoring</u>, ed. David M. Nielsen, 77–84. Worthington, Ohio: National Well Water Association, 1982.

Torgul, Dwight, Director of Research and Development, Sterling Corporation. Letter to the author, 10 August 1990.

U.S. Environmental Protection Agency. <u>Is Your Drinking Water Safe?</u> Washington, D.C.: U.S. Government Printing Office, 1977.

Figure B–3   *(continued)*

b. An article by Ida Brambilla, Alcide Bertani, and Remo Reggiani in volume 123, issue number 5 of the *Journal of Plant Physiology*. This issue was published in June 1986, and the article is entitled, "Effects of Inorganic Nitrogen Nutrition (Ammonium and Nitrate) on Aerobic and Anaerobic Metabolism in Excised Rice Roots." Pages 419–28.

c. An article entitled "Space Goals for 21st Century Depicted in Report to White House," which appears on pages 16 and 17 of the May 26, 1986, issue of *Aviation Week*. This issue is number 21 of volume 24. The journal does not identify the author of the article. *Aviation Week* is published by McGraw-Hill, which has its headquarters in New York City.

d. An article entitled "Genes Could Be the Enemy in Fighting the Battle of the Bulge," which was published on Tuesday, August 5, 1986, in the *Atlanta Constitution*, a newspaper. The article was written by Jane C. Allison, staff writer, and it appears on page A-20.

e. An article entitled "An Evaluation of the Sulfite-AQ Pulping Process," which appears on pages 102 through 105 of the August 1986 issue of the *Tappi Journal*. The journal identified the authors as I. B. Sanborn and K. D. Schwieger. The *Tappi Journal* is published by the Technical Association of the Pulp and Paper Industry, which has its headquarters in New York City. This is number 8 of volume 69.

f. An essay entitled "Identifying Potentially Active Faults and Unstable Slopes Offshore." It appeared in an essay collection, edited by J. I. Ziony. The title of the collection is *Evaluating Earthquake Hazards in the Los Angeles Region— An Earth Science Perspective*. The book was published by the U.S. Geological Survey in Washington, D.C. The essay appeared in pages 347 through 373. It was published in 1985. The authors are S. H. Clarke, H. G. Greene, and M. P. Kennedy.

C

PROJECTS AND CASES

I n this appendix you will find writing and speaking assignments that your instructor may ask you to complete. These assignments share this important feature: they all involve communicating to particular people for specific purposes that closely resemble the purposes you will have for writing in your career. To succeed with these assignments, you will need to keep both your readers and your purpose constantly in mind. For this reason, the best way to start your work on any of the assignments is to complete the Worksheet for Defining Objectives (Figure 3–5).

The fifteen assignments in this appendix fall into two groups. The first consists of nine projects in which you will write or speak to real people about situations that are drawn from your own experience or that you learn about while working on your assignment. These assignments enable you to prepare communications that might actually help you or someone else.

The next six assignments are case studies in which you communicate in fictitious (but realistic) situations. Several of the cases require you to write two communications about the subject, each to a different audience; doing so will help you learn how you must adapt your communication strategies to the particular audience you are addressing.

You will notice that many of these assignments contain specifications about such details as length and format. Your instructor may change these specifications to tailor the assignments to your writing course.

Note to the instructor: You will find helpful notes about these and other possible assignments in the *Instructor's Manual.* The manual includes suggestions about how you might adapt these assignments to your course.

PROJECTS

PROJECT 1 RESUME AND LETTER OF APPLICATION

Write a resume and letter of application addressed to some *real* person in an organization with which you might actually seek employment. If you will graduate this year, you will probably want to write for a full-time, permanent position. If you aren't about to graduate, you may want to apply for a summer position or an internship. If you are presently working, imagine that you have decided to change jobs, perhaps to obtain a promotion, secure higher pay, or find more challenging and interesting work.

To complete this project, you may need to do some research. Among other things, you will have to find an organization that really employs people in the kind of job you want, and you will need to learn something about the organization so that you can persuade your readers in the organization that you are knowledgeable about it. Many employers publish brochures about themselves; your campus placement center or library may have copies. Your public library is also a good source of information. If the publications you find don't give the name of some specific person to whom you can address your letter, call the organization's switchboard to ask for the name of the employment director or the manager of the particular department in which you would like to work. While you work on this assignment, keep this real person in mind—even if you will not actually send your letter to him or her.

Your letter should be an original typed page, but your resume may be a high-quality photocopy of a typed original. Remember that the appearance of your resume, letter, and envelope will affect your readers, as will your attention to such details as grammar and spelling. Enclose your letter and resume in an envelope complete with your return address and your reader's name and address. You will find information about the formats for letters and envelopes in Appendix A.

As part of your package, include the names, addresses, and phone numbers of three or four references. These may be included within your resume. If you choose instead to have your resume say, "References available upon request," enclose a copy of the list of references you would send if the list were requested by an employer. Throughout your work on this project, you should carefully and creatively follow the advice given in Chapter 2 on resumes and letters of application.

PROJECT 2

UNSOLICITED RECOMMENDATION

This assignment is your chance to improve the world—or at least one small corner of it. You are to write a letter of 400 to 800 words in which you make an unrequested recommendation for improving the operation of some organization with which you have personal contact—perhaps the company that employed you last summer, a club you belong to, or your sorority or fraternity.

There are four important restrictions on the recommendation you make:

1. Your recommendation must concern a real situation in which your letter can *really* bring about change. As you consider possible topics, focus on situations that can be improved by the modest measures you can argue for effectively in a relatively brief letter. It is not necessary, however, that your letter aim to bring about a complete solution. In your letter, you might aim to persuade *one* of the key people in the organization that your recommendation will serve the organization's best interests.

2. Your recommendation must be unrequested; that is, it must be addressed to someone who has not asked for your advice.

3. Your recommendation must concern the way an organization operates, not just the way one or more individuals think or behave.

4. Your recommendation may *not* involve a problem that would be decided in an essentially political manner. Thus, you are not to write on a problem that would be decided by elected officials (such as members of Congress or the city council), and you may not address a problem that would be raised in a political campaign.

Of course, you will have to write to an *actual* person, someone who, in fact, has the power to help make the change you recommend. You may have to investigate to learn who that person is. Try to learn also how that person feels about the situation you hope to improve. Keep in mind that most people are inclined to reject advice they haven't asked for; that's part of the challenge of this assignment. From time to time throughout your career, you will find that you want to make recommendations your reader hasn't requested.

In the past, students have completed this assignment by writing on such matters as the following:

● A no-cost way that the student's summer employer could more efficiently handle merchandise on the loading dock

- A detailed strategy for increasing attendance at the meetings of a club the student belonged to

- A proposal that the Office of the Dean of Students establish a self-supporting legal-aid service for students

Bear in mind that one essential feature of a recommendation is that it compares two alternatives: keeping things the way they are now and changing them to the way you think they should be. You will have to make the change seem to be the better alternative *from your reader's point of view*. To do this, you will find it helpful to understand why the organization does things in the present way. By understanding the goals of the present method, you will probably gain insight into the criteria your reader will apply when comparing the present method with the method you recommend.

When preparing this project, follow the advice about the letter format given in Appendix A. Figure 5–15 shows an unsolicited recommendation written by a student in response to this assignment.

PROJECT
3

BROCHURE

Write a brochure about some academic major or student service on your campus. Alternatively, write a brochure for a service organization in your community.

Begin by interviewing people at the organization to learn about their aims for such a brochure. Then follow the advice given in Chapter 3 ("Defining Your Objectives") to learn about the target audience for the brochure. Remember that to be effective the brochure must meet the needs of both the organization and the readers.

Use a folded $8\frac{1}{2} \times 14''$ sheet of paper so that there are three columns (or panels) on each side of the sheet. When the brochure is folded, the front panel should serve as a cover. Use press-on letters, Kroy lettering, or some other form of large type for your cover and headings. In addition to any artwork you may decide to include on the cover, use at least one visual aid (such as a table, flow chart, drawing, or photograph) in the text of the brochure. Note that your cover need not have any artwork; it may consist solely of attractively lettered and arranged words that identify the topic of your brochure. Along with your brochure, turn in copies of any existing brochures or other printed material you used while working on this project.

Turn in a photocopy made from a typed original. In this way, you can create a professional-looking mixture of text and graphics by pasting up a master copy that you then use to print the finished work. Alternatively, you may turn in an original prepared by using desktop-publishing equipment.

Your success in this project will depend largely upon your ability to predict the questions your readers will have about your subject—and upon your ability to answer those questions clearly, concisely, and usefully. Also, think very carefully about how you want your brochure to alter your readers' attitudes about your subject. Remember that, along with the prose, the neatness and visual design of your brochure will have a large effect on your readers' attitudes.

PROJECT
4

INSTRUCTIONS

Write a set of instructions that will enable your readers to operate some device or perform some process used in your major. The procedure must involve at least twenty-four steps.

With the permission of your instructor, you may also choose from topics that are not related to your major. Such topics might include the following:

- Changing an automobile tire
- Developing a roll of film
- Making a pizza from scratch
- Planting a garden
- Rigging a sailboat
- Making homemade yogurt
- Starting an aquarium
- Some other procedure of interest to you that includes at least two dozen steps

Your instructions should guide your readers through some specific process that your classmates or instructor could actually perform. Do not write generic instructions for performing a general procedure. For instance, do not write instructions for "Operating a Microscope" but rather for "Operating the Thompson Model 200 Microscope."

Be sure to divide the overall procedure into groups of steps, rather than presenting all the steps in a single list. Use headings to label the groups of steps.

When preparing your instructions, pay careful attention to the visual design of your finished communication. You must include at least one illustration, and you must use press-on letters, Kroy lettering, or some other form of large type for your title, headings, and other appropriate portions of your instructions. You may rely heavily on figures if they are the most effective way for you to achieve your objectives; in fact, your instructions need not contain a single sentence.

Finally, your instructor may require you to use a page design that has two or more columns (rather than having a single column of type that runs all the way from the left-hand margin to the right-hand margin). In a two-column design, you might put all of your steps in the left-hand column and all of your accompanying illustrations in the right-hand column. Alternatively, you might mix both text and figures in both columns. Large figures and the title for the instructions can span both columns.

You may use the format for your instructions that you believe will work best—whether it is a single sheet of 8½ × 11″ paper, a booklet printed on smaller paper, or some other design. Turn in a photocopy made from a typed original. (In this way, you can create a professional-looking mixture of text and graphics in the way that many professional writers do: by pasting up a master copy that is then used to print the finished work.) Alternatively, you may turn in an original prepared by using desktop-publishing equipment.

Don't forget that your instructions must be accurate.

Note to the instructor: If you wish to make this a larger project, increase the minimum length of the procedure about which your students write. In this case, you may also want to specify that they create an instruction manual rather than, for example, an instruction sheet.

PROJECT 5   PROJECT PROPOSAL

Write a proposal seeking your instructor's approval for a project you will prepare later this term.

Your work on this proposal serves three important purposes. First, it provides an occasion for you and your instructor to agree about what you will do for the later project. Second, it gives you experience at writing a proposal, a task that will be very important to you in your career. Third, it gives you a chance to demonstrate your mastery of the material in Chapters 3 ("Defining Your Objectives"), 4 ("Planning to Meet Your Readers' Informational Needs"), and 5 ("Planning Your Persuasive Strategies").

Notice that while working on this assignment, you will have to define the objectives of two different communications: (1) the *proposal* you are writing now, which is addressed to your instructor, and (2) the *project* you are seeking approval to write, whose purpose and audience you will have to describe to your instructor in the proposal.

When writing your proposal, you may think of your instructor as a person who looks forward with pleasure to working with you on your final project and wants to be sure that you choose a project from which you can learn a great deal and on which you can do a good job. However, until your instructor learns from your proposal some details about your proposed project, his or her attitude toward it will be neutral. While reading your proposal, your instructor will seek to answer many questions, including the following:

- What kind of communication do you wish to prepare?
- Who will its readers be?
- What is its purpose?
  What is the final result you want it to bring about?
  What task will it enable its readers to perform?
  How will it alter its readers' attitudes?
- Is this a kind of communication you will have to prepare at work?
- Can you write the communication effectively in the time left in the term using resources that are readily available to you?

For additional insights into the questions your instructor (like the reader of any proposal) will ask, see Chapter 24 ("Proposals").

Your proposal should be between 400 and 800 words long. Write it in the memo format (see Appendix A), use headings, and include a schedule chart (see Chapter 14, "Creating Twelve Types of Visual Aids").

PROJECT 6   PROGRESS REPORT

Write a report of between 400 and 800 words in which you tell your instructor how you are progressing on the writing project you are currently preparing. Be sure to give your instructor a good sense not only of what you have accomplished but also of what problems you have encountered or anticipate. Use the memo format (see Appendix A). Figure 23–5 shows a sample progress report written by a student to the instructor of her technical writing class.

PROJECT 7   FORMAL REPORT OR PROPOSAL

Write an empirical research report, feasibility report, or proposal. Whichever form of communication you write, it must be designed to help some organization—real or imaginary—solve some problem or achieve some goal, and you must write it in response to a request (again, real or imaginary) from the organization you are addressing.

A real situation is one you have actually encountered. It might involve your employer, your major department, or a service group to which you belong—to name just a few of the possibilities. Students writing on real situations have prepared projects with such titles as:

- **Feasibility of Using a Computer Data Base to Catalog the Art Department's Slide Library.** The students wrote this feasibility report at the request of the chair of the Art Department.

- **Attitudes of Participants in Merit Hotel's R.S.V.P. Club.** The student wrote this empirical research report at the request of the hotel, which wanted to find ways of improving a marketing program that rewarded secretaries who booked their companies' visitors at that hotel rather than at one of the hotel's competitors.

- **Expanding the Dietetic Services at the Campus Health Center: A Proposal.** The student wrote this proposal to the college administration at the request of the part-time dietitian employed by the Health Center.

An imaginary situation is one you create to simulate the kinds of situations that you will find yourself in once you begin your career. You pretend that you have begun working for an employer who has asked you to use your specialized training to solve some problems or answer some questions that face his or her organization. You may imagine that you are a regular employee or that you are a special consultant. Students writing about imaginary situations have prepared final projects with titles such as:

- **Improving the Operations of the Gift Shop of Sea World of Ohio.** The student who wrote this proposal had worked at this shop for a summer job; she imagined that she had been hired by the manager to study its operation and recommend improvements.

- **Performance of Three Lubricants at Very Low Temperatures.** The student wrote this empirical research report about an experiment he had conducted in a laboratory class. He imagined that he worked for a company that wanted to test the lubricants for use in manufacturing equipment used at temperatures below $-100°F$.

- **Upgrading the Monitoring and Communication System in the Psychology Clinic.** The student who wrote this report imagined that she had been asked by the Psychology Clinic to investigate the possibility of purchasing equipment that would improve its monitoring and communication system. All of her information about the clinic and the equipment were real.

For this project, use the book format (see Appendix A). Remember that your purpose is to help your readers make a practical decision or take a practical action in a real or imaginary organization. The body of your report should be between twelve and twenty pages long (not counting cover, executive summary, title page, table of contents, appendixes, and similar parts). Be sure to use headings within your sections (or chapters) where appropriate, and to use desktop publishing, press-on letters, Kroy lettering, or some similar technique to create larger type for your cover, title page, and chapter titles. You may also want to use larger lettering for headings.

PROJECT
8

ORAL BRIEFING I: PROJECT PLANS

At work, you will sometimes be asked to report in brief talks about projects upon which you are working. For this assignment, you are to give an oral briefing to the class about your final project. Here are the things you should cover:

- **What kind of communication are you writing?** Who will your readers be? What role will you be playing? Identify your readers by telling what organization they are in and what positions your key readers hold. Describe your role by saying whether you are imagining that you work for the company as a regular employee or have been hired as a consultant. Tell who you report to.

- **What organizational problem will your communication help your readers solve?** What need or goal will it help them satisfy or reach? Provide full background so your classmates can understand the situation from your readers' point of view.

- **What are you doing to solve the problem?**
  Your research activities: what kind of information are you gathering and how, or what kind of analysis are you providing and why?
  Your writing activities: how do you plan to organize and present your information? What will your communication look like?

- **What is the gist of your message to your readers?** What are the main points you are planning to make?

As you prepare and deliver your oral briefing, pretend that you are interviewing for a job (or for a new job) and that the prospective employer has asked you to give an oral briefing about a project of yours for which you are now writing a report or proposal. The members of your class can play the role of the people your employer has asked to attend your presentation. Pretend that your classmates have not heard about your project as yet, even though you may already have discussed it in class several times. This means that you will have to provide all the background information that will enable your listeners to understand the organizational situation in which you are writing.

As the name implies, a "briefing" is a brief presentation. Make yours between 4 and 5 minutes long—no longer. Gauge the time by making timed rehearsals. In your briefing, you are to use at least one visual aid. It might show an outline for your project or it might be one of the visual aids you will use within your final project. You may present this visual aid as a poster, overhead transparency, or handout, and you may use more than one visual aid if doing so will increase the effectiveness of your briefing.

PROJECT
9

ORAL BRIEFING II: PROJECT RESULTS

At work, people often present the results of their major projects twice, once in a written communication and a second time in an oral briefing that covers the major points of the written document. In some ways, this briefing is like an executive summary—an overview of all the important things presented in more detail in writing.

For this assignment, you are to give an oral briefing on one of your writing projects. Address the class as if it were the same audience as you address in writing, and imagine that the audience has not yet read your communication.

Limit your briefing to 4 or 5 minutes—no longer (see the last paragraph of Project 8). Use at least one visual aid.

CASES

| CASE 1 | THE FRIENDLY CO-WORKER PROGRAM |

THE FRIENDLY CO-WORKER PROGRAM
Case developed by Mary L. Mason

For the past year and a half, you have been working for Management Systems Laboratories (MSL), a consulting firm that advises medium-sized companies (50 to 500 employees) about ways to improve profitability. MSL employees work as members of consulting teams. The teams generally consist of two management specialists, one accounting specialist, and one person specializing in some product or service. For example, for a team that offers advice to companies that develop computer software, the fourth team member would be a systems analyst. Similarly, for a team that offers advice to department stores, the fourth person would be a retailing specialist. Recently, MSL has decided to diversify, acquiring larger businesses as clients. Teams that work for these larger clients have the same four-person composition as do the original teams. On all teams, each member contributes his or her expertise in creating recommendations for the clients MSL serves. Because the clients are from across a five-state area, the job involves some travel to visit client sites to gather information, but much of the work (including report writing) is done at the MSL office.

Recently, MSL established a "friendly co-worker" program in which employees are assigned to orient new employees to the company and, if necessary, to the city and area in which MSL is situated. Although it would be easy enough for MSL simply to mail a new employee the available printed material (such as brochures and maps) describing the area, someone has decided that a letter from a present employee would be more personal and more effective.

You are the first person on your team to be assigned an incoming employee. All you know about the new employee is her name, Margo Reese, and that she is single, twenty-five years old, and has been working for the past three years. She starts work in just a month, and you're anxious to get your letter in the mail as soon as possible.

Assignment

First write the letter to Margo. In it you should:

- Introduce yourself.
- Describe the area in which MSL is located.
- Describe the work atmosphere at MSL, as well as any additional facilities and programs provided by the company (all of which you will make up).
- Provide any other information that you think Margo will find useful (for example, maps and entertainment information).

For the purpose of this case, imagine that MSL has one team serving businesses in your field. Margo also specializes in your field and will work on a new team. However, whereas your team serves medium-sized clients, hers will serve larger ones.

Finally, assume that Margo made such a favorable impression upon the President of MSL, Jesse Davis, that he hired her on the spot at an out-of-state interview. She accepted the offer even without visiting the office, apparently because she was impressed by Mr. Davis and because her parents recently moved from another part of the country to within 50 miles of the MSL office. As a result of the way she was hired, Margo is vaguely aware of the basic team structure of the organization but doesn't yet know which type of team she will be on. Feel free to make up sufficient, *realistic* details about what the atmosphere in the organization is like, what the job pace is, and any other details that you think would address a new employee's questions and concerns.

Locate MSL in the city where you go to college. Alternatively, locate it in your hometown or another city with which you are very familiar. Locate Margo Reese in a city that differs significantly from yours in size, climate, and region of the country. You have learned that she does not yet have housing in your city.

Second, at the request of your boss, send Jesse Davis, the President, a copy of your letter to Margo Reese, along with a memo in which you explain to him what you did in your letter and why you think your doing so will benefit the company. Your boss supports the Friendly Co-worker Program and wants your memo to Davis to persuade him that the program is worthwhile.

According to your boss, Davis is not thrilled by the new program. Some of the PR specialists apparently pushed it through, but Davis thinks the whole idea is pretty silly and a waste of time. Time is a particularly sensitive issue with Davis because at MSL everyone's working hours must be billed either to specific client accounts or to the firm's overhead account. If they are billed to the overhead account, they eat up profits rather than making them. The time you spend on Margo's letter will be billed to overhead.

Davis was in favor of a form letter that could be sent to any entering employee. He approved the program with the idea that it wouldn't be long before it died a natural death.

(For advice about using the memo and letter formats, see Appendix A.)

CASE 2

THE CAREER COUNSELOR
Case developed by Mary L. Mason

You are employed at Open Options, a career counseling firm, and your main function is to help clients in their career choices. Today a client came in who looked familiar to you. After exchanging greetings, you discovered, to your extreme surprise and delight, that this person was in a couple of your college classes. Your friend is younger than you are—just now graduating but not really looking forward to it. You asked why and found that this person has a pretty big problem—your fellow classmate can't decide between two career options. You're just the person who can help.

Assignment

Interview someone in your class to determine two different careers that he or she is interested in. Do some research on the two careers and then prepare a short report that your "client" can study and use to make his or her decision about which career path to pursue. Your report should contain a career recommendation. For this assignment, the career options might be much different (for example, working as a research engineer or as a technical sales person) or they might be similar, yet distinct (such as working in the public relations department of a large company or working for an independent public relations firm).

In your report, compare at least five major areas for the two careers, such as salary, career path, working conditions, travel requirements, working hours required, stress associated with the job, and education required, just to name a few possibilities. During your interview, or perhaps in a follow-up second interview, you should confirm your client's personal preferences regarding each of the topics you might discuss in your report. For example, what is the minimum starting salary the client will accept? By comparing the client's preferences with the research material you collect, such as salary data, you will have a basis for concluding which of the two careers is likely to be better suited to your client.

Write a letter between 400 and 800 words long in which you describe and compare the two career paths. Make a tentative recommendation, but focus on explaining the alternatives clearly for your client. Remember that in this case you are a professional career counselor and that the reader is a friend.

<table>
<tr><td>CASE
3</td><td>

EMPLOYEE TRAINING AT EXCEL

Case developed by Paula Castrogiovani (DuPont)
</td></tr>
</table>

You have worked as a management trainee for Excel Incorporated for the past several months. Your year-long management training program requires you to rotate through various departments at Excel. Currently, you work in Excel's Human Resource Department under Martin Nelson.

Excel is a medium-sized company that sells business supplies and office equipment in a region that covers the corners of three states. The company holds the high regard of its clients, partly due to the high quality of the products it sells but largely because of the courteous, client-centered image it has projected during the 82 years since its inception.

Recently though, clients, as well as people within the corporation, have been commenting on the discourteous treatment they've received from Excel employees, both on the phone and in other oral and written communications. You've noticed this yourself, so you aren't surprised when Mr. Nelson approaches you on Tuesday morning about this problem.

He seems rather upset, but you understand why this is so when he says, "The communication problem has gone too far— it threatens to cost Excel one of its largest clients."

He continues, "This cannot happen again. Our hard-earned reputation is at stake. Mr. Pytash from the Sales and Marketing Department contacted me yesterday and asked me to arrange an Office Etiquette Training Program. The program will teach basic phone skills, as well as appropriate form, tone, and wording for written communications. It's designed to be a brush-up for persons who *do* have the skills, and it will instruct those persons who *don't* have the skills. We hope this will alleviate any future problems—we simply can't afford them."

You agree, and are beginning to wonder what you have to do with all this. So you ask, "Will I help run the course?"

Mr. Nelson replies, "No, that really wouldn't be too practical; you don't have the experience, nor do you have the credibility. But, I want you to write a memo to all those who will be participating in the Training Program. Let me give you the rest of the details, then we can talk about what you'll write to them."

He proceeds, "According to research on training programs, the programs are more effective if a vertical slice of the corporation is involved in the training, rather than, say, solely entry-level persons or solely senior managers. So, persons from top to bottom in the company have been chosen to participate; it was a mostly random selection, with the exception of certain individuals who were recommended for training by their superiors.

"The program will be administered in three two-hour sessions; they will take place from 9 to 11 on Tuesday for three consecutive weeks. If the first program works, ten additional three-week sessions will be scheduled to be sure everyone is included. The participants have been selected for the first session, and changes can be made only in dire emergencies."

Mr. Nelson continues, "That's about it. So, write the memo—for my signature—including details on the program, its scope, and its schedule. Remember, this is a company-wide effort, so the memo goes to participants in all areas, including Sales and Marketing, Data Processing, Accounting, Technical Services, and Customer Relations. Also, participants will be not only from the company's headquarters buildings but also its several branch offices and warehouses.

"Any questions?"

You reply, "No, it seems easy enough—just tell them the facts, right?"

Mr. Nelson hesitates, "Yes . . . and remember not to step on any toes. We don't want this to seem like a punishment. And we don't want people to feel insulted that we're requiring them to attend what they may regard as a Mickey-Mouse workshop on being polite. So play it up; be creative. Present it as a sort of "excel-at-Excel" program. I know you can think of a way to make it sound good, right?"

"Sure," you answer, knowing you don't have any choice.

Mr. Nelson seems reassured, "Great, then I'll expect it on my desk first thing in the morning. Good luck!"

Assignment

1. Write the memo to the persons selected for training at Excel. Inform them of the course, its content and scope, and its schedule.

2. Assume that one of Excel's most valued clients, DynaCorp Computers, has recently threatened to discontinue its business relationship with Excel because of discourteous treatment its purchasing staff received over the telephone when calling Excel's Sales and Marketing Department to make inquiries or place orders. This has happened not just once, but several times. Write a letter to Mr. Fred Z. Taylor, DynaCorp's Purchasing Manager (you may make up an address), apologizing for past problems and *persuading* him (and other readers) that things will be different in the future.

3. Assume that Mr. Nelson has decided to hire an outside consultant to design and conduct the training program he has in mind. Write a letter for Mr. Nelson's signature to a consultant who could conduct the program, inviting him or her to or-

ganize and conduct the workshops. You may choose someone you know whom you think is qualified or make up a hypothetical consultant. Be sure to let him or her know exactly what is needed, who the audience for the program will be, what sort of schedule you want the program to follow, and any other information you think the prospective consultant should know. There probably are some questions you'll want to ask the consultant as well. Keep in mind that one of your tasks is to persuade the consultant to conduct the program for you.

(For advice about using the memo and letter formats, see Appendix A.)

CASE
4

BELCORP SAVINGS PLAN
Case developed by Peter C. Hall

Belcorp Corporation offers its employees a retirement savings plan in which workers can have a portion of their pay deducted from their payroll checks and placed in various investment funds. Taxes on these savings are deferred until the workers withdraw the funds after retirement. For this assignment, you will write, edit, and design a small brochure to help Belcorp employees understand their savings plan and each of its funds. This report is to be a supplement to the personalized account statement employees receive each spring. The brochure should highlight employee participation trends, past growth rates, and other related information through the use of tables, charts, and graphs.

The Facts of the Case

The Belcorp Corporation's Employee Savings Plan (BCESP) was established in 1981. That year there were 1,896 members. In 1990 there are over 3,900 employees taking advantage of the BCESP (Table 1).

1981	47%	1986	74%
1982	55%	1987	76%
1983	67%	1988	82%
1984	73%	1989	84%
1985	69%	1990	89%

Table 1   *Employee participation, 1981–1990*

In 1990 members of BCESP saved an average of 5.5 percent of their pay through the plan. The basic employee contributions totaled $3,694,000 for the year.

Belcorp Corporation matches part or all of the employee contributions. The matching fund's rate is determined by the company's yearly profit. Table 2 shows the percentages matched since 1981:

1981	65%	1986	85%
1982	45%	1987	60%
1983	60%	1988	60%
1984	85%	1989	50%
1985	85%	1990	50%

Table 2   *Portion of employee contribution matched by Belcorp, 1981–1990*

The plan allows members to make full or partial withdrawals after retirement. Retirement withdrawals totaled $3.4 million during 1990. Over $3 million of this was taken under the full-withdrawal option.

At the end of 1990, total funds in the BCESP had a market value of $51,547,214. In the period from the inception of the plan in 1981 through the year 1984, employees could choose to invest their ESP savings in either a stock market fund or a bond fund. During the last year when there were only two options, 73 percent chose the stock market fund and 27 percent chose the bond fund. Starting in July of 1985, the BCESP offered a guaranteed-income fund. In 1990 the distribution of savings in the three funds was as follows: bond fund, 3.5 percent; stock market fund, 25 percent; and guaranteed-income fund, 71.5 percent.

The *bond fund* invests only in securities backed or issued by the Federal Government. Interest income is consistent, and the face value of the bond is paid at maturity. The bond fund is managed by the Magellan Bank of New York, which typically deals in U.S. Treasury Bonds and mortgage-backed government agency securities. Last year (1990) was the first year the bond fund showed a net annual loss. This has been attributed to the many requests by employees to transfer funds to the guaranteed-income fund, which required many bonds to be sold on the market before maturity.

The *stock market fund* is managed by Sloan and Babcock, a Houston investment firm. It holds a wide variety of publicly traded stocks. Its 1990 investments included stocks of 38 different companies ranging from IBM to K-Mart. This fund aims for solid, long-term growth through capital appreciation and modest dividend income. Profits have the potential to be large, but there is always the risk of a market drop.

Returns from the *guaranteed-income fund* are always from interest income. As each year's interest is earned, it is locked in and cannot be lost through future market fluctuations. This allows continued and steady growth limited to each year's contracted rate of return. Table 3 shows the average rates of return for each of the funds.

Bond fund	7.30%
Stock market fund	14.90%
Guaranteed income fund	6.00%

Table 3 *Average yearly percentage return, 1981–1990*

Past performance is not always an indicator of future yields. Bond and stock market yields will depend on bond and stock market changes and on future interest rates. The guaranteed income fund will be offering a guaranteed contract rate of 7.5 percent for the July 1991 to July 1992 year. Table 4 shows the 1990 rates of return for each fund.

Bond fund	−6.30%
Stock market fund	17.84%
Guaranteed income fund	6.07%
1990 all-fund average	9.39%

Table 4 *Rates of return for all funds in 1990*

Discussion

This brochure will be distributed to all the members of the Belcorp Employee Savings Plan. Since it will be read by employees on the loading dock as well as in the executive suites, you must devise ways to present the key information in an easily understood manner to a very broad audience.

You are expected to present the information with both text and illustrations. Using the facts and figures provided, you must present the following information with appropriate chart and graphs:

- The growth in employee participation in BCESP during the ten years of its existence

- The percentages of matching funds paid into BCESP by Belcorp Corporation in past years

- The past and present distribution of BCESP savings in the various funds

The text of the brochure should help the readers interpret the charts and graphs as well as explain the various funds and the Belcorp Corporation matching-fund program.

You will also need to make it clear that this brochure is not meant to be an exhaustive report on the Employee Savings Plan, and if members have questions about specific plan details or withdrawal options they should consult the *Employee Benefits Manual*.

Assignment

You are to design a brochure that is as informative and attractive as possible. All features of the design are up to you. You are, however, expected to provide as complete a report on the performance of the Employee Savings Plan as you can. All the pertinent details are provided in the preceding "Facts of the Case" section. You may take some creative liberties as long as you do not make substantive changes in the facts.

Turn in a photocopy made from a typed original. In this way, you can create a professional-looking mixture of text and graphics by pasting up a master copy that you then use to print the finished work. Use press-on letters, Kroy lettering, or some similar technique to create larger type for your brochure's cover and headings. Alternatively, you may turn in an original prepared by using desktop-publishing equipment.

CASE
5

COMPANY DAY CARE

Case developed by Martin Tadlock

After graduating from college, you were hired as the Special Assistant to the President of PrimeCare, Inc., which owns and operates twelve nursing homes throughout your state. The central office, where you work, oversees all twelve PrimeCare facilities.

Your job is to conduct special research projects for the company president, Walter Henocker, and other executive officers of PrimeCare. This morning, Jan Debliss, Personnel Director, has called you into her office.

"Well, how do you think things are going—after being here for one whole month?" she asks.

"Not bad," you reply. "I've really learned a lot. Everyone has been very helpful getting me on my feet and used to the way things work around here. I'm happy."

"Good," Jan says. "I hope you're feeling ready to help me solve a rather pressing problem. Usually, I'd handle it myself, but I've been snowed under with all the problems we're having on the new record-keeping system and the opening of the new nursing home. In addition, tomorrow I begin a two-week vacation. So I thought I would ask for your help.''

''I'd be glad to help,'' you reply.

''Good,'' Jan answers. ''Let me tell you what I'm looking for and then turn you loose on it.

''We've been having problems finding employees for the day shift. We've also been having problems with absenteeism among the day-shift employees we now have. In fact, the absenteeism has gotten so bad that a lot of the time we aren't fully staffed, and supervisors must call off-duty or second- and third-shift employees to fill in. We're concerned that patients aren't getting the full care they should because we're always shorthanded. Also, these staffing problems are causing added workloads for the employees who do show up consistently as scheduled, and that's killing staff morale.''

Your mind is racing along at top speed now, trying to figure out how to solve such a problem. Jan continues.

''During last month's meeting of the directors of Primecare's twelve homes, the directors agreed that the biggest reason employees give for absenteeism is problems with child care. As you know, most of PrimeCare's employees are young women who work as nursing assistants. Many have young children—preschoolers—and they leave their children with family or friends who act as unpaid sitters. But if the sitter is busy or the kid's sick, the mother stays with the child and won't show up for work. And we can't pay high enough wages for our employees to go out and pay $2.00 an hour for paid sitters, which is what they have to do if no family or friends are available. This makes it hard to find employees for the day shift.

''One of the center directors suggested that we look into providing child care services for our employees. Not only would it cut down on absenteeism, it would boost morale and give us a benefit to offer, helping to make up for the low salary we pay. It won't cost the company much, if anything, since we get a matching state grant for any money we spend on such a service.''

Jan pauses, which gives you a chance to jump into the conversation.

You ask, ''You mean setting up a nursery in an empty room right in each of the nursing homes?''

''Yes, that's one possibility,'' Jan replies, ''but there could be several ways to provide the service. That's where you come in. I'd like you to find out what you can about corporate child care. I read somewhere that over 2,500 employers offer some kind of child-care assistance to their employees now, and the number is growing.''

''It sounds like a good idea,'' you interject. ''What should I do with the information once I find it?''

''Well, I'd like you to write it up in a report that tells me what you find out and also recommends a child-care service that you think would work for us.''

You are wondering if there is more to this than you think. So you ask, ''You mean like a report summarizing what other places have done with child care for their employees, with a recommendation included?''

''Yes,'' Jan answers. ''Just send me a memo summarizing how other places provide child care for their employees and recommend the service that would work best for us. But don't report everything you find out about the services other places provide. We can't handle some huge, expensive thing like some corporations can. However, we can provide up to $800 per facility each month and have it matched by the state. We don't have any start-up money, and the state won't provide any, but at least the monthly funds are available. So focus on services that we could afford, perhaps by adapting them to our situation.''

As Jan speaks, you realize that she probably has some information that could help you, and maybe even an opinion about what PrimeCare should do. You ask, ''Is there any type of child care service that you think would be especially good for us?''

''Well, frankly, I think that setting up a day-care service in each of our 12 homes would be an excellent way to meet our needs. Although I haven't had much time to investigate the matter, I have found that we could probably do it cheaply. Our present insurance would cover an in-house service, so we wouldn't have to face any additional expense there. Also, each of our homes has one room that could be used for this purpose. Though they aren't outfitted with the furniture or equipment needed for day care, they all meet the state's requirements, so we could get a day-care license without any additional expense.''

Though this sounds good, you wonder about one thing: ''But if we don't have any furniture or equipment, how could we get it? You said we won't have any start-up money.''

''Well, that could be a problem,'' Jan responds. ''We might be able to get donations from community people. The senior citizens' center near one of our facilities has already offered some things. And once we have the rooms outfitted, our only operating expenses would be pay for staff; supplies like paper, crayons, and kleenex; and food—snacks and lunches—for the children. We may even be able to staff the centers partly with volunteers.''

''It all sounds great,'' you observe. ''Why not save some time by having me study ways to set up in-house services without investigating other alternatives?''

''Well, there are two reasons. First, some directors think such a service would cost more money than we will have. And, to tell the truth, I haven't had a chance to calculate the actual costs myself, though I'm pretty sure I'm right. Second, some directors feel uneasy about settling on a specific method of providing day-care assistance until all reasonable possibilities have been investigated. They fear that if we focus immediately on one alternative, we may overlook others that would better serve both the employees and the company. So, I need to have you get information on several possible ways of providing our employees with child-care assistance. That's what the directors want, and after all, they are the people who will decide what we will do.''

''I see,'' you reply. After a short pause, you ask, ''Is there anything you can tell me about the number of employees who would use a child-care service at each facility?''

''Yes,'' Jan says. ''We talked informally about all of this at the last directors' meeting. The director who brought it up gave me a sheet with a few figures based on his facility. Maybe it will help you.''

Jan stands up and hands you a sheet scribbled with some figures about the Charleston Avenue PrimeCare facility. She starts toward the door, your signal that the conversation is over.

Jan says, ''Have the memo on my desk when I return from vacation in two weeks. I'd like to be able to use your memo to decide what to recommend to the next directors' meeting. In fact, if I agree with your recommendation, I'll just pass your memo along with a note from me.''

''Okay,'' you reply. ''The memo will be waiting for you when you return.''

On Jan's note sheet you find the following information about the Charleston Avenue facility.

Number of Day-Shift Employees

Nurses	17	Day-shift jobs vacant	4 Nurses' Aides
Nurses' Aides	41		1 Housekeeping
Housekeeping	15		2 Nurses
Kitchen	13		
Maintenance	2	Absent days last month	18
Secretaries	2		
Therapists	3		
Social Relief	1		
Management	2		
	96		

Day Shift
72 employees paid less than $5.00/hour
24 with children under 6 years of age (28 total children)

24 employees paid more than $5.00/hour
6 with children under 6 years of age (7 total children)

Childcare Budget (monthly)
$ 800 local from PrimeCare for each facility
$ 800 state match for each facility
$1600

1) What can we provide?
   —within budget?
   —22 employees say they will use/need in-house day-care services
       —together they have 27 kids under 6
   —volunteers?

Assignment

1. Write the report Jan Debliss has requested. Make it between 700 and 1200 words long.

2. Use the general superstructure for reports (see Chapter 20). Include the following sections:
    ● **Introduction.** In it, review your assignment and describe the most important features of the problem that your research is intended to help the company solve. If you think you can make your memo more effective by briefly stating your recommendation here, do so. Do not give this section a heading.
    ● **Method.** Tell how you got your information. Do not include full bibliographic citations, but name the journals and other sources you consulted. This will be a short section, perhaps only one sentence long.
    ● **Findings.** Discuss each of the major alternatives you think will most interest Jan Debliss and the directors. This section should include the alternative(s) you will recommend. Be sure to describe each alternative in a way that lets your readers know what it involves. Also, be sure to evaluate each of the alternatives from PrimeCare's point of view. This will be your longest section.

● **Recommendation.** State what you think PrimeCare should do (even though your recommendation may already be evident from your discussion of the alternatives). Your recommendation may involve a combination of two or more of the alternatives you described in the preceding section, or it may focus on just one alternative. Be sure that your readers understand how your recommended service would be implemented at PrimeCare; to do that, you may need to provide information not already given in your "Findings" section.

3. Use the memo format, including centered headings (all caps, underlined). You may include subheadings (initial caps, flush left, underlined).

CASE 6

PROPOSE YOUR OWN BUSINESS

For this assignment, you are to write a proposal that seeks a loan to start a student-created business in your city.

Imagine that in the hope of fostering the entrepreneurial spirit of students, a dozen prosperous alumni have persuaded your school to set aside space in a building near the center of your campus to provide rent-free space for several student-created businesses. Calling themselves the Delphi Group, these alumni will also make interest-free loans to the students who set up these businesses. The only stipulation is that repayment begin no later than one year after the students receive the money.

Because the Delphi Group expects a large number of requests for the space and the loans, they have decided to establish a competition in which interested students submit proposals describing the businesses they would like to start. After reading these proposals, the Delphi Group will select the four that seem most likely to succeed.

You are to write a formal proposal to submit in this competition. To assure that you have an appropriate business to propose, you will need to submit a brief description of it for your instructor's review and approval.

Proposal Content

The Delphi Group has specified that the proposals should be between fourteen and twenty double-spaced pages (not counting front matter or appendixes). Proposals are to include the following chapters. Note that this organization is simply a variation on the superstructure for proposals that is described in Chapter 24.

Introduction   Briefly explain that you are responding to the Delphi Group's request for proposals and name the kind of business you would like to start. Tell how much money you want to borrow. Forecast the contents of the rest of your proposal. This chapter should not exceed one page.

Market Analysis   Provide persuasive evidence that a market exists for the business you are proposing. Who will purchase your products or services? What are the important characteristics of these people from the point of view of your proposed business? [*Note to the instructor:* the *Instructor's Manual* discusses guidance you can give your students concerning this research.]

Also, discuss any competition that exists and explain your reasons for thinking that you can compete effectively against it. If there isn't any competition, tell whether a similar business has been tried in your community but has failed. Explain what makes you think your business is more likely to succeed.

This chapter should be as long as is required to persuade your readers that a substantial market exists for the business you are proposing.

Proposed Business   Because the Delphi Group will be extremely interested in the soundness of your business plan, this chapter should be the longest in your proposal. Cover the following topics:

- **Products or services.** Tell specifically and in detail what your business will sell. Relate these products or services directly to your market analysis.

- **Business site.** Describe the physical layout of your business and its decor in a way that persuades that they will make your business efficient and appealing to its target market. Identify the equipment and furniture you will need to purchase. Include a floor plan that shows the overall dimensions of the area you will use and the arrangement of items within that area.

- **Marketing plan.** Tell how you will attract customers. Include both your plans for attracting attention to your business as it first opens and the on-going marketing plan you will follow thereafter.

- **Staffing and management.** Indicate how many people you will hire and what their duties will be. If your employees will need to have any special qualifications, name them. Be sure that you have someone to perform each of the functions required to run a business even if you don't assign one person to each function. For example, even if you don't plan to hire a full-time accountant, tell how your accounting will be done. Your staff may be as small as two people (yourself and one other person), provided that the two of you can perform all the required activities and provide coverage for all the hours the business will be open.

Also, describe your business's management structure, keeping the structure as simple as possible.

Schedule   Provide a prose overview and a Gantt chart of the major events in the development of your business. Begin with the approval of your proposal and continue to the point where your business is fully established and you can start making payments on your loan. Include such events as outfitting and decorating your business location, lining up suppliers, hiring employees, preparing your marketing materials, and opening for the first day of business. Indicate the duration of each event.

This chapter should not exceed three pages, including the Gantt chart.

Qualifications   Explain your qualifications to run the business you propose. If you would need to gain additional knowledge, tell what it is and describe how you will get it (for instance by taking a certain course). This chapter should not exceed one page.

Budget   Explain the projected expenses and income for your business. When discussing *initial expenses,* list major categories (such as furniture, equipment, initial supplies); you do not need to identify each separate item. You may provide rough estimates of costs as long as you have a basis for believing your estimates to be reasonable.

When discussing *operating expenses,* consider only the cost of replenishing supplies and paying salaries. Assume that utilities, janitorial services, and other miscellaneous operating needs will be provided for free by your school; do not include them in your budget.

When discussing *projected income,* state your assumptions about how much business you will do, what your profit margin will be, and so on. Here, too, rough estimates are acceptable, provided that you have a basis for believing them to be reasonable.

Remember that your business may build slowly so that it may be months before your income can cover operating expenses completely. If this is the case, your loan request should include the money you need to operate during this period.

Present your financial plans in a way that will persuade the Delphi Group that you can begin repaying the loan one year after receiving it.

Your budget chapter should not exceed three pages, including prose and tables. If you feel that additional details about your budget would be persuasive, include them in an appendix (which does not count as part of the fourteen to twenty pages required for the body of the report).

Conclusion Briefly summarize your proposal and bring it to an appropriate end. This chapter should not exceed one page.

Format

The Delphi Group has specified that all proposals should use the format described in the style guide distributed by your instructor. [*Note to the instructor:* the *Instructor's Manual* includes suggestions about this style guide.]

Chapter 1: Writing, Your Career, and This Book

1. Lester Faigley and Thomas P. Miller, "What We Learn from Writing on the Job," *College English* 44 (1982): 557–69.
2. Paul V. Anderson, "What Survey Research Tells Us about Writing at Work," in *Writing in Nonacademic Settings,* ed. Lee Odell and Dixie Goswami (New York: Guilford Press, 1985), 3–85.
3. Richard M. Davis, "How Important Is Technical Writing?—a Survey of Opinions of Successful Engineers," *The Technical Writing Teacher* 4 (1977): 83–88.
4. No generally accepted term exists for the overall structural patterns conventionally used for communications. For this book, I have borrowed the term *superstructures* from cognitive psychologists, who study how these conventional patterns help people comprehend the things they read. See, for example, Teun A. van Dijk, "Semantic Macrostructures and Knowledge Frames in Discourse Comprehension," in *Cognitive Processes in Comprehension,* ed. Marcel Adam Just and Patricia A. Carpenter (Hillsdale, N.J.: Erlbaum, 1977), 3–32; Walter Kintsch and Teun A. van Dijk, "Toward a Model of Text Comprehension and Production," *Psychology Review* 85 (1978): 363–94; and Teun A. van Dijk, *Macrostructures* (Hillsdale, N.J.: Erlbaum, 1980).

Chapter 2: Examples of Audience-Centered Writing: Resumes and Letters of Application

1. Baron Wells, Nelda Spinks, and Janice Hargrave, "A Survey of the Chief Personnel Officers in the 500 Largest Corporations in the United States to Determine Their Preferences in Job Application Letters and Personal Resumes," *ABCA Bulletin* 14:2 (June 1981): 3–7.

Chapter 5: Planning Your Persuasive Strategies

1. Richard E. Petty and John T. Cacioppo, "The Elaboration Likelihood Model of Persuasion," *Communication and Persuasion: Central and Peripheral Routes to Attitude Change* (New York: Springer-Verlag, 1986), 1–24.
2. I have drawn this classification of attitudinal changes from Gerald R. Miller, "On Being Persuaded: Some Basic Distinctions," in *Persuasion: New Directions in Theory and Research,* ed. M. Roloff and G. Miller (Beverly Hills: Sage, 1980), 11–28.
3. My general description of the "cognitive response theory" of persuasion paraphrases the excellent description of it by Richard E. Petty and John T. Cacioppo, *Attitudes and Persuasion: Classic and Contemporary Approaches* (Dubuque, Iowa: Wm C Brown, 1981), 225. See also Richard E. Petty, Thomas M. Ostrom, and Timothy C. Brock, *Cognitive Responses in Persuasion* (Hillsdale, N.J.: Erlbaum, 1981); and Petty and Cacioppo, *Communication and Persuasion.*
4. A. Maslow, *Motivation and Personality* (New York: Harper & Row, 1970), and Frederick Herzberg, *Work and the Nature of Man* (Cleveland: World, 1968).

5. J. Weger, *Motivating Supervisors* (New York: American Management Association, 1971), 53–54. For leads to this and several other sources cited in Chapter 5, I am indebted to Mary Munter, *Business Communication: Strategy and Skill* (Englewood Cliffs, N.J.: Prentice-Hall, 1987).

6. E. K. Warren, L. Roth, and M. Devanna, ''Motivating the Computer Professional,'' *Faculty R&D,* a publication of the Columbia Business School (Spring 1984): 8.

7. Petty and Cacioppo, ''Self-Persuasion Approaches,'' *Attitudes and Persuasion,* 213–54.

8. Petty and Cacioppo, ''Motivational Approaches,'' *Attitudes and Persuasion,* 126–61.

9. Stephen Toulmin, Richard Rieke, and Allan Janik, *An Introduction to Reasoning,* 2nd ed. (New York: Macmillan, 1984).

10. Brian Sternthal, Ruby Dholakia, and Clark Leavitt, ''The Persuasive Effect of Source Credibility: Tests of Cognitive Response,'' *Journal of Consumer Research* 4 (1978): 252–60.

11. H. C. Kelman and C. I. Hovland, '' 'Reinstatement' of the Communicator in Delayed Measurement of Opinion Change,'' *Journal of Abnormal and Social Psychology* 48 (1953): 327–35.

12. Robert N. Bostrom, *Persuasion* (Englewood Cliffs, N.J.: Prentice-Hall, 1981), 71–73. Bostrom has speculated that power and dynamic appeal might be different aspects of the same factor; I have chosen to treat them separately.

13. Carl R. Rogers, ''Communication: Its Blocking and Its Facilitation,'' *Harvard Business Review* 30 (1952): 46–50.

Chapter 6: Using the Library

1. For a more detailed introduction to the library, one excellent source is Mary G. Hauer, Ruth C. Murray, Doris B. Dantin, and Myrtle S. Bolner, *Books, Libraries, and Research,* 3rd ed. (Dubuque, Iowa: Kendall/Hunt, 1987).

Chapter 7: Writing Paragraphs, Sections, and Chapters

1. Robert W. Kelton, ''The Internal Report in Complex Organizations,'' in *Proceedings of the 30th International Technical Communication Conference* (Washington, D.C.: Society for Technical Communication, 1984), RET54–57.

2. Linda Flower and John R. Hayes, ''The Dynamics of Composing: Making Plans and Juggling Constraints'' in *Cognitive Processes in Writing,* ed. Lee W. Gregg and Edwin R. Steinberg (Hillsdale, N.J.: Erlbaum, 1980), 31–50.

3. J. D. Bransford and M. K. Johnson, ''Contextual Prerequisites for Understanding: Some Investigations of Comprehension and Recall,'' *Journal of Verbal Learning and Verbal Behavior* 11 (1972): 717–26.

Chapter 9: Beginning a Communication

1. Richard E. Petty and John T. Cacioppo, ''Consequences of the Route to Persuasion,'' *Communication and Persuasion: Central and Peripheral Routes to Attitude Change* (New York: Springer-Verlag, 1986), 173–95.

2. J. C. Mathes and Dwight W. Stevenson, ''The Problematic Context: The Purpose of the Report,'' *Designing Technical Reports: Writing for Audiences in Organizations* (Indianapolis: Bobbs-Merrill, 1976), 24–42.

3. Paul V. Anderson, ''What Survey Research Tells Us about Writing at Work,'' in *Writing in Nonacademic Settings,* ed. Lee Odell and Dixie Goswami (New York: Guilford Press, 1985), 3–85.

Chapter 10: Ending a Communication

1. Harold R. Mancusi-Ungaro, Jr., and Norman H. Rappaport, ''Preventing Wound Infections,'' *American Family Physician* 33 (April 1986): 152.

2. Raymond L. Murray, *Understanding Radioactive Waste* (Columbus, Ohio: Battelle Press, 1982), 100.

3. Spartaco Astolfi Filho et al., ''Stable Yeast Transformants That Secrete Functional a-Amylase Encoded by Cloned Mouse Pancreatic cDNA,'' *Biotechnology* 4 (April 1986): 311–15.

4. General Electric Company, *How To Get the Best from Your Dryer* (Louisville, Ky.: General Electric Company, 1983), 9.

5. Office of Cancer Communications, National Cancer Institute, *Taking Time: Support for People with Cancer and the People Who Care about Them* (Bethesda, Md.: National Cancer Institute, 1983), 53.

6. K. M. Foreman, *Preliminary Design and Economic Investigations of Diffuser Augmented Wind Turbines (DAWT)* (Golden, Col.: Solar Energy Research Institute, 1981), 23.

Chapter 11: Writing Sentences

1. Joseph M. Williams, *Style: Ten Lessons in Clarity and Grace* (Glenview, Ill.: Scott Foresman, 1981), 109.

Chapter 12: Choosing Words

1. Raymond W. Kulhavy and Neil H. Schwartz, ''Tone of Communications and Climate Perceptions,'' *Journal of Business Communication* 18 (Winter 1981): 17–24.

2. George R. Klare, ''The Role of Word Frequency in Readability,'' *Elementary English* 45 (1968): 12–22.

3. For a fuller discussion of sexist and discriminatory language in business, see International Association of Business Communicators, *Without Bias: A Guidebook for Nondiscriminatory Communication,* 2nd ed. (New York: Wiley, 1982).

4. For example, see Maxine Hairston, *Contemporary Composition,* 4th ed. (Boston: Houghton Mifflin, 1986), 251–55.

Chapter 15: Designing Pages

1. Philippa J. Benson, ''Writing Visually: Design Considerations in Technical Publications,'' *Technical Communication* 32 (Fourth Quarter 1985): 37.

2. J. Hartley, *Designing Instructional Text* (New York: Nichols Publishing Co., 1978).

3. M. A. Tinker. *Legibility of Print* (Ames, Iowa: University of Iowa Press, 1969).

Chapter 16: Checking

1. John R. Anderson, ''Attention and Sensory Information Processing,'' *Cognitive Psychology and Its Implications,* 2nd ed. (New York: W. H. Freeman, 1985), 40–48.
2. Frank Smith, ''Word Identification,'' *Understanding Reading,* 3rd ed. (New York: Holt Rinehart and Winston, 1982), 120–34.
3. George R. Klare, ''Readable Technical Writing: Some Observations,'' *Technical Communication* 24:2 (Second Quarter, 1977): 2.
4. Jack Selzer, ''What Constitutes a 'Readable' Technical Style?'' in *New Essays in Technical and Scientific Communication,* ed. Paul V. Anderson, R. John Brockmann, and Carolyn R. Miller (Farmingdale, N.Y.: Baywood, 1983), 71–89.

Chapter 18: Testing

1. Linda Flower, John R. Hayes, and Heidi Swarts, ''Revising Functional Documents,'' in *New Essays in Technical and Scientific Communication,* ed. Paul V. Anderson, R. John Brockmann, and Carolyn R. Miller (Farmingdale, N.Y.: Baywood, 1983), 41–58.

Chapter 21: Empirical Research Reports

1. This report is an adaptation of material from Roy E. Anderson, Richard L. Frey, and James R. Lewis, *Satellite-Aided Mobile Communications Limited Operational Test in the Trucking Industry* (Schenectady, N.Y.: General Electric Company, 1980).
2. Robert B. Hays, ''A Longitudinal Study of Friendship Development,'' *Journal of Personality and Social Psychology* 48 (1985): 909–24. Copyright 1985 by the American Psychological Association.

Chapter 25: Instructions

1. International Business Machines, (IBM), *DPPX/SP Migration Guide* (Kingston, N.Y.: International Business Machines, 1983), 24. Copyright © 1983 by International Business Machines Corporation. Reprinted by permission.

Chapter 26: Writing Collaboratively

1. C. L. Kleinke, A. A. Bustos, F. B. Meeker, and R. A. Staneski, ''Effects of Self-Attributed Gaze in Interpersonal Evaluations between Males and Females,'' *Journal of Experimental Psychology* 9 (1973): 154–63.
2. R. E. Kleck and W. Nuessle, ''Congruence between the Indicative and Communicative Functions of Eye-Contact in Interpersonal Relations,'' *British Journal of Social and Clinical Psychology* 6 (1967): 256–66.
3. Alan Mehrabian, *Nonverbal Communication* (Chicago: Aldine, 1972).

ILLUSTRATION CREDITS

Figure 4–1: Westinghouse Electric Corporation.

Figure 5–1: Welch's, Concord, Mass.

Figure 5–3: Courtesy of The Procter & Gamble Company. Used with permission.

Figure 6–5: *Biological & Agricultural Index,* vol. 74, 1989. Copyright © 1989 by the H. W. Wilson Company. Material reproduced with permission of the publisher.

Figure 7–16: Courtesy J. Vdoviak, General Electric Company.

Figure 7–17: Used with permission of Thomson Consumer Electronics.

Figure 8–7: Boyce Rensberger © 1986 *Science Digest*.

Figure 13–15 (bottom): This graph is based on work performed under contract number DE-AC05-760R00033 between the U.S. Department of Energy and Oak Ridge Associated Universities.

Figure 13–12C: Pergamon Press Incorporated.

Figure 14–6: Mitsubishi Motor Sales of America.

Figure 14–11: Reprinted from August 1984 *Data Communications Magazine.* Copyright 1984 McGraw-Hill, Incorporated. All rights reserved.

Figure 14–17: Reprinted by permission from *Nature,* Vol. 315, p. 237. Copyright © 1990 Macmillan Magazines Ltd.

Figure 14–22: Reprinted with permission © 1990 Society of Automotive Engineers, Incorporated.

Figure 14–23: Courtesy of Miami University, Oxford, OH 45056.

Figure 14–24 (p. 414, bottom): © Nissan Motor Corporation USA.

Figure 14–26: Courtesy APV Crepaco/Food Engineering.

Figure 15–1C: Reprinted with permission from *Electronic Design,* June 26, 1986. Copyright 1986 Penton Publishing.

Figure 15–1D: Reprinted with permission of *Solid State Technology*.

Figure 15–14: Professor Joseph L. Cox III, Professor of Art/Associate Provost.

Figure 25–1: Akron Standard Company.

Figure A–14: Reproduced by permission of the *Canadian Journal of Criminology,* 27(1):1–13, 1985. Copyright by the Canadian Criminal Justice Association.

INDEX

guidelines
  begin by announcing topic, 206–12, 214
  move from most important to least important, 214–15
  present generalizations before details, 212–14
  reveal organization, 215–241
  types of segments, 205–6
Parallel construction
  examples, 53–54, 241
  headings, 223–30
  lists, 238–41
  resumes, 53, 54
Partnership
  between writers and reviewers, 481
  with readers
    creating as a persuasive strategy, 163, 166–67
    creating at the beginning, 282–86, 292
  See also Problems a communication will help readers solve
Passive voice
  cause of wordiness, 320
  converting to active voice, 321
  identifying, 23, 318–19
  when to use, 320–21, 325
  why avoid, 319–20
Patterns of development
  cause and effect
    aims, 267–69
    example, 270, 271–72
    guidelines for describing, 269
    guidelines for persuading, 269–70
  classification
    aims, 250
    creating hierarchies, 249
    defined, 249
    examples, 252–255
    formal and informal methods, 250
    guidelines for formal, 250–51
    guidelines for informal, 251–52
  comparison
    alternating and divided, 78–79, 261–67, 588
    defined, 261
    example, 267, 268
    guidelines, 267
    how it works, 261–67
    when to use it, 261
  mixing, 248–49
  partition
    aims, 257
    defined, 256
    example, 258, 259
    guidelines, 257–58
    how it works, 256
  problem and solution
    aims, 270
    defined, 270
    example, 273, 274
    guidelines for describing, 270–73
    guidelines for persuading, 273
  segmentation

    aim, 258
    defined, 258
    examples, 261, 263–65, 266
    guidelines, 260–61
    two purposes for, 258
  use only as *guides,* 248
Paul, Ronald S., 405
Pergamon Press, 808
Periodical indexes, 189–92
Persona. *See* Role you create
Personal data, describing in resume, 54–55
Persuasion
  aimed at altering attitudes, 142–43
  as a general objective, 142
  benefits to emphasize
    growth needs, 146–48
    organizational objectives, 143–46
  claims made in a resume, 42
  concerns of audience
    how they arise, 148
    importance of addressing, 148–49
  counterarguments by audience
    how they arise, 148
    how to address, 148–49
    importance of addressing, 148
  guidelines
    address concerns and counterarguments, 148–49
    create an effective role for yourself, 159–68
    emphasize benefits for readers, 143–48
    organize to create a favorable response, 154–59
    show sound reasoning, 149–54
  initial generalizations to increase, 213
  resistance to changing attitudes, 148
  three kinds of change in attitudes, 80, 142
  See also Persuasive element of purpose; Readers' attitudes; Reasoning
Persuasive element of purpose
  defined, 80, 109
  identifying, 80–81
  relationship to enabling element, 76
  See also objectives, defining; Persuasion
Petty, Richard E., 804n.1, 805nn.7, 8, 1
Phantom readers. *See* Readers
Photographs. *See* Visual aids, types of
Pictographs. *See* Visual aids, types of
Pie charts. *See* Visual aids, types of
Pihil, Alexander, 407
Pitler, R. K., 405
Planning
  benefits, 108
  for collaborative writing, 682–85
  importance of making flexible, 123
  job application letters, 60–64
  place in composing process, 19–20, 108
  resumes, 42–46
  stages, 123
  using model communications, 121–23
  using outlines, 169–75
  worksheets. *See* Planning guides and worksheets